Bioactive Compounds from Functional Foods

Bioactive Compounds from Functional Foods

Editors

Michał Halagarda
Sascha Rohn

Basel • Beijing • Wuhan • Barcelona • Belgrade • Novi Sad • Cluj • Manchester

Editors
Michał Halagarda
Krakow University of Economics
Krakow
Poland

Sascha Rohn
Technische Universität Berlin
Berlin
Germany

Editorial Office
MDPI
St. Alban-Anlage 66
4052 Basel, Switzerland

This is a reprint of articles from the Special Issue published online in the open access journal *Molecules* (ISSN 1420-3049) (available at: https://www.mdpi.com/journal/molecules/special_issues/8QJ9MB32AL).

For citation purposes, cite each article independently as indicated on the article page online and as indicated below:

Lastname, A.A.; Lastname, B.B. Article Title. *Journal Name* **Year**, *Volume Number*, Page Range.

ISBN 978-3-7258-0375-0 (Hbk)
ISBN 978-3-7258-0376-7 (PDF)
doi.org/10.3390/books978-3-7258-0376-7

Contents

About the Editors . . . . **vii**

Preface . . . . **ix**

Nawal K. Z. AlFadhly, Nawfal Alhelfi, Ammar B. Altemimi, Deepak Kumar Verma, Francesco Cacciola and Arunaksharan Narayanankutty
Trends and Technological Advancements in the Possible Food Applications of Spirulina and Their Health Benefits: A Review
Reprinted from: *Molecules* **2022**, *27*, 5584, doi:10.3390/molecules27175584 . . . . **1**

VS Shilpa, Rafeeya Shams, Kshirod Kumar Dash, Vinay Kumar Pandey, Aamir Hussain Dar, Shaikh Ayaz Mukarram, et al.
Phytochemical Properties, Extraction, and Pharmacological Benefits of Naringin: A Review
Reprinted from: *Molecules* **2023**, *28*, 5623, doi:10.3390/molecules28155623 . . . . **41**

Afifa Aziz, Sana Noreen, Waseem Khalid, Fizza Mubarik, Madiha khan Niazi, Hyrije Koraqi, et al.
Extraction of Bioactive Compounds from Different Vegetable Sprouts and Their Potential Role in the Formulation of Functional Foods against Various Disorders: A Literature-Based Review
Reprinted from: *Molecules* **2022**, *27*, 7320, doi:10.3390/molecules27217320 . . . . **59**

Janaina Sánchez-García, Sara Muñoz-Pina, Jorge García-Hernández, Amparo Tárrega, Ana Heredia and Ana Andrés
In Vitro Digestion Assessment (Standard vs. Older Adult Model) on Antioxidant Properties and Mineral Bioaccessibility of Fermented Dried Lentils and Quinoa
Reprinted from: *Molecules* **2023**, *28*, 7298, doi:10.3390/molecules28217298 . . . . **77**

Katarzyna Najman, Anna Sadowska, Monika Wolińska, Katarzyna Starczewska and Krzysztof Buczak
The Content of Bioactive Compounds and Technological Properties of Matcha Green Tea and Its Application in the Design of Functional Beverages
Reprinted from: *Molecules* **2023**, *28*, 7018, doi:10.3390/molecules28207018 . . . . **95**

Karolina M. Wójciak, Karolina Ferysiuk, Paulina Kęska, Małgorzata Materska, Barbara Chilczuk, Monika Trząskowska, et al.
Reduction of Nitrite in Canned Pork through the Application of Black Currant (*Ribes nigrum* L.) Leaves Extract
Reprinted from: *Molecules* **2023**, *28*, 1749, doi:10.3390/molecules28041749 . . . . **118**

Agnieszka Ciurzynska, Magdalena Trusinska, Katarzyna Rybak, Artur Wiktor and Malgorzata Nowacka
The Influence of Pulsed Electric Field and Air Temperature on the Course of Hot-Air Drying and the Bioactive Compounds of Apple Tissue
Reprinted from: *Molecules* **2023**, *28*, 2970, doi:10.3390/molecules28072970 . . . . **136**

Michał Halagarda and Paweł Obrok
Influence of Post-Harvest Processing on Functional Properties of Coffee (*Coffea arabica* L.)
Reprinted from: *Molecules* **2023**, *28*, 7386, doi:10.3390/molecules28217386 . . . . **160**

Karolina M. Wójciak, Paulina Kęska, Monika Prendecka-Wróbel and Karolina Ferysiuk
Peptides as Potentially Anticarcinogenic Agent from Functional Canned Meat Product with Willow Extract
Reprinted from: *Molecules* **2022**, *27*, 6936, doi:10.3390/molecules27206936 . . . . **169**

Paulina Strugała-Danak, Maciej Spiegel and Janina Gabrielska
Malvidin and Its Mono- and Di-Glucosides Forms: A Study of Combining Both In Vitro and Molecular Docking Studies Focused on Cholinesterase, Butyrylcholinesterase, COX-1 and COX-2 Activities
Reprinted from: *Molecules* **2023**, *28*, 7872, doi:10.3390/molecules28237872 . . . . **183**

N. Afzal Ali, Kshirod Kumar Dash, Vinay Kumar Pandey, Anjali Tripathi, Shaikh Ayaz Mukarram, Endre Harsányi and Béla Kovács
Extraction and Encapsulation of Phytocompounds of Poniol Fruit via Co-Crystallization: Physicochemical Properties and Characterization
Reprinted from: *Molecules* **2023**, *28*, 4764, doi:10.3390/molecules28124764 . . . . **197**

Muhammad Abdul Rahim, Adeela Yasmin, Muhammad Imran, Mahr Un Nisa, Waseem Khalid, Tuba Esatbeyoglu and Sameh A. Korma
Optimization of the Ultrasound Operating Conditions for Extraction and Quantification of Fructooligosaccharides from Garlic (*Allium sativum* L.) via High-Performance Liquid Chromatography with Refractive Index Detector
Reprinted from: *Molecules* **2022**, *27*, 6388, doi:10.3390/molecules27196388 . . . . **212**

Paweł Sroka, Tomasz Tarko and Aleksandra Duda
The Impact of Furfural on the Quality of Meads
Reprinted from: *Molecules* **2024**, *29*, 29, doi:10.3390/molecules29010029 . . . . **223**

Weiyue Zhang, Nana Zhang, Xinxin Guo, Bei Fan, Shumei Cheng and Fengzhong Wang
Potato Resistant Starch Type 1 Promotes Obesity Linked with Modified Gut Microbiota in High-Fat Diet-Fed Mice
Reprinted from: *Molecules* **2024**, *29*, 370, doi:10.3390/molecules29020370 . . . . **235**

About the Editors

Michał Halagarda

Michał Halagarda, born 1984, is an Associate Professor and a Head of the Department of Food Product Quality at the Krakow University of Economics (Poland). He graduated from the University of Abertay in Dundee, Scotland with a BSc in Food Product Design in 2007, and from Krakow University of Economics with an MSc in Commodity Science, in 2008, with his specialty being Product Quality Management. In 2014, he obtained his Ph.D. in Commodity Science from the Faculty of Commodity Science at Krakow University of Economics, Poland, working on new food product development process modelling for bakery industry products. He did his habilitation in 2019 at the Faculty of Commodity Science and Product Management of Krakow University of Economics, Poland. He did internships in the Institute of Food Chemistry at the University of Hamburg, Germany; in the Department of Chemical Engineering (Area of Food Technology) of the Faculty of Science at the University of Vigo, Spain; in the Department of Grain Processing and Baking Technologies at the Institute of Agricultural and Food Biotechnology in Warszawa, Poland; and in the Seidman College of Business, Grand Valley State University, USA. He is a member of the Polish Commodity Science Society and the International Society of Commodity Science and Technology (IGWT).

His research mainly covers food science and consumer economics. His scientific interests include the authentication of traditional, regional, and organic food products, bioactive compounds in foods, food product quality development and assessment, consumer market behavior, and factors affecting success of new product development processes.

Sascha Rohn

Sascha Rohn, born 1973, is a full professor of Food Chemistry at the Technische Universität Berlin (Germany). He graduated from the University of Frankfurt/Main, Germany, with the first and second state examinations in Food Chemistry in 1999. In 2002, he obtained his Ph.D. in Food Chemistry from the Institute of Nutritional Science at the University of Potsdam, Germany, working on the interactions of polyphenols with food proteins. After two years as a postdoc, he left Potsdam for Berlin, where he did a habilitation at the Institute of Food Technology and Food Chemistry of the Technische Universität Berlin. From October 2009 to October 2020, he was a full professor at the Hamburg School of Food Science, Institute of Food Chemistry, University of Hamburg, Germany. His group is dealing with the analysis of bioactive food compounds. In particular, they are characterizing the reactivity and stability of bioactive compounds. The aim is to identify degradation products that serve as quality parameters, as process markers during food/feed processing, or as biomarkers in nutritional physiology. The results of their work have been presented in more than 250 publications so far (they have a Scopus h-index of 60 and more than 10,000 citations). More than 30 well-known scientific journals regularly ask Prof. Rohn to review scientific manuscripts. From 2006 to 2012, he was the chairman of the Northeastern branch of the German Food Chemical Society (LChG). From 2015 to 2023, he also headed the Institute of Food and Environmental Research in Bad Belzig, Germany, dealing with applied research in the fields of new natural raw materials for new food/feed/non-food products. He is actually a member of the board of the German Nutrition Society (DGE) and the chair of the scientific advisory board of the Max Rubner-Institut (MRI), Germany.

Preface

Foods with certain health-beneficial effects are still attracting widespread consumer interest. Their action is primarily connected with the content of bioactive compounds, which are, besides vitamins and minerals, mainly secondary plant metabolites (flavonoids, carotenoids, phenolic acids, and many more) that are applied to traditional foods or recipes as extracts, or sometimes as pure compounds. However, primary plant metabolites, and sometimes those originating from animal sources (e.g., selected peptides, indispensable amino acids, and fatty acids), as well as compounds from other raw materials such as single cells (e.g., yeasts, bacteria, and algae), can be considered functional ingredients. Biologically active compounds demonstrate various positive physiological functions. The most prominent mechanism discussed is the ability of biological compounds to act as antioxidants and thus, diminish the risk of various diseases, including cancer. Although this topic has been discussed, not without controversy, for quite a while, there is not still a satisfying conclusion, and novel aspects are emerging in research, making this still an interesting research approach. Other functions described are the stimulation of defense mechanisms, in order to e.g., prevent widespread damage or enhance cell repair. One of the limitations in the application of functional ingredients is their stability, which is closely connected to the way the (functional) food or its ingredients are processed along the whole value-added chain.

This compilation of scientific publications aims to bring together the latest knowledge, novel ideas, considerations, and overviews on bioactive compounds that are related to identification of the validity of functional foods. In particular, this compilation contains three review articles and eleven original research contributions, which provide a broad coverage of the progress concerning functional biologically active food and food ingredients. The review chapters present the trends and technological advancements in the possible food applications of spirulina and their health benefits, in discussed phytochemical properties, in extraction optimization, and in the pharmacological benefits of naringin, a quite prominent citrus flavonoid, in addition to characterizing the potential role of bioactive compounds from different vegetable sprouts in the formulation of functional foods. The presented results of recent research aim to investigate the influence of age-related digestive conditions on plant phenolic stability, antioxidant activity, and the bioaccessibility of minerals (Ca, Fe, and Mg) in two types of unfermented, fermented, and fermented dried quinoa and lentils. Another study deals with the content of bioactive compounds and technological properties of matcha green tea and its application in the design of functional beverages. Another study analyses the effects of reducing the amount of sodium-(III) nitrite added to canned meat by enriching it with freeze-dried blackcurrant leaf extract. Further research studies include the investigation of the effect of pulsed electric field pretreatment and air temperature on the course of hot air drying and selected chemical properties of the apple tissue of 'Gloster' apples, the evaluation of the influence of four selected postharvest coffee fruit treatments on the antioxidant and psycho-active properties of Arabica coffee, the discussion of the anti-cancer potential of willow herb extract enhanced canned pork product, the determination of relationships between structural derivatives of malvidin and their anti-cholinergic and anti-inflammatory activity, and many more topics, all highlighting the diversity of this quite complex research field, which still bears so many novel aspects to be discovered.

We highly appreciate the great effort of all authors to prepare their excellent contributions to this compilation.

Michał Halagarda and Sascha Rohn
Editors

molecules

Review

Trends and Technological Advancements in the Possible Food Applications of Spirulina and Their Health Benefits: A Review

Nawal K. Z. AlFadhly [1,*], Nawfal Alhelfi [1], Ammar B. Altemimi [1,2], Deepak Kumar Verma [3], Francesco Cacciola [4,*] and Arunaksharan Narayanankutty [5]

1 Department of Food Science, College of Agriculture, University of Basrah, Basrah 61004, Iraq
2 College of Medicine, University of Warith Al-Anbiyaa, Karbala 56001, Iraq
3 Agricultural and Food Engineering Department, Indian Institute of Technology Kharagpur, Kharagpur 721302, West Bengal, India
4 Department of Biomedical, Dental, Morphological and Functional Imaging Sciences, University of Messina, 98125 Messina, Italy
5 Division of Cell and Molecular Biology, PG and Research Department of Zoology, St. Joseph's College (Autonomous), Devagiri, Calicut 673008, Kerala, India
* Correspondence: nawal.zben@uobasrah.edu.iq (N.K.Z.A.); cacciolaf@unime.it (F.C.)

Abstract: Spirulina is a kind of blue-green algae (BGA) that is multicellular, filamentous, and prokaryotic. It is also known as a cyanobacterium. It is classified within the phylum known as blue-green algae. Despite the fact that it includes a high concentration of nutrients, such as proteins, vitamins, minerals, and fatty acids—in particular, the necessary omega-3 fatty acids and omega-6 fatty acids—the percentage of total fat and cholesterol that can be found in these algae is substantially lower when compared to other food sources. This is the case even if the percentage of total fat that can be found in these algae is also significantly lower. In addition to this, spirulina has a high concentration of bioactive compounds, such as phenols, phycocyanin pigment, and polysaccharides, which all take part in a number of biological activities, such as antioxidant and anti-inflammatory activity. As a result of this, spirulina has found its way into the formulation of a great number of medicinal foods, functional foods, and nutritional supplements. Therefore, this article makes an effort to shed light on spirulina, its nutritional value as a result of its chemical composition, and its applications to some food product formulations, such as dairy products, snacks, cookies, and pasta, that are necessary at an industrial level in the food industry all over the world. In addition, this article supports the idea of incorporating it into the food sector, both from a nutritional and health perspective, as it offers numerous advantages.

Keywords: spirulina algae; chemical composition; health and nutritional value; functional foods; food formulation; biological activity

Citation: AlFadhly, N.K.Z.; Alhelfi, N.; Altemimi, A.B.; Verma, D.K.; Cacciola, F.; Narayanankutty, A. Trends and Technological Advancements in the Possible Food Applications of Spirulina and Their Health Benefits: A Review. *Molecules* **2022**, *27*, 5584. https://doi.org/10.3390/molecules27175584

Academic Editors: Michał Halagarda and Sascha Rohn

Received: 11 August 2022
Accepted: 28 August 2022
Published: 30 August 2022

1. Introduction

Spirulina algae, also known as *Arthrospira platensis*, are members of the class of cyanobacteria (also named blue-green algae) that are classified under the phylum of multicellular organisms. These filaments are unbranched and spiral in shape. Algae are a diverse group of aquatic organisms that have the ability to conduct photosynthesis. In subtropical and tropical climates, such as Hawaii, Mexico, Asia, and Central Africa, they flourish naturally in water tanks that contain high levels of salt and alkaline. GRAS stands for "generally regarded as safe," which is the designation that the Food and Drug Administration (FDA) has bestowed upon it. Research on humans in clinical trials, as well as studies on animals carried out in the most recent decade, provide credence to this assertion. *A. platensis*, *A. maxima*, and *A. fusiformis* are three of the species of spirulina that have been put to use in food, and have been the subject of a significant amount of research [1–4].

Spirulina algae have high nutritional value. As a result of their high protein content (60–70% on a dry weight basis), vitamins, minerals, essential fatty acids, and other nutrients, the FDA has designated them as the ideal food for mankind and a "super food," containing high concentrations of beta(β)-carotene, vitamin B12, iron, trace elements, and the extremely rare essential gamma(γ)-linolenic acid. The Food and Agriculture Organization (FAO) of the United Nations have referred to spirulina as a "highly digestible protein product," and the US space agency has utilized it as a dietary supplement for astronauts. Because of this, spirulina deserves the title of "the food of the future" more than any other food on Earth [1,2].

It has been demonstrated that spirulina is both biologically and economically significant due to the numerous applications that have been developed for it in the food, pharmaceutical, biofuel, cosmetics, and agricultural industries. These algae are readily accessible for purchase and have a significant geographic distribution. This is because the manufacturers want to obtain the biomass of spirulina in order to make use of its important biologically active compounds, such as phycocyanins, phenols, polysaccharides, polyunsaturated fatty acids (PUFAs), carotenoids, vitamins, and sterols. This is why there is such a high demand for spirulina. The majority of these compounds play an important therapeutic role in the treatment of cardiovascular diseases (CVDs), high cholesterol, high blood sugar, obesity, high blood pressure, tumors, and inflammatory diseases. In addition to bolstering the immune system, the presence of these compounds is associated with a reduced risk of developing neurodegenerative conditions, such as Parkinson's disease, Alzheimer's disease, and multiple sclerosis, in particular. Spirulina is regarded as a natural medicine and is utilized in the manufacturing of functional foods and nutritional supplements all over the world due to the qualities that have been described [5,6].

Spirulina algae can be produced in the form of powder, liquid, oil, tablets, or capsules, and are used in many food industries, including the manufacture of sweets, snacks, and pastries. This helps the market meet the demand for variety while also providing highly nutritious food that can aid in the feeding of children and the fight against malnutrition [7]. In addition to the introduction of spirulina in the production of functional beverages, such as fruit juices, which have gained a great deal of relevance in terms of health, it is also employed in the production of dairy products, pasta, oil derivatives, and nutritional supplements [8–10]. In addition to being used as a coloring agent in the food industry, spirulina has a wide range of uses in the areas of human nutrition, animal feed, and fish feed [11–13].

The production of spirulina algae, which are a rich source of protein, has increased in recent years. This coincides with an increase in the demand for protein, which has led to the development of the industry. Because of this, food companies have begun marketing proteins derived from a range of sources, including those derived from animals, plants, single-celled organisms, and spirulina. Pasta, sushi, and jerky are just a few examples of the new food products that have been produced for consumers that are based on spirulina [14,15].

In spite of the facts on the nutritional, environmental, and social significance of spirulina that have been acquired from a broad spectrum of the published literature, it is still possible to draw the conclusion that the production of spirulina is restricted to a select number of natural places. As a result, a group of researchers and scientists from throughout the world are campaigning for extensive spirulina production everywhere in the world.

The objective of this review paper is to shed light on emerging tendencies and technological developments in a variety of facets of spirulina. This includes providing a concise introduction to spirulina as important algae for human food and health, as well as spirulina's various nutritional and biochemical components. In addition, this article offers an insightful discussion on the use of spirulina in the food industry for the purpose of the formulation of a variety of food products. The biological and therapeutic significance of spirulina has also received a lot of attention. This includes its importance in weight control, intestinal flora, and immunological activities, as well as its application for the treatment of various diseases such as diabetes, cancer, cardiovascular, and so on.

2. A Brief Overview on Spirulina as Important Algae for Human Food and Health

The term "algae" refers to a wide collection of organisms that produce their own food via the process of photosynthesis and may be found in a variety of habitats, including marine and freshwater environments [3]. They are found in almost every part of the world and may be divided into two categories. Microalgae are the most basic and fundamental members of the plant kingdom. The bulk of their cells are rather thin, measuring between 3 and 20 μm, and some species form simple colonies. Macroalgae are typically multicellular, expand at a quick rate, and can reach widths of up to 20 m. When compared to the growth rates of terrestrial plants, the rates of growth of macroalgae are significantly higher. Production of macroalgae in maritime habitats, also known as seaweed, does not need the usage of arable land or fertilizer and can take place without either of those factors being present. Seaweeds have the capacity to generate more biomass per hectare than vascular plants do, develop at a far faster rate, and make use of the light energy and carbon dioxide that is taken in from the environment. In the field of applied botany, the tiny algae known as cyanobacteria were once known as cyanophyceae. Cyanobacteria are some of the earth's oldest primitives. They are one of the prokaryotes that have certain properties in common with plants, such as the capacity to carry out photosynthesis, and their cytoskeleton is similar (phototrophic nutrition). The cellular forms of cyanobacteria have undergone several transformations during the course of their evolution, ranging from unicellular to multicellular structures. They can be found in ecosystems containing fresh water, marine life, and terrestrial life, as well as certain severe or harsh habitats, such as hot springs, dry soils, some saline environments, and glaciers [3,16–18].

Arthrospira platensis is the species of spirulina that is multicellular, filamentous, heterogeneous, non-branching, and does not fix nitrogen. It is also capable of photosynthesis and the production of chemical compounds that are necessary for existence. It is grown in liquid farms that are located within open ponds, and flourishes naturally in brackish waters, salt lakes, and warm conditions that are rich in bicarbonate and carbonate [19–23]. In Iraq, many species of spirulina, such as *A. jenner*, were discovered, identified as novel algae, and listed in the inventory of Iraq's algal flora [24,25].

Before 1962, spirulina was considered to be a type of algae. However, in that year, it was reclassified as a member of the prokaryotic kingdom, and the name "cyanobacteria" was suggested for it [18,26]. Different-sized filaments or spiral trichomes can be produced by organisms belonging to the genus *Arthrospira*. Spirulina may fold and bend to varying degrees, taking on shapes that range from a tightly coiled form to a shape that is straight and unwound. Solitary in nature, filaments reproduce by a process known as binary fission. The lengths of the filaments typically range from 2 to 12 μm but can go as high as 16 μm at times [27,28]. The diameter of the thread ranges anywhere from 3 to 12 μm, and the cells that make up the filament contain gas vacuoles that aid in floating [19,29,30].

3. Nutritional and Biochemical Components of Spirulina

Food is the primary means through which the body receives the myriad of vital nutrients that are required for development, the performance of essential biological activities, and the preservation of overall health. On the one hand, considering that our bodies are unable to produce some nutrients, it is necessary to receive them through this diet. On the other hand, several diseases have been related to an imbalance in the human diet, which can be caused by the presence of certain unsuitable nutritional components or the body's incapacity to absorb them [31]. The overall composition of spirulina changes depending on the source of the algae used to cultivate it, the environmental conditions of the manufacturing facility, and the season of the year. Proteins make up between 55% and 70% of the body of spirulina, while carbohydrates make up between 15% and 25%, fats make up between 6% and 8%, minerals make up between 7 and 13%, moisture (dried algae) makes up between 3% and 7%, and dietary fibers make up between 8% and 10% [32]. Figure 1 presents a description of the components that make up spirulina. The proportion of PUFAs is between 1.5% and 2% of the total fat content, and it is rich in linolenic acid, which accounts for 36% of the total PUFAs, as well as vitamins (B1, B2, B3, B6, B9, B12, C, D, and E) and minerals (K, Ca, Cr, Cu, Fe, Mg, Mn, P, Se, Na, and Zn), as well as the pigments (chlorophyll A, xanthophylls, β-carotene, echinenone, myxoxanthophyll, zeaxanthin, canthaxanthin, diatoxanthin, 3-hydroxychininone, β-cryptoxanthin oscillaxanthin, phycobiliproteins, C-phycocyanin, allophycocyanin) and enzymes (such as lipase) [33]. The components of spirulina's chemical makeup are summarized in Table 1.

Table 1. The value of proximate composition of spirulina from different reported research.

Proximate Composition (%)						Food Energy	References
Moisture	Fat/Lipid	Protein	Ash	Fiber	Carbohydrate		
4–5	4–7	65–72	6–12%	3–7	15–25	2.90 cal/g	[34]
3–7	6–8	55–70	7–13	8–10	15–25	–	[32]
5.37	7.19	61.57	7.10	7.93	16.21	–	[35]
5.45–9.92	6.61–6.84	52.85–65.00	9.55–9.93	9.79–11.37	15.29–13.62	329.89–379.58	[36]
5.27	1.27	71.90	3.50	9.70	13.63	353.55	[37]
4.74	6.93	62.84	7.47	8.12	–	–	[38]
–	7.16	52.95	–	–	13.20	–	[39]
1	6	63	8	–	22	–	[40]
–	4	65	3	3	19	–	[41]
4–6	5–7	55–70	3–6	5–7	–	–	
6	6	61	9	–	14	–	
9	7	60	11	–	–	–	

– Not reported.

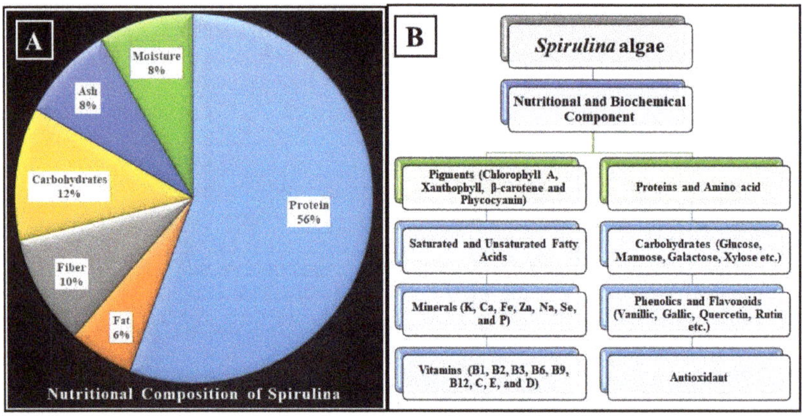

Figure 1. (**A**) Nutritional composition and (**B**) biochemical components of spirulina [36].

3.1. Carbohydrates

According to the findings of various studies, the proportion of carbohydrates present in *spirulina* spp. is around 13.6% [42,43]. On the other hand, a number of additional studies came to the conclusion that the total carbohydrate content of spirulina ranged from 15% to 25% dry weight [32,33,42,44–46]. There is no cellulose present in spirulina algae's carbohydrates; instead, they are made up of a variety of sugars, such as glucose, mannose, galactose, and xylose, in addition to glycogen. As a result, the carbohydrates included in spirulina are simple to digest, as well as nutrient-dense, and may be consumed by elderly individuals and those who have intestinal malabsorption. In addition to that, it has a polysaccharide with high molecular weight known as immolina. Rhamnose is the primary component in it, accounting for around 52.3% of the total sugars generated by spirulina. In another variety, rhamnose accounts for roughly 49.7% of the total sugars produced. Spirulina has a biomass of 1.22 g/L, its polysaccharide content is 2.590% of its biomass, and the total sugars that it contains are 17.275% of its polysaccharides [6,43,47–49]. Polymers, such as glucosamine (1.9%), rhamnosamine (9.7%), and glycogen (0.5%), as well as small amounts of glucose, fructose, sucrose, glycerine, mannitol, and sorbitol, are the primary components of virtually all absorbable carbohydrates. Spirulina has sugars in its cell wall that are analogous to the sugars found in the cell walls of Gram-negative bacteria. These are composed of glucosamine, muramic acid, and glucosamine that have bound to peptides. Due to the fact that these cell walls are relatively thin, digestive enzymes are able to access the contents of the cell with relative ease [42]. Of the various culture media that are utilized during the production of spirulina, each has an impact on the total amount of carbohydrates that are produced. According to the findings of Madkour et al. [39], the percentage of carbohydrate content in spirulina algae grown in low-cost culture media varied depending on the type of nitrogen source present in the culture medium. To accomplish this, all of the nutrients found in the standard medium are swapped out for more affordable and readily available commercial chemicals and fertilizers in the region. The percentage of carbohydrates present in the medium with the standard nitrogen source was 13.20%, but this percentage increased to 16.01% when the nitrogen source was replaced with a medium containing urea. In the ammonium nitrate (NH4NO3) medium, the concentration of carbohydrates rose to 24.50% on a dry weight basis; however, other researchers discovered that the amount of carbohydrates varied depending on the region of production and the kind of product being made [32].

3.2. Lipids/Fats and Fatty Acids

According to the findings of several researchers, the lipid content of *S. platensis* ranges from 5% to 10% of the dry weight. Other research that used more effective extraction techniques found that the percentage was greater than 11%. In most cases, it will contain fats that are necessary for human survival, and free fatty acids will make up between 70% and 80% of the total fat. These total lipids may be divided into a saponified fraction that makes up 83% of the total and an unsaponifiable fraction that makes up 17%, with the unsaponifiable fraction mostly consisting of paraffin, pigments, terpene alcohols, and sterols. Omega-6 fatty acids make up the majority of the total fat, and there is just a trace quantity of cholesterol (less than 0.1 mg/100 g dry mass) present [33,42]. Adults need 1–2% of their total energy intake to come from essential fatty acids, whereas children need 3% of their total energy intake [50].

The location of the closest polyunsaturated point in the MTG is used to describe the optimal omega-6 to omega-3 ratio that some nutritionists advocate, which falls between 4 and 5 [31,51]. It was discovered that the total fatty acid concentration of *A. platensis* is 81.2 mg/g on a dry weight basis, which demonstrates that spirulina is an excellent source of fatty acids [52]. However, Sharoba [38] discovered that the proportion of total saturated fatty acids was 44.21 mg/100 g, but the proportion of total essential unsaturated fatty acids was 55.79 mg/100 g. When looking at the nutritional value of spirulina, researchers found that it has a significant amount of palmitic acid (16:0), which makes up more than 60% of the lipids in *S. maxima* and 25% in *S. platensis*, respectively. While the proportion of saturated palmitic acid in the total fatty acids was 25.8%, the percentage of γ-linolenic acid in the total fatty acids was 40.1%. Spirulina is an excellent dietary supplement for essential fatty acid deficits as a result [42].

Spirulina was discovered to have a significant quantity of PUFAs, with levels ranging from 1.5% to 2.0% fat. This has piqued the curiosity of many researchers, who have been doing studies on PUFAs to determine how much of this nutrient is contained in spirulina [23]. According to the findings of another study, PUFAs made about 30% of the total fats [6], while other researchers reported that the proportion of these fatty acids ranged between 19.4% and 21.9% of the total fatty acids [53]. Its primary fatty acid, 15,12.9-octadecatrienoic acid, accounted for 10.1% of its total fatty acid content, whereas the omega-3 content accounted for less than 1% of its total fatty acid content. Additionally, it had some omega-6 type fatty acids. In addition to this, a significant amount of saturated hexadecanoic acid was discovered (37.6%). The concentration of monounsaturated fatty acids (MUFAs) was low, with the octadec-9-enoic acid (18:1) omega-9 type falling below 2.0%. The quantity of γ-linolenic acid, which came in at 16 mg, had the greatest content, followed by palmitic acid, which had the highest percentage (23%), while myristic acid had the lowest percentage (0.2%) [52].

According to Matos et al. [53], the amount of fatty acids in spirulina algae might vary depending on a variety of parameters, including the growing circumstances and development stage at the time of harvest. The total fatty acid content was estimated to be 4.25 mg/100 g, and it was discovered to include sapienic acid at a level of 2.25 mg/100 g, linoleic acid at a level of 16.7%, and γ-linolenic acid at a level of 14% [51]. According to the findings of Alyasiri et al. [54], one gram of spirulina has a high concentration of linolenic acid of the omega-6 type; specifically, the concentration was 29.1 mg/g, which corresponds to a rate of 2.91%. Additionally, it has PUFAs, which are saturated with 18 carbon atoms and include omega-6. When it comes to the most significant biologically active compounds found in spirulina, phytol had the highest percentage (100%), followed by monolinoleoylglyceroltrimethylsilylether-1b (71.31%), steroid and cholestan-3-ol (2-methylene-3β, 5α) (54.62%), and 9, 12, 15-octadecatrienoic acid, 2, 3-dihydroxypropyl ester (28.21%), hexadecanoic acid methyl ester (23.23%), and methenamine (23.21%) [55]. According to Legezynska et al. (2014), spirulina algae are one of the primary sources of omega-3 fatty acids that fish feed on. Examples of these fatty acids are docosahexaenoic acid (DHA) and eicosapentaenoic acid (EPA). As a result, a higher percentage of these essential fatty acids might be found in fish oils [56]. Linoleic acid, which belongs to the omega-6 group, and alpha (α)-linolenic acid, which belongs to the omega-3 group, is found in the fats of marine algae, suspended algae, and fish oils, respectively. EPA, DHA, α-linolenic acid, and docosapentaenoic acid are the four essential fatty acids (omega-3) that are considered to be of the utmost importance [57]. According to the research conducted by Liestianty et al. [52], the fatty acids contained in spirulina include myristic, heptadecanoic, stearic, oleic, palmitoleic, omega-3, omega-6, linoleic acid, and palmitic acid. Omega-6 kinds, the most significant of which are palmitoleic, oleic, linoleic, and γ-linolenic, and omega-3 types, including α-linoleic acid, are among the most essential types that may be found [38]. Linolenic acid, stearidonic acid, EPA, DHA, and arachidonic acid are found in high concentrations in it [23]. The omega-6 family, which includes γ-linolenic acid and arachidonic acid, and the omega-3 family, which includes EPA and DHA, are the most essential long-chain PUFAs that algae can produce [6]. Utilizing gas chromatography–mass spectrometry (GC-MS) and high-performance liquid chromatography (HPLC), Al-Dhabi and Valan Arasu [51] were able to identify PUFAs in 37 different commercialized spirulina species. Myristic acid, stearic acid, and eicosadienoic acid were identified as the three saturated fatty acids that were present in the spirulina samples. It was found that ten of the unsaturated fatty acids in the spirulina samples were substantially different from one another.

The accumulation of toxic compounds in fish, as well as the odor, strange taste, and oxidative instability of the oils extracted from fish, has a negative impact on the total dependence on the synthesis of long-chain PUFAs (especially omega-3 type) from fish oil. This has a negative impact on the total dependence on the synthesis of long-chain PUFAs from fish oil. As a result, the focus shifted toward the possibility of employing spirulina algae in a commercial environment as a different source to produce these fatty acids [56].

Spirulina algae are a potential source of polyunsaturated fatty acids (PUFAs). Essential fatty acids, such as omega-3 and omega-6, are unable to be produced by humans and, as a result, must be received through the consumption of food. They play a significant role in preserving health and warding off disease. Even though the human gut microbiota is capable of synthesizing long-chain fatty acids, such as linoleic and α-linolenic acids, the synthesis of these acids is controlled by various variables, which makes the consumption of these fatty acids vital for the maintenance of good health [6]. Because it is not commonly found in foods that people eat on a regular basis, despite it having a high nutritional value, the presence of -linolenic acid is interesting. This acid is typically generated in humans from γ-linolenic acid (18:2 omega-6), which comes from vegetable sources [42].

Spirulina is the only food source that contains large amounts of essential fatty acids, especially γ-linolenic acid, which is an omega-6 type that helps regulate all hormones and has anti-inflammatory properties. Comparatively, breast milk is the only food source that contains large amounts of essential fatty acids [19,39,54]. The other supply comes from the oil that is derived from borage, black currant, and evening primrose seeds. In comparison, an evening primrose oil intake of 500 mg has just 45 mg of γ-linolenic acid, whereas 10 g of spirulina has 135 mg of γ-linolenic acid. Comparatively, evening primrose oil only contains 9% linoleic acid, whereas the lipids of spirulina contain around 20–25% of γ-linolenic acid [58].

3.3. Protein

The structure and function of the body, as well as the organization of tissues and organs, are all significantly impacted by the presence of protein. If all of the essential amino acids (EAAs) are present in the food that is ingested, the body will be able to generate the protein that it requires. A protein is considered to be of excellent quality if it includes all of the essential amino acids (EAAs) in the amounts required by the body while retaining its bioavailability. The fundamental components of life are referred to as proteins and amino acids. When comparing various sources of protein, the following criteria should be taken into account: the quantity of protein, the quality of the amino acids, the amount of protein that can be consumed, the ease with which the protein can be digested, and the amount of fat, calories, and cholesterol that the protein contains [57].

Spirulina has a protein level that is quite high, reaching from 60% to 70% of its dry weight (compared to 22% in beef). This type of algae contains an extraordinarily high amount of protein for a plant source; in fact, it contains twice as much protein as the finest source of protein found in vegetables. This amount is significantly higher than the percentages found in animal meat and fish (15–25%), soybeans (35%), powdered milk (35%), peanuts (25%), eggs (12%), cereals (8–14%), and whole milk (3%). Table 2 compares the amount of protein found in various dietary sources to that which may be found in spirulina algae. The amount of protein that these algae contain fluctuates by between 10% and 15% depending on the time of harvest, with the most protein being present in the algae during the early morning [42,57].

Spirulina is a great food source of proteins because it has a high percentage of essential amino acids (EAAs), which account for around 38.81–47.00% of the total weight of proteins. The quantity, proportion, and quality of the amino acid contents of a protein are used to evaluate the protein's overall quality [2,42]. Leucine, valine, and isoleucine are the three amino acids in spirulina that have the greatest concentrations, and complete spirulina proteins include all of the EAAs. Spirulina is superior to all plant proteins, including legume proteins, despite having lower concentrations of the amino acids methionine, cysteine, and lysine than the conventional dietary proteins that come from meat, eggs, or milk [33,45].

Table 2. A comparison of the relative protein content of spirulina algae with other food and food products based on the literature.

Food and Food Products		RPC [a] (%)	References
Spirulina		55.70	[59]
		55–70	[23]
		30–55	[60]
Beef		20.71	
Chicken		21.96	[61]
		22.25%	[62]
Fish	Carp	16.70	[63]
	Cod	17.40	
	Herring	18.10	
	Salmon	18.40	
Whole egg		12.60	[64]
Sausage		14.43	[65]
Milk	Buffalo	4.17	[66]
	Camel	3.38	
	Cow	3.56	
	Goat	3.44	
	Sheep	4.35	
Whey Protein	Buffalo	0.72	
	Camel	0.58	
	Cow	0.53	
	Goat	0.54	
	Sheep	0.74	
Whey Proteins		54.8 3.0–74.8 4.1%	[67]
Whey Protein Concentrate Powder		33.30	[68]
White Cheese		16.20	[69]
Organic hard cheese		21.53–25.70	[70]
Soy bean		38.30–40.30	[71]
		35.35–39.80	[72]
Common Oat		11.61	[73]
Oat grains		9.70	[74]
Black Bean (Organically produced)		25.20	[75]
Maize		12.65–12.45	[76]
Rice		7.76–8.31	[77]
Wheat		11.88	[78]

[a] Relative Protein Content.

Undernutrition is a problem that affects public health, particularly in developing nations. This has led to a trend toward the utilization of spirulina algae, which has been utilized as a functional food for decades. Developing countries are more at risk of undernutrition. According to Salmeán et al. [79], spirulina has adequate sensory properties and has not shown any sign of toxicity, which means that it is safe for human ingestion [79]. Because it has such a high proportion of macro- and micronutrients, this algae was initially utilized in the health food and nutritional supplement sector as a protein supplement. In many countries outside Europe and North America, it is often referred to as supplementary food [23,48,80,81]. According to the findings of El-Chaghaby et al. [82], the protein content of spirulina platensis was 53.30% (dry weight), which was much higher than that of Chlorella vulgaris (20.67%) and Scenedesmus obliquus (31.07%).

Batista et al. [83] were able to boost the protein availability in the biscuit samples by adding *A. platensis* as a high protein food source. On a dry weight basis, *A. platensis* is composed of 68.9% protein; therefore, this allowed them to achieve their goal. The greatest levels of IVPD (in vitro digestibility protein) were reported in samples that included 6% algae. These samples contained 14.3% protein and had IVPD values of 83%. The quantity of protein that could be digested in the treated samples was 11.9 g/100 g of biscuits. This proportion is much greater than the one found in the control sample, which had a protein content of 9.8% and a digestible protein content of 7.3 g/100 g of biscuits. Protein insufficiency is a prevalent condition that affects more than 300 million individuals throughout the world [79]. Protein deficiency may be treated, but it requires a dependency on the ingestion of protein from various sources.

The plant proteins known as phycobiliprotein, which are C-phycocyanin and allophycocyanin in a ratio of around 10:1, are responsible for the majority of spirulina's beneficial effects on human health [32,45]. C-phycocyanin is one of the primary proteins that may be discovered in moss, and it accounts for around 20% of the total dry weight of all protein fractions. This particular pigment is a molecule that is similar to biliverdin in that it contains phycocyanobilin [3,30,43,79,84].

Proteins included in spirulina are not difficult to digest, even for elderly individuals who have difficulties absorbing complex proteins via their intestines and who adhere to certain diets. This is because the cell walls of algae are composed of mucopolysaccharides, which are simple sugars that are easy to digest and absorb. Cellulose, on the other hand, is indigestible for humans. Additionally, it has a high digestibility of its proteins, ranging from 85% to 95%, and the process of acquiring proteins through proteolytic enzymes is straightforward in comparison to the process of obtaining enzymes from vegetables. Patients suffering from malnutrition, such as kwashiorkor, in which there is a reduction in the capacity of the gut to absorb nutrients, would be good candidates for this treatment. It was discovered that this algae is more beneficial in children who are suffering from malnutrition than milk powders, which contain lactic acid that is difficult to digest and absorb. This alga was shown to be more effective than milk powders [3,43,57,85].

3.4. Amino Acids

Proteins are composed of amino acids, which are the core structural units of proteins and also the fundamental components of numerous coenzymes, hormones, and nucleic acids. Foods that include all of the EAAs are essential for overcoming a wide variety of dietary and health issues, since these EAAs play a variety of structural and functional roles in the body. It is essential to have access to a sufficient amount of protein in one's diet in order to preserve the structure and function of one's cells, as well as one's health and capacity [10]. There is a correlation between the presence of EAAs and the quality of proteins. Animal proteins are the only source of complete proteins and are an abundant source of EAAs, which the human body is unable to produce on its own for biological reasons. Plant proteins are considered to be incomplete proteins because they are missing one or more of the EAAs. These essential amino acids include histidine, isoleucine, lysine, methionine, phenylalanine, threonine, and valine [12].

Proteins derived from spirulina have a comprehensive profile thanks to the presence of EAAs and non-EAAs in enough quantities in each molecule. It has EAAs, which account for 47% of the algae's total protein weight, so it is really good. These include leucine, tryptophan, methionine, phenylalanine, lysine, thionine, and valine; the corresponding levels of these amino acids were 55, 10, 14, 28, 30, 33, 36, and 45 mg/g. In terms of the non-EAAs, they contribute to the synthesis of the proteins that are required by the cells of the body alongside the EAAs, which is how they play a vital part in the body. Any non-EAAs that are present in excess can be turned into glucose, which serves as a source of energy for the body. In general, it was discovered that the non-EAAs contained in spirulina include cysteine, histidine, proline, tyrosine, glycine, serine, arginine, alanine, aspartic, and glutamate acid at levels of 7, 10, 27, 30, 32, 33, 44, 47, 60, and 92 mg/g, respectively. These amino acids were detected in spirulina [42,52].

Spirulina contains the amino acids methionine, lysine, threonine, tryptophan, isoleucine, leucine, phenylalanine, valine, alanine, arginine, cysteine, glutamine, glycine, histidine, proline, serine, and threonine [46]. According to research conducted by Siva et al. [18], the highest levels of the EAAs leucine and valine were found in spirulina at 5400 and 4000 mg/100 g, respectively. On the other hand, the highest levels of glutamic acid and aspartic acid, which are not EAAs, were found at 9100 and 6100 mg/100 g, respectively. Salmeán et al. [79] reported that EAAs were found in the highest amounts of leucine (5380 mg/100 g), valine (3940 mg/100 g) and isoleucine (3500 mg/100 g), while the non-EAAs were reported as glutamic acid, aspartic acid, and alanine with values of 9130 mg/100 g, 5990 mg/100 g, and 4590 mg/100 g, respectively.

Isoleucine, leucine, lysine, valine, arginine, alanine, aspartic acid, glutamic acid, and glycine were the amino acids that were discovered in dry spirulina algae; their quantities were 3.209, 4.947, 3.025, 3.512, 4.147, 4.515, 5.793, 8.386, 3.099 g/100 g, respectively [43]. According to the research conducted by [86], proteolytic enzymes were utilized in order to extract the amino acids from dry extracts of spirulina algae. It was discovered that the extracts were abundant in free amino acids as well as short peptides. Furthermore, it was discovered that the Alcalase enzyme extract contained the highest percentage of amino acids, which was 45% (weight-to-weight) dry extract, whereas the extract without enzymatic aid produced only 34%. This value was in line with the range of amino acid content in spirulina, which was between 50% and 65% (weight-to-weight). The quantity of amino acids that could be extracted from spirulina using the Alcalase enzyme approach was 1426 μmol/g, while the amount that could be extracted using the traditional method was only 573 μmol/g [86].

Because of the high nutritional value of EAAs, their application in the production of nutritional supplements has become more widespread in recent years. EAAs make up around 35% of the total amino acids found in spirulina [6]. Due to the fact that spirulina and its extracts include a diverse assortment of EAAs and non-EAAs, they can serve as a source of nutrients and nutritional supplements. It was discovered that glutamic acid and aspartic acid are the most prevalent non-EAAs found in dry algae, whereas isoleucine and phenylalanine are the most common EAAs. Glutamic acid was shown to be the most abundant of the three. The level of the amino acid arginine was found to be greater in the bulk of spirulina, with 8.153 μmol/100 mg, compared to the extract, which had 7.89 μmol/100 mg of dry weight. It was discovered that the biological value of the proteins in spirulina is very high, complete, and contains all of the EAAs, despite having a low percentage of methionine and cysteine when compared to the standard proteins egg albumin and milk casein. This was discovered when comparing the proteins of spirulina to those of egg albumin and milk casein. However, according to Salmeán et al. [79], the proteins found in spirulina are of better quality than the proteins found in any other vegetable protein source, including legumes and soybeans.

Fortifying foods with spirulina, which leads to an increase in the amount of amino acids in treated foods, has been shown to be beneficial by a significant number of researchers who have published their findings. This has a beneficial impact on increasing the nutritional content of foods that have been fortified. Spirulina in its dry form is an excellent source of amino acids, particularly the amino acids alanine, arginine, aspartic acid, glutamic acid, leucine, and valine, which are the most plentiful. The addition of spirulina to biscuits at several levels (0%, 5%, 10%, and 15%) was the subject of an experiment, and the results showed that the amount of amino acids present in the finished product grew as the percentage of addition increased [34]. Aljobair et al. [10] found that the addition of spirulina to two different types of date juice boosted the amount of essential amino acids (EAAs). The total amount of EAAs found in spirulina algae, as well as in two other types of date juice that were supplemented with spirulina at a concentration of 10%, were, respectively, 38.46%, 48.69%, and 46.02%. In addition, spirulina has a high proportion of total non-EAAs; specifically, 61.54%, which is followed by date juice fortified with 10% spirulina, which has 53.98% and 51.32% respectively. In comparison to the proteins found in dates, spirulina algae proteins are of a far higher quality, which contributes to the juice's high concentration of amino acids.

3.5. Vitamins

Vitamins are necessary micronutrients, but since humans are unable to produce them in adequate quantities, they need to be received from the food that they eat. Their absence is linked to a wide variety of ailments, and foods derived from algae are particularly rich in vitamins [3,81]. Of these algae, spirulina is employed in the development of functional foods. As an alternative source of vitamin production, it generates a significant quantity of spirulina at an affordable price [86]. It was demonstrated that spirulina contains a high concentration of vitamins, and the addition of these algae to food products, such as drinks and juices, can boost the vitamin content of the product while also improving its nutritional and health benefits [10]. These algae contain all vitamins, including vitamin A (β-carotene), vitamin D, vitamin E, vitamin K, vitamin C, and vitamin B complexes, such as thiamine (vitamin B1), riboflavin (vitamin B2), niacin (vitamin B3), pantothenic acid (vitamin B5), pyridoxine (vitamin B6), folic acid (vitamin B9), and cobalamin (vitamin B12) [6,10,32,85,86]. Edelmann et al. [87] investigated the amount of various vitamins that are present in dry spirulina algae. These vitamins include riboflavin (vitamin B2), cobalamin (vitamin B12), and folic acid (vitamin B9), and their respective amounts were 36.3, 2.4–0.6, and 3.5 µg/g, while the amount of niacin (vitamin B3) was 0.16 mg/g. The researchers came to the conclusion that the dry spirulina had a low quantity of folic acid and that the majority of vitamin B12 was inactive. Therefore, they suggested that spirulina powder in the amount of 4–5 g per day is what one ought to consume on a daily basis in order to fulfill one's needs for folic acid and vitamin B12 [87]. According to Choopani et al. [31], dried spirulina includes a wide variety of vitamins, the levels of which range from 100–200, 1.5–4.0, 0.5–0.7, and 5.0–20 mg/100 g, respectively, for vitamins A, B1, B6, and E, respectively.

According to the findings of a number of studies that referred to the analysis of vitamins, dried spirulina contains 3.6 mcg/g of vitamin B12 and is an excellent source of β-carotene with a content of 5.8 mg/g. β-carotene is absorbed by the body and transformed into vitamin A. Consuming between 1 and 2 g of spirulina per day is adequate to fulfill the body's daily requirements for vitamin A, which are around 1 mg/day [52]. It is crucial to remember that 100 g of spirulina includes 1100 IU of vitamin A, which is necessary for maintaining healthy immunity, eyesight, and reproduction [86]. It was described that dried spirulina is abundant in beta-carotene, which makes up nearly half of the carotenoids, and 1 g of spirulina has 0.9 mg of all-trans β-carotene. It was also reported that spirulina includes a significant amount of vitamins, including vitamin A (β-carotene) at 211 mg/100 g, vitamin K at 1090 µg, and vitamin B12 at 162 µg [88]. According to Seyidoglu et al. [43], dry spirulina is an excellent source of vitamins, such as vitamin E, vitamin B1, vitamin B2, vitamin B3, vitamin B6, vitamin B12, vitamin K, folic acid, biotin, and pantothenic acid. The amounts of these vitamins in dry spirulina were 5, 3.5, 4.0, 14.0, 0.8, 0.32, 2.2, 0.01, 0.005, 0.1 mg/100 g. According to Stanic-Vucinic et al. (2018), 3 g of spirulina contains 11,250 IU, 75 µg, and 9 µg of vitamin A, vitamin K, and vitamin B12, respectively. Spirulina is a good source of these vitamins. A sample of 10 g of spirulina contains vitamin A (23000 IU), vitamin B1 (0.35 mg), carotene (14 mg), vitamin B2 (0.40 mg), vitamin C (0.8 mg), vitamin B3 (1.4 mg), vitamin D (1200 IU), vitamin B6 (60 mg), vitamin E (1.0 mg), folic acid (1.0 mg), vitamin K (200 mg), vitamin B12 (20.0 mg, biotin (0.5 mg), pantothenic acid (10.0 mg), and inositol (6.4 mg) [32]. It was demonstrated by Falquet and Hurni [42] that the β-carotene found in spirulina can be converted into vitamin A in mammals. The daily requirement for vitamin A in mammals is estimated to be less than 1 mg, and it only takes 1–2 g of spirulina to fulfill this requirement [79]. Spirulina in its dry form also has vitamin E (190–50 mg/kg), which is enough to meet the recommended daily allowance of 12 mg. Spirulina is enriched with vitamin B12, a nutrient that is hard to come by in a vegetarian or vegan diet because it is only found in animal products. It is essential for the production of red blood cells as well as DNA, and it is present in quantities that are four times higher than those found in raw liver [6,86,89]. In addition to that, it has the bioavailable form of vitamin B12 known as methylcobalamin, which is found in it. Spirulina, which has 35–38 μg/100 g, is regarded as the food that contains the highest concentration of this vitamin [33]. Because the daily needs of vitamins B1, B2, B3, and B12 are met by consuming 20 g of spirulina, it is considered a complete source of these vitamins [45]. Mogale [90] found that the vitamins found in spirulina algae play an important role in the metabolism of cells. Additionally, he found that the vitamin A, vitamin E, and vitamin C present in the aqueous extract of spirulina are nonenzymatic antioxidants that protect membrane lipids from oxidative damage. They had a vitamin A content of 18.1 mg/g, a vitamin E content of 3.91 mg/g, and a vitamin C content of 17.2 mg/g, whereas the recommended daily allowances for these vitamins are 0.9 mg/day, 90 mg/day, and 15 mg/day, respectively [54].

There are various environmental variables (cultivation conditions), harvesting methods, and cell drying methods that can have a significant impact on the vitamin content of microalgae [6]. These factors can also substantially modify the vitamin content of microalgae In addition to this, it was discovered that the process of extracting vitamin B12 from spirulina algae has an impact on the total quantity of vitamin that is produced, There were six different extraction procedures employed. The extraction using potassium cyanide (KCN), which recovered 92–95% of the vitamin, was shown to be the most effective approach [91]. Rats were given a B12-deficient diet for the duration of six weeks in a comparative study that was carried out by Usharani et al. [23]. The rats were also given spirulina as a source of vitamins to supplement their diet. Rats given a diet consisting of spirulina had a significantly increased amount of cobalamin in their livers. El-Nakib et al. [34] also found that an increase in the proportion of algae led to an increase in the number of vitamins present in enriched biscuits (0%, 5%, 10%, and 15%). Dry spirulina has significantly higher vitamin content than other food sources, such as liver, carrots, spinach, and vegetables. Aljobair et al. [10] came to the conclusion that enriching date juice with spirulina led to an increase in the quantity of vitamins already present in the juice as well as a reduction in the deficiency of vitamins, such as vitamin B3, vitamin B5, vitamin B6, vitamin B12, vitamin E, and vitamin K in the juice. This was one of the main findings of the study.

3.6. Minerals

Spirulina is an excellent source of a wide variety of minerals, such as potassium (K), calcium (Ca), chromium (Cr), copper (Cu), iron (Fe), magnesium (Mg), manganese (Mn), phosphorous (P), selenium (Se), sodium (Na), boron (B), molybdenum (Mo), and zinc. Other nutritional components include boron (B), phosphorous (P), and selenium (Zn). It also has a very high proportion of both macro- and micronutrients, in addition to other nutritious components, and its products are utilized in agriculture, the food industry, the pharmaceutical industry, the perfumery industry, and medical practice [23,33]. The quality of the nutritional supplement offered by two different species of spirulina, *A. platensis* and *A. maxima*, was analyzed, and the mineral components were identified for each of the four concentrations [92]. Using microwave-induced plasma atomic emission spectroscopy, the levels of 15 elements that were found in 11 different dietary supplement products were investigated and studied. These elements included Al, Ba, Ca, Cd, Cr, Cu, Fe, K, Mg, Mn, Na, Ni, P, V, and Zn [92]. In all of the analyses conducted, the results showed that the levels of the mineral elements in the spirulina samples fell below or were within the level of the recommended daily intake (mg/daily) that was established by the Codex Alimentarius

Commission (CODEX). This information was provided by the FDA. However, there was an exception for the concentration of Cd, which was beyond allowed limits; this was explained by the capacity of algae to bioaccumulate this element. The Cd concentration was above permissible values.

The Fe content in spirulina is 10 times higher than that of other foods that are rich in Fe. Spirulina is an iron-rich diet. Spirulina's Fe content is absorbed by the body at a rate that is approximately 60% higher than that of ferrous sulfate, which is often included in Fe supplements. Due to the high percentage of Fe in spirulina and the relevance of Fe in the treatment of Fe deficiency (anemia), many studies have been conducted on the topic. Pregnant women and children are particularly at risk of Fe insufficiency. It was revealed that it does not cause any toxicity in comparison to Fe supplements that are administered in the form of ferrous sulfate, which is known to produce toxicity issues and frequently leads to diarrhea. Additionally, cereals have a high concentration of phytic and oxalic acids, both of which greatly inhibit the bioavailability and absorption of Fe (as happens in spinach). For spirulina, the bioavailability of Fe was proven in both people and rats. Many dietitians indicate that most food sources have relatively low ratios of K to Na; however, spirulina has a high K content. Spirulina contains Ca, P, and Mg in proportions that are comparable to those found in milk. Because of this, there is no danger of decalcification, which can occur when the amount of P in a food source is increased [33,42]. According to Salmeán et al. [79], the Fe content of spirulina algae is roughly 580–1800 mg/kg, which is a reasonably significant amount when compared to cereals, which are a rich source of Fe and contain 150–250 mg/kg [79].

According to the findings of El-Nakib et al. [34], the incorporation of spirulina during the production of biscuits led to an increase in the amount of minerals present, hence improving the products' overall nutritional value. This is because attending school has been shown to improve the health of children who are undernourished. The same study found that the Fe concentration of spirulina, which is 0.522 mg/g, is greater than the Fe content of spinach, which is 0.109 mg/g, and the Fe content of soybeans, which is 0.115 mg/g. The amount of Ca, Mg, Fe, P, K, and Na that was present in dry spirulina was 168, 2.55, 0.52, 9.18, 18.30, and 10.98 mg/g, respectively; meanwhile, the amount of Mn, Zn, B, Cu, Mo and Se that was present in spirulina was 19, 2, 30, 3, 30, and 5 g/g, respectively. According to Verdasco-Martín et al. [93], the percentage of minerals found in spirulina is 15%, and the most important minerals are Fe, Ca, P, and K. These minerals are significant and necessary for the construction of the body as well as the performance of its numerous essential functions [94]. When researching the role of spirulina as a nutritional supplement, Siva et al. [18] found that the algae contained minerals that were within the recommended dietary allowance (RDA). These minerals included Ca (1300 mg), Fe (10 mg), iodine (150 mcg), P (700 mg), Mg (420 mcg), Zn (11 mg), Se (0.055 mg), Cu (0.9 mg), Mn (2.3 mg), B (1000–10,000 mcg), and germanium (1.5 mg). The majority were within the range of 100 g of spirulina, although Cr (35 mcg) and Mo (45 mcg) were safe with 10 g/day of spirulina consumption (DRIs, 2004). According to the findings of a number of studies, the most important minerals found in spirulina are Fe, Ca, P, and K, with percentages of 0.058–0.18%, 0.13–1.4%, 0.67–0.9%, and 0.64–1.54% of the dry weight, respectively [45].

Spirulina contains all of the necessary minerals, including Ca, K, Mg, Na, P, Cu, Fe, Mn, Zn, Cr, Se, B, and Mo, which may be obtained in the following proportions: 922.278, 2085.28, 1.1902, 1540.46, 2191.71, 1.2154, 273.197, 5.6608, 3.6229, 0.325, 0.394, 0.325, and 0.394 mL/g [39]. Spirulina is a nutritional supplement that is appropriate for vegetarians, because it is able to absorb the mineral elements that are present in the culture medium in high concentrations while it is growing. These elements include Fe, Ca, P, and K in the following proportions: 0.58–1.8, 1.3–14, 6.7–9.0, and 6.4–15.4 g/kg, respectively [33]. It was revealed that the quantity of minerals in dried spirulina algae is as follows: 700 mg/100 g for 700 mg/100 g for Ca, 0.28 mg/100 g for Cr, 1.2 mg/100 g for Cu, 100 mg/100 g for Fe, 400 mg/100 g for Mg, 5.0 mg/100 g for Mn, 800 mg/100 g for P, 1400 mg/100 g for K, 900 mg/100 g for Na, and 3 mg/100 g for Zn [43]. The mineral composition of spirulina is determined by the source of the algae used to make it as well as the conditions under which it was grown [28,33]. The production of spirulina algae, which may absorb heavy metals if they are present in the culture medium, has been shown to raise concerns about the presence of heavy metals, which has been reported to be a reason to be concerned. Therefore, it is utilized in the removal of heavy metals from contaminated water in the cultivation of spirulina. According to EU Commission Regulation (EC) 1881/2006 for the maximum levels of some pollutants in foodstuffs, the allowable amounts of lead (Pb), Cd, and mercury (Hg) for food supplements made from algae are 3, 3, and 0.1 mg/kg, respectively. This regulation establishes the maximum levels of certain pollutants that can be present in foodstuffs.

3.7. Pigments

One of the greatest dietary sources rich in pigments are spirulina algae, particularly C-phycocyanin, which has 14% of the element Fe in it. Additionally, it has the greatest value of chlorophyll (1%) Chlorophyll A pigment is a type of phytonutrient that assists the body in cleansing and detoxifying itself [43]. Spirulina included chlorophyll and phycocyanin pigments, both of which are potent antioxidants; the percentages of these pigments were 1.472% and 14.18%, respectively [2]. Researchers from all around the world are interested in examining the efficacy of these pigments, as well as their uses, applications, and quantitative effects. It was confirmed that the alcoholic extract of spirulina algae contains many important pigments. These pigments include chlorophyll and β-carotene, as well as the protein pigments phycocyanin, allophycocyanin, and phycoerythrin, with their quantities being, respectively, 0.301 mg/g, 0.372 mg/g, and 0.247 mg/g. This is due to the fact that the majority of the antioxidant activity comes from these pigments [95]. There are three primary categories for the pigments found in phycobiliproteins. The primary pigment is called C-phycocyanin, and it accounts for around 20% of the dry weight of the substance. It is also sold in the commercial world as an antioxidant and an anti-inflammatory agent, in addition to its usage as a natural colorant. Phycoerythrins are water-soluble protein compounds that have many beneficial effects on one's health [45,96–99]. Phycoerythrins and allophycocyanins are typically present in lower amounts in a 1:10 ratio.

Spirulina contains a number of essential plant pigments, including total carotenoids (400–650 mg/100 g), β-carotene (150–250 mg/100 g), xanthophylls (250–470 mg/100 g), zeaxanthin (125–200 mg/100 g), chlorophyll (1300–1700 mg/100 g), and phycocyanin (15,000–19,000 mg/100 g) [89,100]. It was also reported that the overall concentration of carotenoids in dried spirulina was 6.928 mg/kg, with the total content of xanthophylls accounting for 83.6% (5.787 mg/kg) of the total carotenoid content [43,101]. The levels of phycocyanin, chlorophyll, and carotene found in spirulina were measured and found to be 180, 11, and 6 mg/g of spirulina, respectively [52]. Phycocyanin is a member of an important group of pigments that are found in spirulina algae. Patel et al. (2005) conducted research on the quantitative assessment of the phycobiliprotein pigments C-phycocyanin, allophycocyanin, and phycoerythrin in three distinct species of cyanobacteria. These cyanobacteria species were *spirulina* spp. and *Phormidium* spp. and *Lyngbya* spp. The amounts of phycocyanin pigment were 17.5%, 4.1%, and 3.9% weight to weight for the three different algae, respectively. The amounts of allophycocyanin pigment were 3.8%, 1%, and 0.8% weight to weight, while the quantity of phycoerythrin pigment was 1.2, 0.3, and 0.4% weight to weight. Usharani et al. [23] also reported that spirulina algae have a variety of photosynthetic pigments, including chlorophyll A, xanthophyll, β-carotene, echinenone, myxoxanthophyll, and phycobiliprotein pigments, in addition to lutein, zeaxanthinoxant echinenone, and β-cryptoxanthin pigments [43,101].

Protein pigments are responsible for converting the light energy that falls within the visible wavelength range, which ranges from 400 to 700 nanometers, into chemical energy [102–105]. At a pH of 7, phycocyanin had the greatest amount of solubility; however, the protein quickly denatures at a pH lower than 3, which causes the color to precipitate [48,102]. It has been shown that the use of an ultrasonic water bath is the most effective way for extracting phycocyanin from spirulina algae. This method produced 43.75 mg/g of pigment at a concentration of 0.21 mg/mL [106]. There are a few different approaches that may be taken to accomplish this task.

Pigments are sensitive to a wide variety of environmental conditions, including temperature, light, oxygen, pH, and oxidizing agents, such as ascorbic acid and trace metal ions, among others. C-phycocyanin is a protein pigment that is photosensitive, heat-unstable, and vulnerable to the oxidation caused by free radicals [48]. Stability in the pigment is provided by the sugars glucose (20%), sucrose (20%), or sodium chloride (2.5%). The acids citric and benzoic can further reduce the heat degradation rate of the pigment [103,107,108]. Abd El-Monem et al. [103] conducted research to investigate how the pigment content was affected by the pH function. The highest pigment content was found at pH 10, with a concentration of 2.8 μg/mL chlorophyll A and 2.6 μg/mL carotenoids, whereas the highest pigment stability was found at pH 5–6 [103,109]. The impact of medium salinity on the pigments was also tested at concentrations of 15, 20, 25, and 30 parts per thousand (ppt), and it was seen from the quantity of chlorophyll A and phycobiliprotein pigments that salinity did not influence these pigments [110]. It was discovered that the lowest possible light intensity resulted in the best production of pigments when exposed to 80 $\mu mol/m^2/second$. This was the case regardless of the influence that light intensity had on the exposure length. On the other hand, photodegradation of pigments happens by photooxidation when exposed to very intense light intensity over an extended period of time. The amount of phycocyanin that was present increased as the light intensity rose. The light had an intensity of 160 $\mu mol/m^2/second$. Once more, the production of pigment was made significantly better. The longer the darkness lasted, the higher the pigment production as well as the amount of energy and biomass that was produced [111]. Sandeep et al. [59] conducted research on the influence that the type of culture medium has on the quantity of pigments. They discovered that the contents of phycocyanin pigment in the treated seawater medium and the m-NRC standard medium were comparable, coming in at 50.9 and 50.95 mg/g, respectively, but that it dropped to 49.82 mg/g in the mixture medium consisting of a combination of m-NRC standard medium and mixture medium (treated seawater: shrimp wastewater). When compared with the standard medium, the content of total carotenoids in the mixed medium was 20.3% lower than what it was in the standard medium. The chlorophyll pigment level was comparable between the two media.

3.8. Phenols and Flavonoids

Phenolic chemicals, also known as polyphenolics and present in abundant amounts in algae, are often regarded as being among the most significant naturally occurring antioxidant molecules. They are by-products of the metabolism and are associated with the systems of chemical protection mechanisms that algae have against a variety of biological stimuli, such as ultraviolet rays, pathogens, and mineral pollution [6]. In general, phenolic acids make up one-third of the phenolic compounds, while flavonoids make up the remaining two-thirds. Flavonols and anthocyanins make up the majority of the flavonoid compounds that are found in the diet. Among the phenolic compounds found in algae is the fluorotannin compound, which is primarily found in brown algae, but can also be found in some red algae in smaller quantities. This compound participates in the formation of the cell wall, plays a role in algal proliferation, and acts as a protective mechanism against biological factors [112]. According to Hidayati et al. [95], spirulina algae have a total of phenolic compounds that equate to 26.64 mg of gallic acid per gram of extract. Due to the redox characteristics that they possess, they play a significant role in antioxidant activity and are very efficient in scavenging harmful free radicals. The total quantity of phenols recorded in the spirulina extract was determined to be 2238.46 mg of gallic acid/kg of the extract, while the total amount of flavonoids found in the extract was determined to be 142.23 mg of quercetin/kg of extract [82]. Matos et al., [53] found that the alcoholic extract of spirulina algae (*A. platensis*), included the highest concentration of polyphenols. This concentration was 205 mg gallic acid/100 g dry weight, although the percentage was greater in the aqueous extract, which contained 334 mg gallic acid/100 g. According to the findings of Michael et al. [36], dry spirulina algae contained a significant quantity of total phenols, flavonoids, and carotenoids. The amount of phenols in spirulina extract was found to be 409.28 mg gallic acid/g, and the amount of flavonoids was found to be 13.25 mg rutin/g extract. Another study found that fresh and dried spirulina algae extract both contained a significant quantity of total flavonoids, with values of 22.10 mg/g and 10.91 mg/g, respectively [113]. According to the findings of Gabr et al. [114], the highest concentration of total phenols was found in spirulina biomass (51.20 μg/mL), then in the ethanolic extract (49.48 μg/mL), and the lowest concentration was found in the aqueous extract (15.26 μg/mL). When calculating the total flavonoids, it was observed that the amount of flavonoids acquired from algal biomass was the largest (97.73 μg/mL), followed by the amount received from the ethanolic extract (69.07 μg/mL), and the amount obtained from the aqueous extract was the lowest (4.67 μg/mL). When compared to algae and the extract made with ethanol, it was noted that the amount of phenolic and flavonoid chemicals present in the aqueous extract was significantly lower.

According to Salamatullah, [115], the methanolic extract of algae resulted in 3.46 mg/g of phenols, while the ethanolic extract resulted in 2.28 mg/g of phenols. The acetone extract of algae resulted in 7.37 mg/g of phenols, which was the highest value among the three extracts. In comparison to the other extracts, the quantity of total flavonoids that were produced by the methanolic extract, both with and without the addition of hydrochloric acid, was the greatest, reaching 6.37 and 6.05 mg/g, respectively. In contrast to the dried sample, fresh spirulina was shown to contain a greater number and higher quality of beneficial chemicals, according to the findings of another study. It was reported that the total flavonoids of fresh spirulina had 469.96 quercetin/g extracted in dry weight, whereas dried spirulina had 119.43 quercetin/g extracted in dry weight [116]. It may be deduced from the fact that the phenolic acids measured 38.64 and 7.50 mg gallic acid/g extract on a dry weight basis, respectively, that this is because of the variation in the amount of water present in the sample. Pyrogallol, gallic, chlorogenic caffeine, vanillic, p-coumaric, naringin, hespirdin, rutin, quercetrin, naringenin, catechin, and hespirtin are some of the phenolic chemicals that might be found in spirulina algae. Pyrogallol was found to contain the highest amount of phenolic compounds in both the biomass and the aqueous extract of spirulina, with a concentration of 638.50 and 12.33 mg/100 g, respectively. On the other hand, the compound E-vanillic acid contained the highest amount of phenolic compounds in the ethanolic extract, with 18.20 mg/100 g. Flavonoids such as catechein, epicatechein naringin, hespirdin, rutin, quercetrin, quercetin, naringenin, hespirtin, kampferol, and apigenin can be found in spirulina algae. In the spirulina biomass, the flavonoid component with the greatest concentration was hespirdin, which had a value of 9.013 mg/100 g. This was followed by ethanol and aqueous extract, which had concentrations of 0.652 and 0.359 mg/100 g, respectively [114].

The bioavailability of various phenols is affected differently depending on the type of food consumed [117]. Because it is found in enough quantities, spirulina may be regarded as a reliable source of phenolic compounds, In addition to being primarily found in fruits and beverages, such as tea, wine, and coffee, polyphenols can also be found in vegetables, leguminous plants, and grains [112]. The amount of polyphenols present in different foods can be attributed to a number of factors, including genetic, environmental, and industrial A variety of parameters, including the acidity function of the growth media, can have an effect on the total quantity of phenolic compounds that are present in spirulina. Abd El-Monem et al. [103] cultivated spirulina in a conventional growth medium by adjusting the pH values, and extracts of algae were made using a variety of solvents at a concentration of 70% (acetone, methanol, and ethanol). It was observed that the optimal pH was 10, which led to the maximum amount of flavonoids in the ethanol extract (7.6 mg/g) and the highest quantity of phenols in the acetone extract (0.52 mg/g) [103]. Therefore, they claimed that algae may produce phenols as a defense mechanism against disease, stress, depletion of nutrients, and an increase in the pH of the culture media. This resulted in an enhanced generation of phenols to reduce oxidative stress induced by high pH. Additionally, it was shown that the polyphenols in these algae include a greater quantity of total phenol compounds and a smaller quantity of total flavonoid compounds [36].

The determination of phenolic compounds in the alga *A. platensis* and their application in food preservation to increase the shelf life was the subject of just a few investigations. In order to achieve this goal, the treated product's oxidative deterioration is decreased, and the product's structural and organoleptic features are improved. The risk of developing illnesses that are brought on by oxidative stress is decreased by eating foods that are high in antioxidants. Even when utilizing a concentration that was greater than 40 mg/mL, it was determined that the extract of spirulina did not demonstrate any inhibitory effect against the stable radicals caused by the substance DPPH (1,1-diphenyl-2-picrylhydrazyl) [118]. In addition, Batista et al. [83] revealed that when estimating the amount of total phenols in biscuits enriched with spirulina algae, the addition of spirulina by 2% did not give any significant differences compared to the control sample, as the amount of phenols was 1.4 mg gallic acid/g, while the content of phenolics increased significantly to 1.6–1.7 mg gallic acid/g when spirulina was increased to 6%. El-Beltagi et al. [8] studied that the aqueous extract of spirulina contained a significant amount of total flavonoids, which was 79.6 mg/g dry weight, and that the amount of total phenols was 16.0 mg/g dry weight. Consequently, these were used to enhance pomegranate juice because of the positive effects that these chemical compounds have on one's health. According to Martelli et al. [97], the presence of phenolic compounds in spirulina extract was associated with antimicrobial activity. Martelli et al. [97] also noted that the amount of phenolic compounds that showed the highest level of inhibitory activity against the bacteria that cause transmitted diseases was 3.18–3.40 mg gallic acid/g. This paves the way for the use of adding spirulina as a food preservative, as there are numerous phenolic chemicals that have the ability to inhibit the growth of microorganisms and are present in different kinds of algae.

4. Potential Health Benefits of Spirulina from a Human Nutrition, Biological, and Medicinal Standpoint

4.1. Spirulina for Formulation of Various Food Products in the Food Industries

4.1.1. Snack Food Formulation

Because of its great nutritional value as a source of proteins and minerals, the biscuit and snack industry has been increasingly interested in incorporating algae, particularly spirulina, into its products. One option to increase and improve the nutritional content of this snack is to produce it with the addition of spirulina. This is one technique. The results revealed that the addition of spirulina at any of the following concentrations: 0%, 2.5%, 5%, 7.5%, 10%, or 12.5% led to an increase in the percentage of protein (9.43–18.11%) and ash (1.31–2.67%). After undergoing the microbiological testing, all of the samples were found to be risk-free, and the conclusions drawn from the sensory analysis pointed to the exclusion of the addition at the 12.5% concentration [2]. There is a growing trend in the manufacture of biscuits that involves the use of beneficial ingredients. Biscuits are a type of snack food that is consumed by a large number of people. Algae, such as spirulina and Chlorella, are used as a source of proteins, antioxidants, and biologically active molecules in the process of promoting wheat cookies. These algae are included among the additions. There were two different addition percentages, which were 2% and 6% by weight/weight respectively. The incorporation of 6% each of spirulina and Chlorella resulted in significantly increased protein content (13.2–13.5%) [83]. As a result, these additions are regarded as a source of protein, since the biscuit that was loaded with the spirulina alga *A. platensis* had the strongest antioxidant activity and received the highest score in the sensory evaluation [83].

The effect of adding spirulina to biscuits during the fortification process at varying rates of 0%, 5%, 10%, and 15% in comparison to regular biscuits has been researched in terms of its impact on the nutritional content of the biscuits [119]. Comparing the scores acquired by the control sample and the samples that had spirulina added to them, the findings of the sensory assessment revealed that the control sample had a lower score for the majority of the sensory qualities. Additionally, it was found that the incorporation of spirulina into biscuits resulted in an increase in both the percentage of protein (9.09 and 14.73%) and the amount of energy (554, 672 kCal), respectively, that they contained. The functional biscuits that are supplemented with spirulina algae are shown in Table 3A On the other hand, research carried out by El Nakib et al. [34] demonstrated that the nutritional content of biscuits might be improved by adding certain percentages of spirulina to the foods that schoolchildren eat for snacks. The percentages tested were 0%, 5%, 10%, and 15%. It also revealed an increase in fatty acids, such as omega-3, omega-6, omega-7, and omega-9, with omega-6 containing the highest percentage of unsaturated fatty acids, specifically linolenic and γ-linolenic acid, with values of 33.0 and 30.0 mg/g, respectively [3,83,84]. Spirulina is included in the formulation of the snacks and functional baby food listed in Table 3B.

Table 3. Various developed functional food products with incorporation of spirulina biomass.

Type of Functional Food Products	*Spirulina* Biomass	Other Ingredients	References
	(A) Functional Biscuits		
Cookies	10–15%	–	[34]
	5%, 10%, and 15%	–	[119]
Wheat Crackers	2% and 6%		[83]
	(B) Functional Snacks and Baby Food		
Snack	0%, 2.5%, 5%, 7.5%, 10%, and 12.5%	–	[2]
Snakes for the elderly	750 mg/100 g	–	[120]
Baby food formulation	0%, 2.5%, 5%, and 7.5%	–	[39]
	(C) Functional Pasta		
Pasta	1–15	–	[121]
	0%, 0.25%, 0.5%, 0.75%, and 1%	–	[122]
	0%, 0.6%, 2.0%, 3.4%, and 4.0%	–	[123]
	0%, 0.25%, 0.5%, and 0.75%	–	[124]
Novel Food Products Pasta, Maki-Sushi, Jerky Pasta,	Spirulina filled	–	[15]
	(D) Functional Ice Cream		
Ice cream	5%	–	[125]
	0.15%	Fructo oligosaccharides (FOS) and Probiotic Vegetable Milk	[126]
	0%, 0.6%, and 1.2%	–	[116]
	0.075%, 0.15%, 0.23%, and 0.3%	–	[40]
	(E) Functional Dairy Products		
Ayran Yoghurt	0%, 0.25%, 0.5%, and 1%	*Bifidobacterium lactis, Lactobacillus acidophilus, L. bulgaricus* and *Streptococcus thermophilus*	[127]
Low-fat Probiotic Yogurt	0.1–1%	Algae: *Spirulina platensis* and *Fexulago angulata*	[128]
		Bacteria: *L. delbruckii* subsp. *bulgaricus, L. acidophilus* ATCC 4356, and *S. thermophiles*	

Table 3. *Cont.*

Type of Functional Food Products	*Spirulina* Biomass	Other Ingredients	References
	(E) Functional Dairy Products		
Probiotic Fermented Milk Product	–	–	[127]
Soft cheese	0%, 1%, and 1.5%	–	[116]
Yoghurt	0%, 0.5%, 0.75%, 1%, 2%, and 3%	*L. acidophilus, L. delbruckii* subsp. *bulgaricus* and *S. thermophilus*	
	0%, 0.5%, and 1%	*Bifidobacterium animalis* ssp. *Lactis* (BB-12) and *L acidophilus* (ha-5)	[129]
	0.3%, 0.5%, and 0.8%	*L. acidophilus, L. delbruckii subsp. bulgaricus, S. thermophiles,* and spinach (10–13 *w*/*w*%)	
	0.5% and 1%	*L. acidophilus, L. delbruckii* subsp. *bulgaricus* and *S. thermophilus*	[130]
Soy Yoghurt	0.80%	*L. delbruckii* subsp. *Bulgaricus* and *S. thermophilus*	[131]
Functional Yogurt	0.1%, 0.2%, 0.3%, and 0.5%	–	[40]
Probiotic Fermented Milks	–	–	[132]
Fermented Symbiotic Lassi	0%, 0.05%, 0.1%, 0.2%, 0.3%, 0.4%, and 0.5%	*L. helveticus* MTCC 5463 (V3) and *Streptococcus thermophilus* MTCC 5460 (MD2)	[133]
Acidophilus Milk	0.5% and 1%	*L. acidophilus*	[130]
Acidophilus bifidus-thermophiles Fermented (ABT) Milk	3 g/L	*Bifidobacteria, L. acidophilus,* and *S. thermophiles*	[134]

– Not reported.

4.1.2. Pasta Formulation

Algae from the genus *Spirulina* have been put to use in the pasta-manufacturing industry in order to produce enriched pasta with enhanced nutritional, sensory, and therapeutic benefits [122]. It has developed a wide variety of new food products based on spirulina in order to cater to the requirements of the customers, such as sushi, jerky, and pasta [15]. Spirulina was added to wheat flour at a rate of 5% and 10% in order to prepare macaroni. As a result, the protein and energy content of the macaroni increased to 10.32% and 14.50%, respectively, while the caloric value increased to 322.94 and 327.60 kcal/g in the pasta that had been enriched with algae at a rate of 5% and 10%, respectively [124]. The chemical characteristics of the pasta were significantly improved as a result of the addition of spirulina at several percentages (0%, 0.25%, 0.5%, 0.75%, and 1% weight to weight), which resulted to a considerable improvement in the nutritional value of the pasta. In comparison to the sample that served as the control, the macaroni that had a concentration of 0.25% moss added to it received the highest score on the sensory evaluation. For this reason, this ratio was utilized throughout the production process of pasta in order to generate the greatest possible product that had been fortified in terms of its nutritional content, sensory value, and functional therapeutic capabilities [122]. In addition, Pagnussatt et al. [123] observed that the addition of spirulina and oats to dry pasta resulted in an increase in the product's nutritive content, which was a result of the product being fortified. The amount of soluble solids included in the pasta as well as its color was impacted by spirulina, whereas the acidity value of the pasta was impacted by oats, which also enhanced the values of crude protein and overall food flavor (13.06%). These additions, when compared to the amount of fiber that is often found in commercial pasta (2.40%), can be regarded as a source of fiber, which lends them a healthful and nutritious value [123]. The functional pasta that has been enhanced with spirulina algae is shown in Table 3C.

4.1.3. Formulation of Dairy Food Products

The production of a wide variety of dairy products includes the use of spirulina algae. Mocanu et al. [127] employed *S. platensis* to improve the nutritional content of fermented dairy products. During incubation and storage, the impact of *S. platensis* on the probiotics Bifidobacterium animalis ssp. lactis and Lactobacillus acidophilus was investigated. According to the findings, the consumption of spirulina over the course of the full storage time had a positive impact on the number of initiator bacteria B. animalis ssp. lactis and L. acidophilus that survived. Spirulina, which has a protein content of between about 55% and 70%, can be added to food products in order to increase the nutritional value of those products. When it came to the preparation of the soft-cheese product, the findings indicated that adding 1% of *S. platensis* was the optimal concentration from both the physicochemical and the sensory points of view. This addition had a considerable and favorable influence on the levels of protein, water, fat, and β-carotene, as well as on the tissues [116].

Ice cream

Because algae include various antioxidants, such as polycarotene phenols, which have the power to scavenge free radicals, the usage of algae in the making of ice cream has many positive impacts on one's health. The ice cream product that was added to algae extract demonstrated an increase in antioxidant activity and gave the highest inhibition level of 39.7% in mint ice cream samples. This was indeed in comparison to the control sample, which did not include algae and resulted in an inhibition level of 32.8% [125]. In order to make a functional ice cream that had a high nutritional value, *S. platensis* was included in the production process. The optimum concentration of algae was 1.2%, which had a beneficial impact on the amount of protein in the product and had sensory and physical attributes that were satisfactory [116]. Although spirulina was employed as a stabilizer in the production of ice cream by Malik [40] with several replacement percentages (25, 50, 75, and 100%), they found that the optimal replacement percentage was 50% with a concentration of 0.15%. In comparison to the control sample, the treated product maintained the highest possible level of microbiological quality throughout the duration of the storage period. This treatment also had a favorable influence on the nutritional content of the product and improved its sensory characteristics. The spirulina-enriched functional ice cream is presented in Table 3D.

Yogurt and Acidophilic Milk

Functional yogurt was made using dried spirulina at different concentrations (0.1%, 0.2%, 0.3%, and 0.5%), with the concentration of 0.3% giving the best positive effect in enhancing the nutritional value and improving the sensory characteristics. In addition, it showed the highest survivability of S. thermophilus and L. bulgaricus compared with the control throughout the storage period [40]. Spirulina was also used in the preparation of yogurt at concentrations of 0.5% and 1% *w*/*w*, as it was discovered that adding 1% spirulina to yogurt had a positive effect on the numbers of lactic acid bacteria, which decreased in the control sample, while adding 0.5% was superior to adding 1% for the sensory characteristics of the product. Therefore, yoghurt that has been supplemented with spirulina is an excellent approach to keeping the viability of lactic acid bacteria for the thirty days that the product is stored in the refrigerator [130]. *S. platensis* was added to milk that had been fermented by probiotic bacteria using a starter (ABT-4) that contained L. acidophilus (A), Bifidobacteria (B), and Streptococcus thermophilus (T). The effect of this addition was evaluated. According to the findings, including *S. platensis* in the mix had a beneficial impact on increasing the number of initiator bacteria that made it through the storage period [134,135]. When making soy yogurt that was enriched with spirulina, it was observed that the optimal addition ratio was 0.80% (*w*/*w*), which produced the ideal yogurt at a temperature of 40 °C and a fermentation time of 12 h. This was found when the yogurt was being prepared [131]. The dairy products that have been fortified with spirulina are listed in Table 3E.

4.2. Biological and Therapeutic Significance of Spirulina

Spirulina, especially *A. platensis*, has a high nutritional value, does not harm the environment, and possesses remarkable curative capabilities. According to studies, spirulina plays a role in the treatment of a number of diseases, and it can be utilized in the reduction of blood sugar levels, the reduction of blood pressure, the modification of dysbiosis caused by an unbalanced intestinal flora, anticancer, antioxidant, antibacterial, anti-allergic, antiaging, and anti-inflammatory activities, as well as in the reduction of HIV immunity [112]. The chemical composition of spirulina, which includes minerals (especially Fe), phenols, phycocyanin, and polysaccharides, is primarily responsible for its beneficial effects on human health. The clinical data that are now available do not show that ingesting spirulina poses a threat to one's health, as stated by the US Food Supplements Convention [45]. The World Health Organization (WHO) verified that not only is spirulina a good diet due to its high Fe and protein content, but it may also be given to children without any adverse effects [39]. Figure 2 demonstrates the biological as well as therapeutic benefits of the spirulina algae. In addition to that, it has been described in further detail in Table 4.

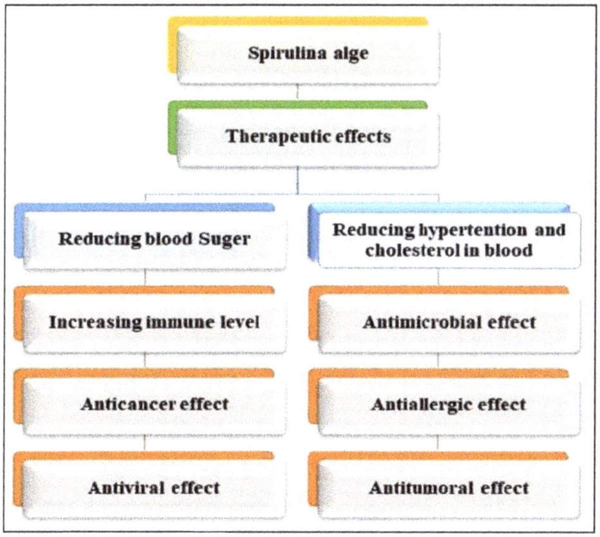

Figure 2. A diagrammatic explanation of the therapeutic value conferred by spirulina algae.

Table 4. Biofunctional effect of spirulina algae biomass application.

Biomass of Spirulina and Associated Products	Remarkable Observations		References
	Nutritional and Biochemical Component	Health and Other Beneficial Remarks	
Powdered/Dried solution 10%	Phenolic compound (phycocyanins, β-carotene), fatty acid composition (PUFAs), polysaccharide, vitamins, and mineral	Antioxidant properties	[130,136]
Spirulina extracts 70% (acetone, methanol and ethanol)	Phenolic and flavonoid compounds	Antimicrobial activity	[103,137]
10%, 20%, 30%, 40%, and 50%	Polysaccharides, phenolic compounds, phytopigments (carotenoids, chlorophyll phycocyanin) and phycocyanobilin	Anticancer and antioxidant properties	[8]
–	Phycocyanin	Anticancer, antidiabetic, and anti-inflammatory properties	[138]
–	Polyphenols and phycocyanin	Antioxidant and anti-inflammatory properties	[139]
–	Polysaccharides, protein, polyunsaturated fatty acid, vitamins, minerals, phenolic, pigments, sterols, and volatile compound	Antioxidant, anticancer, anti-inflammatory, antidepressing, antihypertensive, antiaging activities as well as arthritis, cardiovascular effusions, hypertension, increasing HDL cholesterol and reducing triglyceride properties	[140,141]
–	Protein, amino acid, unsaturated fatty acids, vitamins (A, B2, B6, B8, B12, E, and K), minerals (Fe and Ca) and antioxidant compound	Antiinflammatory, antitumoral, antivirial activities as well as properties of reducing blood lipid profile, blood sugar, body weight, and wound healing time	[142,143]
–	Protein, essential fatty acids (omega-3 and omega-6 type), vitamins (A, B1, B2, B12, and B3), minerals (Cu, Fe, Mg, Mn, and K), and pigments (xanthophyll and carotenoids)	Decrease LDL cholesterol, triglyceride levels, blood pressure, and blood sugar, while, increase hemoglobin level of red blood cells. Furthermore, it improved immune system	[144–149]
–	Polysaccharides, phycobiliproteins, omega-3 (EPA and DHA), omega-6 (γ-linolenic acid), vitamins (A, C, E, B, B1, B2, B3, B5, B6, B9, B12), and phenolic compounds (polyphenolics, *p*-coumaric, and ferulic acid)	Antioxidant, antitumor, anti-inflammatory, immunomodulatory, antifungal, antiviral, as well as radical scavenging properties. Furthermore, it prevents diseases such as atherosclerosis, cardiovascular, and heart related. In addition to this, it has also played key role in healthy skin, blood circulation, preventing clots forming in the blood as well as synthesis DNA and cholesterol in body.	[6]

Table 4. *Cont.*

Biomass of Spirulina and Associated Products	Remarkable Observations		References
	Nutritional and Biochemical Component	Health and Other Beneficial Remarks	
–	Polysaccharide, protein, essential amino acid, essential fatty acids, minerals, vitamins (vitamin E and carotenoids), phenolics and phycocyanins	Antioxidant, anti-inflammatory, immunostimulatory, antihypertensive, hypoglycemic as well as hypolipidemic properties. Furthermore, it has promoted the activity of natural killer cells as well as improve the growth of beneficial intestinal microbiota.	[150]
–	Carbohydrates, lipids, protein, vitamins, minerals, pigments and phytonutrients	Anticancer, antihypertension, antihypercholesterolemia, antidiabetes, antianemia properties as well as responsible growth of Lactobacilli	[151]
–	Polyunsaturated fatty acid (γ-linoleic acid)	Positive effect on chronic diseases, cancer, diabetes, heart disease, arthritis, Al Zheimer's disease and inflammatory property	[31]
Spirulina extract 4%	Phenolic and flavonoids compounds as well as high free radical scavenging activity	Antioxidant properties and phagocytosis inhibition	[87]
–	Carbohydrate, protein, vitamins (B1, B2, B3, and E) and minerals (Fe)	Antioxidant, anticancer, antiviral properties as well as control of obesity, allergies, arthritis, inflammation, diabetes, hyperlipidemia, cholesterol and immune system	[48]
–	Polysaccharide, allophycocyanin, carotene, xanthophyll and *C*-phycocyanin	Antioxidant, antivirial, anticancer as well as anti-immunostimulant effects	[33]
–	Polysaccharides, γ-linolenic acid, β-carotene and *C*-phycocyanin	Antioxidant, anticancer, antiviral as well as anti-allergic properties	[152]
–	Acid development and lower pH	Promoted the growth of lactic acid bacteria	[153]
Tablets	Polyunsaturated fatty acids (omega-3 and omega-6 type)	Increased omega-3 omega-6 in human blood lipids as well as cell membrane lipids	[154]
–	Polysaccharide, protein, vitamins, mineral, phycocyanin, and other nutritional supplements	Improved immunity and immune system, activity of superoxide dismutase (SOD) as well as lowering blood lipid level	[41]

– Not reported.

4.2.1. Importance in Weight Control

Several different studies have pointed to the relevance of spirulina in maintaining a healthy weight. Spirulina may help regulate body weight by inhibiting the migration of macrophages into visceral fat, preventing the buildup of fat in the liver, lowering the levels of oxidative stress in the body, and increasing insulin sensitivity and satiety. These are the hypothesized mechanisms of action. In their study, DiNicolantonio et al. [155] validated the effects of spirulina on weight loss. They found that obese adults who consumed 1–2 g of spirulina per day for a period of three months saw a reduction in their body weight, body mass index (BMI), and waist circumference. A considerable rise in body weight, total protein, albumin, and hemoglobin levels were seen in diabetic rats after they were given an aqueous extract of spirulina to consume for a period of fifty days. This points to an improvement in general health conditions as well as metabolic mechanisms brought about by efficient glycemic management. On the other hand, the weight of diabetic animals that were not treated dropped [55].

According to the findings of the research conducted by Huang et al. [156], consuming spirulina supplements in doses ranging from 1 to 19 g per day for a period of time ranging from 2 to 48 weeks had a beneficial impact on a variety of key indicators of the body. It was noticed that there had been an increase in overall body weight as well as in BMI. When compared to the rats who were fed the control diet, the rats that were given either pomegranate juice, spirulina extract, or both together had a considerable rise in their body weight. These liquids stimulated hunger, which ultimately resulted in weight gain [8]. In addition, Moradi et al. [157] found that supplementation with spirulina significantly lowered body weight from −1.98 to −1.14 kg, notably in obese adults at a dosage of 4 g/day from −2.45 to −1.68 kg. These results were found when the participants were given spirulina. The decline was larger than the people who were overweight (−1.62 to −0.93 kg), the percentage of body fat reduced from −1.02 to −0.54 kg, and the waist circumference decreased from −1.40 to −1.39 cm. It has been observed that spirulina has beneficial effects on weight and waist circumference when used for at least 12 weeks, and it has a positive effect on BMI when used for a longer period of time. Chronic diseases are associated with obesity and overweight, and it has been observed that spirulina has these beneficial effects [158]. Supplements containing spirulina contain antioxidants, which are known to have a significant role in the management of weight for diabetics and those who are obese. They increase the body's need for energy while simultaneously preventing the production of adipocytes and the enzyme lipase. The usage of these supplements resulted in a substantial reduction in body weight from −1.76 to −0.88 kg and a significant reduction in waist circumference from −1.40 to −1.39 cm; however, there was no significant effect on BMI observed. Spirulina was found to reduce lipid accumulation in the liver by inhibiting macrophage infiltration into visceral fat. Additionally, the phenylalanine content of spirulina was found to be reduced, which may have led to an increase in the secretion of cholecystokinin, a hormone that suppresses appetite [159]. Figure 3A is a diagrammatic illustration of the function that spirulina algae play in the process of weight reduction.

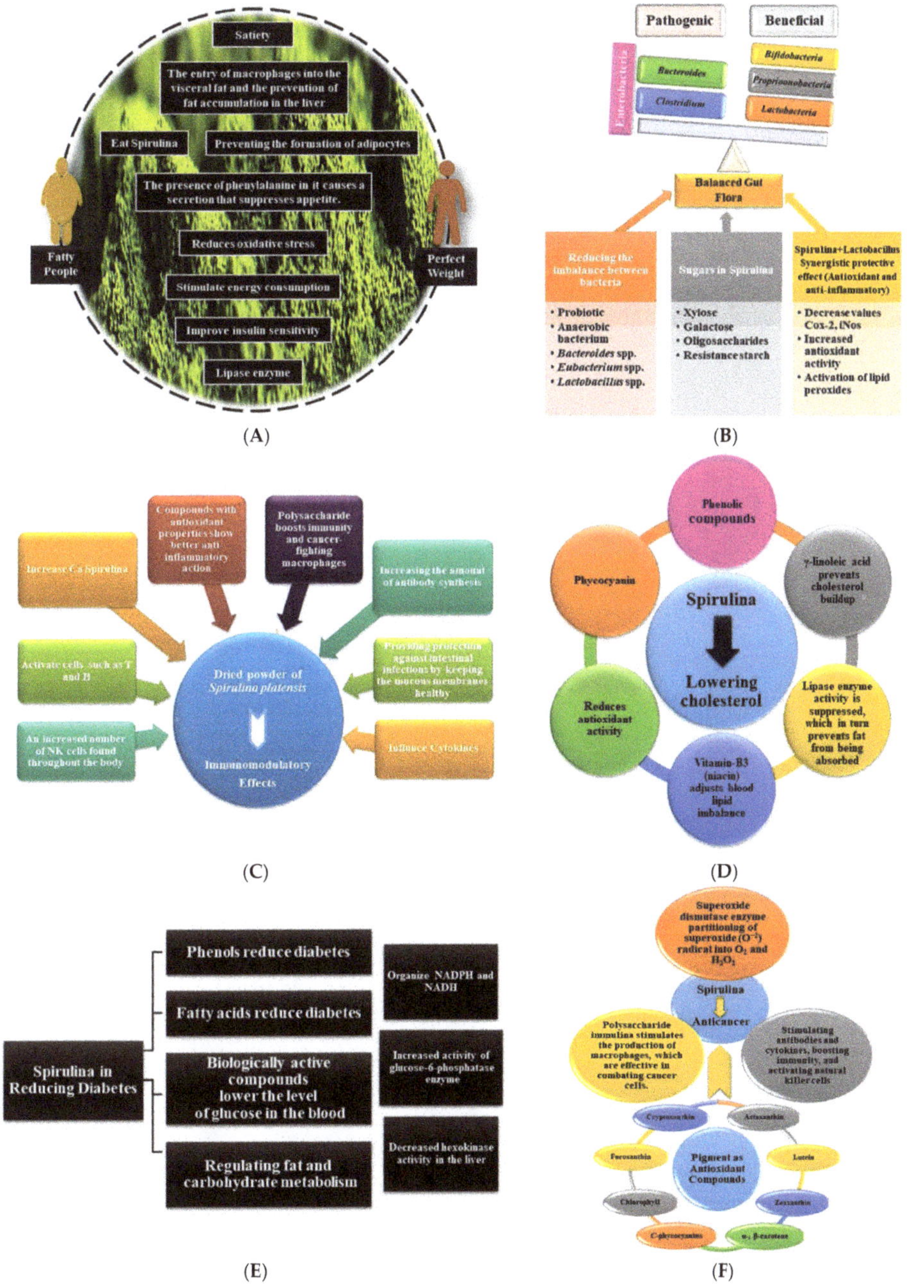

Figure 3. A pictorial representation of the numerous beneficial health effects that have been derived from consuming spirulina algae and their associated bioactive components. (**A**) Role of spirulina algae play in the process of weight loss; (**B**) How consuming spirulina algae might alter the pattern of the intestinal balance; (**C**) Role played by spirulina algae in the regulation of immunity; (**D**) Function that spirulina algae play in reducing cholesterol levels; (**E**) Role that spirulina algae play in lowering the risk of developing diabetes; and (**F**) Anticancer properties of the spirulina algae.

4.2.2. Importance in Intestinal Flora

The findings of studies on spirulina suggest that it plays a role in the modulation of dysbiosis, which is an imbalance in the intestinal flora. This suggests that the combination of spirulina and probiotics could represent a new strategy for improving the growth of beneficial intestinal microorganisms [150,160]. The aqueous extract of S. platensis was shown to be the greatest source of the biostimulators xylose, galactose, oligosaccharides, and resistant starch. These biostimulators had a tremendous effect on boosting the development of probiotic bacteria. The biostimulators encouraged the growth of beneficial bacteria in the colon, such as lactic acid bacteria, Bifidobacteria, and Lactobacilli, while also reducing the number of dangerous bacteria, such as Enterobacteria and Clostroids [161]. Spirulina has been shown to have a positive impact on human and animal health by altering the composition of gut bacteria and boosting the growth of useful bacteria. This has the knock-on effect of enhancing overall health. Changes in the composition of the bacteria in the intestinal tract are a contributing factor in the development of a wide variety of disorders, including those affecting immunity and metabolism. There are a wide variety of disorders that are linked to microorganisms, one of which is inflammatory bowel disease (IBD). It has been observed that consuming spirulina brings about a reduction in the imbalance that exists between the genera of beneficial probiotic bacteria and the genera of natural anaerobic bacteria, particularly *Bacteroides* spp., *Eubacterium* spp., and *Lactobacillus* spp. [150,162–164]. According to the findings of research that was conducted by Abdel-Moneim et al. [165], alcoholic spirulina extracts (methanol, acetone, and hexane) and biosynthesized Se nanoparticles (SeNPs) both possessed significant antioxidant activity and antibacterial activity. The pathogens included three strains of Gram-positive bacteria, three strains of Gram-negative bacteria, and three strains of *Candida* spp. and *Aspergillus* spp. The spirulina methanolic extract had the highest antibacterial and antioxidant activity levels, as well as the highest total phenol content. The total phenol content of spirulina and SeNPs, as well as their biological activity, were found to have a correlation with one another. In the realm of the effectiveness of antibacterial and antioxidant agents, with the potential for safe medicinal uses as alternatives to traditional chemical medications and antibiotics. The capacity of functional foods to change the composition of the gut microbiota is the source of their many health advantages. These foods also have the potential to interfere with patients' existing health conditions [166]. It was observed that the antioxidant and anti-inflammatory actions of spirulina and Lactobacillus bacteria had a complementary impact on the treatment of ulcerative colitis (UC). When the mice were given Lactobacillus at a rate of 1×10^9 CFU per mouse per day and spirulina at a rate of 500 mg/kg per day, the standard effect of both was determined. It has been demonstrated that the protective benefits are caused by its capacity to lower levels of the inflammatory markers iNOS and COX-2, as well as boost antioxidant activity and significantly prevent lipid peroxidation [167]. Figure 3B is a diagrammatic description of how ingesting spirulina algae may affect the pattern of the balance in the digestive tract.

4.2.3. Immunological Importance

In terms of the immune system, spirulina has a major role to play in boosting immunity due to the fact that it possesses a wide range of biological activities and is important to nutrition due to the high concentration of natural nutrients it contains [117]. In addition, it can modulate the immune system and the bioactivity of macrophages by activating T and B cells, promoting the production of antibodies and cytokines, increasing the concentration of NK cells in tissues, and encouraging the generation of antibodies [84]. Spirulina is a rich source of antioxidants and anti-inflammatory substances, which are able to not only support and develop physiological functions of the nervous system and brain, as well as compensate for nutritional deficiencies, but also promote a beneficial immune response [5]. Several studies have also demonstrated the positive effects of spirulina, including the immunological and antioxidant effects that are attributable to its carbohydrate content, particularly polysaccharides. These effects are due to the fact that spirulina has a high concentration of polysaccharides [44,100].

The chemical composition of spirulina (*A. platensis*), which is rich in minerals and carbohydrates and consists of polymers including glucose and branching polysaccharides comparable to glycogen, is largely responsible for the positive effects on human health that these microalgae have. These high molecular weight compounds of negatively charged sugars are referred to as "Immulina," and they have significance in the pharmaceutical industry. It was discovered that after ingestion at a rate of 400 mg per day for seven days, they enhanced the activity of natural degrading cells against cancer cells. This finding gives them importance in the pharmaceutical industry. In addition to this, the antiviral activity of these drugs has been demonstrated against types I and II of the herpes simplex virus [6,43,168,169]. In most cases, this immunoreactivity is brought on by Ca spirulan, which is made up of the following saccharides: rhamnose, methylrhamnose (acofriose)-3-O, methylrhamnose di-O-2,3, methylxylose-3-O, uronic acids and sulfates [27,33,45,170,171].

Spirulina is utilized to produce phycobiliproteins, which are then put to use as colorants. According to the findings of a large number of studies, these pigments offer a wide range of potential uses. In clinical medicine and immunological analysis, as well as for therapeutic and diagnostic reasons, they are utilized as fluorescent material. It was also revealed that it had a considerable effect on lowering the quantity of cholesterol detected in the blood and that it acted as a protective agent against hepatitis [120,172,173].

Spirulina is a stimulant of immune system cells because it increases anti-inflammatory resistance, stimulates the synthesis of antibodies and cytokines, activates macrophages, stimulates natural killer cells, and activates T and B cells. Consuming spirulina helps to keep the intestinal epithelium, which is the first line of defense of the mucosal barrier against infections, in good working order [44,89]. Figure 3C presents a diagrammatic description of the role that spirulina algae play in the control of immunity.

4.2.4. Blood Pressure Treatment

It was observed that supplementation with spirulina led to a substantial drop in diastolic blood pressure (DBP), but at the same time, it led to a rise in systolic blood pressure, which is one of the indications of cardiovascular safety and metabolism in humans [156]. This finding suggests that spirulina has a beneficial influence on blood pressure. Vasodilation is the mechanism that has been demonstrated to be responsible for spirulina's ability to alleviate hypertension in mice [94]. When taken orally at a dose of 4.5 g per day for a period of six weeks, spirulina is related to a drop in both systolic and diastolic blood pressure. Because it contains the pigment C-phycocyanin, it was also reported to have a lipid-lowering effect. The high K and relatively low Na content of spirulina have beneficial effects on blood pressure. It was shown that C-phycocyanin stops platelets from adhering to one another by preventing Ca mobilization and moderating free radicals that are generated by platelets. Spirulina also helps prevent atherosclerosis [89]. Lipoproteins are not only used as a coloring agent but they have also been associated with a wide variety of beneficial impacts on one's health [99]. When rats with naturally high blood pressure were given phycocyanin pigment in their regular diet at doses of 2500, 5000, or 10,000 mg/kg over a period of 26 weeks, hypotension was found. The ACE-I inhibitor peptide, which has been shown to have an effect on lowering blood pressure, was discovered during the process of phycocyanin synthesis [174].

4.2.5. Cholesterol Treatment

Spirulina seems to have a positive effect on metabolic risk factors, particularly elevated blood lipid, according to a number of studies. It was noted that spirulina has an effect on lowering blood lipids, particularly triglycerides and cholesterol linked to low-density lipoprotein (LDL), as well as an indirect effect on total cholesterol and cholesterol linked to high-density lipoprotein (HDL), due to the presence of a pigment called phycocyanin [89]. Phenolic compounds are antioxidants that protect cells and natural chemicals in the body from the damage caused by free radicals. Free radicals are responsible for causing tissue damage in the body as well as oxidizing LDL cholesterol, which can lead to heart disease if it accumulates in the arteries [6]. Consuming spirulina can help lower levels of dangerous LDL and triglycerides, and it can also play a part in the indirect modification of high cholesterol and total cholesterol levels. This is because the components of spirulina have antioxidant action, and the aqueous extract of spirulina reduces the absorption of fats from foods by reducing the activity of the lipase enzyme that is released by the pancreas [94]. This is why spirulina is so beneficial. When investigating the impact of the ethanolic extract of spirulina on lipid contents in the blood of artificially hypercholesterolemic male laboratory rabbits at two different dosages (33 and 66 mg/kg) over the course of four weeks, the researchers found that the extract had no effect [175]. It was revealed that there was a reduction in the levels of cholesterol, triglycerides, LDL, and dangerous LDL, while there was an increase in the levels of HDL, which was favorable.

Over one month, hypercholesterolemic rabbits had their total cholesterol, triglycerides, and HDL levels monitored to determine how they were affected by a diet that included 0.5 g of spirulina per day as a dietary supplement [176]. An increase in the level of HDL in the blood was shown to be associated with a decrease in the levels of total cholesterol found in the blood, but there was no discernible change in the levels of triglycerides found in the blood. According to the findings of the study, this is because spirulina contains natural antioxidants such as phenolic compounds, γ-linolenic acid, and phycocyanin, all of which work to prevent hypercholesterolemia and the negative consequences of the condition. Spirulina algae, when consumed as a functional food, has been shown to aid in the treatment of cardiovascular disorders such as high cholesterol and atherosclerosis, as validated by Wang et al. [162]. In addition to this, it regulates the level of LDL and HDL in the blood, which results in favorable effects on high blood lipids. It was noted that taking spirulina supplements had a beneficial impact on blood lipid parameters, as it was observed that the values of triglycerides, total cholesterol, LDL, and very LDL decreased, while the value of HDL increased. This was found to be the case because it was found that the values of HDL increased [156]. Spirulina can contribute to in avoiding the production of cholesterol, which is mediated by γ-linolenic acid. A lack of γ-linolenic acid can cause the artery wall to become thicker, which can lead to high blood pressure and an increase in blood lipids. In addition, spirulina includes vitamin B3 known as niacin, which corrects an imbalance in the lipids of the blood [155]. Figure 3D provides a graphic of the role that spirulina algae play in lowering cholesterol levels.

4.2.6. Diabetes Treatment

One of the metabolic illnesses, diabetes is also one of the most common diseases and a major cause for concern all over the world owing to the toll it takes on public health. Diabetes is a metabolic ailment [117]. Spirulina demonstrated both glucose and lipid modulation activities, indicating that it may have a regulatory function in the metabolism of carbohydrates and lipids in both experimental animals and diabetics [84]. Gheda et al. [55] found the presence of a variety of bioactive compounds when they conducted an analysis of the methanolic extract of S. platensis using GC-MS. These bioactive compounds included phytols, phenolic compounds, and methyl esters of fatty acids. These bioactive compounds worked together to produce a synergistic effect. These are what give spirulina its antioxidant effects, as well as its ability to decrease cholesterol and blood sugar levels. In addition, it was realized that administering the methanolic extract to experimental mice at doses of 15 and 10 mg/kg body weight resulted in hypoglycemic activity and improved the histological disorders of the liver and pancreas that are associated with diabetes. This was found to be the case when the mice were given the extract. According to the findings of the study, this algae should be included in the formulation of medicinal products intended for the treatment of diabetes and the symptoms associated with it [55].

During the in vivo phase of the research, diabetic mice were given an extract of spirulina to cure their condition. The methanolic extract of spirulina caused hypoglycemic activity when it was administered at concentrations of 15 and 10 mg/kg by body weight. This activity was characterized by a reduction in a variety of liver and kidney functions, an increase in fats, and an improvement in the histological disorders of the liver and pancreas that are associated with diabetes [55]. The presence of these chemical molecules with biological activity, which may have the potential to serve as antioxidants in a synergistic way and may also have an impact that is diabetic-lowering. Spirulina has also been demonstrated to be useful in decreasing the level of glucose that is present in the blood when the individual is fasting [94]. Diabetic people who consume spirulina have lower blood sugar levels and are less likely to experience problems. Spirulina also controls cholesterol levels. Due to the fact that spirulina possesses antioxidant and immunomodulatory qualities, diabetic individuals might benefit from consuming it as a functional food in order to better manage their type 2 diabetes. Spirulina supplementation can be used to maintain nutritional balance, reduce blood lipids and modify carbohydrates, and drop blood sugar levels by a large amount. It can also be used to minimize inflammatory stress levels [44,156].

The majority of spirulina's effect can be attributed to the upregulation of NADPH and NADH. NADH is a cofactor in lipid metabolism that is responsible for the high activity of glucose-6 hydrogen ion phosphatase (H+). This enzyme binds to NADP+ in the form of NADPH. Spirulina's effect can also be attributed to the high activity of glucose-6 hydrogen ion phosphatase (H+). This help is necessary for the synthesis of lipids from carbohydrates, and spirulina may be able to oxidize NADPH. Spirulina is a blue-green alga. One animal investigation revealed that the activity of hexokinase in the liver of diabetic control mice was greatly reduced, whilst the activity of glucose-6-phosphatase was significantly enhanced. Both of these changes were observed in the liver. The administration of spirulina to diabetic mice led to an increase in hexokinase activity and a reduction in glucose-6-phosphatase activity. Mice that were given spirulina had higher levels of hexokinase activity, which showed that the mice's hepatocytes took in more glucose from the blood [89]. Figure 3E presents a schematic representation of the function that spirulina algae perform in reducing the likelihood of an individual getting diabetes in the future.

4.2.7. Cancer Treatment

There is a tremendous opportunity for the potential new natural anticancer compounds from BGA, particularly spirulina, in the treatment of cancer. It has been demonstrated that the high concentration of antioxidant molecules found in algae is what makes it so effective as an alternative treatment for a variety of diseases, including cancer, diabetes, and inflammation [3,114]. Spirulina has been shown to have therapeutic effects, such as protection against some types of cancer, enhancement of the immune system, protection against radiation, and reduction of high blood fats and obesity, according to several studies. Figure 3F presents a schematic representation of the anticancer effects that may be attributed to the consumption of spirulina algae. It was reported by Hosseini et al. [33] that the spirulina algae product possesses many pharmacological activities, such as anticancer, antiviral, antimicrobial, anti–heavy metal, immunomodulatory, and antioxidant due to its high content of protein and PUFAs, particularly the high content of γ-linolenic acid. These pharmacological activities are attributed to the high content of γ-linolenic acid [45]. Spirulina has been shown to boost the stimulation of antibody and cytokine production, as well as NK activation. Additionally, it has the capacity to augment human immunity and plays a role in inhibiting the growth of tumors. Spirulina exerts its effects on human myeloid progenitors and natural killer cells in either a direct or indirect manner, depending on the context [94].

The astaxanthin tincture has potent anti-inflammatory and antioxidant capabilities, which have the capacity to prevent or lessen the severity of a wide variety of ailments, including cancer. Lutein, zeaxanthin, and β-carotene are known to reduce the risk of developing premenopausal breast cancer. Cryptoxanthin and α-carotene are known to reduce the risk of developing cervical cancer. Chlorophyll tincture has been demonstrated to have both antioxidant and anticarcinogenic action, while fucoxanthin pigment has been proven to have antiobesity and anticarcinogenic effects. As a result of this, the food industry has begun to make use of spirulina pigments as a coloring ingredient [98,177,178].

At a concentration of 50 μM for up to 48 h, C-phycocyanin and β-carotene had an anticancer impact on the human chronic myeloid leukemia (K562) cell line, lowering 49% of cell growth [44]. This feature of spirulina can be related to the presence of antioxidants with a high level of β-carotene and superoxide dismutase enzyme in the food. Spirulina has been shown to have a protective impact against cancer cells [89]. Spirulina extract was tested at five different concentrations to see whether or not it has an anticancer effect: 6.25, 12.5, 25.50, and 100 μg/mL. The cytotoxicity test revealed that cells were inhibited by 50% at a concentration of 19.18 μg/mL after 72 h. The results demonstrated that the extract had a highly toxic impact against cancer cells, and this effect increased with increasing concentration [179].

The cytotoxicity of spirulina extract was demonstrated against cancer cell lines when it was tested on colon cancer (HCT-116), liver cancer (HepG2), and CACO colon cancer cell lines. The LC50 values for each of these cancer cell lines were 21.8, 14, and 11.3 mcg/mL, respectively. This effect was attributed to the presence of antioxidant plant pigments (carotenoids, chlorophyll, and phycocyanin), as well as the presence of polysaccharides, as the use of spirulina extract has been suggested in the development of anticancer drugs. Phycocyanin is a pigment found in blue-green algae [8]. When employing algal extract at a concentration of 1000 μg/mL, the findings of Matos et al. [53] indicated that spirulina was cytotoxic to Hela cells by 61.4% and reduced cell viability by 39%. This was the case when the cells were exposed to spirulina.

4.2.8. Cardiovascular Disease Treatment

Cardiovascular disease and other circulatory disorders are the main causes of death and illness on a global scale [3,117]. The World Health Organization (WHO) says that cardiovascular diseases (CVDs) are responsible for the deaths of 17.3 million people a year, and it is anticipated that this figure would increase to more than 23.6 million by the year 2030. The primary contributors to CVDs are the formation of lipids in the blood and prolonged exposure to high levels of stress [156]. Spirulina, which is classified as a BGA, can be consumed either as a whole meal or as a dietary supplement. It is commonly taken in many Asian countries, where the incidence of several ailments that are prevalent in Western society, such as coronary heart disease, cancer, and arthritis, is far lower than in those countries [115].

According to a number of studies, taking spirulina supplements may help reduce the risk of CVDs and other conditions linked to atherosclerosis. It was shown that it plays a function in lowering triglycerides, total cholesterol, and LDL cholesterol while simultaneously raising HDL cholesterol levels. The consumption of spirulina by the experimental animals helped to keep the level of lipids in their blood unchanged to a substantial degree. Because of these features, which are a result of the presence of phycocyanin molecules, phenolic compounds, and PUFAs, spirulina was regarded to be a functional food capable of reducing cholesterol levels and avoiding atherosclerosis [180]. Patients who suffer from both high blood pressure and high blood lipids have an increased chance of experiencing a heart attack. It is necessary to bring LDL and DBP levels, as well as weight and blood sugar, down to normal levels in order to both prevent and cure this condition. Because of the significant antioxidant activity that spirulina supplements possess in human beings, their consumption holds a great deal of promise for the treatment and prevention of CVDs. It has been demonstrated that using these supplements has a beneficial impact on a number of metabolic and cardiovascular health markers in people, including triglycerides, total cholesterol, LDL, extremely LDL, fasting blood glucose, and blood pressure, without causing any negative side effects [156].

Consumption of PUFAs is linked to a reduced risk of death from a variety of illnesses, including cardiovascular disease and other conditions. EPA (20:5 n3) and DHA (22:6 n3) are two of the PUFAs that can be found in the spirulina [53]. The pigment phycocyanin in spirulina, which has properties comparable to those of the pigment bilirubin in terms of its structure, is thought to be responsible for the preventive effect that spirulina has against CVDs. It has been proved that the pigment bilirubin, which is present in bile, contains powerful antioxidants. These antioxidants prevent oxidative stress and the production of radical byproducts in plasma proteins and aromatic amino acid residues [162].

4.2.9. Other Disease Treatment

Spirulina has been demonstrated to have a beneficial impact on the activation of glial cells as well as the treatment of neurodegenerative disorders. These conditions manifest themselves as a result of a deficiency in the body's naturally occurring antioxidants and anti-inflammatory defensive mechanisms. This makes the brain more susceptible to the damaging effects of free radicals, such as ROS and RNS, which play an important role in the majority of neurological disorders, such as Alzheimer's disease, Parkinson's disease, multiple sclerosis, inflammatory lesions, and aging [5]. Piovan et al. [181] found that spirulina extract, which blocks lipopolysaccharides, helped control the activation of microglia and prevent the occurrence of neuroinflammation when they studied the neuroprotective effects before and after early treatment with the extract. This was the case when they compared the effects before and after early treatment with the extract. This is because there are a large number of bioactive chemicals present. Chlorophyll, pheophytin, carotenoids, β-carotene, and zeaxanthin are the components that have a positive impact on one's health. In addition, Abdullahi et al. [182] pointed out the neuroprotective role of spirulina in mitigating the effects of spinal cord injury and its protective ability for the spinal cortical tracts and behavioral recovery in laboratory injured rat models. It was noticed that the optimal concentration of spirulina for rats was 180 mg/kg, and this concentration was found to be the most effective for rats [182].

Linolenic acid has several health benefits, including the ability to develop and maintain healthy bones, alleviate back pain, prevent arthritis and osteoporosis, avoid kidney stones, and protect teeth by ensuring that jaw bones remain strong. The benefits of γ-linolenic acid include enhanced brain function, alleviation of osteoporosis symptoms, enhancement of insulin metabolism, facilitation of the elimination of kidney stones, promotion of metabolic processes, defense against oxidative stress, and prevention of vitamin D shortage [183–186].

Some disorders that are linked to food may be traced back to a lack of certain nutrients or an inability to absorb them, including PUFAs. Being a rich source of γ-linolenic acid, in addition to linoleic and oleic acids, γ-linolenic acid is the most vital of the three. This particular acid makes for around 20% of the overall PUFA content that can be found in spirulina. The fact that unsaturated fatty acids have a beneficial effect on the majority of the chronic diseases plaguing modern civilization, such as cancer, diabetes, heart disease, arthritis, and Alzheimer's disease, is one of the primary reasons for their significance. Because of the influence that they have on prostaglandins and leukotrienes, they are essential for the proper functioning of the cardiovascular system as well as the immunological system. As a result, the food and pharmaceutical industries have begun using these algae as a complementary food in order to treat a variety of health conditions [31,43,48,79,187,188].

Chlorophyll, carotenoids, and phycobilins are the three pigments in algae that are considered to be the most significant. These pigments have been shown to have a beneficial impact on a variety of health and therapeutic uses [2,189,190]. Spirulina has several benefits and is essential for the development of young children. It is also an excellent choice for adolescents and adults who are still developing their bodies. Because it is high in Ca and Fe and contains a little amount of selenium, spirulina is beneficial in situations of general weakness and poverty. It also protects against osteoporosis, and blood illnesses, as its Ca, Fe, and Se concentrations are, respectively, 1043.62, 338.76, and 0.0488 mg/100 g.

5. Concluding Remarks

Spirulina is the cyanobacterium species that has received the most interest from researchers in the pharmaceutical and food sectors. Researchers, analysts, and scientists from all around the world have conducted numerous studies and research projects on rising trends and new technological breakthroughs. As a result of this, a number of papers have been published during the last several years. As a consequence, we wished to gather all of the scattered material and consolidate it into a single source that will give crucial information to researchers and scientists in their ongoing research, scientific pursuits, and industrial initiatives in the future. As an outcome of the facts offered in this article, we can conclude that consuming spirulina algae has the potential for monetary gain as well as health advantages. This is mostly due to the contents of these algae, as well as their capacity to synthesize a wide range of chemical compounds that are both biologically and commercially important. As a natural byproduct, a substantial amount of biomass may be produced at a low cost for food processing, with the certainty of getting naturally occurring components with high nutritional value. Polyunsaturated fatty acids, carotenoids, phycobilins, polysaccharides, vitamins, sterols, and a wide range of bioactive compounds, such as antioxidants and cholesterol reducers, are among the naturally occurring substances that can be utilized in the development of functional foods. Despite the fact that the nutritional, environmental, and social advantages of spirulina have been gathered from a variety of published literature, it can be extrapolated that spirulina production is still limited to a few natural regions. As a result, a growing number of researchers and scientists worldwide are advocating for the wider production of spirulina.

Author Contributions: Conceptualization, N.K.Z.A. and N.A.; methodology, A.B.A.; software, D.K.V.; validation, N.A. and A.B.A.; formal analysis, N.K.Z.A.; investigation, N.A.; resources, F.C.; data curation, N.K.Z.A.; writing—original draft preparation, N.K.Z.A.; writing—review and editing, D.K.V. and F.C.; visualization, A.B.A.; supervision, N.A.; project administration, A.B.A. and A.N.; funding acquisition, A.B.A. and A.N. All authors have read and agreed to the published version of the manuscript.

Funding: The study was not supported by any funding. AN acknowledges infrastructural development support by DBT-STAR to St. Joseph's College, Devagiri, Calicut.

Institutional Review Board Statement: Not applicable.

Informed Consent Statement: Not applicable.

Data Availability Statement: The data may be shared upon valid request.

Acknowledgments: The authors are grateful to the University of Basrah for providing opportunity and support for this research work.

Conflicts of Interest: The authors declare no conflict of interest.

Sample Availability: Not applicable.

Abbreviations

B	Boron
BGA	Blue-green algae
BMI	Body mass index
Ca	Calcium
CODEX	Codex Alimentarius Commission
Cr	Chromium
Cu	Copper
CVDs	Cardiovascular diseases
DBP	iastolic blood pressure
DHA	Docosahexaenoic acid

dw	Dry weight
EAAs	Essential amino acids
EPA	Eicosapentaenoic acid
FBG	Fasting blood glucose
FDA	Food and Drug Administration
Fe	Iron
GCMS	Gas chromatography–mass spectrometry
HDL	High-density lipoprotein
IBD	Inflammatory bowel disease
IVPD	In vitro digestibility protein
K	Potassium
KCN	Potassium cyanide
LDL	Low-density lipoprotein
Mg	agnesium
Mn	Manganese
Mo	Molybdenum
MUFAs	Monounsaturated fatty acids
Na	Sodium
P	Phosphorous
PUFAs	Polyunsaturated fatty acids
Se	Selenium
SeNPs	Se nanoparticles
WHO	World Health Organization
Zn	Zinc

References

1. Carlson, S. *Spirulina platensis (Conventional and Organic), Spirulina, Organic Spirulina, or Arthrospira Platensis*; Division of Biotechnology and Gras Notice Review, Office of Food Additive Safety-CFSAN: Dauphin Island, AL, USA, 2011; p. 36.
2. Haoujar, I.; Haoujar, M.; Altemimi, A.B.; Essafi, A.; Cacciola, F. Nutritional, sustainable source of aqua feed and food from microalgae: A mini review. *Int. Aquat. Res.* **2022**, *14*, 1–9.
3. Singh, S.; Verma, D.K.; Thakur, M.; Tripathy, S.; Patel, A.R.; Shah, N.; Utama, G.L.; Srivastav, P.P.; Benavente-Valdés, J.R.; Chávez-González, M.L.; et al. Supercritical Fluid Extraction (SCFE) as Green Extraction Technology for High-value Metabolites of Algae, Its Potential Trends in Food and Human Health. *Food Res. Int.* **2021**, *150 Pt A*, 110746. [CrossRef]
4. Nascimento, R.Q.; Deamici, K.M.; Tavares, P.P.L.G.; de Andrade, R.B.; Guimarães, L.C.; Costa, J.A.V.; Guedes, K.M.; Druzian, J.I.; de Souza, C.O. Improving water kefir nutritional quality via addition of viable Spirulina biomass. *Bioresour. Technol. Rep.* **2022**, *17*, 100914. [CrossRef]
5. Trotta, T.; Porro, C.; Cianciulli, A.; Panaro, M.A. Beneficial Effects of Spirulina Consumption on Brain Health. *Nutrients* **2022**, *14*, 676. [CrossRef]
6. Andrade, L.M.; Andrade, C.J.; Dias, M.; Nascimento, C.; Mendes, M.A. Chlorella and spirulina microalgae as sources of functional foods. *Nutraceuticals Food Suppl.* **2018**, *6*, 45–58.
7. Gogna, S.; Kaur, J.; Sharma, K.; Prasad, R.; Singh, J.; Bhadariya, V.; Kumar, P.; Jarial, S. Spirulina-An Edible Cyanobacterium with Potential Therapeutic Health Benefits and Toxicological Consequences. *J. Am. Nutr. Assoc.* 2022, *in press*. [CrossRef] [PubMed]
8. El-Beltagi, H.S.; Dhawi, F.; Ashoush, I.S.; Ramadan, K. Antioxidant, anti-cancer and ameliorative activities of *Spirulina platensis* and pomegranate juice against hepatic damage induced by CCl4. *Not. Bot. Horti Agrobot. Cluj-Napoca* **2020**, *48*, 1941–1956. [CrossRef]
9. Manjula, R.; Vijayavahini, R.; Lakshmi, T.S. Formulation and quality evaluation of spirulina incorporated ready to serve (RTS) functional beverage. *Int. J. Multidiscip. Res. Arts Sci. Commer.* **2021**, *1*, 29–35.
10. Aljobair, M.O.; Albaridi, N.A.; Alkuraieef, A.N.; AlKehayez, N.M. Physicochemical properties, nutritional value, and sensory attributes of a nectar developed using date palm puree and spirulina. *Int. J. Food Prop.* **2021**, *24*, 845–858. [CrossRef]
11. Shahidi, F.; Alasalvar, C. *Handbook of Functional Beverages and Human Health*; CRC Press: Boca Raton, FL, USA, 2016; Volume 11, 886p.
12. Bleakley, S.; Hayes, M. Algal Proteins: Extraction, Application, and Challenges Concerning Production. *Rev. Foods* **2017**, *6*, 33. [CrossRef] [PubMed]
13. Tavakoli, M.; Habibi Najafi, M.B.; Mohebbi, M. Effect of The Milk Fat Content and Starter Culture Selection on Proteolysis and Antioxidant Activity of Probiotic Yogurt. *Heliyon* **2019**, *5*, e01204. [CrossRef]
14. Li, D.M.; Qi, Y.Z. Spirulina industry in China: Present status and future prospects. *J. Appl. Phycol.* **1997**, *9*, 25–28. [CrossRef]
15. Grahl, S.; Strack, M.; Weinrich, R.; Mörlein, D. Consumer-Oriented Product Development: The Conceptualization of Novel Food Products Based on Spirulina (*Arthrospira platensis*) and Resulting Consumer Expectations. *J. Food Qual.* **2018**, *2018*, 1919482. [CrossRef]

16. Fogg, G.E.; Thake, B. *Algal Cultures and Phytoplankton Ecology*; University of Wisconsin Press: Madison, WI, USA, 1987.
17. Schirrmeister, B.E.; Antonelli, A.; Bagheri, H.C. The origin of multicellularity in cyanobacteria. *BMC Evol. Biol.* **2011**, *11*, 45. [CrossRef]
18. Siva Kiran, R.R.; Madhu, G.M.; Satyanarayana, S.V. Spirulina in combating protein energy malnutrition (PEM) and protein energy wasting (PEW)—A review. *J. Nutr. Res.* **2015**, *1*, 62–79.
19. Ciferri, O. Spirulina, the edible microorganism. *Microbiol. Rev.* **1983**, *47*, 551–578. [CrossRef]
20. Vonshak, A. Spirulina: Growth, physiology and biochemistry. In *Spirulina platensis (Arthrospira): Physiology, Cell Biology and Biotechnology*; Taylor and Francis Ltd.: London, UK, 1997; pp. 43–65.
21. Pelizer, L.H.; Danesi, E.D.G.; Rangel, C.O.; Sassano, C.E.; Carvalho, J.C.M.; Sato, S.; Moraes, I.O. Influence of inoculum age and concentration in *Spirulina platensis* cultivation. *J. Food Eng.* **2003**, *56*, 371–375. [CrossRef]
22. Costa, J.A.V.; Colla, L.M.; Filho, P.D. *Spirulina platensis* growth in open raceway ponds using fresh water supplemented with carbon, nitrogen and metal ions. *Z. Nat. C -A J. Biosci.* **2003**, *58*, 76–80.
23. Usharani, G.; Saranraj, P.; Kanchana, D. Spirulina cultivation: A review. *Int. J. Pharm. Biol. Arch.* **2012**, *3*, 1327–1341.
24. Maulood, B.K.; Hassan, F.M.; Al-Lami, A.A.; Toma, J.J.; Ismail, A.M. *Checklist of Algal Flora in Iraq*; Ministry of Environment: Baghdad, Iraq, 2013.
25. Al-Yassiry, T.M.H. Ecological Assessment of the Sewage in The City of Hilla/Iraq Using the Canadian Model and the Study of Phytoplankton. Master's Thesis, University of Babylon, Hillah, Iraq, 2014.
26. Stanier, R.Y.; Van Niel, C.B. The concept of a bacterium. *Arch. Mikrobiol.* **1962**, *42*, 17–35. [CrossRef]
27. Gershwin, M.E.; Belay, A. (Eds.) *Spirulina in Human Nutrition and Health*; CRC Press: Boca Raton, FL, USA; Taylor and Francis Group: London, UK; New York, NY, USA, 2007.
28. Heinsoo, D. Cultivation of Spirulina on Conventional and Urine Based Medium in a Household Scale System. Master's Thesis, KTH School of Biotechnology, Stockholm, Sweden, 2014; 46p.
29. Tomaselli, L. Morphology, ultrastructure and taxonomy of Arthrospira (Spirulina) maxima and Arthrospira (Spirulina) platensis. In *Spirulina platensis (Arthrospira): Physiology, Cell Biology and Biotechnology*; Vonshak, A., Ed.; Taylor and Francis: London, UK, 1997; pp. 1–15.
30. Belay, A.; Kato, T.; Ota, Y. Spirulina (Arthrospira): Potential application as an animal feed supplement. *J. Appl. Phycol.* **1996**, *8*, 303–311. [CrossRef]
31. Choopani, A.; Poorsoltan, M.; Fazilati1, M.; Latifi, A.M.; Salavati, H. Spirulina a source of gamma-linoleic acid. *J. Appl. Biotechnol. Rep.* **2016**, *3*, 483.
32. Jung, F.; Krüger-Genge, A.; Waldeck, P.; Küpper, J.-H. *Spirulina platensis*, a super food? *J. Cell. Biotechnol.* **2019**, *5*, 43–54. [CrossRef]
33. Hosseini, S.M.; Shahbazizadeh, S.; Khosravi-Darani, K.; Reza Mozafari, M. Spirulina paltensis: Food and function. *Curr. Nutr. Food Sci.* **2013**, *9*, 189–193. [CrossRef]
34. El Nakib, D.M.; Ibrahim, M.M.; Mahmoud, N.S.; Abd El Rahman, E.N.; Ghaly, A.E. Incorporation of Spirulina (*Athrospira platensis*) in traditional Egyptian cookies as a source of natural bioactive molecules and functional ingredients: Preparation and sensory evaluation of nutrition snack for school children. *Eur. J. Nutr. Food Saf.* **2019**, *9*, 372–397. [CrossRef]
35. Bahlol, H.E.M. Utilization of Sprulina Algae to Improve the Nutritional Value of Kiwifruits and Cantaloupe Nectar Blends. *Ann. Agric. Sci. Moshtohor* **2018**, *56*, 315–324. [CrossRef]
36. Michael, A.; Kyewalyanga, M.S.; Mtolera, M.S.; Lugomela, C.V. Antioxidants activity of the cyanobacterium, Arthrospira (Spirulina) fusiformis cultivated in a low-cost medium. *Afr. J. Food Sci.* **2018**, *12*, 188–195.
37. Saharan, V.; Jood, S. Nutritional composition of *Spirulina platensis* powder and its acceptability in food products. *Int. J. Adv. Res.* **2017**, *5*, 2295–2300. [CrossRef]
38. Sharoba, A.M. Nutritional value of spirulina and its use in the preparation of some complementary baby food formulas. *J. Agroaliment. Process. Technol.* **2014**, *20*, 330–350. [CrossRef]
39. Madkour, F.F.; Kamil, A.E.W.; Nasr, H.S. Production and nutritive value of *Spirulina platensis* in reduced cost media. *Egypt. J. Aquat. Res.* **2012**, *38*, 51–57. [CrossRef]
40. Malik, P. Utilization of Spirulina Powder for Enrichment of Ice Cream and Yoghurt. Master's Thesis, Karnataka Veterinary, Animal and Fisheries Sciences University, Bida, India, 2011; 151p.
41. Habib, M.A.B.; Parvin, M.; Huntington, T.C.; Hasan, M.R. *A Review on Culture, Production and Use of Spirulina as Food for Humans and Feeds for Domestic Animals and Fish*; FAO Fisheries and Aquaculture Circular No. 1034; FAO: Rome, Italy, 2008.
42. Falquet, J.; Hurni, J.P. The Nutritional Aspects of Spirulina. Antenna Foundation. 1997. Available online: https://www.antenna.ch/wp-content/uploads/2017/03/AspectNut_UK (accessed on 10 August 2022).
43. Seyidoglu, N.; Inan, S.; Aydin, C. A prominent superfood: *Spirulina platensis*. In *Superfood and Functional Food the Development of Superfoods and Their Roles as Medicine*; IntechOpen: London, UK, 2017; Volume 22, Chapter 1; pp. 1–27.
44. Ravi, M.; De, S.L.; Azharuddin, S.; Paul, S.F. The beneficial effects of Spirulina focusing on its immunomodulatory and antioxidant properties. *Nutr. Diet. Suppl.* **2010**, *2*, 73–83.
45. Sotiroudis, T.G.; Sotiroudis, G.T. Health aspects of Spirulina (Arthrospira) microalga food supplement. *J. Serb. Chem. Soc.* **2013**, *78*, 395–405. [CrossRef]
46. Jamil, A.B.M.R.; Akanda, M.R.; Rahman, M.M.; Hossain, M.A.; Islam, M.S. Prebiotic competence of spirulina on the production performance of broiler chickens. *J. Adv. Vet. Anim. Res.* **2015**, *2*, 304–309. [CrossRef]

47. Walter, P. Effects of vegetarian diets on aging and longevity. *Nutr. Rev.* **1997**, *55*, S61–S65. [CrossRef] [PubMed]
48. Mishra, P.; Singh, V.P.; Prasad, S.M. Spirulina and its Nutritional Importance: A Possible Approach for Development of Functional Food. *Biochem. Pharmacol.* **2014**, *3*, e171.
49. Rachidi, F.; Benhima, R.; Kasmi, Y.; Sbabou, L.; Arroussi, H.E. Evaluation of microalgae polysaccharides as biostimulants of tomato plant defense using metabolomics and biochemical approaches. *Sci. Rep.* **2021**, *11*, 930. [CrossRef] [PubMed]
50. Pascaud, M.; Brouard, C. Acides gras polyinsaturés essentiels ω6 et ω3. Besoins nutritionnels, équilibres alimentaires. *Cah. Nutr. Diététique* **1991**, *26*, 185–190.
51. Al-Dhabi, N.A.; Valan Arasu, M. Quantification of phytochemicals from commercial Spirulina products and their antioxidant activities. *Evid. -Based Complement. Altern. Med.* **2016**, *2016*, 7631864.
52. Liestianty, D.; Rodianawati, I.; Andi Arfah, R.; Asma Assa, A.; Patimah; Sundari; Muliadi. Nutritional analysis of spirulina sp. to promote as superfood candidate. In Proceedings of the 13th Joint Conference on Chemistry (13th JCC), Semarang, Indonesia, 7–8 September 2018; pp. 1–6.
53. Matos, J.; Cardoso, C.L.; Falé, P.; Afonso, C.M.; Bandarra, N.M. Investigation of nutraceutical potential of the microalgae Chlorella vulgaris and *Arthrospira platensis*. *Int. J. Food Sci. Technol.* **2020**, *55*, 303–312. [CrossRef]
54. Alyasiri, T.; Alchalabi, S.; AlMayaly, I. In vitro and In vivo antioxidant effect of *Spirulina platensis* against Lead induced toxicity in rats. *Asian J. Agric. Biol.* **2018**, *6*, 66–77.
55. Gheda, S.F.; Abo-Shady, A.M.; Abdel-Karim, O.H.; Ismail, G.A. Antioxidant and Antihyperglycemic Activity of *Arthrospira platensis* (*Spirulina platensis*) Methanolic Extract: In vitro and in vivo Study. *Egypt. J. Bot.* **2021**, *61*, 71–93. [CrossRef]
56. Viso, A.C.; Marty, J.C. Fatty-acids from 28 marine microalgae. *Phytochemistry* **1993**, *34*, 1521–1533. [CrossRef]
57. Cottin, S.C.; Sanders, T.A.; Hall, W.L. The differential effects of EPA and DHA on cardiovascular risk factors. *Proc. Nutr. Soc.* **2011**, *70*, 215–231. [CrossRef]
58. Henrikson, R. A nutrient rich super food for super health. In *Earth Food Spirulina*; Ronore Enterprises, Inc.: Hana, HI, USA, 2009; pp. 25–41.
59. Sandeep, K.P.; Shukla, S.P.; Vennila, A.; Purushothaman, C.S.; Manjulekshmi, N. Cultivation of *Spirulina* (*Arthrospira*) *platensis* in low cost seawater based medium for extraction of value added pigments. *Indian J. Geo-Mar. Sci.* **2015**, *44*, 1–10.
60. Biel, W.; Czerniawska-Piątkowska, E.; Kowalczyk, A. Offal Chemical Composition from Veal, Beef, and Lamb Maintained in Organic Production Systems. *Animals* **2019**, *9*, 489. [CrossRef]
61. Pambuwa, W.; Tanganyika, J. Determination of chemical composition of normal indigenous chickens in Malawi. *Int. J. Avian Wildl. Biol.* **2017**, *2*, 86–89.
62. Probst, Y. *Nutrient Composition of Chicken Meat*; Rural Industries Research and Development Corporation: Wagga, Australia, 2009; pp. 4–16.
63. Usydus, Z.; Szlinder-Richert, J.; Adamczyk, M.; Szatkowska, U. Marine and farmed fish in the Polish market: Comparison of the nutritional value. *Food Chem.* **2011**, *126*, 78–84. [CrossRef]
64. Kusum, M.; Verma, R.C.; Renu, M.; Jain, H.K.; Deepak, S. A review: Chemical composition and utilization of egg. *Int. J. Chem. Stud.* **2018**, *6*, 3186–3189.
65. De Lima, A.L.; Guerra, C.A.; Costa, L.M.; de Oliveira, V.S.; Lemos Junior, W.J.F.; Luchese, R.H.; Guerra, A.F. A Natural Technology for Vacuum-Packaged Cooked Sausage Preservation with Potentially Postbiotic-Containing Preservative. *Fermentation* **2022**, *8*, 106. [CrossRef]
66. Yasmin, I.; Iqbal, R.; Liaqat, A.; Khan, W.A.; Nadeem, M.; Iqbal, A.; Chughtai, M.F.J.; Rehman, S.J.U.; Tehseen, S.; Mehmood, T.; et al. Characterization and comparative evaluation of milk protein variants from pakistani dairy breeds. *Food Sci. Anim. Resour.* **2020**, *40*, 689. [CrossRef] [PubMed]
67. Pehlivanoğlu, H.; Bardakçi, H.F.; Yaman, M. Protein quality assessment of commercial whey protein supplements commonly consumed in Turkey by in vitro protein digestibility-corrected amino acid score (PDCAAS). *Food Sci. Technol.* **2021**, *42*, 1–8. [CrossRef]
68. Barone, G.; Moloney, C.; O'Regan, J.; Kelly, A.L.; O'Mahony, J.A. Chemical composition, protein profile and physicochemical properties of whey protein concentrate ingredients enriched in α-lactalbumin. *J. Food Compos. Anal.* **2020**, *92*, 103546. [CrossRef]
69. Yaman, H.; Aykas, D.P.; Jiménez-Flores, R.; Rodriguez-Saona, L.E. Monitoring the ripening attributes of Turkish white cheese using miniaturized vibrational spectrometers. *J. Dairy Sci.* **2022**, *105*, 40–55. [CrossRef]
70. Popović-Vranješ, A.; Paskaš, S.; Kasalica, A.; Jevtić, M.; Popović, M.; Belić, B. Production, composition and characteristics of organic hard cheese. *Biotechnol. Anim. Husb.* **2016**, *32*, 393–402. [CrossRef]
71. Karr-Lilienthal, L.K.; Bauer, L.L.; Utterback, P.L.; Zinn, K.E.; Frazier, R.L.; Parsons, C.M.; Fahey, G.C. Chemical Composition and Nutritional Quality of Soybean Meals Prepared by Extruder/Expeller Processing for Use in Poultry Diets. *J. Agric. Food Chem.* **2006**, *54*, 8108–8114. [CrossRef] [PubMed]
72. Ciabotti, S.; Silva, A.C.B.B.; Juhasz, A.C.P.; Mendonça, C.D.; Tavano, O.L.; Mandarino, J.M.G.; ConÇAlves, C.A.A. Chemical composition, protein profile, and isoflavones content in soybean genotypes with different seed coat colors. *Int. Food Res. J.* **2016**, *23*, 621–629.
73. Youssef, M.K.E.; Nassar, A.G.; EL–Fishawy, F.A.; Mostafa, M.A. Assessment of proximate chemical composition and nutritional status of wheat biscuits fortified with oat powder. *Assiut J. Agric. Sci* **2016**, *47*, 83–94.

74. Sterna, V.; Zute, S.; Brunava, L. Oat grain composition and its nutrition benefice. *Agric. Agric. Sci. Procedia* **2016**, *8*, 252–256. [CrossRef]
75. Barreto, N.M.B.; Pimenta, N.G.; Braz, B.F.; Freire, A.S.; Santelli, R.E.; Oliveira, A.C.; Bastos, L.H.P.; Cardoso, M.H.W.M.; Monteiro, M.; Diogenes, M.E.L.; et al. Organic black beans (*Phaseolus vulgaris* L.) from Rio de Janeiro state, Brazil, present more phenolic compounds and better nutritional profile than nonorganic. *Foods* **2021**, *10*, 900. [CrossRef]
76. Ignjatovic-Micic, D.; Vancetovic, J.; Trbovic, D.; Dumanovic, Z.; Kostadinovic, M.; Bozinovic, S. Grain nutrient composition of maize (Zea mays L.) drought-tolerant populations. *J. Agric. Food Chem.* **2015**, *63*, 1251–1260. [CrossRef]
77. Sangwongchai, W.; Tananuwong, K.; Krusong, K.; Thitisaksakul, M. Yield, Grain Quality, and Starch Physicochemical Properties of 2 Elite Thai Rice Cultivars Grown under Varying Production Systems and Soil Characteristics. *Foods* **2021**, *10*, 2601. [CrossRef] [PubMed]
78. Yousefian, M.; Shahbazi, F.; Hamidian, K. Crop Yield and Physicochemical Properties of Wheat Grains as Affected by Tillage Systems. *Sustainability* **2021**, *13*, 4781. [CrossRef]
79. Salmeán, G.G.; Castillo, L.H.F.; Chamorro-Cevallos, G. Nutritional and toxicological aspects of Spirulina (Arthrospira). *Nutr. Hosp. Organo Of. Soc. Española Nutr. Parenter. Enter.* **2015**, *32*, 34–40.
80. Layam, A.; Lekha, C.; Reddy, K. Antidiabetic property of Spirulina. *Diabetol. Croat.* **2006**, *35*, 29–33.
81. Wells, M.L.; Potin, P.; Craigie, J.S.; Raven, J.A.; Merchant, S.S.; Helliwell, K.E.; Smith, A.G.; Camire, M.E.; Brawley, S.H. Algae as nutritional and functional food sources: Revisiting our understanding. *J. Appl. Phycol.* **2017**, *29*, 949–982. [CrossRef]
82. El-Chaghaby, G.A.; Rashad, S.; Abdel-Kader, S.F.; Rawash, E.S.A.; Abdul Moneem, M. Assessment of phytochemical components, proximate composition and antioxidant properties of Scenedesmus obliquus, Chlorella vulgaris and *Spirulina platensis* algae extracts. *Egypt. J. Aquat. Biol. Fish.* **2019**, *23*, 521–526. [CrossRef]
83. Batista, A.P.; Niccolai, A.; Bursic, I.; Sousa, I.; Raymundo, A.; Rodolfi, L.; Biondi, N.; Tredici, M.R. Microalgae as Functional Ingredients in Savory Food Products: Application to Wheat Crackers. *Foods* **2019**, *8*, 611. [CrossRef]
84. Khan, Z.; Bhadouria, P.; Bisen, P. Nutritional and Therapeutic Potential of Spirulina. *Curr. Pharm. Biotechnol.* **2005**, *6*, 373–379. [CrossRef]
85. Vakarelova, M. Microencapsulation of Bioactive Molecules from *Spirulina platensis* and *Haematococcus pluvialis*. Ph.D. Thesis, University of Verona, Verona, Italy, 2017; 119p.
86. Mahmoud, A.; Sabae, S.A.; Helal, A.M. Culture and Biorefinary of Two Freshwater Microalgae; *Spirulina platensis* and Chlorella vulgaris As Vitamins Sources. *Biosci. Res.* **2018**, *15*, 4584–4589.
87. Edelmann, M.; Aalto, S.; Chamlagain, B.; Kariluoto, S.; Piironen, V. Riboflavin, niacin, folate and vitamin B12 in commercial microalgae powders. *J. Food Compos. Anal.* **2019**, *82*, 103226. [CrossRef]
88. Tang, G.; Suter, P.M. Vitamin A, nutrition, and health values of algae: Spirulina, Chlorella, and Dunaliella. *J. Pharm. Nutr. Sci.* **2011**, *1*, 111–118.
89. Mohan, A.; Misra, N.; Srivastav, D.; Umapathy, D.; Kumar, S. Spirulina, the nature's wonder: A review. *Lipids* **2014**, *5*, 7–10.
90. Mogale, M. Identification and Quantification of Bacteria Associated with Cultivated Spirulina and Impact of Physiological Factors. Master's Thesis, University of Cape Town, Cape Town, South Africa, 2016; 164p.
91. Kumudha, A.; Sarada, R. Effect of different extraction methods on vitamin B12 from blue green algae, *Spirulina platensis*. *Pharm. Anal. Acta* **2015**, *6*, 1000337. [CrossRef]
92. Neher, B.D.; Azcarate, S.M.; Camiña, J.M.; Savio, M. Nutritional analysis of Spirulina dietary supplements: Optimization procedure of ultrasound-assisted digestion for multielemental determination. *Food Chem.* **2018**, *257*, 295–301. [CrossRef] [PubMed]
93. Verdasco-Martín, C.M.; Echevarrieta, L.; Otero, C. Advantageous preparation of digested proteic extracts from *Spirulina platensis* biomass. *Catalysts* **2019**, *9*, 145. [CrossRef]
94. Ghaeni, M.; Roomiani, L. Review for application and medicine effects of Spirulina, microalgae. *J. Adv. Agric. Technol.* **2016**, *3*, 114–117.
95. Hidayati, J.R.; Yudiati, E.; Pringgenies, D.; Oktaviyanti, D.T.; Kusuma, A.P. Comparative study on antioxidant activities, total phenolic compound and pigment contents of tropical *Spirulina platensis*, Gracilaria arcuata and Ulva lactuca extracted in different solvents polarity. *E3S Web Conf.* **2020**, *147*, 03012. [CrossRef]
96. Scheer, H.; Zhao, K.H. Biliprotein maturation: The chromophore attachment. *Mol. Microbiol.* **2008**, *68*, 263–276. [CrossRef]
97. Martelli, F.; Cirlini, M.; Lazzi, C.; Neviani, E.; Bernini, V. Edible seaweeds and spirulina extracts for food application: In vitro and in situ evaluation of antimicrobial activity towards foodborne pathogenic bacteria. *Foods* **2020**, *9*, 1442. [CrossRef]
98. García, J.L.; De Vicente, M.; Galán, B. Microalgae, old sustainable food and fashion nutraceuticals. *Microb. Biotechnol.* **2017**, *10*, 1017–1024. [CrossRef]
99. Lafarga, T.; Fernández-Sevilla, J.M.; González-López, C.; Acién-Fernández, F.G. Spirulina for the food and functional food industries. *Food Res. Int.* **2020**, *137*, 109356. [CrossRef] [PubMed]
100. Thomas, S.S. *The Role of Parry Organic Spirulina in Health Management*; Parry Nutraceuticals, Division of EID Parry (India) Ltd.: Crawley, UK, 2010.
101. Anderson, D.W.; Tang, C.S.; Ross, E. The xanthophylls of Spirulina and their effect on egg yolk pigmentation. *Poult. Sci.* **1991**, *70*, 115–119. [CrossRef]

102. Ting, C.S.; Rocap, G.; King, J.; Chisholm, S.W. Cyanobacterial photosynthesis in the oceans: The origins and significance of divergent light-harvesting strategies. *Trends Microbiol.* **2002**, *10*, 134–142. [CrossRef]
103. Chaiklahan, R.; Chirasuwan, N.; Bunnag, B. Stability of phycocyanin extracted from *Spirulina* sp.: Influence of temperature, pH and preservatives. *Process Biochem.* **2012**, *47*, 659–664. [CrossRef]
104. Rybner, T.V. Improving the biomass productivity and phycocyanin concentration by mixotrophic cultivation of *Arthrospira platensis*. Master's Thesis, Section for Sustainable Biotechnology, Aalborg University Copenhagen, Aalborg, Denmark, 2016; 57p.
105. Gao, X.; Sun, T.; Pei, G.; Chen, L.; Zhang, W. Cyanobacterial chassis engineering for enhancing production of biofuels and chemicals. *Appl. Microbiol. Biotechnol.* **2016**, *100*, 3401–3413. [CrossRef] [PubMed]
106. Moraes, C.C.; Sala, L.; Cerveira, G.P.; Kalil, S.J. C-phycocyanin extraction from *Spirulina platensis* wet biomass. *Braz. J. Chem. Eng.* **2011**, *28*, 45–49. [CrossRef]
107. Kannaujiya, V.K.; Sinha, R.P. Thermokinetic stability of phycocyanin and phycoerythrin in food-grade preservatives. *J. Appl. Phycol.* **2016**, *28*, 1063–1070. [CrossRef]
108. Abd El-Monem, A.M.; Gharieb, M.M.; Hussian, A.M.; Doman, K.M. Effect of pH on phytochemical and antibacterial activities of *Spirulina platensis*. *Int. J. Appl. Environ. Sci.* **2018**, *13*, 339–351.
109. Duangsee, R.; Phoopat, N.; Ningsanond, S. Phycocyanin extraction from *Spirulina platensis* and extract stability under various pH and temperature. *Asian J. Food Agro-Ind.* **2009**, *2*, 819–826.
110. Widawati, D.; Santosa, G.W.; Yudiati, E. Pengaruh Pertumbuhan *Spirulina platensis* terhadap Kandungan Pigmen beda Salinitias. *J. Mar. Res.* **2022**, *11*, 61–70. [CrossRef]
111. Buso, D.; Zissis, G.; Prudhomme, T. Influence of light intensity and photoperiod on energy efficiency of biomass and pigment production of Spirulina (*Arthrospira platensis*). *OCL* **2021**, *28*, 37.
112. Machu, L.; Misurcova, L.; Vavra Ambrozova, J.; Orsavova, J.; Mlcek, J.; Sochor, J.; Jurikova, T. Phenolic content and antioxidant capacity in algal food products. *Molecules* **2015**, *20*, 1118–1133. [CrossRef]
113. Dianursanti; Prakasa, M.B.; Nugroho, P. The effect of adding microalgae extract *Spirulina platensis* containing flavonoid in the formation of Sunscreen towards cream stability and SPF values. *AIP Conf. Proc.* **2020**, *2255*, 040022.
114. Gabr, G.A.; El-Sayed, S.M.; Hikal, M.S. Antioxidant activities of phycocyanin: A bioactive compound from *Spirulina platensis*. *J. Pharm. Res. Int.* **2020**, *32*, 73–85. [CrossRef]
115. Salamatullah, A. Characterization of Extraction Methods to Recover Phenolic-Rich Antioxidants from Blue Green Algae (Spirulina) Using Response Surface Approaches. Master's Thesis, University of Nebraska, Lincol, NE, USA, 2014; 82p.
116. Agustini, T.W.; Maâ, W.F.; Widayat, W.; Suzery, M.; Hadiyanto, H.; Benjakul, S. Application of *Spirulina platensis* on ice cream and soft cheese with respect to their nutritional and sensory perspectives. *J. Teknol.* **2016**, *78*, 245–251. [CrossRef]
117. Banwo, K.; Olojede, A.O.; Adesulu-Dahunsi, A.T.; Verma, D.K.; Thakur, M.; Tripathy, S.; Singh, S.; Patel, A.R.; Gupta, A.K.; Aguilar, C.N.; et al. Functional importance of bioactive compounds of foods with Potential Health Benefits: A review on recent trends. *Food Biosci.* **2021**, *43*, 101320. [CrossRef]
118. Jerez-Martel, I.; García-Poza, S.; Rodríguez-Martel, G.; Rico, M.; Afonso-Olivares, C.; Gómez-Pinchetti, J.L. Phenolic profile and antioxidant activity of crude extracts from microalgae and cyanobacteria strains. *J. Food Qual.* **2017**, *2017*, 2924508. [CrossRef]
119. Sonam, K.; Neetu, S. A comparative study on nutritional profile of spirulina cookies. *Int. J. Food Sci. Nutr.* **2017**, *2*, 100–102.
120. Santiago-Santos, M.C.; Ponce-Noyola, T.; Olvera-Ramírez, R.; Ortega-López, J.; Cañizares-Villanueva, R.O. Extraction and purification of phycocyanin from *Calothrix* sp. *Process Biochem.* **2004**, *39*, 2047–2052. [CrossRef]
121. Fradinho, P.; Niccolai, A.; Soares, R.; Rodolfi, L.; Biondi, N.; Tredici, M.R.; Sousa, I.; Raymundo, A. Effect of *Arthrospira platensis* (spirulina) incorporation on the rheological and bioactive properties of gluten-free fresh pasta. *Algal Res.* **2020**, *45*, 101743. [CrossRef]
122. Mostolizadeh, S.; Moradi, Y.; Mortazavi, M.S.; Motallebi, A.A.; Ghaeni, M. Effects of incorporation *Spirulina platensis* (Gomont 2020, 1892) powder in wheat flour on chemical, microbial and sensory properties of pasta. *Iran. J. Fish. Sci.* **2020**, *19*, 410–420.
123. Pagnussatt, F.A.; Spier, F.; Bertolin, T.E.; Vieira Costa, J.A.; Gutkosk, L.C. Technological and nutritional assessment of dry pasta with oatmeal and the microalga *Spirulina platensis*. *Braz. J. Food Technol.* **2014**, *17*, 296–304. [CrossRef]
124. Lemes, A.C.; Takeuchi, K.P.; Carvalho, J.C.M.; Danesi, E.D.G. Fresh Pasta Production Enriched with *Spirulina platensis*. *Biomass Braz. Arch. Biol. Technol.* **2012**, *55*, 741–750. [CrossRef]
125. Szmejda, K.; Duliński, R.; Byczyński, L.; Karbowski, A.; Florczak, T.; Żyła, K. Analysis of The Selected Antioxidant Compounds In Ice Cream Supplemented With *Spirulina* (*Arthrospira platensis*) Extrac. *Biotechnol. Food Sci.* **2018**, *82*, 41–48.
126. Patil, A.G.; Banerjee, S. Variants of ice creams and their health effects. *MOJ Food Process. Technol.* **2017**, *4*, 58–64.
127. Mocanu, G.; Botez, E.; Nistor, O.V.; Andronoiu, D.G.; Vlăscean, G. Influence of *Spirulina platensis* Biomass over Some Starter Culture of Lactic Bacteria. *J. Agroaliment. Process. Technol.* **2013**, *19*, 474–479.
128. Aghajani, A.; Mortazav, S.A.; Tabtabai Yazdi, F.; Shafafi Zenosian, M.; Saeidi Asl, M.R. Color, microbiological and sensory properties of low-fat probiotic yogurt supplemented with *Spirulina platensis* and Ferulago angulata hydroalcoholic extracts during cold storage. *Banat. J. Biotechnol.* **2019**, *10*, 20–34. [CrossRef]
129. Fadaei, V.; Mohamadi-Alasti, F.; Khosravi-Darani, K. Influence of *Spirulina platensis* powder on the starter culture viability in probiotic yoghurt containing spinach during cold storage. *Eur. J. Exp. Biol.* **2013**, *3*, 389–393.
130. Guldas, M.; Gurbuz, O.; Cakmak, I.; Yildiz, E.L.İ.F.; Sen, H. Effects of honey enrichment with *Spirulina platensis* on phenolics, bioaccessibility, antioxidant capacity and fatty acids. *LWT-Food Sci. Technol.* **2022**, *153*, 112461. [CrossRef]

131. Sengupta, S.; Bhowal, J. Optimization of Ingredient and Processing Parameter for the Production of *Spirulina platensis* Incorporated Soy Yogurt Using Response Surface Methodology. *J. Microbiol. Biotechnol. Food Sci.* **2017**, *6*, 1081–1085. [CrossRef]
132. Beheshtipour, H.; Mortazavian, A.M.; Mohammadi, R.; Sohrabvandi, S.; Khosravi-Darani, K. Supplementation of *Spirulina platensis* and Chlorella vulgaris Algae into Probiotic Fermented Milks. *Compr. Rev. Food Sci. Food Saf.* **2013**, *12*, 144–154. [CrossRef]
133. Chatterjee, M. Development of Spirulina Containing Fermented Synbiotic Lassi. Master's Thesis, Anand Agricultural University, Anand, India, 2012.
134. Varga, L.; Szigeti, J.; KováCs, R.; FoLdes, T.; Buti, S. Influence of A *Spirulina platensis* Biomass on the Microflora of Fermented Abt Milks During Storage (R1). *Am. Dairy Sci. Assoc. (J. Dairy Sci.)* **2002**, *85*, 1031–1038. [CrossRef]
135. Gouveia, L.; Coutinho, C.; Mendonça, E.; Batista, A.P.; Sousa, I.; Bandarra, N.M.; Raymundo, A. Functional biscuits with PUFA-ω3 from Isochrysis galbana. *J. Sci. Food Agric.* **2008**, *88*, 891–896. [CrossRef]
136. Wu, Q.; Liu, L.; Miron, A.; Klímova, B.; Wan, D.; Kuča, K. The antioxidant, immunomodulatory, and anti-inflammatory activities of Spirulina: An overview. *Arch. Toxicol.* **2016**, *90*, 1817–1840. [CrossRef] [PubMed]
137. Jabber, M.M. The Inhibitory Efficacy of Extracts of Some Alge Species aginst Some Pathogenic Bacteria. Master's Thesis, Department of Biology, Collage of Education for pur Science, University of Thi-Qar, Thi-Qar, Iraq, 2021; 118p.
138. Prabakaran, G.; Sampathkumar, P.; Kavisri, M.; Moovendhan, M. Extraction and characterization of phycocyanin from *Spirulina platensis* and evaluation of its anticancer, anti diabetic and anti inflammatory effect. *Int. J. Biol. Macromol.* **2020**, *153*, 256–263. [CrossRef]
139. Niccolai, A.; Bažec, K.; Rodolfi, L.; Biondi, N.; Zlatić, E.; Jamnik, P.; Tredici, M.R. Lactic acid fermentation of *Arthrospira platensis* (Spirulina) in a vegetal soybean drink for developing new functional lactose-free beverages. *Front. Microbiol.* **2020**, *11*, 560684. [CrossRef]
140. Camacho, F.; Macedo, A.; Malcata, F. Potential industrial applications and commercialization of microalgae in the functional food and feed industries: A short review. *Mar. Drugs* **2019**, *17*, 312. [CrossRef]
141. Omar, H.H.; Dighriri, K.A.; Gashgary, R.M. The Benefit Roles of Micro- and Macro-Algae in Probiotics. *Nat. Sci.* **2019**, *17*, 258–279.
142. Merchant, R.E.; Andre, C.A. A review of recent clinical trials of the nutritional supplement Chlorella pyrenoidosa in the treatment of fibromyalgia, hypertension, and ulcerative colitis. *Altern. Ther. Health Med.* **2001**, *7*, 79–91.
143. Gyenis, B.; Szigeti, J.; Molnár, N.; Varga, L. Use of dried microalgal biomasses to stimulate acid production and growth of Lactobacillus plantarum and Enterococcus faecium in milk. *Acta Agrar. Kaposváriensis* **2005**, *9*, 53–59.
144. Koyande, A.K.; Chew, K.W.; Rambabu, K.; Tao, Y.; Chu, D.T.; Show, P.L. Microalgae: A potential alternative to health supplementation for humans. *Food Sci. Hum. Wellness* **2019**, *8*, 16–24. [CrossRef]
145. Sathasivam, R.; Radhakrishnan, R.; Hashem, A.; Abd_Allah, E.F. Microalgae metabolites: A rich source for food and medicine. *Saudi J. Biol. Sci.* **2019**, *26*, 709–722. [CrossRef] [PubMed]
146. Parikh, P.; Mani, U.; Iyer, U. Role of Spirulina in the control of glycemia and lipidemia in type 2 diabetes mellitus. *J. Med. Food* **2001**, *4*, 193–199. [CrossRef]
147. Mazokopakis, E.E.; Starakis, I.K.; Papadomanolaki, M.G.; Mavroeidi, N.G.; Ganotakis, E.S. The hypolipidaemic effects of Spirulina (*Arthrospira platensis*) supplementation in a Cretan population: A prospective study. *J. Sci. Food Agric.* **2014**, *94*, 432–437. [CrossRef] [PubMed]
148. Torres-Duran, P.V.; Ferreira-Hermosillo, A.; Juarez-Oropeza, M.A. Antihyperlipemic and Antihypertensive Effects of Spirulina maxima in an Open Sample of Mexican Population: A Preliminary Report. *Lipids Health Dis.* **2007**, *6*, 33. [CrossRef]
149. Selmi, C.; Leung, P.S.; Fischer, L.; German, B.; Yang, C.-Y.; Kenny, T.P.; Cysewski, G.R.; Gershwin, M.E. The effects of Spirulina on anemia and immune function in senior citizens. *Cell. Mol. Immunol.* **2011**, *8*, 248–254. [CrossRef]
150. Finamore, A.; Palmery, M.; Bensehaila, S.; Peluso, I. Antioxidant, Immunomodulating, and Microbial-Modulating Activities of The Sustainable and Ecofriendlyspirulina. *Oxidative Med. Cell. Longev.* **2017**, *2017*, 3247528. [CrossRef]
151. Gupta, S.; Gupta, C.; Garg, A.B.; Prakash, D. Prebiotic Efficiency of Blue Green Algae on Probiotics Microorganisms. *J. Microbiol. Exp.* **2017**, *4*, 00120. [CrossRef]
152. Karkos, P.D.; Leong, S.C.; Karkos, C.D.; Sivaji, N.; Assimakopoulos, D.A. Spirulina in clinical practice: Evidence-based human applications. *Evid. -Based Complement. Altern. Med.* **2011**, *2011*, 531053.
153. Bhowmik, D.; Dubey, J.; Mehra, S. Probiotic efficiency of *Spirulina platensis*-stimulating growth of lactic acid bacteria. *World J. Dairy Food Sci.* **2009**, *4*, 160–163.
154. Diraman, H.; Koru, E.; Dibeklioglu, H. Fatty acid profile of *Spirulina platensis* used as a food supplement. *Isr. J. Aquac. -Bamidgeh* **2009**, *61*, 134–142. [CrossRef]
155. DiNicolantonio, J.J.; Bhat, A.G.; OKeefe, J. Effects of spirulina on weight loss and blood lipids: A review. *Open Heart* **2020**, *7*, e001003. [CrossRef] [PubMed]
156. Huang, H.; Liao, D.; Pu, R.; Cui, Y. Quantifying the effects of spirulina supplementation on plasma lipid and glucose concentrations, body weight, and blood pressure. *Diabetes Metab. Syndr. Obes.* **2018**, *11*, 729. [CrossRef] [PubMed]
157. Moradi, S.; Ziaei, R.; Foshati, S.; Mohammadi, H.; Nachvak, S.M.; Rouhani, M.H. Effects of Spirulina supplementation on obesity: A systematic review and meta-analysis of randomized clinical trials. *Complement. Ther. Med.* **2019**, *47*, 102211. [CrossRef] [PubMed]

158. Zarezadeh, M.; Faghfouri, A.H.; Radkhah, N.; Foroumandi, E.; Khorshidi, M.; Rasouli, A.; Zarei, M.; Honarvar, N.M.; Karzar, N.H.; Ebrahimi Mamaghani, M. Spirulina supplementation and anthropometric indices: A systematic review and meta-analysis of controlled clinical trials. *Phytother. Res.* **2021**, *35*, 577–586. [CrossRef]
159. Bohórquez-Medina, S.L.; Bohórquez-Medina, A.L.; Zapata, V.A.B.; Ignacio-Cconchoy, F.L.; Toro-Huamanchumo, C.J.; Bendezu-Quispe, G.; Pacheco-Mendoza, J.; Hernandez, A.V. Impact of spirulina supplementation on obesity-related metabolic disorders: A systematic review and meta-analysis of randomized controlled trials. *NFS J.* **2021**, *25*, 21–30. [CrossRef]
160. Niamah, A.K.; Al-Sahlany, S.T.G.; Ibrahim, S.A.; Verma, D.K.; Thakur, M.; Singh, S.; Patel, A.R.; Aguilar, C.N.; Utama, G.L. Electro-hydrodynamic processing for encapsulation of probiotics: A review on recent trends, technological development, challenges and future prospect. *Food Biosci.* **2021**, *44*, 101458. [CrossRef]
161. Hadebe, N. Isolation and Characterization of Prebiotic Oligosaccharides from Algal Extracts and Their Effect on Gut Microflora. Ph.D. Thesis, Durban University of Technology, Durban, South Africa, 2016.
162. Wang, Y.; Ocampo, M.F.; Rodriguez, B.; Chen, J. Resveratrol and Spirulina: Nutraceuticals that potentially improving cardiovascular disease. *J. Cardiovasc. Med. Cardiol.* **2020**, *7*, 138–145.
163. Davis, C.D. The gut microbiome and its role in obesity. *Nutr. Today* **2016**, *51*, 167–174. [CrossRef]
164. Schroeder, B.O.; Bäckhed, F. Signals from the gut microbiota to distant organs in physiology and disease. *Nat. Med.* **2016**, *22*, 1079–1089. [CrossRef]
165. Abdel-Moneim, A.M.E.; El-Saadony, M.T.; Shehata, A.M.; Saad, A.M.; Aldhumri, S.A.; Ouda, S.M.; Mesalam, N.M. Antioxidant and antimicrobial activities of *Spirulina platensis* extracts and biogenic selenium nanoparticles against selected pathogenic bacteria and fungi. *Saudi J. Biol. Sci.* **2022**, *29*, 1197–1209. [CrossRef] [PubMed]
166. Catinean, A.; Neag, M.A.; Muntean, D.M.; Bocsan, I.C.; Buzoianu, A.D. An overview on the interplay between nutraceuticals and gut microbiota. *Peer J.* **2018**, *6*, e4465. [CrossRef]
167. Ghazy, E.W.; Mokhbatly, A.A.A.; Keniber, S.S.; Shoghy, K.M. Synergistic ameliorative effect of Lactobacillus and *Spirulina platensis* against experimintal colitis in albinorats: Antioxidant, histopathological and molecular studies. *Slov. Vet. Res.* **2019**, *56* (Suppl. S22), 553–569. [CrossRef]
168. Nielsen, C.H.; Balachandran, P.; Christensen, O.; Pugh, N.D.; Tamta, H.; Sufka, K.J.; Wu, X.; Walsted, A.; Schjørring-Thyssen, M.; Enevold, C.; et al. Enhancement of natural killer cell activity in healthy subjects by Immulina®, a Spirulina extract enriched for Braun-type lipoproteins. *Planta Med.* **2010**, *76*, 1802–1808. [CrossRef] [PubMed]
169. Pugh, N.; Ross, S.A.; ElSohly, H.N.; ElSohly, M.A.; Pasco, D.S. Isolation of three high molecular weight polysaccharide preparations with potent immunostimulatory activity from *Spirulina platensis*, Aphanizomenon flos-aquae and Chlorella pyrenoidosa. *Planta Med.* **2001**, *67*, 737–742. [CrossRef] [PubMed]
170. Cohen, Z. *The Chemicals of Spirulina in Spirulina platensis (Arthrospira): Physiology, Cell-Biology, and Biotechnology*; Vonshak, A., Ed.; Taylor and Francis: London, UK, 1996; p. 175.
171. Parages, M.L.; Rico, R.M.; Abdala-Díaz, R.T.; Chabrillón, M.; Sotiroudis, T.G.; Jiménez, C. Acidic polysaccharides of Arthrospira (Spirulina) platensis induce the synthesis of TNF-α in RAW macrophages. *J. Appl. Phycol.* **2012**, *24*, 1537–1546. [CrossRef]
172. González, R.; González, A.; Remirez, D.; Romay, C.; Rodriguez, S.; Ancheta, O.; Merino, N. Protective effects of phycocyanin on galactosamine-induced hepatitis in rats. *Biotechnol. Apl.* **2003**, *20*, 107–110.
173. Sekar, S.; Chandramohan, M. Phycobiliproteins as a commodity: Trends in applied research, patents and commercialization. *J. Appl. Phycol.* **2008**, *20*, 113–136. [CrossRef]
174. Anekthanakul, K.; Senachak, J.; Hongsthong, A.; Charoonratana, T.; Ruengjitchatchawalya, M. Natural ACE inhibitory peptides discovery from Spirulina (*Arthrospira platensis*) strain C1. *Peptides* **2019**, *118*, 170107. [CrossRef]
175. Muna, W. Studying the Effect of the Ethanolic Extract of *Spirulina platensis* on Lipid Concentrations and Some Physiological Parameters in Male Laboratory Rabbits Induced by Hypercholesterolemia. Master's Thesis, College of Education for Pure Sciences, University of Basrah, Basrah, Iraq, 2018; 138p.
176. Colla, L.M.; Muccillo-Baisch, A.L.; Costa, J.A.V. *Spirulina platensis* effects on the levels of total cholesterol, HDL and triacylglycerols in rabbits fed with a hypercholesterolemic diet. *Braz. Arch. Biol. Technol.* **2008**, *51*, 405–411. [CrossRef]
177. Hosikian, A.; Lim, S.; Halim, R.; Danquah, M.K. Chlorophyll extraction from microalgae: A review on the process engineering aspects. *Int. J. Chem. Eng.* **2010**, *2010*, 391632. [CrossRef]
178. Oo, Y.Y.N.; Su, M.C.; Kyaw, K.T. Extraction and determination of chlorophyll content from microalgae. *Int. J. Adv. Res. Publ.* **2017**, *1*, 298.
179. Najem, A.M.; Abed, I.J.; Fadhel, L.Z. Assessment the Activity of *Spirulina platensis* Alcoholic Extract in MCF-7 Breast Cancer Cells Inhibition and P-53 gene induction. *Public Health* **2021**, *24*, 442–449. [CrossRef]
180. Pandey, R.; Singh, S. Spirulina and Herbal Combination on Metabolic Alterations of Cardiovascular Diseases (CVDs). *Int. J. Res. Rev.* **2022**, *9*, 308–317. [CrossRef]
181. Piovan, A.; Battaglia, J.; Filippini, R.; Dalla Costa, V.; Facci, L.; Argentini, C.; Pagetta, A.; Giusti, P.; Zusso, M. Pre- and Early Post-treatment With *Arthrospira platensis* (Spirulina) Extract Impedes Lipopolysaccharide triggered Neuroinflammation in Microglia. *Front. Pharmacol.* **2021**, *12*, 724993. [CrossRef] [PubMed]
182. Abdullahi, D.; Ahmad Annuar, A.; Sanusi, J. Improved spinal cord gray matter morphology induced by *Spirulina platensis* following spinal cord injury in rat models. *Ultrastruct. Pathol.* **2020**, *44*, 359–371. [CrossRef]

183. Bandarra, N.M.; Pereira, P.A.; Batista, I.; Vilela, M.H. Fatty acids, sterols and α-tocopherol in *Isochrysis galbana*. *J. Food Lipids* **2003**, *10*, 25–34. [CrossRef]
184. Donato, M.; Vilela, M.H.; Bandarra, N.M. Fatty acids, sterols, α-tocopherol and total carotenoids composition of Diacronema vlkianum. *J. Food Lipids* **2003**, *10*, 267–276. [CrossRef]
185. Swanson, D.; Block, R.; Mousa, S. Omega-3 fatty acids EPA and DHA: Health benefits throughout life. *Adv. Nutr.* **2012**, *3*, 1–7. [CrossRef]
186. Stonehouse, W. Does consumption of LC omega-3 PUFA enhance cognitive performance in healthy school-aged children and throughout adulthood? Evidence from clinical trials. *Nutrient* **2014**, *6*, 2730–2758. [CrossRef]
187. Quoc, K.P.; Dubacq, J.P.; Demandre, C.; Mazliak, P. Comparative effects of exogenous fatty acid supplementations on the lipids from the cyanobacterium *Spirulina platensis*. *Plant Physiol. Biochem.* **1994**, *32*, 501–509.
188. Ramírez-Moreno, L.; Olvera-Ramírez, R. Uso tradicional y actual de *Spirulina* sp. (*Arthrospira* sp.). *Interciencia* **2006**, *31*, 657–663.
189. Dufossé, L.; Galaup, P.; Yaron, A.; Arad, S.M.; Blanc, P.; Murthy, K.N.C.; Ravishankar, G.A. Microorganisms and microalgae as sources of pigments for food use: A scientific oddity or an industrial reality? *Trends Food Sci. Technol.* **2005**, *16*, 389–406. [CrossRef]
190. Koller, M.; Muhr, A.; Braunegg, G. Microalgae as versatile cellular factories for valued products. *Algal Res.* **2014**, *6*, 52–63. [CrossRef]

molecules

MDPI

Review

Phytochemical Properties, Extraction, and Pharmacological Benefits of Naringin: A Review

VS Shilpa [1], Rafeeya Shams [1,*], Kshirod Kumar Dash [2,*], Vinay Kumar Pandey [3,4], Aamir Hussain Dar [5], Shaikh Ayaz Mukarram [6], Endre Harsányi [7] and Béla Kovács [6,*]

1 Department of Food Technology & Nutrition, Lovely Professional University, Phagwara 144001, Punjab, India
2 Department of Food Processing Technology, Ghani Khan Choudhury Institute of Engineering and Technology Malda, Malda 732141, West Bengal, India
3 Department of Bioengineering, Integral University, Lucknow 226026, Uttar Pradesh, India
4 Department of Biotechnology, Axis Institute of Higher Education, Kanpur 209402, Uttar Pradesh, India
5 Department of Food Technology, Islamic University of Science and Technology, Awantipora 192122, Kashmir, India
6 Faculty of Agriculture, Food Science and Environmental Management Institute of Food Science, University of Debrecen, 4032 Debrecen, Hungary
7 Faculty of Agriculture, Food Science and Environmental Management, Institute of Land Utilization, Engineering and Precision Technology, University of Debrecen, 4032 Debrecen, Hungary
* Correspondence: rafiya.shams@gmail.com (R.S.); kshirod@tezu.ernet.in (K.K.D.); kovacsb@agr.unideb.hu (B.K.)

Abstract: This review describes the various innovative approaches implemented for naringin extraction as well as the recent developments in the field. Naringin was assessed in terms of its structure, chemical composition, and potential food sources. How naringin works pharmacologically was discussed, including its potential as an anti-diabetic, anti-inflammatory, and hepatoprotective substance. Citrus flavonoids are crucial herbal additives that have a huge spectrum of organic activities. Naringin is a nutritional flavanone glycoside that has been shown to be effective in the treatment of a few chronic disorders associated with ageing. Citrus fruits contain a common flavone glycoside that has specific pharmacological and biological properties. Naringin, a flavone glycoside with a range of intriguing characteristics, is abundant in citrus fruits. Naringin has been shown to have a variety of biological, medicinal, and pharmacological effects. Naringin is hydrolyzed into rhamnose and prunin by the naringinase, which also possesses l-rhamnosidase activity. D-glucosidase subsequently catalyzes the hydrolysis of prunin into glucose and naringenin. Naringin is known for having anti-inflammatory, antioxidant, and tumor-fighting effects. Numerous test animals and cell lines have been used to correlate naringin exposure to asthma, hyperlipidemia, diabetes, cancer, hyperthyroidism, and osteoporosis. This study focused on the many documented actions of naringin in in-vitro and in-vivo experimental and preclinical investigations, as well as its prospective therapeutic advantages, utilizing the information that is presently accessible in the literature. In addition to its pharmacokinetic characteristics, naringin's structure, distribution, different extraction methods, and potential use in the cosmetic, food, pharmaceutical, and animal feed sectors were discussed.

Keywords: naringin; flavonoid; extraction; bioactive potential; pharmaceutical

Citation: Shilpa, VS.; Shams, R.; Dash, K.K.; Pandey, V.K.; Dar, A.H.; Ayaz Mukarram, S.; Harsányi, E.; Kovács, B. Phytochemical Properties, Extraction, and Pharmacological Benefits of Naringin: A Review. *Molecules* **2023**, *28*, 5623. https://doi.org/10.3390/molecules28155623

Academic Editor: Adele Papetti

Received: 4 June 2023
Revised: 15 July 2023
Accepted: 19 July 2023
Published: 25 July 2023

1. Introduction

Numerous phytochemicals, such as flavonoids (such as hesperidin and naringin), limonoids (such as limonin and nomilin), carotenoids (such as beta-carotene and lutein), and vitamin C are abundant in citrus fruits. Citrus fruits' vivid colors, distinctive flavors, and distinctive scents are all influenced by these phytochemicals. Citrus fruits include a variety of phytochemicals that have many health advantages [1]. They have antioxidant capabilities that assist the body in fighting off dangerous free radicals and guarding against oxidative stress and cellular damage. Citrus phytochemicals have also been associated with anti-inflammatory effects, which can help reduce the risk of chronic diseases like cardiovascular disease and certain types of cancer. Citrus fruit polyphenols have also been linked to stronger immune systems, better cardiovascular health, and potential anti-diabetic effects. According to some research, these substances may assist with healthy weight management, lowering cholesterol levels, and lowering blood pressure. Citrus fruits include a wide variety of phytochemicals that are essential for overall health and wellbeing, thus including them in the diet is crucial. Beyond what can be achieved by a single vitamin, these chemicals act synergistically to promote health. Regular citrus fruit consumption can support optimal health and lower the risk of chronic diseases by promoting a balanced and nutrient-rich diet [2].

Citrus species are among the most frequently cultivated fruit crops around the world, used to make both fresh juice and food products. The plant genus Citrus, which comprises several varieties of oranges, sour and sweet oranges, tangors, lemons, and tangerines, is a member of the Rutaceae family [3]. Citrus fruits are an excellent supplier of secondary metabolites like terpenoids and polyphenols. These are rich in vitamins A, vitamin C, vitamin E, dietary fibers, and essential minerals [4]. Natural phenolic molecules called flavonoids have a wide variety of bioactivities. Three rings, including two benzene rings and 15 carbon atoms, make up the basic flavonoid structure [5]. Any flavonoid's antioxidant potential is decided by the presence of hydroxyl groups in positions 3,5, an O-dihydroxy structure in the B-ring, a 2,3-double bond conjugated with 4-oxo function, and a 2,3-double bond. A total of 4000 flavonoids have already been isolated, mostly in fruits, herbs, and vegetables. The concentrations and profiles of citrus flavonoids differ greatly between species [6].

The peels, pulp, seeds, and juice of citrus fruits contain a variety of bioactive compounds [1]. Citrus peel has a wealth of bioactive chemicals, including natural antioxidants like flavonoids, and accounts for 50 percent to 65 percent of the total mass of the fruits [7] Citrus flavonoids that have been isolated were discovered to have anti-inflammatory, anti-bacterial, anti-aging, anti-cancer, cardiovascular protective, and hepatoprotective properties in several investigations [8]. Naringin, scientifically known as 5,7-trihydroxyflavonone-7-rhamnoglucoside, comes under the category of flavanone glycoside and it is found in grapes and citrus fruits. The quantity of naringin in fruit is usually determined by its maturity. The immature fruit has a higher concentration of naringin. The fruit maturity is an essential consideration in juice processing, particularly in grapefruit juices, which have a high level of bitterness [9].

Numerous scientific studies have stated that naringin alters the blood levels of some medications, when taken concurrently, by interfering with the operations of enzymatic proteins and transporters in the intestines [10]. Naringin is a powerful inhibitor of transporter proteins such as the multidrug resistance protein (MDR) and organic anion transporting polypeptide (OATP) isoforms, as well as sulfotransferase (SULT). Naringin inhibits a number of cytochrome isoenzymes (CYP) as well [11]. It has been demonstrated that the anticancer medication naringin reduces the expression of p-glycoprotein, which is a membrane-associated drug efflux pump whose increased expression causes anticancer medication resistance for doxorubicin. The flavanone naringin also prevents CYP isoenzymes by inhibiting the production of carcinogens, indicating a potential role in the mitigation of cancer [12]. The bitterness brought by naringin in the manufacturing of commercial grapefruit juice can be removed using a specific enzyme known as naringinase. Two rhamnose units are attached to its aglycon portion, naringenin, at the 7-carbon position. Both naringin and naringenin are strong antioxidants [13]. Naringin is less potent compared with naringenin because the sugar moiety in the former causes steric hindrance of the scavenging group. Naringin is moderately soluble in water. The gut microflora breaks down naringin to its aglycon naringenin in the intestine; it is then absorbed from the gut [1]. However, the enzyme naringinase, which is present in the stomach of humans, transforms naringin into aglycone naringenin. The objective of this review was to assess the various innovative approaches implemented for the extraction of naringin and to study its chemical structure, chemical components, and potential food sources. The pharmacological properties of naringin, including its potential as an anti-diabetic, anti-inflammatory, and hepatoprotective substance, were also discussed. This review also explored the wide range of biological, medicinal, and pharmacological effects associated with naringin along with its applications in various industries such as cosmetic, food, pharmaceuticals and animal feed [14].

2. Chemical Composition of Naringin

The flavonoid substance naringin is mostly present in grapefruits and other citrus fruits. In chemical terms, it is a glycoside made up of the disaccharide neohesperidose and the flavone naringenin. The chemical structure of naringin consists of a flavonoid backbone, two phenolic rings, and a heterocyclic pyran ring. Its molecular weight per mole is 580.54 g and its chemical formula is $C_{27}H_{32}O_{14}$. Pharmaceutical and nutraceutical research is interested in naringin because of its bitter taste and its variety of biological qualities, such as antioxidant, anti-inflammatory, anticancer, and cardioprotective properties [1].

2.1. Significance of Flavonoids

In plants, animals, and microbes, flavonoids have a variety of biological effects. Long known to be synthesized at specific locations in plants, flavonoids are also important for the color and scent of flowers, the ability of fruits to draw pollinators and, as a result, fruit dispersion, the germination of seeds and spores, and the development and growth of seedlings. Plants are protected from various biotic and abiotic challenges by flavonoids, which also serve as special UV filters, allopathic substances, signal molecules, phytoalexins, antimicrobial defensive components, and detoxifying agents. Flavonoids have protective effects against frost drought resistance and hardiness, and they may serve to help plants adapt to heat and tolerate freezing temperatures. There are six types of flavonoids [14]. The major classes of flavonoids, their examples, chemical structures, and main dietary sources are listed in Table 1.

Table 1. The major classes of flavonoids, examples, chemical structures, and main dietary sources.

Flavonoids	Examples	Chemical Structure with Molar Mass (g/mol)	Food Sources	Reference
Anthocyanin	Cyanidin, pelargonidin, peonidin	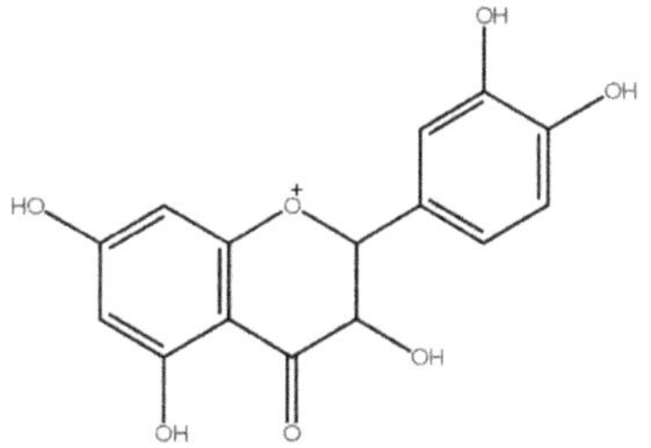 Cyanidin (287.24)	*Solanum melongena, Rubus fruticosus, Ribes nigrum, Vaccinium sect. Cyanococcus*	[15,16]
Flavan-3-ol	Catechin, epicatechin, epigallocatechin	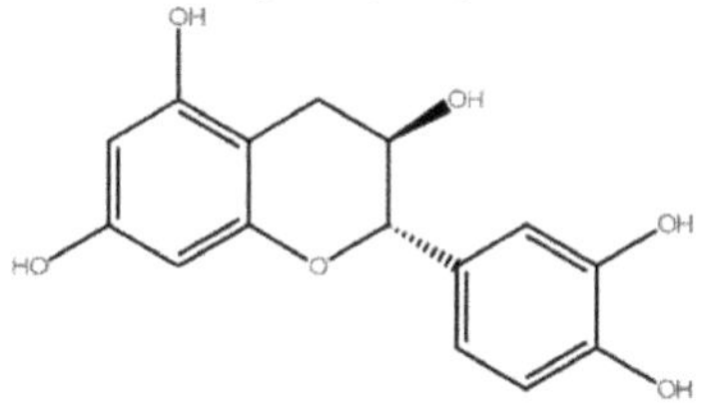 Catechin (290.26)	Green tea, Chocolate, *Phaseolus vulgaris* L., *Prunus avium*	[16]

Table 1. *Cont.*

Flavonoids	Examples	Chemical Structure with Molar Mass (g/mol)	Food Sources	Reference
Flavanones	Hesperidin, Naringin, Eriodictyol	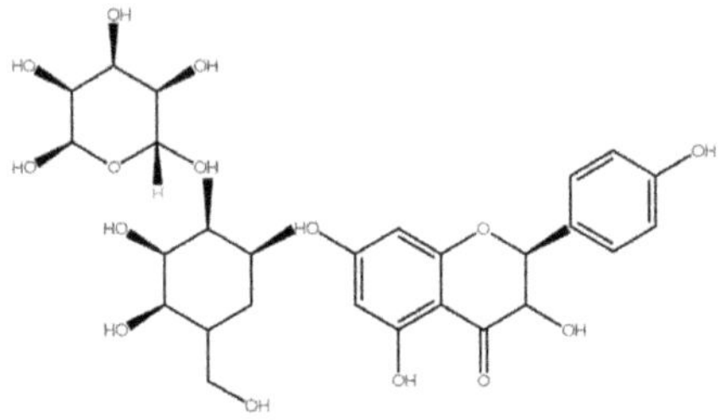 Naringin (580.54)	Orange juice, grapefruit juice, lemon juice	
Flavanones	Apigenin, luteolin	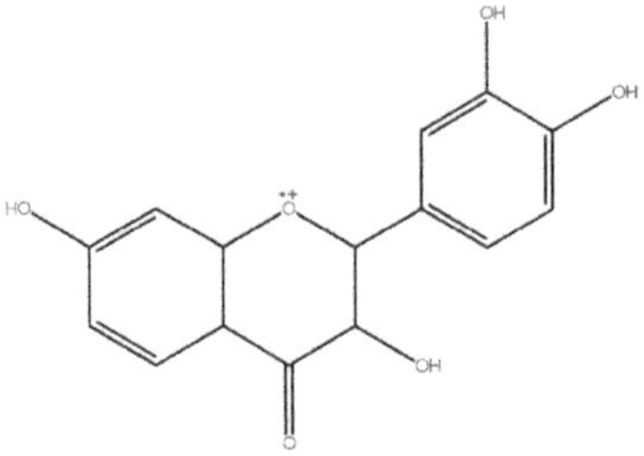 Apigenin (270.05)	*Petroselinum crispum, Apium graveolens, Capsicum annuum*	[17]
Flavonols	Quercetin, kaempferol, myricetin	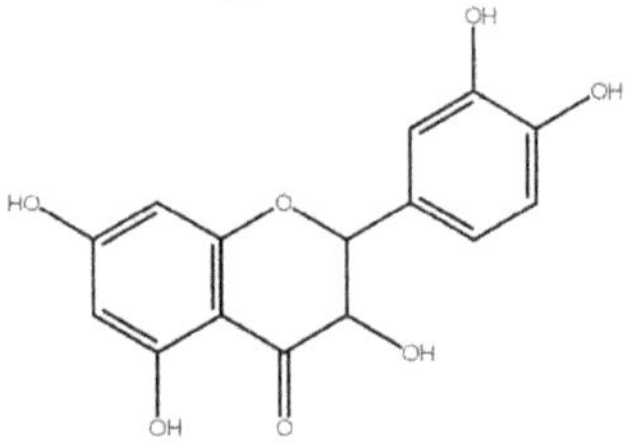 Quercetin (302.23)	*Allium cepa, Malus domestica, Brassica oleracea* var. *sabellica, Allium porrum*	[6]
Isoflavones	Genistein, daidzein, glycitein	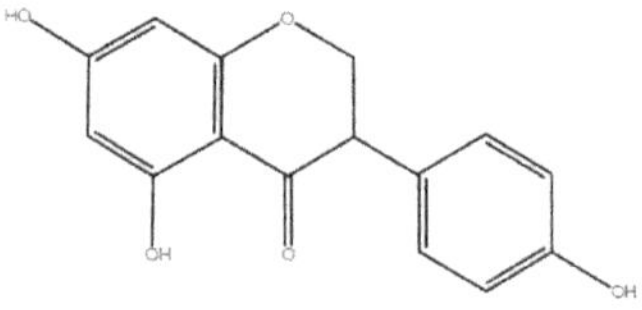 Genistein (270.24)	Soyflour, soymilk, *Glycine max.*	[18,19]

Flavones, which are distinguished by a flavone backbone, are often present as glucosides in a variety of fruits and vegetables, including parsley, celery, chamomile, red peppers, and mint [20,21]. Proanthocyanins are created from flavonols, which are found in large quantities in a variety of foods such as tomatoes, grapes, apples, and berries. Flavonols are identified by a ketone group and a hydroxyl group in the third position of the C ring [22–24]. Isoflavones have a flavone-like structure and are mostly found in legumes like soybeans and leguminous plants. They are frequently found as glycosides, which the gut bacteria can change into aglycones [3,25]. Anthocyanins are phenolic chemicals that are members of the flavonoid family and make up the biggest category of water-soluble pigments. Flowers, plants, and fruits are given bright colors by them, which are distinguished by a flavylium cation structure [18,26]. Citrus fruits including lemons, oranges, and grapes contain flavanones, which are dihydroflavones with an unsaturated C ring [27]. Flavanols, often referred to as flavan-3-ols or flavanonols, are substances that have a hydroxyl group in the third position of the C ring on the flavan skeleton. These substances, which provide a wide range of health advantages, are found in many different plant-based sources [24,28].

The two primary groups of phenolic chemicals present in citrus fruits are flavonoids and phenolic acids. Citrus flavonoids have been proven to have anti-cancer, anti-inflammatory, anti-aging, anti-bacterial, hepatoprotective, and cardiovascular protective effects, according to several studies. The primary class of phytochemicals found in citrus fruits, particularly in the pulp, peels, and seeds, are called flavonoids. Flavones, flavanones, and flavonols are the three main categories of citrus flavonoids. Table 2 depicts the classification of citrus flavonoids isolated in citrus species, major fruit sources, C-ring structures, and substitution patterns [29].

Table 2. Flavonoids isolated in citrus sp. their structure, molecular weight, fruit sources, C-ring structure and substitution pattern (FLA: flavanone FLO: flavone FOL: flavonol).

Flavonoid	Molecular Weight	C-Ring Structure	Fruit Sources	Substitution Pattern	Reference
Naringin	580.541 g/mol	FLA FLA	*Citrus paradisi* *Citrus aurantium*	5,4′-OH 7-O-Neo	[8,21]
Neoeriocitrin	596.5 g/mol	FLA	*Citrus aurantium*	5,3′,4′-OH 7-O-Neo	[6,8]
Diosmin	608.54 g/mol	FLO	*Citrus sinensis* *Citrus limonia*	5,3′-OH 4′-OMe 7-O-Rut	[29]
Hesperidin	610.1898 g/mol	FLA	*Citrus sinensis*	5,3′-OH, 4′-OMe 7-O-Rut	[28]
Rutin	610.517 g/mol	FOL	*Citrus limonia*	5,7,3′,4′-OH 3-O-Rut	[4,28]
Naringenin	272.257 g/mol	FLA	*Citrus paradisi*	5,7,4′-OH	[30,31]
Hesperetin	302.27 g/mol	FLA	*Citrus sinensis*	5,7,3′-OH 4′-OMe	[3,21]
Kaempferol	286.23 g/mol	FOL	*Citrus paradisi*	5,7,3,4′-OH	[8]
Quercetin	302.236 g/mol	FOL	*Citrus limonia*	5,7,3,3′,4′-OH	[28]
Tangeretin	372.37 g/mol	FLO	*Citrus aurantium* *Citrus paradisi* *Citrus limonia*	5,6,7,8,4′-OMe	[5]
Luteolin	286.24 g/mol	FLO	*Citrus limonia* *Citrus aurantium*	5,7,3′,4′-OH	[10]

2.2. Structure of Naringin

Asahina and Inubuse identified and characterized the chemical structure and molecular formula of naringin in 1928. A 2-O-(alpha-L-rhamnopyranosyl)-beta-D-glucopyranosyl moiety is substituted at position 7 of the disaccharide derivative naringin by an alpha-L-rhamnopyranosyl group via a glycosidic bond. The melting point of naringin is 83 °C at a solubility of 1 mg/mL at 40 °C and the molecular weight of naringin is 580.5 g/mol [24]. The molecular structure of naringin is shown in Figure 1. With a rise in temperature, naringin and naringenin, its aglycon equivalent, become more soluble in various solvents. In the order of methanol, ethyl acetate, n-butanol, isopropanol, petroleum ether, and hexane, naringin was soluble in the six solvents [30]. Naringin complexes are 15 times more soluble in water at 37 ± 0.1 °C than free naringin. It starts to degrade at temperatures above 100 °C or when light is present [31]. The presence of a carboxylic group is suggested by the wide, strong -OH stretching absorption from 3300 per cm to 2500 per cm. Alcohols and phenols are represented by the strong and wide hydrogen-bonded O-H stretching bands centered at 3300 cm^{-1} and 3400 cm^{-1} [17]. The C=C stretching bands for aromatic rings typically appear outside the typical region where C=C emerges for alkenes (1650 cm^{-1}) between 1600 and 1450 cm^{-1}. These peaks only occur with naringin. When researching flavonoid-cyclodextrin inclusion complexes, it was discovered that the characteristic peaks for aromatic rings and phenols in naringin at 1519 cm^{-1} and 1361 cm^{-1} had disappeared.

The amount of naringin measured and the correlation coefficient (r) for sensory bitterness was 0.97IBU [32].

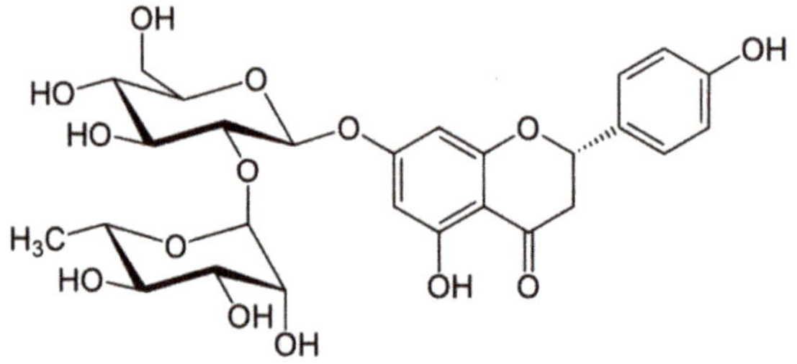

Figure 1. Molecular structure of naringin.

3. Sources of Naringin

Plants contain a variety of flavonoids, which are widely dispersed and have significant biological functions. Since the quantity of naringin is comparatively higher at the immature stage, citrus fruits are typically used in studies to determine the amount of naringin in fruits [33]. Citrus fruits provide a large number of flavonoids in the diet. Naringin is mostly found in the peel of grapefruit, lime, and their variations; it has several biological functions and is frequently used in food, cosmetics, and medicine. Naringin is a glycoside flavanone seen in grapes and citrus fruits. Naringin was first discovered by DeVry in 1857 [34]. It has been reported that the pith contains a higher quantity of naringin in grapefruit, followed by the peel with the membrane, the seeds, and the juice [35]. The amount of naringin in the seeds of grape fruit is 200 μg/mL and 2300 μg/mL is found in the peel of grape fruit [36]. Pummelo has plenty of naringin in it. Compared to the juice, the quantity of naringin was higher in the peel; the naringin content of the juice of pummelo is 220 μg/mL and in the peel it is 3910 μg/mL. The amount of naringin in lime is very low when compared with pummelo. In both species, a high amount of naringin content is present in the skin of the fruits. The amount of naringin found in skin, juice, and seed is 517.2 μg/mL, 98 μg/mL, and 29.2 μg/mL, respectively [37]. The distribution of naringin based on the calculations of various studies in *Citrus aurantiifolia* is shown in Figure 2. Naringin content in sour orange is 47.1 μg/mL. In sour orange flower, the amount of naringin in the receptacle, ovary, and stigma is 1.3444 μg/mL, 9.036 μg/mL, and 2.554 μg/mL, respectively [38]. Phenol is a chemical compound with a hydroxyl group attached to an aromatic ring. Tannin is a type of phenol compound found in plants, known for its astringent properties. Naringin is a flavonoid compound found in citrus fruits that exhibits antioxidant and anti-inflammatory effects.

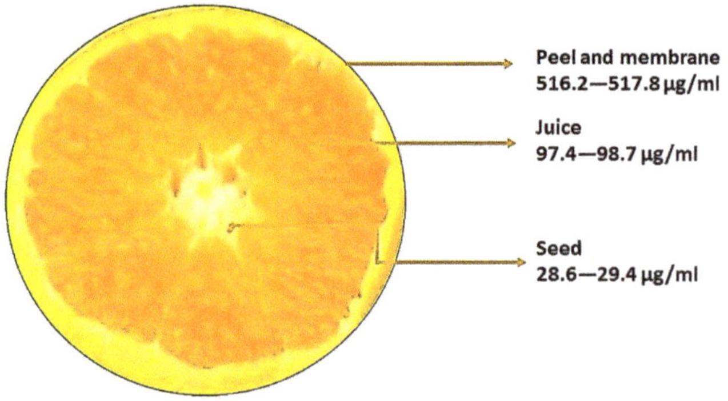

Figure 2. The distribution of naringin in *Citrus aurantiifolia*.

4. Extraction of Naringin

The citrus peel contains significant levels of the flavanones neohesperidin, hesperidin, naringin, and narirutin as well as polymethoxylated flavones tangeretin, sinensetin, and nobiletin. Flavonols, glycosylated flavones, and hydrocinnamic acid are present in very minor amounts [39]. There are three main steps for naringin isolation from fruits including extraction, separation, and purification [29]. Only after utilizing the proper extraction process can flavonoids be isolated, recognized, and classified. The amount of naringin in fruit is determined by numerous factors which include the harvesting time of the fruit, the section of fruit utilized, and whether the peel is a source of naringin. The most prevalent method of extraction is conventional solvent extraction [40]. The various non-traditional techniques include high hydrostatic pressure extraction [41], ultrasound assisted extraction [42], microwave assisted extraction [43], and subcritical [44] and supercritical [45] extraction. The first step in the process involves pre-treating or preparing the sample, during which centrifugation, filtration, drying, and other techniques may be performed. Naringin is extracted, isolated, and purified from various plant materials in the second stage. Naringin is extracted utilizing techniques like soxhlet, maceration, water infusion, microwave extraction, ultrasound extraction, supercritical fluid extraction, auto-hydrolysis, and solid micro-phase extraction, etc. in this step, as given in Figure 3. The final phase often involves the identification, quantification, and recovery of flavonoid components using chromatography techniques on the purified and extracted extracts.

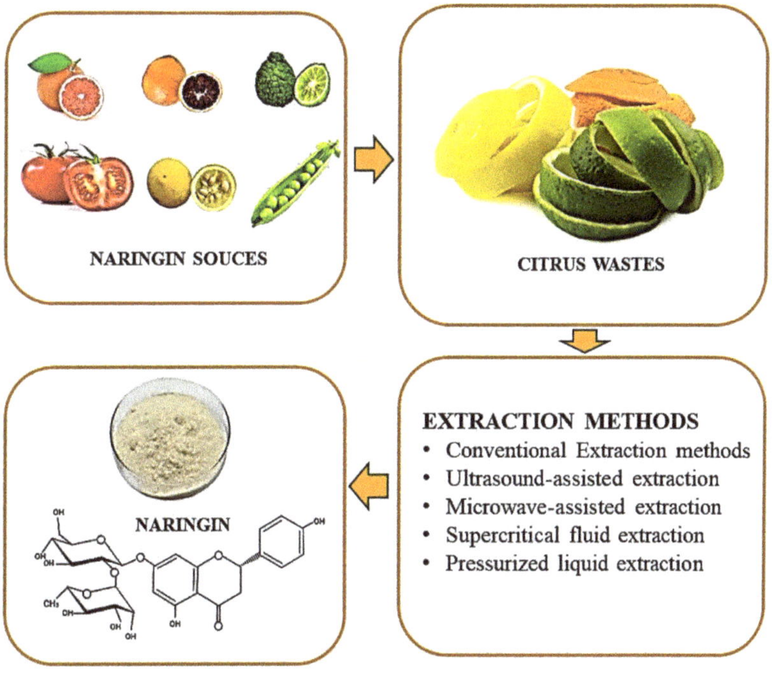

Figure 3. Extraction of naringin using different techniques.

4.1. Conventional Techniques for Naringin Extraction

The most extensively used conventional method for extracting flavonoids is liquid–liquid or solid–liquid extraction. For the extraction of bioactive chemicals, these approaches have incorporated the use of solvents such as ethanol (C_2H_6O), methanol (CH_3OH), and acetone $(CH_3)_2CO$ rather than only water. Because of the significantly greater yields obtained in the recovery of flavonoids, ethanol (C_2H_6O) and methanol (CH_3OH) are the most extensively utilized solvents for flavonoid extraction [40]. In a study, soxhlet method was used for extraction on fresh grapefruit peel samples; 40.0 g was grounded for 1 min in a blender with 100 mL Ethyl alcohol before the mixture was filtered and the filtrates and residue were separated. For removing the alcohol, the residue was air dried. A portion of the filtrate was put in a solvent extraction flask with 50 mL of ethanol and extraction was completed in three hours. After that, the filtrate was evaporated and dried at 50 °C and kept at room temperature.

4.2. Novel Techniques for Naringin Extraction

Numerous innovative extraction techniques have been developed to provide more environmentally friendly extraction processes that utilize the minimum amount of energy and solvent by generating high amounts of yields. Many of these techniques made use of accelerators like microwave or ultrasound, as well as supercritical fluid or subcritical fluid and the use of high pressure. To accelerate the extraction operation and improve the extraction kinetics and yield, several authors recommended using two processes in sequence, like ultrasonic aided extraction and instant controlled pressure drop technology, or a combination of techniques, like enzyme-assisted extraction [20].

Naringin from the peel of *Citrus paradisi* L. can be extracted using supercritical fluid extraction (SFE) techniques with the highest yield of 14.4 g/kg. In this technique, SC modified with 15% ethanol and fresh peels at 1377.86 psi and 58.6 °C can also be used which will result in a lower consumption of solvent and time (45 min) [42]. Naringin from grapefruit seeds has been extracted using the supercritical fluid extraction process. For this, the extraction process was split into two halves. SC-CO_2 was used to extract less polar limonin in stage one, whereas SC-CO_2 was modified with C_2H_6O as a co-solvent in stage two for the extraction of high polar limonin-17-b-D-glucopyranoside (LG) and naringin. The highest quantity of naringin, which is 0.2 mg/g from seeds, was obtained with the optimized conditions of 40 min at 41.4 MPa pressure, 50 °C temperature, and 20% ethanol concentration. The flow rate of the given mobile phase was kept constant at 5.0 L/min throughout the trials. The results showed that SC-CO_2 extraction of naringin from grapefruit seeds is an ecologically friendly and feasible method [46].

Ultrasonic extraction was found to be effective for extracting naringin from ripe pomelo peels. The effectiveness of this method was determined by the agent concentration, the sample-to-solvent ratio, and the ultrasonic duration. In this experiment, 0.6 L of 70% of aqueous ethanol was mixed with 80 g of powdered material in a flask. The flask was kept in the ultrasonic bath with the frequency of 0.04 MHz for 30 min, which was followed by a filtering process that was performed once more. The naringin concentration from peels of ripe *Citrus maxima* is 2.20 % and the yield of purified naringin is 77.26 % under optimal purification conditions, according to the data [4]. The ultrasound-assisted aqueous two-phase extraction (UA-ATPE) method is used for extracting synephrine, neohesperidin, and naringin from *Citrus aurantium* L. fruitlets and is also used for their preliminary purification. Five distinct forms of ethanol or salts from an aqueous two-phase system (ATPS) were used for response surface methodology (RSM) and single-factor studies to further tune the extraction conditions. The following are the optimal process parameters: 20.60% (*w*/*w*) K_2CO_3, 27% (*w*/*w*) ethanol, 45.17:1 (g:g) solvent to material ratio, the 120-mesh particle size of fruit powder, 50 °C temperature, 30 min extraction period, and 80 W ultrasonic power. Under the given conditions, the yields of naringin, synephrine, and neohesperidin were 7.39 mg/g, 11.17 mg/g, and 89.27 mg/g, respectively [46].

Deep eutectic solvents can also be considered promising green and efficient solvents for the extraction of naringin, hesperidin, and neohesperidin from citrus fruits such as *Aurantii Fructus*. In a study, a series of tunable deep eutectic solvents was prepared and investigated by mixing choline chloride or betaine with different hydrogen-bond donors, and betaine/ethanediol was found to be the most suitable extraction solvent. The optimum extraction conditions were 40% of water in betaine/ethanediol (1:4) at 60 °C for 30 min extraction time with solid/liquid ratio 1:100 g/mL. Under these conditions, the extraction yield of narirutin, naringin, hesperidin, and neohesperidin was found to be 8.39 ± 0.61,

83.98 ± 1.92, 3.03 ± 0.35 and 35.94 ± 0.63 mg/g, respectively, which was comparatively higher than when using methanol as extraction solvent [47]. It has also been reported that deep eutectic solvents or aqueous glycerol can replace the traditional solvents for the extraction of polyphenols (naringin) from citrus peels such as grapefruit peels. It can increase the extraction of polyphenols and especially naringin flavonoid from grapefruit peels as compared to water [48].

5. Schematic Overview of the Possible Health Benefit Based on the Literature Review

Since ancient times, citrus fruits have been utilized as natural herbal treatments in traditional medicine. Citrus peel has been utilized in traditional Chinese medicine to enhance digestion, minimize gastric gas, bloating, and clear congestion [12]. Clinical and epidemiologic research states that eating citrus fruits lowers the risk of lifestyle-related disorders like cancer, cardiovascular disease, diabetes (type-2), and osteoporosis [49]. Naringin has been shown to have anticancer, antiapoptotic, cholesterol-lowering, antiatherogenic, and metal binding capabilities, as well as antioxidant qualities, as shown in Figure 4. Naringin is also said to enhance medication absorption and metabolism [50].

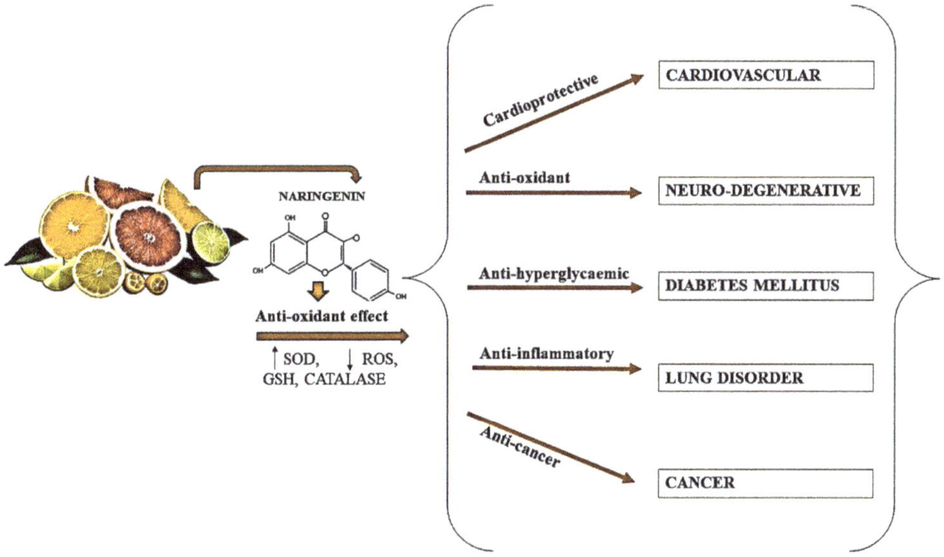

Figure 4. Potential health benefits of naringin; SOD: superoxide dismutase, ROS: Reactive oxygen species, GSH: γ-l-glutamyl-l-cysteinyl-glycine (glutathione).

5.1. Anticancer Properties of Naringin

Naringin has been reported to inhibit many malignancies through the regulation of various cellular signaling cascades, including the inhibition of malignant cell growth, the induction of apoptosis and also the arresting of the cell cycle and the regulation of oxidative stress, inflammatory processes, and angiogenesis [51]. It was discovered that naringin at concentrations of 250–2000 M promoted cell apoptosis in cervical cancer cells (SiHa) in a dose-dependent way. This impact of naringin is thought to have contributed to the suppression of cell growth as well and also increase in apoptosis [29].

Naringin in the concentrations of 1 M, 5 M, and 10 M has reduced cell mortality caused by rotenone in human neuroblastoma cells (SH-SY5Y). In 4, 6-diamidino-2-phenylindol (DAPI) staining and terminal deoxynucleotidyl transferase dUTP nick end labeling (TUNEL) tests, naringin prevents condensation of chromatin and breakage of DNA strand production by rotenone [52]. Naringin also decreases rotenone-induced phosphorylation of the mitogene-activated kinase (MAPK) family members p38 and Jun NH2-terminal protein kinase (JNK) [53]. According to one study, naringin inhibits the growth of cells, and apoptosis was induced in K562, HL-60, and Kasumi-1 human myeloid leukemia cells in a concentration- and time-dependent manner by downregulating Mcl-1 expression and activating the caspase and PARP pathways. In U937 and THP-1 human leukemia cells, naringin therapy increased cell death and lowered cell cervical proliferation and expansion [54]. The most dangerous and prevalent brain tumors are gliomas, and they are still fatal despite advances in therapeutic care. As a result, a variety of therapy techniques are required to combat this deadly disease. In glioblastoma cells, naringin was able to block FAKp-Try397 and limit focal adhesion kinase (FAK) activity and its downstream pathway. Naringin treatment of U251 glioblastoma cells and U87 glioblastoma cells restricts their growth by inhibiting the cyclin D1 pathway or FAK pathway and induces the death of cells by inhibiting the BAD pathway or FAK pathway. By inhibiting the MMP or FAK pathways, it prevents metastasis of cells and invasion of cells. In the case of U251 glioma cells, naringin treatment restricts the proliferation of cells and its viability [55]. Naringin also inhibits the migration of cells and invasion of cells by modulating the matrix metallopeptidase-2 (MMP-2) expression and MMP-9. As a result, by decreasing the p38 signal transduction pathways, naringin has a potential and therapeutic effect on the regulation of invasive malignant gliomas [56].

Breast cancer is a term that refers to various types of cancers. A vast variety of individualized treatments for breast cancer have recently been offered, all of which have been shown to be effective [57]. Chemotherapy and cancer chemoprevention are both carried out with natural products containing bioactive chemicals. In MCF-7 cell lines, naringin treatment reduced proliferation and growth while also increasing apoptosis. In canine mammary cancer cells (CMT-U27), naringin oxime treatment decreased cell proliferation and viability [53]. Cervical cancer is the second most common cancer in women around the globe and continues to be difficult. At a dose of 750 M, naringin displayed a 50% suppression of SiHa human cervical cancer cells. Apoptosis, intra nucleosomal DNA fragmentation, morphological abnormalities, and mitochondrial transmembrane potential reduction were seen in SiHa cells. The findings imply that naringin works well in human cervical cancer treatment [58]. The common extra-cranial solid tumor in children is neuroblastoma. Plant-derived nutritional chemicals are gaining popularity as a treatment for a variety of solid tumors, including malignant neuroblastoma [59]. Naringin therapy reduced the viability of cells and induced apoptosis in rotenone-treated SH-SY5Y human neuroblastoma cells via inhibiting P38 and JNK phosphorylation and activating caspase-9 and caspase-3 [54]. The common malignant tumor in the endocrine system is thyroid cancer. In SW1736 cells, a TPC-1 naringin dose dependently raised the expression of Bax, cleaved Caspase3, and Caspase3, when the expression of c-Myc, cyclin D1, Bcl-2, and survivin decreased. It also decreased AKT pathway or P13K pathway activation in thyroid cancer (TC) cells. The naringin showed anti-tumor actions in TC cells by limiting cell division of TC and cell death promotion by controlling the gene expression associated with cell apoptosis and division and activating the AKT pathway or PI3K pathway [60].

5.2. Antidiabetic Properties of Naringin

It has been demonstrated that naringin enhances insulin sensitivity. Naringin has shown that it can improve insulin action and cell uptake of glucose. Insulin resistance is a major contributor to the onset of type 2 diabetes. This can enhance overall glycemic control and help control blood sugar levels. Insulin moves sugars from the bloodstream into cells, where they are used or stored as energy [61]. Diabetes is when the body does not produce enough insulin or cannot utilize the insulin it makes efficiently [5]. Diabetes is divided into two types. Type 1 Diabetes is a condition that is autoimmune. Cells in the pancreas, which make insulin, are attacked and destroyed by the immune system and what generates this attack remains an enigma. Approximately 10% of diabetics have this type of diabetes [32]. In type 2 diabetes, blood sugar levels increase if the body becomes resistant to insulin [62]. Some enzymes involved in the metabolism of carbohydrates may be inhibited by naringin.

It has the ability to block glucosidase, an enzyme that converts complex carbs into simple sugars. Naringin can lower postprandial glucose levels by blocking this enzyme, which will slow down the digestion and absorption of carbohydrates. While naringin has the potential to be an effective anti-diabetic substance, it should not be used as a stand-alone treatment for diabetes. Along with traditional diabetes treatment techniques like a good diet, frequent exercise, and prescribed medications, it might be considered as a complimentary strategy.

Naringin is a powerful biomolecule that has the potential to help people with diabetes and its consequences [63]. Naringin restricts the secretion and sensitivity of insulin, PPAR, glucose transporters, blood lipids, hepatic glucose production, peripheral glucose uptake, intestinal glucose absorption, biosynthesis of cholesterol, oxidative stress, and inflammation [64]. Inflammatory cytokines are elevated and insulin resistance and hyperglycemia are generated by a high-fat diet. Naringin's hypoglycemic impact has been thoroughly documented. Vitamin C (50 mg/kg) with naringin co-treatment improved insulin concentration and oxidative stress reduction in rats with streptozotocin-induced diabetes [5]. In diabetic nephropathic rats, naringin minimized streptozotocin-induced kidney dysfunction and damage, inhibited streptozotocin-induced oxidative stress in vivo, and prevented high glucose-induced apoptosis and ROS levels in vitro [65]. Naringin has anti-inflammatory and antioxidant benefits in diabetic nephropathic rats, as evidenced by the downregulation of IL-1, proinflammatory cytokines TNF, and IL-6 and the upregulation of antioxidants SOD, GSH, and CAT [6].

5.3. Anti-Inflammatory Properties of Naringin

The process by which the body's white blood cells and the substances they make protect against bacterial and viral illness is known as inflammation. Flavanone-rich plants, such as naringin, hesperidin, and neohesperidin, have long been known to have anti-inflammatory properties [3]. Inflammation is divided into two types. Acute inflammation is the body's reaction to a quick injury, such as cutting your finger. Your body sends inflammatory cells to the wound to help it heal. The healing process begins with these cells [66]. Chronic inflammation occurs when your body sends inflammatory cells even when there is no external threat. Inflammatory cells and chemicals assault joint tissues in rheumatoid arthritis, for example, causing an inflammation that comes and goes and can cause serious damage to joints, including pain and deformity [67].

The anti-inflammatory process controlled by nuclear factor-erythroid 2–related factor 2 (Nrf2) regulates cellular antioxidant synthesis and thus plays a very important role in preventing various degenerative illnesses [68]. In 3-nitropropionic acid-induced rats, naringin upregulates the expression of mRNA in HO-1, GST P1, NAD(P)H:quinone oxidoreductase 1, and g-glutamylcysteine ligase; this is followed by activating Nrf2 and the reduced expression of proinflammatory mediators like TNF-a, cyclooxygenase-2, and inducible NO synthase [69]. Naringin did not inhibit cell proliferation, but it did inhibit RANTES (regulated upon activation of normal T-cell expressed and secreted) production in a human epidermal keratinocytes cell line (HaCaT cells) by restricting nuclear translocation of NF-JB [70]. In animal models of inflammation, naringin has been demonstrated to reduce the production of inflammatory signaling factors like interleukin-8 (IL-8), interleukin-6 (IL-6), inducible nitric oxide synthase (iNOS), TNF-a, and nuclear factor erythroid 2-related factor 2 (Nrf2). Naringin treatment prevented an improvement in serum IL-6 during aging-related inflammation in 20-month-old male Wistar rats [54]. Naringin, neohesperidin, paeoniflorin, and platycodin-D are all found in "painopowder", a traditional Chinese medicine. The four-ingredient combination had the greatest anti-inflammatory impact in a model of acute inflammation, while naringin was shown to have the most important role among the four substances [71].

5.4. Hepatoprotective Properties of Naringin

The capability of a chemical compound to inhibit liver toxicity is known as hepatoprotection [72]. Naringin is suggested to enhance the functioning of the hepatic antioxidant system as well as the metabolism of hepatotoxic substances [73]. Naringin exhibits protection against naturally occurring genotoxins in food, like PhIP (2-Amino-1-methyl-6-phenylimidazo[4,5-b] pyridine) and other cooked food mutagens, by lessening PHIP induced genotoxicity in human liver segments at a concentration of 1000 M [69]. Naringin (0.05–0.125 g/L) increased ethanol and lipid metabolism in rats, alleviating the adverse effects of ethanol consumption. It also reduced necrosis, steatosis, and fibrosis in rat models of alcoholic liver disease, as demonstrated by reduced expression of PGC1α (Peroxisome proliferator-activated receptor-gamma coactivator) or Sirt1; it is an enzyme involved in regulating energy metabolism in response to calorie restrictions at a dosage of 100 mg/day [71]. Table 3 describes various medical conditions and the therapeutic effect of the action of naringin.

Table 3. The therapeutic effect of naringin in different medical condition.

Medical Condition	Therapeutic Effects of Naringin	Other Possible Treatments	References
Diabetes	Improved insulin sensitivity, enhanced glucose uptake, inhibition of α-glucosidase enzyme, antioxidant, and anti-inflammatory properties.	Conventional antidiabetic medications, lifestyle modifications.	[5]
Cardiovascular Health	Reduction of cholesterol levels, prevention of LDL oxidation, anti-inflammatory effects, improvement of endothelial function.	Statins, blood pressure medications, lifestyle modifications.	[74]
Cancer	Anticancer properties, inhibition of tumor growth, induction of apoptosis, antioxidant and anti-inflammatory effects.	Chemotherapy, radiation therapy, targeted therapies.	[75]
Neurodegenerative Diseases	Neuroprotective effects, reduction of oxidative stress and inflammation, potential improvement of cognitive function.	Symptomatic treatments, rehabilitation therapies.	[76]
Liver Health	Protection against liver damage, reduction of liver fibrosis, antioxidant effects, improvement of liver enzyme levels.	Lifestyle modifications, hepatoprotective medications.	[77]
Obesity	Suppression of adipogenesis, reduction of body weight gain, improvement of metabolic parameters.	Diet and exercise interventions, weight loss medications.	[2]

5.5. Pharmacokinetics of Naringin

Studies were conducted with help of rats to understand the pharmacokinetic properties of naringin. The study of the absorption, distribution, metabolism, and excretion of drugs is known as pharmacokinetics [78]. Proton-coupled active transport and passive diffusion are used to absorb flavanone aglycones into the enterocytes. The low molecular weight, high lipophilicity, and slightly acidic character of aglycones cause passive diffusion. Once within the cells, naringin is expected to go through phase I metabolism, such as oxidation or demethylation by cytochrome P450 monooxygenases, then passing to phase II metabolism, such as sulfation, glucuronidation, or methylation, in intestinal cells or liver cells [79,80]. Naringin is rapidly absorbed in the blood, with the initial concentration peaking at 15 min and the second peaking at 3 h after naringin monomer oral administration; 480 min later, it is undetectable [81]. The affinity of these food chemicals for serum albumin, the primary transport protein, coincides with their tissue distribution and elimination. To begin with, the bound medication works as a reservoir and has a prolonged half-life, whilst the unbound portion is responsible for the biological impact. Second, drug binding to this protein reduces drug filtration by the kidney [82].

In terms of tissue distribution, the liver had the largest quantities of flavanone conjugates after repeated or single dose flavanone treatment in rats. By partially undergoing breakage of the bacterial ring and then the three bridges of carbon to dihydrochalcone moiety, naringin is eliminated by the kidneys into the urine and by the liver into bile According to Fuhr and Kummert's findings (1995). Urine excretion ranges from 5 to 57 percent of total intake. Sulfates were the most common naringenin type detected in the tissues of rats. Only the liver and kidney had glucuronide concentrations that could be measured [83]. The average Cmax of naringin in portal plasma was 18.83.8 min (determined by the concentration reached at tmax in portal plasma), whereas the absorption ratios of naringin in portal plasma and lymph fluid were approximately 95.9 and 4.1, respectively, after naringin administration via a duodenal cannula (600 and 1000 mg/kg). This suggests that naringin is absorbed largely by portal blood rather than mesenteric lymph fluid and that it is excreted primarily by bile, with just a tiny quantity entering systemic circulation following hepatic metabolism [84]. Grapefruit juice, a key factor in pharmacokinetic drug interactions, possesses a higher amount of naringin, hence it was previously thought to be moderating these interactions. Despite extensive investigation, the constituents in grapefruit juice responsible for drug interaction, particularly P-glycoprotein and CYP3A4 levels, are unknown; however, it appears that certain coumarins, rather than flavanones, are likely involved [85]. However, naringin has also been shown to possess therapeutic properties and has potential as a therapeutic agent to treat numerous diseases such as various liver, heart, and metabolic disorders [86].

6. Application of Naringin

Naringin has many benefits for numerous industries including the food, pharmaceutical, cosmetics, and animal feed industries.

6.1. In Cosmetic Industry

The flavonoid naringin has anti-cancer, anti-oxidative, anti-aging, antibacterial, anti-inflammatory, cholesterol-lowering, and free radical scavenging properties [87]. Studies show that naringin reduces the risk of toxicity caused by other sunscreen ingredients like TiO_2 when it is added to sunscreen formulations because of its antioxidant activity. It also scavenges free radicals produced by UV radiation and by the photocatalytic activity of ZnO and TiO_2, which further lowers the risk of toxicity [88]. Essential oils like eucalyptus oil, lavender oil, and peppermint oil, are known to have strong antioxidant and antibacterial qualities that could be employed as environmentally friendly substitutes for synthetic antioxidants and preservatives in skin care formulations [70,89]. Naringin-loaded microemulsions were created using essential oils as the oil phase; these microemulsions showed antioxidant and antibacterial effects that were comparable to or outperformed those of synthetic ones. In comparison to their unformulated counterparts, naringin-loaded microemulsion-gel formulations demonstrated improved stability and release profiles. Skin care formulations have experienced advantages by utilizing microemulsions of essential oils. These microemulsions enhance the release and permeation of active ingredients into the skin, improve their stability, and serve as environmentally friendly alternatives to synthetic antioxidants [90].

6.2. Pharmaceutical Application

The area of the wound and the length of the epithelization phase significantly decreased during treatment with naringin ointment formulation, whilst the velocity of wound contraction dramatically increased. Naringin ointment formulation modulates collagen-1 expression to promote angiogenesis, which in turn promotes wound healing. This is accomplished by down-regulating the expression of inflammatory (ILs, NF-Jb, and TNF-a), apoptotic (pol-g and Bax), and growth factor (TGF-b and VEGF) genes [91]. An investigation was carried out into the potential protective benefits of hesperidin and naringin against diclofenac-induced liver damage and also the involvement of oxidative stress, inflammation, and apoptosis modulation. The study found that, when hesperidin and naringin were given to diclofenac-injected rats, the raised levels of blood LDH, GGT, ALP, AST, IL-17 levels, total bilirubin, and TNF level, liver p53 and caspase-3 mRNA expression, liver lipid peroxidation all significantly decreased. Through anti-inflammatory, antioxidant, and anti-apoptotic effects, hesperidin, naringin, and their combination proved effective for reversing diclofenac-induced liver damage. The liver adverse effects of medications like diclofenac can be treated with naringin, hesperidin, or a combination of the two [92].

6.3. In Livestock Sector

Naringin and quercetin reduce protozoa and methanogen populations in the rumen and suppress methane production without negatively affecting the parameters of ruminal fermentation. Daily diets containing hesperidin and naringin have proved successful in enhancing milk's oxidative stability while having no negative impacts on the substance's chemical compositions, coagulation abilities, or fatty acid profile [93]. According to studies, adding Macleaya cordata extract at a dosage of 120 mg/kg of diet and naringin at a dosage of 50 mg/kg of diet to post-weaning piglet diets enhanced growth performance and nutrient digestibility while having no effect on the villi and crypts' histo-morphological status in the jejunum. These results show that these items have the potential to be utilized as feed supplements to improve the growth performance of weaned piglets [94]. Hesperidin and naringin, when consumed, had a substantial impact on the antioxidative capacity of broiler breast and thigh meat, presumably indicating that these bioflavonoids had entered the cell phospholipid membranes of the broiler muscles. Broiler meat's slower lipid oxidation rate has a good impact on meat shelf life, which benefits both the consumer and the poultry meat industry. Hesperidin and naringin have a good effect on the antioxidative qualities of meat without having a negative impact on the development capacity and meat quality features of poultry; for this reason, they can be introduced as a significant additive to poultry feed [95].

6.4. Food Industry

The use of naringin microspheres in yogurt demonstrated their ability to effectively reduce whey precipitation and to slow pH drop. According to a study, naringin-encapsulated microspheres could extend the shelf life of this bioactive product and offer a fresh concept for functional yogurt [96]. Hesperidin, naringin, and coumarins have been found to inhibit xanthine oxidase, which directly reduces cellular free radical production. When compared to dietary citrus pulp and control diets, feeding dietary citrus pulp prolonged the shelf life of beef during retail display by increasing antioxidant activity, lowering coliforms, and reducing lipid and protein oxidation [97]. Naringin's incorporation caused significant UV blocking, plasticizing, and antioxidant and antibacterial effects. The biological oxygen demand (BOD) in saltwater was used to test the biodegradability of these films, showing excellent disintegration under these circumstances [98]. It was observed that naringin could prevent browning in soybean and mung bean and could keep their appearance and quality intact for a period of six days while they were stored. Since naringin treatment enhanced the quantity of p-coumaric acid and gallic acid in mung bean sprouts while also increasing the quantity of rutin and daidzein in soybean sprouts, the utilization of naringin for the postharvest preservation of soybean sprouts and mung bean sprouts will maintain good consumer quality [99].

7. Conclusions and Future Prospects

Flavonoids are a large group of polyphenolic compounds that are present in almost every part of the plant, including the leaves, flowers, stems, roots, fruits, and seeds. Although flavonoids are classed differently in the study, they all share a similar structural foundation. Flavonoids are categorized into six different groups based on the variations in their substitution and the activity of the carbon skeletons: flavones, flavanones, flavan-3-ols, flavonols, anthocyanins, and isoflavones. There are almost 10,000 identified flavonoids, and several studies have shown their antioxidant, pro-oxidant, anti-inflammatory, antiviral/bacterial, antidiabetic, cardioprotective, anticancer, and anti-aging properties. Naringin (bitter flavonoid) has been reviewed in detail. Commercial grapefruit juice manufacturing can benefit from the use of the enzyme naringinase to get rid of the bitterness brought on by naringin. The stomach-based enzyme naringinase transforms naringin in humans into aglycone naringenin. Naringin has been demonstrated to possess anti-cancer, anti-apoptotic, cholesterol-lowering, anti-atherogenic, antioxidant, and metal binding properties. Naringin is also reported to improve the metabolism and absorption of medications. It has been reported that exposure to naringin in-vitro and in-vivo in a variety of test animals and cell lines has actions that may be used to treat osteoclast genesis, hyperthyroidism, hyperlipidemia, diabetes, tumors, and asthma. Recent research indicates that naringin may potentially be utilized to treat COVID-19. The value of naringin for both therapeutic and commercial applications has recently been the focus of greater investigation. Naringin is a promising research subject with several applications.

Author Contributions: V.S.: writing—original draft preparation; R.S.: writing—review and editing; K.K.D. and V.K.P.: writing—review and editing, visualization, supervision; A.H.D., S.A.M., E.H. and B.K.: supervision. All authors have read and agreed to the published version of the manuscript.

Funding: Project No. TKP2021-NKTA-32 has been implemented with support from the National Research, Development, and Innovation Fund of Hungary, financed under the TKP2021-NKTA funding scheme for conducting this research.

Institutional Review Board Statement: Not applicable.

Informed Consent Statement: Not applicable.

Data Availability Statement: Not applicable.

Conflicts of Interest: The authors declare no conflict of interest.

Sample Availability: Not applicable.

References

1. Alam, M.A.; Subhan, N.; Rahman, M.M.; Uddin, S.J.; Reza, H.M.; Sarker, S.D. Effect of citrus flavonoids, naringin and naringenin, on metabolic syndrome and their mechanisms of action. *Adv. Nutr.* **2014**, *5*, 404–417. [CrossRef]
2. Jakab, J.; Miškić, B.; Mikšić, Š.; Juranić, B.; Ćosić, V.; Schwarz, D.; Včev, A. Adipogenesis as a potential anti-obesity target: A review of pharmacological treatment and natural products. *Diabetes Metab. Syndr. Obes.* **2021**, *14*, 67–83. [CrossRef] [PubMed]
3. Su, W.; Wang, Y.; Li, P.; Wu, H.; Zeng, X.; Shi, R.; Zheng, Y.; Li, P.; Peng, W. The potential application of the traditional Chinese herb Exocarpium Citri grandis in the prevention and treatment of COVID-19. *Tradit. Med. Res.* **2020**, *5*, 160–166. [CrossRef]
4. Tang, D.M.; Zhu, C.F.; Zhong, S.A.; Da Zhou, M. Extraction of naringin from pomelo peels as dihydrochalcone's precursor. *J. Sep. Sci.* **2011**, *34*, 113–117. [CrossRef]
5. Ahmed, O.M.; Mahmoud, A.M.; Abdel-Moneim, A.; Ashour, M.B. Antidiabetic effects of hesperidin and naringin in type 2 diabetic rats. *Diabetol. Croat.* **2012**, *41*, 53–67.
6. Hollman, P.C.H.; Arts, I.C.W. Flavonols, flavones and flavanols—Nature, occurrence and dietary burden. *J. Sci. Food Agric.* **2000**, *80*, 1081–1093. [CrossRef]
7. Gorinstein, S.; Leontowicz, H.; Leontowicz, M.; Krzeminski, R.; Gralak, M.; Delgado-Licon, E.; Martinez Ayala, A.L.M.; Katrich, E.; Trakhtenberg, S. Changes in plasma lipid and antioxidant activity in rats as a result of naringin and red grapefruit supplementation. *J. Agric. Food Chem.* **2005**, *53*, 3223–3228. [CrossRef]
8. Li, Q.; Zhang, N.; Sun, X.; Zhan, H.; Tian, J.; Fei, X.; Liu, X.; Chen, G.; Wang, Y. Controllable biotransformation of naringin to prunin by naringinase immobilized on functionalized silica. *J. Chem. Technol. Biotechnol.* **2021**, *96*, 1218–1227. [CrossRef]
9. Zeng, X.; Su, W.; Zheng, Y.; He, Y.; He, Y.; Rao, H.; Peng, W.; Yao, H. Pharmacokinetics, tissue distribution, metabolism, and excretion of naringin in aged rats. *Front. Pharmacol.* **2019**, *10*, 34. [CrossRef] [PubMed]
10. Sharma, A.; Bhardwaj, P.; Arya, S.K. Naringin: A potential natural product in the field of biomedical applications. *Carbohydr. Polym.* **2021**, *2*, 100068. [CrossRef]
11. Egert, S.; Rimbach, G. Which sources of flavonoids: Complex diets or dietary supplements. *Adv. Nutr.* **2011**, *2*, 8–14. [CrossRef]
12. Bacanli, M.; Başaran, A.A.; Başaran, N. The major flavonoid of grapefruit: Naringin. In *Polyphenols: Prevention and Treatment of Human Disease*; Academic Press: Cambridge, MA, USA, 2018; pp. 37–44.
13. Renugadevi, J.; Prabu, S.M. Naringenin protects against cadmium-induced oxidative renal dysfunction in rats. *Toxicology* **2009**, *256*, 128–134. [CrossRef] [PubMed]
14. Panche, A.N.; Diwan, A.D.; Chandra, S.R. Flavonoids: An overview. *J. Nutr. Sci.* **2016**, *5*, e47. [CrossRef]
15. Nawaz, R.; Abbasi, N.A.; Khan, M.R.; Ali, I.; Hasan, S.Z.U.; Hayat, A. Color development in "Feutrell's early" (Citrus reticulata Blanco) affects peel composition and juice biochemical properties. *Int. J. Fruit Sci.* **2020**, *20*, 871–890. [CrossRef]
16. Tan, J.; Li, Y.; Hou, D. The Effects and Mechanisms of cyanidin-3-glucoside and Its phenolic Metabolites in Maintaining intestinal Integrity. *Antioxidants* **2019**, *8*, 479. [CrossRef] [PubMed]
17. Puri, M. Updates on naringinase: Structural and biotechnological aspects. *Appl. Microbiol. Biotechnol.* **2012**, *93*, 49–60. [CrossRef]

18. Křížová, L.; Dadáková, K.; Kašparovská, J.; Kašparovský, T. Isoflavones. *Molecules* **2019**, *24*, 1076. [CrossRef]
19. Da Silva, F.L.; Escribano-Bailón, M.T.; Pérez Alonso, J.J.; Rivas-Gonzalo, J.C.; Santos-Buelga, C. Anthocyanin pigments in strawberry. *LWT—Food Sci. Technol.* **2007**, *40*, 374–382. [CrossRef]
20. Addi, M.; Elbouzidi, A.; Abid, M.; Tungmunnithum, D.; Elamrani, A.; Hano, C. An overview of bioactive flavonoids from citrus fruits. *Appl. Sci.* **2022**, *12*, 29. [CrossRef]
21. Verma, A.K.; Pratap, R. Chemistry of biologically important flavones. *Tetrahedron* **2012**, *68*, 8523–8538. [CrossRef]
22. Heiss, C.; Keen, C.L.; Kelm, M. Flavanols and cardiovascular disease prevention. *Eur. Heart J.* **2010**, *31*, 2583–2592. [CrossRef] [PubMed]
23. Hackman, R.M.; Polagruto, J.A.; Zhu, Q.Y.; Sun, B.; Fujii, H.; Keen, C.L. Flavanols: Digestion, absorption and bioactivity. *Phytochem. Rev.* **2007**, *7*, 195–208. [CrossRef]
24. Aherne, S.A.; O'Brien, N.M. Dietary flavonols: Chemistry, food content, and metabolism. *Nutr. J.* **2002**, *18*, 75–81. [CrossRef]
25. Wollenweber, E. Flavones and flavonols. The flavonoids. *J. Adv. Res.* **2017**, 259–336.
26. Jung, U.J.; Kim, S.R. Effects of naringin, A flavanone glycoside in grapefruits and citrus fruits, On the nigrostriatal dopaminergic projection in the adult brain. *Neural Regen. Res.* **2014**, *9*, 1514–1517.
27. Felgines, C.; Texier, O.; Morand, C.; Manach, C.; Scalbert, A.; Régerat, F.; Remesy, C.; Felgines, C.; Texier, O.; Morand, C.; et al. Bioavailability of the flavanone naringenin and its glycosides in rats. *Am. J. Physiol. Gastrointest. Liver Physiol.* **2020**, *279*, G1148–G1154. [CrossRef] [PubMed]
28. Billowria, K.; Ali, R.; Rangra, N.K.; Kumar, R.; Chawla, P.A. Bioactive Flavonoids: A Comprehensive Review on Pharmacokinetics and Analytical Aspects. *Crit. Rev. Anal. Chem.* **2022**, 1–15. [CrossRef] [PubMed]
29. Sharma, K.; Mahato, N.; Lee, Y.R. Extraction, characterization and biological activity of citrus flavonoids. *Rev. Chem. Eng.* **2019**, *35*, 265–284. [CrossRef]
30. Zhang, L.; Song, L.; Zhang, P.; Liu, T.; Zhou, L.; Yang, G.; Lin, R.; Zhang, J. Solubilities of naringin and naringenin in Different Solvents and Dissociation Constants of naringenin. *J. Chem. Eng. Data* **2015**, *60*, 932–940. [CrossRef]
31. Aron, P.M.; Kennedy, J.A. Flavan-3-ols: Nature, occurrence and biological activity. *Mol. Nutr. Food Res.* **2008**, *52*, 79–104. [CrossRef]
32. Ioannou, I.; Nouha, M.; Chaaban, H.; Boudhrioua, N.M.; Ghoul, M. Effect of the process, temperature, light and oxygen on naringin extraction and the evolution of its antioxidant activity. *Int. J. Food Sci. Technol.* **2020**, *53*, 2754–2760. [CrossRef]
33. Wang, S.; Yang, C.; Tu, H.; Zhou, J.; Liu, X.; Cheng, Y.; Luo, J.; Deng, X.; Zhang, H.; Xu, J. Characterization and metabolic diversity of flavonoids in Citrus species. *Sci. Rep.* **2017**, *7*, 10549. [CrossRef]
34. Hoffmann, E. Naringin (Hesperidin de Vry). *Arch. Pharm.* **1879**, *214*, 139–145. [CrossRef]
35. Liu, Y.; Heying, E.; Tanumihardjo, S.A. History, global distribution, and nutritional importance of citrus fruits. *Compr. Rev. Food Sci. Food Saf.* **2012**, *11*, 530–545. [CrossRef]
36. Pichaiyongvongdee, S.; Haruenkit, R. Comparative studies of limonin and naringin distribution in different parts of pummelo (*Citrus grandis* (L.) Osbeck) cultivars grown in Thailand. *J. Nat. Sci.* **2009**, *43*, 28–36.
37. Sir, K.A.; Randa, E.; Amro, A.B. Content of phenolic compounds and vitamin C and antioxidant activity in wasted parts of Sudanese citrus fruits. *Food Sci. Nutr.* **2018**, *6*, 1214–1219.
38. Hassan, R.A.; Hozayen, W.G.; Abo Sree, H.T.; Al-Muzafar, H.M.; Amin, K.A.; Ahmed, O.M. Naringin and hesperidin counteract diclofenac-induced hepatotoxicity in male Wistar rats via their antioxidant, anti-inflammatory, and antiapoptotic activities. *Oxidative Med. Cell. Longev.* **2021**, *2021*, 9990091. [CrossRef] [PubMed]
39. Bahorun, T.; Ramful-baboolall, D.; Neergheen-bhujun, V.; Aruoma, O.I.A.; Verma, S.; Tarnus, E.; Robert, C.; Silva, D.; Rondeau, P. Phytophenolic Nutrients in Citrus: Biochemical and Molecular Evidence. In *Advances in Citrus Nutrition*; Springer: Berlin/Heidelberg, Germany, 2012.
40. Zhao, B.T.; Kim, E.J.; Son, K.H.; Son, J.K.; Min, B.S.; Woo, M.H. Quality evaluation and pattern recognition analyses of marker compounds from five medicinal drugs of Rutaceae family by HPLC/PDA. *Arch. Pharm. Res.* **2015**, *38*, 1512–1520. [CrossRef]
41. Grassino, A.N.; Pedisić, S.; Dragović-Uzelac, V.; Karlović, S.; Ježek, D.; Bosiljkov, T. Insight into high-hydrostatic pressure extraction of polyphenols from tomato peel waste. *Plant Foods Hum. Nutr.* **2020**, *75*, 427–433. [CrossRef] [PubMed]
42. Victor, M.M.; David, J.M.; Sakukuma, M.C.K.; França, E.L.; Nunes, A.V.J. A simple and efficient process for the extraction of naringin from grapefruit peel waste. *Green Process. Synth.* **2018**, *7*, 524–529. [CrossRef]
43. Chávez-González, M.L.; Sepúlveda, L.; Verma, D.K.; Luna-García, H.A.; Rodríguez-Durán, L.V.; Ilina, A.; Aguilar, C.N. Conventional and emerging extraction processes of flavonoids. *Processes* **2020**, *8*, 434. [CrossRef]
44. Ciğeroğlu, Z.; Bayramoğlu, M.; Kırbaşlar, Ş.İ.; Şahin, S. Comparison of microwave-assisted techniques for the extraction of antioxidants from Citrus paradisi Macf. biowastes. *J. Food Sci. Technol.* **2021**, *58*, 1190–1198. [CrossRef]
45. Cheigh, C.I.; Chung, E.Y.; Chung, M.S. Enhanced extraction of flavanones hesperidin and narirutin from Citrus unshiu peel using subcritical water. *J. Food Eng.* **2012**, *110*, 472–477. [CrossRef]
46. Yu, J.; Dandekar, D.V.; Toledo, R.T.; Singh, R.K.; Patil, B.S. Supercritical fluid extraction of limonoids and naringin from grapefruit (Citrus paradisi Macf.) seeds. *Food Chem.* **2007**, *105*, 1026–1031. [CrossRef]
47. Giannuzzo, A.N.; Boggetti, H.J.; Nazareno, M.A.; Mishima, H.T. Supercritical fluid extraction of naringin from the peel of Citrus paradisi. *Phytochem. Anal.* **2003**, *14*, 221–223. [CrossRef]
48. Liu, Y.; Zhang, H.; Yu, H.; Guo, S.; Chen, D. Deep eutectic solvent as a green solvent for enhanced extraction of narirutin, naringin, hesperidin and neohesperidin from Aurantii Fructus. *Phytochem. Anal.* **2019**, *30*, 156–163. [CrossRef]

49. El Kantar, S.; Rajha, H.N.; Boussetta, N.; Vorobiev, E.; Maroun, R.G.; Louka, N. Green extraction of polyphenols from grapefruit peels using high voltage electrical discharges, deep eutectic solvents and aqueous glycerol. *Food Chem.* **2019**, *295*, 165–171. [CrossRef] [PubMed]
50. Ma, G.; Zhang, L.; Sugiura, M.; Kato, M. *Citrus and Health. The Genus Citrus*; Elsevier: Amsterdam, The Netherlands, 2020.
51. Yan, Y.; Zhou, H.; Wu, C.; Feng, X.; Han, C.; Chen, H.; Liu, Y.; Li, Y. Ultrasound-assisted aqueous two-phase extraction of synephrine, naringin, and neohesperidin from *Citrus aurantium* L. fruitlets. *Prep. Biochem. Biotechnol.* **2021**, *51*, 780–791. [CrossRef]
52. Gupta, A.K.; Dhua, S.; Sahu, P.P.; Abate, G.; Mishra, P.; Mastinu, A. Variation in phytochemical, antioxidant and volatile composition of pomelo fruit (*Citrus grandis* (L.) osbeck) during seasonal growth and development. *Plants* **2021**, *10*, 1941. [CrossRef]
53. Ghanbari-Movahed, M.; Jackson, G.; Farzaei, M.H.; Bishayee, A. A systematic review of the preventive and therapeutic effects of naringin against human malignancies. *Front. Pharmacol.* **2021**, *12*, 639840. [CrossRef]
54. Li, H.; Yang, B.; Huang, J.; Xiang, T.; Yin, X.; Wan, J.; Luo, F.; Zhang, L.; Li, H.; Ren, G. Naringin inhibits growth potential of human triple-negative breast cancer cells by targeting β-catenin signaling pathway. *Toxicol. Lett.* **2013**, *220*, 219–228. [CrossRef] [PubMed]
55. Chen, R.; Qi, Q.L.; Wang, M.T.; Li, Q.Y. Therapeutic potential of naringin: An overview. *Pharm. Biol.* **2016**, *54*, 3203–3210. [CrossRef]
56. Cells, N.S.; Kim, H.; Song, J.Y.; Park, H.J.; Park, H.; Yun, D.H.; Chung, J. Naringin protects against rotenone-induced apoptosis in human. *J. Physiol. Pharmacol.* **2009**, *13*, 281–285.
57. Aroui, S.; Fetoui, H.; Kenani, A. Natural dietary compound naringin inhibits glioblastoma cancer neoangiogenesis. *BMC Pharmacol. Toxicol.* **2020**, *21*, 46. [CrossRef]
58. Li, J.; Dong, Y.; Hao, G.; Wang, B.; Wang, J.; Liang, Y.; Liu, Y.; Zhen, E.; Feng, D.; Liang, G. Naringin suppresses the development of glioblastoma by inhibiting FAK activity. *J. Drug Target.* **2017**, *25*, 41–48. [CrossRef] [PubMed]
59. Memariani, Z.; Abbas, S.Q.; ul Hassan, S.S.; Ahmadi, A.; Chabra, A. Naringin and naringenin as anticancer agents and adjuvants in cancer combination therapy: Efficacy and molecular mechanisms of action, a comprehensive narrative review. *Pharmacol. Res. Commun.* **2021**, *171*, 105264. [CrossRef] [PubMed]
60. Ramesh, E.; Alshatwi, A.A. Naringin induces death receptor and mitochondria-mediated apoptosis in human cervical cancer (SiHa) cells. *Food Chem. Toxicol.* **2013**, *51*, 97–105. [CrossRef]
61. Ahmed, S.; Khan, H.; Aschner, M.; Hasan, M.M.; Hassan, S.T.S. Therapeutic potential of naringin in neurological disorders. *Food Chem. Toxicol.* **2019**, *132*, 110646. [CrossRef]
62. Zhou, J.; Xia, L.; Zhang, Y. Naringin inhibits thyroid cancer cell proliferation and induces cell apoptosis through repressing PI3K/AKT pathway. *Pathol. Res. Pract.* **2019**, *215*, 152707. [CrossRef]
63. Mahmoud, A.M. Anti-diabetic effect of naringin: Insights into the molecular mechanism. *Int. J. Obes.* **2016**, *8*, 324–328. [CrossRef]
64. Pandey, V.K.; Shams, R.; Singh, R.; Dar, A.H.; Pandiselvam, R.; Rusu, A.V.; Trif, M. A comprehensive review on clove (*Caryophyllus aromaticus* L.) essential oil and its significance in the formulation of edible coatings for potential food applications. *Front. Nutr.* **2022**, *9*, 987674. [CrossRef] [PubMed]
65. Bharti, S.; Rani, N.; Krishnamurthy, B.; Arya, D.S. Preclinical evidence for the pharmacological actions of naringin: A review. *Planta Med.* **2014**, *80*, 437–451. [CrossRef]
66. Chen, F.; Zhang, N.; Ma, X.; Huang, T.; Shao, Y.; Wu, C.; Wang, Q. Naringin alleviates diabetic kidney disease through inhibiting oxidative stress and inflammatory reaction. *PLoS ONE.* **2015**, *10*, e0143868. [CrossRef]
67. Heidary Moghaddam, R.; Samimi, Z.; Moradi, S.Z.; Little, P.J.; Xu, S.; Farzaei, M.H. Naringenin and naringin in cardiovascular disease prevention: A preclinical review. *Eur. J. Clin. Pharmacol.* **2020**, *887*, 173535. [CrossRef] [PubMed]
68. Kumar, R.; Clermont, G.; Vodovotz, Y.; Chow, C.C. The dynamics of acute inflammation. *J. Theor. Biol.* **2004**, *230*, 145–155. [CrossRef]
69. Lawrence, T.; Gilroy, D.W. Chronic inflammation: A failure of resolution. *Int. J. Exp. Pathol.* **2007**, *88*, 85–94. [CrossRef]
70. El-Desoky, A.H.; Abdel-Rahman, R.F.; Ahmed, O.K.; El-Beltagi, H.S.; Hattori, M. Anti-inflammatory and antioxidant activities of naringin isolated from Carissa carandas L.: In vitro and in vivo evidence. *Phytomedicine* **2018**, *42*, 126–134. [CrossRef] [PubMed]
71. Shirani, K.; Yousefsani, B.S.; Shirani, M.; Karimi, G. Protective effects of naringin against drugs and chemical toxins induced hepatotoxicity: A review. *Phytother. Res.* **2020**, *34*, 1734–1744. [CrossRef]
72. Pari, L.; Amudha, K. Hepatoprotective role of naringin on nickel-induced toxicity in male Wistar rats. *Eur. J. Clin. Pharmacol.* **2011**, *650*, 364–370. [CrossRef]
73. Chen, J.C.; Li, L.J.; Wen, S.M.; He, Y.C.; Liu, H.X.; Zheng, Q.S. Quantitative analysis and simulation of anti-inflammatory effects from the active components of Paino Powder in rats. *Chin. J. Integr. Med.* **2011**, 201203, *online ahead of print*.
74. Yadav, M.; Sehrawat, N.; Singh, M.; Upadhyay, S.K.; Aggarwal, D.; Sharma, A.K. Cardioprotective and hepatoprotective potential of citrus flavonoid naringin: Current status and future perspectives for health benefits. *Asian J. Biol. Sci.* **2020**, *9*, 1–5. [CrossRef]
75. Higashi, Y. Endothelial Function in Dyslipidemia: Roles of LDL-Cholesterol, HDL-Cholesterol and Triglycerides. *Cells* **2023**, *12*, 1293. [CrossRef]
76. Kopustinskiene, D.M.; Jakstas, V.; Savickas, A.; Bernatoniene, J. Flavonoids as anticancer agents. *Nutrients* **2020**, *12*, 457. [CrossRef] [PubMed]

77. Franzoni, F.; Scarfò, G.; Guidotti, S.; Fusi, J.; Asomov, M.; Pruneti, C. Oxidative stress and cognitive decline: The neuroprotective role of natural antioxidants. *Front. Neurosci.* **2021**, *15*, 729757. [CrossRef] [PubMed]
78. Gillessen, A.; Schmidt, H.H.J. Silymarin as supportive treatment in liver diseases: A narrative review. *Adv. Ther.* **2020**, *37*, 1279–1301. [CrossRef]
79. Kawaguchi, K.; Maruyama, H.; Hasunuma, R.; Kumazawa, Y. Suppression of inflammatory responses after onset of collagen-induced arthritis in mice by oral administration of the Citrus flavanone naringin. *Immunopharmacol. Immunotoxicol.* **2011**, *33*, 723–729. [CrossRef]
80. Yang, X.; Zhao, Y.; Gu, Q.; Chen, W.; Guo, X. Effects of naringin on postharvest storage quality of bean sprouts. *Foods* **2022**, *11*, 2294. [CrossRef]
81. Bailey, D.G.; Dresser, G.K. Interactions between grapefruit juice and cardiovascular drugs. *Am. J. Cardiol.* **2004**, *4*, 281–297. [CrossRef]
82. Hsiu, S.L.; Huang, T.Y.; Hou, Y.C.; Chin, D.H.; Chao, P.D. Comparison of metabolic pharmacokinetics of naringin and naringenin in rabbits. *Life Sci.* **2002**, *70*, 1481–1489. [CrossRef]
83. Tsai, Y.J.; Tsai, T.H. Mesenteric lymphatic absorption and the pharmacokinetics of naringin and naringenin in the rat. *J. Agric. Food Chem.* **2012**, *60*, 12435–12442. [CrossRef]
84. Tesseromatis, C.; Alevizou, A. The role of the protein-binding on the mode of drug action as well the interactions with other drugs. *Eur. J. Drug Metab. Pharmacokinet.* **2008**, *33*, 225–230. [CrossRef]
85. Zeng, X.; Yao, H.; Zheng, Y.; He, Y.; He, Y.; Rao, H.; Li, P.; Su, W. Tissue distribution of naringin and derived metabolites in rats after a single oral administration. *J. Chromatogr.* **2020**, *1136*, 121846. [CrossRef] [PubMed]
86. Najmanová, I.; Vopršalová, M.; Saso, L.; Mladěnka, P. The pharmacokinetics of flavanones. *Crit. Rev. Food Sci. Nutr.* **2020**, *60*, 3155–3171. [CrossRef] [PubMed]
87. Goyal, A.; Verma, A.; Dubey, N.; Raghav, J.; Agrawal, A. Naringenin: A prospec-tive therapeutic agent for Alzheimer's and Parkinson's disease. *J. Food Biochem.* **2022**, *46*, e14415. [CrossRef]
88. Turfus, S.C.; Delgoda, R.; Picking, D.; Gurley, B.J. Pharmacokinetics. In *Pharmacognosy*; Elsevier: Amsterdam, The Netherlands, 2017.
89. Sudto, K.; Pornpakakul, S.; Wanichwecharungruang, S. An efficient method for the largescale isolation of naringin from pomelo (*Citrus grandis*) peel. *Int. J. Food Sci. Technol.* **2009**, *44*, 1737–1742. [CrossRef]
90. Gollavilli, H.; Hegde, A.R.; Managuli, R.S.; Bhaskar, K.V.; Dengale, S.J.; Reddy, M.S.; Kalthur, G.; Mutalik, S. Naringin Nano-ethosomal novel sunscreen creams: Development and performance evaluation. *Colloids Surf. B Biointerfaces* **2020**, *193*, 111122. [CrossRef]
91. Pandey, V.K.; Islam, R.U.; Shams, R.; Dar, A.H. A comprehensive review on the application of essential oils as bioactive compounds in nano-emulsion based edible coatings of fruits and vegetables. *Appl. Food Res.* **2022**, *2*, 100042. [CrossRef]
92. Goliomytis, M.; Kartsonas, N.; Charismiadou, M.A.; Symeon, G.K.; Simitzis, P.E.; Deligeorgis, S.G. The influence of naringin or hesperidin dietary supplementation on broiler meat quality and oxidative stability. *PLoS ONE* **2015**, *10*, e0141652. [CrossRef]
93. Oliva, J.; French, B.A.; Li, J.; Bardag-Gorce, F.; Fu, P.; French, S.W. Sirt1 is involved in energy metabolism: The role of chronic ethanol feeding and resveratrol. *Exp. Mol. Pathol.* **2008**, *85*, 155–159. [CrossRef]
94. Lee, S.H.; Chow, P.S.; Yagnik, C.K. Developing eco-friendly skin care formulations with microemulsions of essential oil. *J. Cosmet. Sci.* **2022**, *9*, 30. [CrossRef]
95. Simitzis, P.; Massouras, T.; Goliomytis, M.; Charismiadou, M.; Moschou, K.; Economou, C.; Papadedes, V.; Lepesioti, S.; Deligeorgis, S. The effects of hesperidin or naringin dietary supplementation on the milk properties of dairy ewes. *J. Sci. Food Agric.* **2019**, *99*, 6515–6521. [CrossRef]
96. Goodarzi-Boroojeni, F.; Männer, K.; Zentek, J. The impacts of Macleaya cordata extract and naringin inclusion in post-weaning piglet diets on performance, nutrient digestibility and intestinal histomorphology. *Arch. Anim. Nutr.* **2018**, *72*, 178–189. [CrossRef]
97. Kandhare, A.D.; Alam, J.; Patil, M.V.K.; Sinha, A.; Bodhankar, S.L. Wound healing potential of naringin ointment formulation via regulating the expression of inflammatory, apoptotic and growth mediators in experimental rats. *Pharm. Biol.* **2016**, *54*, 419–432. [CrossRef] [PubMed]
98. Wang, H.; Hu, H.; Zhang, X.; Zheng, L.; Ruan, J.; Cao, J.; Zhang, X. Preparation, physicochemical characterization, and antioxidant activity of naringin–silk fibroin–alginate microspheres and application in yogurt. *Foods* **2022**, *11*, 2147. [CrossRef] [PubMed]
99. Tayengwa, T.; Chikwanha, O.C.; Gouws, P.; Dugan, M.E.R.; Mutsvangwa, T.; Mapiye, C. Dietary citrus pulp and grape pomace as potential natural preservatives for extending beef shelf life. *Meat Sci.* **2020**, *162*, 108029. [CrossRef] [PubMed]

molecules

Review

Extraction of Bioactive Compounds from Different Vegetable Sprouts and Their Potential Role in the Formulation of Functional Foods against Various Disorders: A Literature-Based Review

Afifa Aziz [1], Sana Noreen [2], Waseem Khalid [1,*], Fizza Mubarik [2], Madiha khan Niazi [2], Hyrije Koraqi [3], Anwar Ali [4], Clara Mariana Gonçalves Lima [5], Wafa S. Alansari [6], Areej A. Eskandrani [7], Ghalia Shamlan [8] and Ammar AL-Farga [6]

1 Department of Food Science, Faculty of Life Sciences, Government College University, Faisalabad 38000, Pakistan
2 University Institute of Diet and Nutritional Sciences, Faculty of Allied Health Sciences, The University of Lahore, Lahore 54000, Pakistan
3 Faculty of Food Science and Biotechnology, UBT-Higher Education Institution, St. Rexhep Krasniqi No. 56, 10000 Pristina, Kosovo
4 Department of Epidemiology and Health Statistics, Xiangya School of Public Health, Central South University, Changsha 410017, China
5 Department of Food Science, Federal University of Lavras, Lavras 37203-202, Brazil
6 Biochemistry Department, Faculty of Science, University of Jeddah, Jeddah 21577, Saudi Arabia
7 Chemistry Department, Faculty of Science, Taibah University, Medina 30002, Saudi Arabia
8 Department of Food Science and Nutrition, College of Food and Agriculture Sciences, King Saud University, Riyadh 11362, Saudi Arabia
* Correspondence: waseemkhalid@gcuf.edu.pk

Citation: Aziz, A.; Noreen, S.; Khalid, W.; Mubarik, F.; Niazi, M.k.; Koraqi, H.; Ali, A.; Lima, C.M.G.; Alansari, W.S.; Eskandrani, A.A.; et al. Extraction of Bioactive Compounds from Different Vegetable Sprouts and Their Potential Role in the Formulation of Functional Foods against Various Disorders: A Literature-Based Review. *Molecules* **2022**, *27*, 7320. https://doi.org/10.3390/molecules27217320

Academic Editors: Sascha Rohn and Michał Halagarda

Received: 13 September 2022
Accepted: 25 October 2022
Published: 28 October 2022

Abstract: In this review, we discuss the advantages of vegetable sprouts in the development of food products as well as their beneficial effects on a variety of disorders. Sprouts are obtained from different types of plants and seeds and various types of leafy, root, and shoot vegetables. Vegetable sprouts are enriched in bioactive compounds, including polyphenols, antioxidants, and vitamins. Currently, different conventional methods and advanced technologies are used to extract bioactive compounds from vegetable sprouts. Due to some issues in traditional methods, increasingly, the trend is to use recent technologies because the results are better. Applications of phytonutrients extracted from sprouts are finding increased utility for food processing and shelf-life enhancement. Vegetable sprouts are being used in the preparation of different functional food products such as juices, bread, and biscuits. Previous research has shown that vegetable sprouts can help to fight a variety of chronic diseases such as cancer and diabetes. Furthermore, in the future, more research is needed that explores the extraordinary ways in which vegetable sprouts can be incorporated into green-food processing and preservation for the purpose of enhancing shelf-life and the formation of functional meat products and substitutes.

Keywords: sprout; bioactive compound; new extraction; functional product; pharmacological role

1. Introduction

Sprouts are germinated from seeds of crops such as radish, cereals (rice and legumes), soybeans, and trees (*Toonasinensis* and pepper). Sprouts have been a popular dish in China for over 5000 years and have now spread to other Eastern countries. Sprout consumption has increased in Western cultures due to a shift in lifestyle towards convenience and health [1]. Sprouts have become more popular worldwide because of their nutritional value and health advantages. As compared with adult edible plant portions, sprouts are abundant in health-promoting bioactive chemicals, vitamins, and minerals. Sprouting is a food processing method that boosts the nutritional value of cereals, oilseeds, and vegetable seeds [2], by inducing macronutrient breakdown and increasing the amounts of amino acids, simple sugars, and other nutritional components [3]. Sprouting also helps to reduce anti-nutritional components and to improve sprouts' digestibility and sensory aspects. Sprouts aid in the synthesis of new beneficial components such as polyphenols and vitamin C. Recently, sprouts have grasped consumers' consumption interests due to their functional properties and phytonutritional profile. Sprouts have gained popularity as a top healthy food [4]. Due to the presence of biologically active compounds [5], sprouts have an important role in preventing several forms of malignancies. In the literature, sprouts have been shown to have a substantial anti-genotoxic impact against DNA damage [6]. According to Gawlik-Dziki et al. [7], Brassica and vegetable sprouts help to minimize the incidence of lung and colorectal cancer occurrence. According to epidemiological studies, the consumption of broccoli sprout-rich foods has been linked to a lower incidence of many malignancies and chronic degenerative diseases [8]. Cowpea sprouts have been celebrated for reducing cell proliferation and boosting anti-colorectal cancer activity [9]. Isoflavonoids in soybean sprouts defend against cancer and cardiovascular disease [10]. Sprouts include a variety of nutrients that are beneficial to human health and help to avoid a variety of diseases [11]. According to the research, sprouts are an excellent source of a range of phenolic compounds that protect against oxidative reactions. Mung bean sprouts have been reported to lower gastrointestinal issues and heart stroke [12]. Soybean sprouts have been demonstrated to have health-promoting qualities such as lowering cancer and cardiovascular disease risk [13]. Sprouts have been discovered to offer antidiabetic properties. In studies, *Brassica oleracea* sprouts have been shown to have antidiabetic, hepatoprotective, and antioxidant properties. The extracts had a lowering effect on blood glucose levels in the body and hepatoprotective and antioxidant properties. As a consequence, Brassica sprouts have anti-hyperglycemic activity [14]. Antibacterial activity has also been discovered in sprouts. Broccoli and pea sprouts have antibacterial properties against *Helicobacter pylori* (bacteria linked to stomach cancer) [15,16]. The purpose of this article is to explore methods for extracting bioactive components from certain vegetable sprouts to make functional meals and their bioavailability against certain diseases.

2. Bioactive Components in Different Vegetables Sprouts

Sprouts are recognized due to their high concentration of bioactive compounds [17]. Bioactive chemicals are present in high concentrations in food and food waste. Several bioactive compounds have been found with a wide range of functional and structural features [18] (Figure 1).

Natural chemical components present in minute amounts in plants are called bioactive chemicals [19]. These substances can interact with one or more live tissue components, resulting in many conceivable outcomes. Bioactive chemicals are naturally occurring essential and nonessential substances present in the food chain that have been shown to impact human health. Antioxidants and polyphenols have been studied for their many biological actions. Due to their unique properties, polyphenols such as protocatechuic acid, vanillic acid, caffeic acid, quercetin, and kaempferol have been shown to protect against oxidative damage [20]. Polyphenols are abundant in sprouts [21]. Polyphenols are reducing agents, hydrogen-donating antioxidants, and singlet oxygen quenchers formed in response to biotic or abiotic stress. Polyphenols' multifunctionality is due to their dispersion in various tissues and organs of plants at various concentrations. Plants produce phenolic substances that contain at least one aromatic ring with one or more hydroxyl substituents and can be divided into flavonoids and phenolic acids based on their chemical structure. In recent years, phenolic compounds have been extensively studied for their antioxidant, anthelmintic, antiallergenic, anticancer, anti-inflammatory, antiviral, antiulcer, anti-hepatotoxic, antidiarrheal, and antiproliferative properties [3,22]. Micronutrients (vitamins and minerals) are important for tissue maintenance, bone and tooth production, and general health. They help to regulate and coordinate most biological activities and other biochemical and physiological functions by acting as cofactors and coenzymes in diverse enzyme systems. Humans and other creatures require micronutrients at varying levels throughout their lives to coordinate numerous physiological activities and sustain health [23,24]. Vitamins are essential for optimal health and perform vital functions in the human body. Vitamin C, often known as ascorbate, is a micronutrient that humans require. In humans, vitamin C deficiency inhibits the function of several enzymes and can lead to scurvy. Ascorbic acid is a cofactor in a variety of essential enzymatic processes [25], and it is involved in collagen formation. Vitamin E (tocopherols and tocotrienols) protects DNA, low-density lipoproteins, and polyunsaturated fatty acids against oxidative damage. Vitamin E is also involved in hemoglobin production, immune response regulation, and membrane structure stability, and it helps to keep blood coagulation, bone growth, and healing under control. In newborns, vitamin K deficiency can cause hemorrhagic illness and surgical bleeding, muscular hematomas, and intracranial hemorrhages in adults [26].

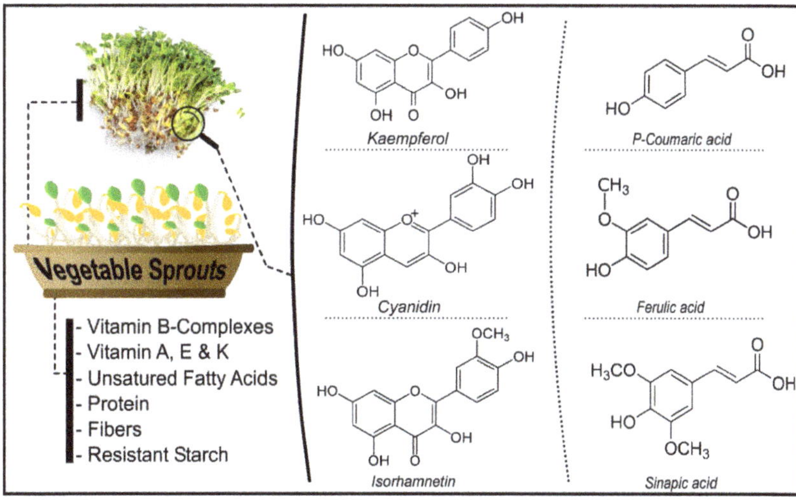

Figure 1. Different bioactive compounds in vegetables sprouts.

Minerals are required to perform processes that are necessary for a healthy life. The human body needs calcium for optimal heart and muscle function, bone production, and blood cell creation and function. Copper, molybdenum, selenium, and zinc are key components of various critical enzymes in the human body. In contrast, iron is required for several protein syntheses, including hemoglobin, which helps to avoid anemia. Magnesium is necessary for ATP processing and bone health. Sodium and potassium are electrolytes found throughout the body and are necessary for the coregulation of ATP. Phosphorus is found in bones and cells, and it also plays a role in energy metabolism, DNA and ATP (as phosphate), and a variety of other functions [27]. The extraction of bioactive compounds from different types of vegetable sprouts is shown in Table 1.

Table 1. Extraction of bioactive compound from vegetables sprouts.

Vegetable	Type	Extraction/Detection/Methods	Bioactive Compounds	Solvent	References
Water spinach	Leafy	–	Carotenoids and phenolic acids	–	[28]
Quinoa	Leafy	Conventional –	Phenolics, flavonoids, carotenoids (β-carotene and lycopene), and chlorophylls (a and b)	Ethanol	[29]
Brassica	Leafy	–	Polyphenols and glucosinolates	–	[30]
Kale and broccoli	Leafy	Ultrahigh-performance liquid chromatography high-resolution mass spectrometry	Polyphenols and glucosinolates	Methanol/water	[31]
Quinoa	Leafy	Spectrophotometry	Total phenolic compounds and antioxidants	Methanol	[32]
Radish, broccoli, leek, and beetroot	Leafy, Root	ABTS, FRAP, and ORAC	Polyphenols, L-ascorbic acid, carotenoids, and chlorophylls	Methanol	[33]
Garden cress	Leafy	HPLC, ABTS, and DPPH	Phenolic compound, antioxidant,, and flavonoids	Methanol	[34]
Onion	Root	HPLC and FTIR	Flavonoids, quercetin, and glucosides	Ethanol	[35]
Broccoli and red radish	Leafy Root	Conventional and HPLC-DAD	Glucosinolates and phenolic compounds	Methanol	[36]
Turnip	Root	Gas chromatography, mass spectroscopy, and DPPH	Phenolics, glucosinolates, and antioxidant,	Water	[37]
Fennel	Root	HPLC	Vitamin C, polyphenols, and antioxidants	Methanol/water	[38]
Brassicaceae	Root and leafy	FRAP	Phenolic compound and antioxidants	Methanol	[39]
Brussels	Leafy	HPLC and spectroscopic analysis	Chlorophyll, vitamin C, polyphenols, flavonoids, and antioxidants	Methanol	[36]
Sweet potato	Root	DPPH, spectrophotometer, LCMS/MS method	Anthocyanin and antioxidant,	Ethanol	[40]
Vegetable	Leafy	–	Glucosinolates, phenolics, and isoflavones	–	[41]
Brassica	Leafy	Conventional	Sulforaphane	–	[42]

3. Conventional and New Extraction Methods Are Used to Extract the Bioactive Compounds

Several traditional extraction procedures can extract bioactive chemicals from plant sources. The majority of these methods rely on a solvent's ability to extract and the use of heat and/or mixing. The three most common procedures for extracting bioactive chemicals from plants are Soxhlet extraction, maceration, and hydro distillation [43]. There are both traditional and modern ways of obtaining isoflavones from plants. Maceration, percolation, decoction, infusion, Soxhlet extraction, and hot reflux extraction are all examples of traditional extraction procedures that have been employed to extract bioactive chemicals [44]. As a result, developing quick, safe, and environmentally acceptable technology for analyzing and separating bioactive chemicals is critical. Isoflavones have been separated via "ultrasound-assisted extraction (UAE), microwave-assisted extraction (MAE), supercritical fluid extraction (SFE), and pressured liquid extraction with green solvents such as ionic water liquids and supercritical carbon dioxide (PLE)". These methods use organic solvents, take less time, and produce higher yields and quality [45].

In recent years, natural deep eutectic solvents (DESs) have received much attention as good green solvents for extracting bioactive chemicals from natural resources [46]. The current research has looked at the feasibility and effectiveness of extracting isoflavones from chickpea sprouts using various polarities of natural deep eutectic solvents (DESs), by testing 20 different DESs that included hydrogen bond acceptors such as "choline chloride, betaine, and L-proline with different hydrogen bond donors (carboxylic acids, alcohols, sugars, and amine). The researchers looked at the yields of four isoflavones (ononin, sissotrin, formononetin, and biochanin A), total flavonoid concentration, and antioxidant activity to estimate extraction efficiency. Using a Box–Behnken design in conjunction with response surface techniques, "the components that contribute to optimal ultrasound-assisted extraction conditions were then examined." The extraction yields of isoflavones were significantly affected by DES water content and extraction temperature. Our findings suggest that DESs might be utilized to extract bioactive chemicals from a variety of biomaterials [47]. Sprouts from peanuts yield trans-resveratrol in accelerated solvent extraction [48].

Soxhlet extraction has been used to recover a large number of phytochemicals from *Azadirachta indica* (Neem) leaf powder, predominantly nonpolar components [49]. Evaluation of Soxhlet extraction for *Moringa oliefera* leaves resulted in lower yield, as well as phenolic and flavonoid contents [50]. *Centella asiatica* extraction was optimized using Soxhlet extraction, which produced the best results at 25 °C, a sample-solvent ratio of 1:45, 200 rpm agitation speed, and 1.5 h [51]. After removing lipoidal components from powdered *Clitorea ternate* flowers with petroleum ether at 60–80 °C, the yield was 2.2 percent *w*/*w* [52].

After more ethanol extraction from the marc, alkaloids and saponins were confirmed to be present. However, the anthocyanin, i.e., the main pigment of *Clitorea ternate* flowers, was not present, indicating that oxidation and degradation had taken place. As compared with other solvents such as petroleum ether, chloroform, and water, the extraction of *Psidium guajava* L. [53] leaves using ethanolic and hydro alcohol extracts (4:1 *v*/*v*) produced the highest extraction yield with the greatest presence of phytoconstituents (alkaloids, saponins, carbohydrates, tannins, and flavonoids) [50]. Nonpolar solvents such as petroleum ether and chloroform revealed no retained active chemicals and very low tannin content in the extracts, respectively. With the exception of the absence of any alkaloids, water was shown to be as effective as ethanol. Polar solvents have been shown to work better for removing bioactive compounds from *Psidium guajava* [54]. As compared with aqueous extracts of *Garnicia atriviridis*, methanol extracts (1:10 *w*/*v*) had stronger antioxidant activities, while the aqueous extracts had better anti-hyperlipidemic activities [50]. Based on total phenolics, maceration with various solvents at a ratio of 1:10 *w*/*v* sample to solvent, for an hour, revealed that 70% acetone was an effective solvent for Portucala oleracea and 70% methanol was an effective solvent for flavonoids in *Cosmos caudatus* [55]. As compared with Soxhlet extraction and percolation using a comparable solvent, maceration with 70% ethanol and powdered dried materials at 1:40 w/v showed the greatest phenolic and flavonoid concentrations for *Moringa oliefera* [56]. Using 100% ethanol as the solvent at 75 °C and an irradiation power of 600 W for four cycles, MAE has been evaluated as a new technique to extract triterpene from *Centella asiatica* and the yield was increased by two times over Soxhlet extraction [50]. Combining MAE with enzyme lysis (such as cellulase) has been shown to increase extraction; the ideal conditions of sample/solvent ratio at 1:36, enzyme pretreatment at 45 °C for 30 min, and irradiation at 650 W for 110 s produced a yield of 27.10 percent. However, Trusheva et al.'s observation that an extra MAE cycle had an impact on the phytochemical degradation was not examined. The best extraction was obtained using MAE with 100 W and 1:12.5 sample/solvent ratios on *Dioscorea hispida*. The UAE has been shown to be the most productive technique for extracting propolis based on its high yield, lengthy (10–30 min) extraction duration, and excellent selectivity. To shorten extraction times and to prevent exposure to high temperatures, UAE was used to extract thermolabile chemicals such as anthocyanin from floral components. When extracting *Withania somnifera* using water as the solvent for 15 min, the yield was at its highest, reaching 11.85 percent as compared with ethanol and water-ethanol at various 5, 15, and 20 min extraction times. A higher effectiveness on phenolics was observed when *Cratoxylum formosum* was extracted using ultrasonic at 45 kHz, 50.33 percent ethanol by volume, at 65 °C for 15 min. Free radical production at irradiation frequencies higher than 20 kHz, however, may need to be taken into account [50].

Sprouts and microgreens are edible seedlings of different vegetables and herbs that have become increasingly popular due to their positive health benefits and are now referred to as functional foods or superfoods. Bioactive components have long been appreciated in broccoli seedlings (*Brassica oleracea* L. var. *Italica*). Secondary metabolites have been linked to several positive health outcomes. In in vitro and animal studies, broccoli seedlings have been shown to have health benefits. A current study has summarized previous research on the bioactive components and bioactivities of various broccoli derivatives, as well as the mechanisms of action associated with them [57]. Conventional and new techniques can be used to extract bioactive compounds from different types of vegetable sprouts, as shown in Figure 2.

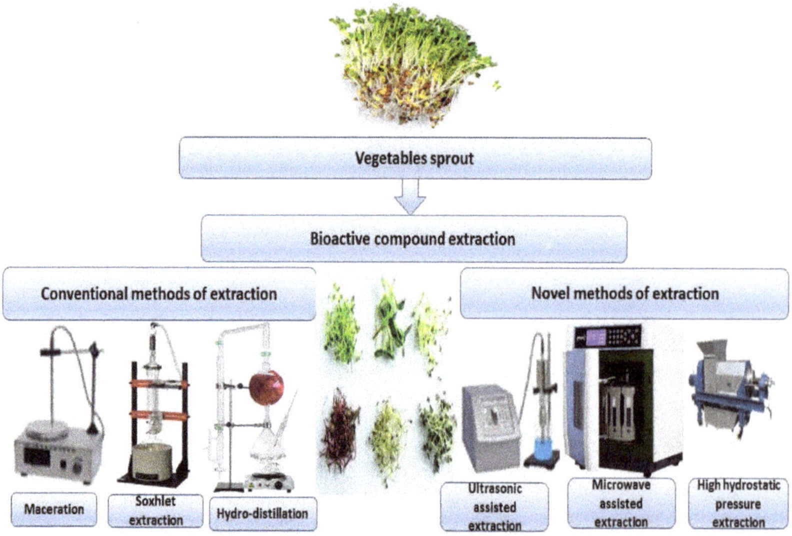

Figure 2. Conventional and new techniques can be used to extract bioactive compounds from different types of vegetable sprouts.

4. Food Applications of Vegetable Sprouts

Nowadays, the food industry has been focused on developing healthier products that are more responsive to changing customer demands. Recently, sprouted grains have become a new element in the culinary world. Sprouted grains have a higher nutritional value, lower antinutrient content, a prime source of bioactive compounds, and a sweeter flavor, making them a potential new food component [58]. Sprouted grains were formerly only used in bread, but they may now be found in tortillas, granola, cookies, crackers, muffins, snacks, bars, morning cereals, side dishes, and salads [59]. Sprouted grains may be used in various culinary applications without requiring any formulation adjustments, and they can assist considerably in differentiating products.

After sprouting and drying, a whole grain kernel can be milled into flour or processed into grits, coarse meals, or flakes, among other granulations. Wheat, rye, spelt, barley, brown rice, oat, sorghum, millet, quinoa, buckwheat, and amaranth may all be sprouted and used in several nutritious applications as long as the germ is intact. Bars, cereals, granola, bread, tortillas, frozen dough, candies, snacks, side dishes, soups, and pasta are common components. Gluten-free foods such as sprouted sorghum, millet, quinoa, amaranth, buckwheat, brown rice, and purity protocol oats are naturally gluten-free foods. They can be added to gluten-free diets to boost nutrition. Because of the wide variety of grains available, bakers, food scientists, and chefs have a lot of creative freedom. The functional differences between sprouted grains and their unsprouted counterparts must be recognized and addressed for optimal formulation, processing, and end-product attributes. Due to their high nutritional content, interesting technical possibilities, and sensory qualities, sprouted grains are being exploited as a component in a variety of food product innovations. The quality of sprouts and their specific culinary behavior depends on the germination circumstances. In this article, we look at two applications of sprouted grains: the effect of wheat sprouting time on the production of innovative baking flours and the microbiological risk of homemade rejuvelac, a sprouted wheat-based fermented beverage [60]. Bakers may notice a shortening of the proofing period or an increase in bread absorption, contributing to higher yields.

In tortillas, flour made of sprouted whole wheat can help to soften them, lengthen their shelf-life, and improve their sensory qualities [40]. The potential of sprouted wheat to improve the likability of bread and tortillas might lead to an increase in whole grain consumption, which would be a tremendous step forward in human health, especially because these staples are frequently consumed on a regular basis [52]. Table 2 shows the food applications of vegetable sprouts.

Table 2. Food applications of vegetables sprouts.

Sprout Source	Application in Food Products	Improvement	References
Vegetable sprout	–	Improve nutritional value of different food product	[61]
Radish, red cabbage, vegetable green, buckwheat and broccoli seeds	–	Vegetable sprouts are rich in nutrients	[62]
Fresh alfalfa and flax sprouts	Rabbit meat	Modified the fat content, fatty acid, and phytochemical profile of the meat	[63]
Wheat seeds	Bread	To examined the profile of phenolic acids and antioxidant properties of wheat bread	[64]
Brown rice	Wheat bread	Sensory acceptance and longer shelf-life	[65]
Broccoli	Broccoli sprout juice	Broccoli sprouts are naturally enriched in glucoraphanin (GR)	[66]

Table 2. *Cont.*

Sprout Source	Application in Food Products	Improvement	References
Brussels	Juice	It may reduce the risk of cancer of the alimentary tract	[67]

5. Bioavailability of Sprout against Different Diseases

With the intake of plant sprouts, bioavailability has long been regarded as crucial. The bioavailability of phytochemicals in various sprout diets varies substantially depending on several parameters. Interindividual factors, including delivery mode, and even intraindividual biochemical variances and the makeup and function of the gut microbiota are all factors to consider. In one study, to test iso-thiocyanate bioavailability, mice were administered either thermally processed broccoli sprout powders or pure isothiocyanate sulforaphane. The greatest quantities of the isothiocyanate metabolite were discovered in slightly cooked broccoli sprout powdered meals. In vivo, nonheated broccoli sprouts were followed by powdered broccoli sprout meals. They identified erusin and sulforaphane interconversion and observed that erusin was the preferred form in the kidney, liver, and bladder even when just sulforaphane was digested. It is worth mentioning that the bioavailability of sulforaphane from broccoli sprouts varies greatly depending on the delivery method [68]. The study suggested the inhibitory effect of broccoli sprout extracts on the properties of two prostate cancer cell lines characterized by low (AT-2) and high (MAT-LyLu) metastatic potential. These effects may be due to the fact that broccoli sprouts contained flavonoids and phenolic acids [7]. A previous study suggested that fava bean sprouts had higher antioxidant activity because they contained more polyphenols and l-3,4-dihydroxyphenylalanine (l-DOPA) than the bean itself [69]. The study suggested that lentil sprouts contained melatonin that is a multifunctional antioxidant neurohormone. The results showed that germination of lentils increased the content of melatonin. In another study, Sprague Dawley rats were used to investigate the pharmacokinetic profile of melatonin after oral administration of a lentil sprout extract and to evaluate plasma and urine melatonin and related biomarkers and antioxidant capacity. The outcomes showed that lentil sprout intake increased melatonin plasmatic concentration and attenuate plasmatic oxidative stress [70]. Figure 3 shows the bioavailability of sprouts against different disorders.

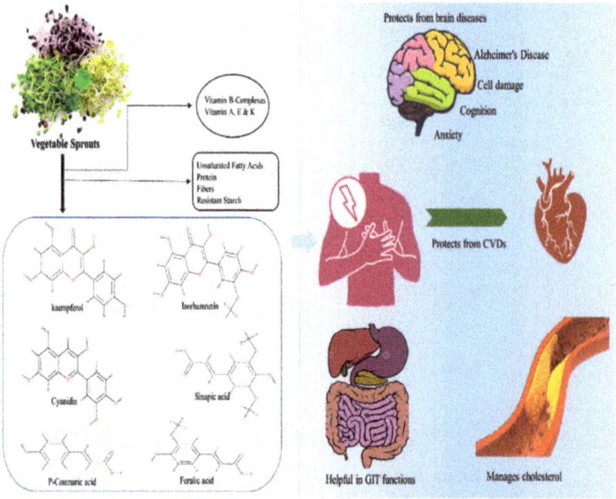

Figure 3. Bioavailability of sprouts against different disorders.

5.1. Bioavailability of Sprouts against Brain Issues

Nervous system diseases are a common ailment that will become more common as populations age. Axonopathy, also known as dying-back axonopathy, is a neurological illness in which axons become disconnected from their destinations, resulting in functional impairment. Axons can renew or sprout in response to several neurologic illnesses to re-establish synaptic function and to reconnect with the target before motor neuron death. Compensatory motor axon sprouting and neuromuscular junction reinnervation has been demonstrated in ALS patients, although the disease's course has typically outpaced these advantages. In ALS and kindred illnesses defined by dying-back axonopathy, potential therapeutics that encourage compensatory sprouting and reinnervation may delay symptom onset and may sustain muscle function for extended periods. Many questions concerning the impact of various disease-causing mutations on axonal outgrowth and regeneration, especially in motor neurons derived from patient-induced pluripotent stem cells, remain unsolved. Researchers must mimic the human neuromuscular circuit using motor neurons created from human-induced pluripotent stem cells to uncover drugs that stimulate axonal regeneration, sprouting, and reinnervation of neuromuscular junctions [71]. Regarding colored flavonoids, anthocyanins, the majority of which are highly acylated, and glycosylated forms of cyanidin are abundant in broccoli, radishes, cabbages, and kale sprouts [72]. Recently, anthocyanins have attracted more attention due to their potential to improve brain function and their role in the prevention and treatment of disorders including diabetes and obesity. One study suggested that two crude juices of broccoli sprouts had a protective effect on SH-SY5Y cells treated with the fragment Aβ25–35 because they contained different amounts of polyphenols and sulforaphane. The sprouts' juices both protected against Aβ-induced cytotoxicity and apoptotic cell death as evidenced by cell viability, nuclear chromatin condensation, and apoptotic body formation measurements [73]. Another study suggested that cruciferous vegetables were a good source of sulforaphane. The results of this study showed that sulforaphane protected against acute brain injuries and neurodegenerative diseases through activating the Nrf2 signaling pathway [74].

5.2. Compensatory Sprouting as a Potential Therapeutic Strategy for Amyotrophic Lateral Sclerosis

Functional motor recovery can be aided by the sprouting of motor axons and the reinnervation of denervated NMJs. Axonal sprouting allows motor units to increase 5–8 times their initial size. In amyotrophic lateral sclerosis, there is evidence of motor axon sprouting [75]. ALS is more common in certain motor neuron subpopulations that are also less prone to sprouting. In people with amyotrophic lateral sclerosis, compensatory sprouting may be employed to slow the onset of muscle denervation and weakness. The global number of ALS cases is expected to rise by 2040. Any drug that can improve the quality of life for ALS patients is badly needed. Axonal sprouting, which involves the functional reinnervation of NMJs, has the potential to improve life quality [76]. Phenolic acids are present in different vegetable sprouts. The diverse neuroprotective effects of phenolic acids make them interesting candidates for better ALS therapies. Study outcomes have shown that protocatechuic acid administration at 100 mg/kg in SOD1G93A mice prolonged survival, recovered motor functions, and decreased gliosis [4,77]. An in vitro study suggested that antioxidant molecules were capable of rescuing NSC34 motor neuron cells expressing an ALS-associated mutation of superoxide dismutase 1 [78].

5.3. Bioavailability of Sprouts against Gastrointestinal Tract (GIT) Health Problems

Sprouts may make it easier for you to digest your diet. According to a study, sprouted seeds increased the amount of fiber in them, making them more accessible. According to one study, cereals sprouted for five days had up to 133 percent more fiber than non-sprouted grains. Another study found that growing beans until the sprouts were 5 mm long boosted the overall fiber content by 226 percent. Sprouting appears to enhance the amount of insoluble fiber, a type of fiber that aids stool creation and passage through the stomach, reducing constipation risk. Finally, sprouted beans, grains, vegetables, nuts, and seeds have lower antinutrient levels than their non-sprouted counterparts. This makes it easier for the body to absorb nutrients during digestion [79]. Fiber that the human gut cannot digest on its own, but some bacteria can digest, is an essential source of nutrients that your gut microbe need to stay healthy. Fiber helps to stimulate the growth of colonic flora, to increase the weight of the stool, and to enhance the number of bacteria in the gut. The growth of bacteria present in the gut enhances the health of the intestines. However, short-chain fatty acids are produced by anaerobic gut bacteria through saccharolytic fermentation of complex resistant carbohydrates, which escape digestion and absorption in the small intestine [80]. In contrast to micro- and macronutritional contents, dietary polyphenols tend to be recognized as xenobiotic by humans during absorption, and therefore, their biological accessibility is significantly low. Furthermore, polymerization and structural complexity influence digestion in the small intestine [81]. The small intestine usually consumes approximately 5–10% of the absorbed polyphenols. The residual polyphenols (90–95%) might develop up to millimolar proportions in the large intestine linked to bile conjugates spilled into lumen in which they are susceptible to the enzymatic reactions of the gut bacteria species [82]. According to current data, dietary polyphenols that penetrate gut microflora, also including volatile compounds produced, manufacture and generate differences in the microbiota community through their prebiotic properties and functioning as an antiseptic towards infectious intestinal microbiota [83].

Onions have been proven to offer digestive system-protective properties, such as preventing stomach ulcers, regulating gut flora, and alleviating colitis. In rats, raw onion sprouts were shown to suppress histamine-induced stomach acid release and to attenuate ethanol-stimulated gastric ulcers. However, boiling the onion was less effective. In common carp juveniles, dietary supplementation with onion sprout powder has been shown to alter gut microbiota by increasing the number of lactic acid bacteria [84]. In rats, bioactive substances produced from onions, such as quercetin and quercetin monoglycosides, were found to boost the enzymatic activity of the gut microbiota. In colitis mice caused by dextran sodium sulfate, quercetin monoglycosides were shown to affect a variety of gut bacteria. Furthermore, onions and other *Allium* species have been demonstrated to protect against upper aerodigestive tract and gastrointestinal tract cancers [85]. Peanut sprout ethanolic extract at a purification of 80% (v/v) has been administered to loperamide-induced constipated SD rats, which revealed its laxative effects [86].

5.4. Bioavailability of Sprouts against Cardiovascular Diseases (CVDs)

In hypercholesterolemic Wistar rats, dietary supplementation with onion reversed high-cholesterol diet-induced changes in lipid mediators such as oxylipin and sphingolipid profiles [87]. Using an animal model to study increased blood pressure, researchers investigated the relationship between oxidative stress and a diet rich in broccoli sprouts with a high quantity of glucoraphanin. After 14 weeks, rats were fed broccoli sprouts that were either low in the chemical or rich in glucoraphanin. After the trial, they observed that rats fed a glucoraphanin-rich diet had lower blood pressure and less heart inflammation. According to the researchers, the benefits were attributed to better antioxidant defense systems and a decreased glucoraphanin-induced inflammatory response. Broccoli and broccoli sprouts contain different antioxidants (vitamin E, β-carotene, α-tocopherol, and ascorbic acid) that may aid in the prevention of cardiovascular diseases. In laboratory rats, the chemical glucoraphanin increased heart function, decreased inflammation, and boosted natural antioxidant defenses. When unstable molecules, called free radicals, react with oxygen in the body, they promote inflammation and cell death, raising the risk of heart disease and cancer. Antioxidants are supposed to help reduce oxidative stress in the body, preventing these detrimental consequences. Glucoraphanin is a chemical that boosts the body's antioxidant defenses by acting as an indirect antioxidant. It is naturally found in broccoli and broccoli sprouts [88].

Onions have been shown in trials to enhance lipid profiles and to prevent platelet aggregation, lowering the risk of heart disease. Onions and their bioactive components have been widely researched for their hypocholesterolemia effects in rats fed high-cholesterol or high-fat diets. Onion sprouts successfully reduced total cholesterol, triglyceride, and low-density lipoprotein cholesterol levels in hyperlipidemic rats [89]. Polyphenol-rich onion extract alleviated hyperlipidemia in Sprague-Dawley rats' livers by upregulating the low-density lipoprotein receptor (LDLR) and downregulating the 3-hydroxy-3-methylglutaryl (HMG)-CoA reductase (HMGCR). In addition, Lee et al. [90] found that quercetin-rich onion peel extract increased fecal cholesterol, reduced the atherogenic index, cardiac risk factor, and activation of LDLR and cholesterol 7-monooxygenase (CYP7A1) in high-cholesterol diet-fed mice, indicating that onion had a cholesterol-lowering effect via fecal excretion. They proved that fecal excretion of onions lowered cholesterol. When onions were added to a high-cholesterol diet supplied to rats, the bile acid levels in their stools changed. Dietary onion increased antioxidant enzyme activity and enhanced anti-inflammatory response and cardiovascular risk markers in rats fed a high-cholesterol diet [66]. An overview of vegetable sprouts' bioavailability against different diseases is given in Table 3.

Table 3. Bioavailability of sprouts against different diseases.

Vegetable Sprout Types	Study Design	Disease	Recovery	References
Cruciferous	Human (male and female)	Cancer	Cruciferous vegetables reduce the risk of cancer by decreasing the damage to DNA	[91]
Broccoli	–	–	The biological properties of broccoli are antioxidant, anticancer, anticancer, antimicrobial, anti-inflammatory, anti-obesity, and antidiabetic activities	[92]
Broccoli	Mice and rats	Alleviate pain	The broccoli sprouts have ability in pain therapy	[36]
Red cabbage, broccoli, Galega kale and Penca cabbage	,–	–	Different vegetables sprouts have antioxidant and anti-carcinogenic properties	[93]
Broccoli	Human	Cancer	Broccoli may reduce the risk of cancer by managing metabolism	[94]
Broccoli	Mice	Prostate tumorigenesis	Broccoli sprouts have significant inhibitory effects on prostate tumorigenesis.	[95]
Alfalfa	Mice	Inflammation	The study suggests that alfalfa supplementation can suppress the production of proinflammatory cytokines and alleviate acute inflammatory hazards.	[96]
Brussels	–	Cancer	Brussels sprouts have cancer preventive effects which may be due to a reduction in oxidative DNA damage	[97]
Spinach, kale, Brussels sprouts, mustard greens, green bell peppers, cabbage, and collards	Human	Binding of bile acids	The results show equal health-promoting potential of spinach, kale, brussels sprouts, mustard greens, green bell peppers, and collards, as indicated by their bile acid binding on dry matter basis	[98]

Table 3. *Cont.*

Vegetable Sprout Types	Study Design	Disease	Recovery	References
Brussels	Human	–	The results show that compounds in cooked and autolysed brussels sprouts can enhance lymphocyte resistance towards H2O2-induced DNA strand breaks in vitro	[99]
Radish, broccoli, leek, and beetroot	In vitro	Diabetic, obesity and cholinergic	Different vegetable sprouts can be used daily as superfoods or functional food	[33]
Turnip, cauliflower, and mustard	In vitro	Cancer	In vitro antiproliferative study supports that sprouts are a good source of anticancer agents	[100]
Broccoli	In vitro	Cancer	In vitro study indicates that broccoli sprouts can reduce prostate cancer	[101]

5.5. Bioavailability of Sprouts against Oxidative Stress-Related Diseases such as Cancer and Diabetes

In addition, secondary metabolites are abundantly present in sprouts, especially the glucosinolates (GLs), as in the case of the *Brassicaceae* family [102]. Gulcosinolates consist of an amino acid group and a thiohydroximate-O-sulfonate attached to the glucose unit [103]. Myrosinase acts to hydrolyse these GLs to thiocyanates and isothiocyanates [89] when the pH is between 6.0 and 7.0 [79], and then it yields anti-mutagenic activity, having a limiting effect on oxidative stress and playing a role in chemoprotection, especially in cancers and diabetes [104]. Glucoraphenin and glucobrassicin are the GLs excessively present in sprouted radish, which readily enhance antioxidant activity, and consequently decrease carcinogenesis in the body [105]. Kale sprouts do not have dehydroerucin but have a better GL profile as compared with sprouted radish due to gluconapoleiferin, glucoiberin, gluconasturtin, gluconapin, progoitrin, glucobrassicin, neoglucobrassicin, 4-hydroxyglucobrassicin, and sinigrin, which potentially reduce oxidative stress, and hence, decrease the risk of related diseases, i.e., diabetes, cancer, and heart diseases [94]. Taniguchi et al. [106] used Japanese radish sprouts in normal and streptozotin-induced diabetic mice to show the benefits of cruciferous sprouts on DM. It was shown that radish sprout consumption decreased plasma levels of fructosamine, glucose, and insulin, suggesting that the hypoglycemia brought on by radish sprout consumption may not be related to an increase in insulin synthesis but rather to enhanced sensitivity or an insulin-like action [107]. It depends on the ktype of sorghum, enzyme-inducing, and anti-proliferative capabilities. The most effective inducer of quinone oxidoreductase, a phase II detoxifying enzyme, has been shown to be an extract from black tea (non-tannin) that is abundant in 3-deoxyanthocyanins. Comparatively speaking, white sorghum extract has been shown to be a relatively potent inducer. Despite not inducing quinone oxidoreductase, tannin sorghum extracts have provided the most potent antiproliferative effects on human esophageal and colon cancer cells.

5.6. Bioavailability of Protein against Malnutrition

Sprouting enhances protein content, as evidenced from a study conducted by Devi et al. [95] on cowpea (lobia) by enhancing its bioavailability and digestibility. Sprouting is an interesting phenomenon that influences metabolic enzymes, especially proteinases, which increase the content of protein [96]. Sprouted chickpeas have more protein content than black gram. The sprouting process decreases the protease inhibitors and even enhances lipase activity, yielding increased content of fatty acids, and this also improves the digestibility of starches [97].

6. Sprout Vegetables as an Ingredient or Substitute for Meat Products

Dried sprouted food ingredients have been a trend for healthy/functional foods to live healthier, particularly by incorporating them into bread making flours or traditional beverages and juices [108]. With continued research on the influence of sprouted dietary feed on animals, it has revealed increased phytochemicals, particularly antioxidants, in the animals' meat, as well as enhanced fatty acid content, particularly when sprouted alfalfa and flax were fed to rabbits [54]. Meat product consumption to achieve protein requirements has increased, and currently, it is difficult to rely on just one livestock source. With advancements in in vitro meat technology, tissue culturing engineers have started to develop lab-grown meat [109].

7. Conclusions

It is concluded that sprouts have been introduced as a new food for some years. Vegetables are basically important plant-based foods. Vegetable sprouts are composed of bioactive compounds, including phenolic compounds, antioxidants, etc. These bioactive compounds are extracted from sprouts by using different conventional methods and new techniques. Plant protein content improvement and enhanced protein bioavailability and digestibility can set up a better opportunity to research and develop plant-based meat protein substitutes. Vegetable sprouts are being used to develop functional foods and they also play an important role in maintaining the stability of food products. Furthermore, pharmaceutically, they aid in the defense of different types of chronic disorders.

8. Future Prospective and Recommendations

Vegetable sprouts' potential involvement in the prevention and treatment of chronic diseases has to be investigated further. The high nutrient content of sprouts may provide extra lipid-lowering advantages. Sprouts are fiber-rich foods that are likely to provide a feeling of fullness. It is also crucial to remember that functional meals must be consumed often in order to offer their somewhat modest benefits. To check the probable positive effect of vegetable sprout foods on chronic diseases or risk factors related to lifestyle, a comprehensive scientific human study is required. The influence of the whole meal, which represents the synergistic effect between components, must be explored by conducting different studies on extracts and components. It is equally crucial to consider the makeup of the background diet because it might bias results and could create challenges in connecting the effects to the fitting dietary elements. The effect of vegetable sprout-based meals on chronic diseases or risk factors related to lifestyle must reflect the whole diet in order to apply the trial's findings in practice. Unfortunately, in vegetable sprout research, because metabolic changes and their link that may affect biological activity in the body after consumption have not been taken into consideration, it is difficult to characterize the direct antioxidant impact of vegetable sprouts. It is necessary to determine the safety of ingesting the quantities of vegetable sprout extracts utilized in these studies by dietary consumption of foods containing vegetable sprouts. However, information gained from many types of experimental studies has contributed to a broader understanding of how the vegetable sprout food matrix may be advantageous. Keeping in mind the gap between protein supply and demand, more time is needed to research and develop sprouted vegetable protein-based products, which would be a better approach because this source would provide a better choice of protein accompanied by phytochemicals.

Author Contributions: Conceptualization, W.K. and A.A. (Afifa Aziz); methodology, W.K.; software, H.K.; validation, A.A. (Afifa Aziz) and S.N.; formal analysis, W.K., A.A. (Anwar Ali) and S.N.; investigation, C.M.G.L., M.k.N. and A.A. (Anwar Ali); data curation, W.S.A., G.S., A.A.E. and S.N.; writing—original draft preparation, M.k.N., W.S.A., G.S., A.A.E. and W.K.; writing—review and editing, M.k.N., F.M., W.K., C.M.G.L. and H.K. supervision, W.K. and A.A. (Afifa Aziz); project administration, M.k.N.; funding acquisition, A.A.-F., W.S.A., G.S. and A.A.E. All authors have read and agreed to the published version of the manuscript.

Funding: This research received no external funding.

Data Availability Statement: Not available.

Acknowledgments: This research did not receive any specific grant from funding agencies in the public, commercial, or non-profit sectors.

Conflicts of Interest: The authors declare no conflict of interest.

References

1. Sikin, A.M.D.; Zoellner, C.; Rizvi, S.S.H. Current Intervention Strategies for the Microbial Safety of Sprouts. *J. Food Prot.* **2013**, *76*, 2099–2123. [CrossRef] [PubMed]
2. Zhang, C.; Zhao, Z.; Yang, G.; Shi, Y.; Zhang, Y.; Shi, C.; Xia, X. Effect of slightly acidic electrolyzed water on natural Enterobacteriaceae reduction and seed germination in the production of alfalfa sprouts. *Food Microbiol.* **2021**, *97*, 103414. [CrossRef] [PubMed]
3. Liu, H.K.; Kang, Y.F.; Zhao, X.Y.; Liu, Y.P.; Zhang, X.W.; Zhang, S.J. Effects of elicitation on bioactive compounds and biological activities of sprouts. *J. Funct. Foods* **2019**, *53*, 136–145. [CrossRef]
4. Chen, F.; Zhang, M.; Yang, C. hui Application of ultrasound technology in processing of ready-to-eat fresh food: A review. *Ultrason. Sonochem.* **2020**, *63*, 104953. [CrossRef] [PubMed]
5. Ki, H.H.; Poudel, B.; Lee, J.H.; Lee, Y.M.; Kim, D.K. In vitro and in vivo anti-cancer activity of dichloromethane fraction of Triticum aestivum sprouts. *Biomed. Pharmacother.* **2017**, *96*, 120–128. [CrossRef] [PubMed]
6. The Role of Sprouts in Human Nutrition. A Review. Available online: https://www.cabdirect.org/cabdirect/abstract/20113263994 (accessed on 23 July 2022).
7. Gawlik-Dziki, U.; Jezyna, M.; Świeca, M.; Dziki, D.; Baraniak, B.; Czyz, J. Effect of bioaccessibility of phenolic compounds on in vitro anticancer activity of broccoli sprouts. *Food Res. Int.* **2012**, *49*, 469–476. [CrossRef]
8. Kensler, T.W.; Ng, D.; Carmella, S.G.; Chen, M.; Jacobson, L.P.; Muñoz, A.; Egner, P.A.; Chen, J.G.; Qian, G.S.; Chen, T.Y.; et al. Modulation of the metabolism of airborne pollutants by glucoraphanin-rich and sulforaphane-rich broccoli sprout beverages in Qidong, China. *Carcinogenesis* **2012**, *33*, 101–107. [CrossRef]
9. Mendoza-Sánchez, M.; Pérez-Ramírez, I.F.; Wall-Medrano, A.; Martinez-Gonzalez, A.I.; Gallegos-Corona, M.A.; Reynoso-Camacho, R. Chemically induced common bean (*Phaseolus vulgaris* L.) sprouts ameliorate dyslipidemia by lipid intestinal absorption inhibition. *J. Funct. Foods* **2019**, *52*, 54–62. [CrossRef]
10. Nakamura, Y.; Kaihara, A.; Yoshii, K.; Tsumura, Y.; Ishimitsu, S.; Tonogai, Y. Content and Composition of Isoflavonoids in Mature or Immature Beans and Bean Sprouts Consumed in Japan. *J. Health Sci.* **2001**, *47*, 394–406. [CrossRef]
11. Teixeira-Guedes, C.I.; Oppolzer, D.; Barros, A.I.; Pereira-Wilson, C. Phenolic rich extracts from cowpea sprouts decrease cell proliferation and enhance 5-fluorouracil effect in human colorectal cancer cell lines. *J. Funct. Foods* **2019**, *60*, 103452. [CrossRef]
12. Tang, D.; Dong, Y.; Ren, H.; Li, L.; He, C. A review of phytochemistry, metabolite changes, and medicinal uses of the common food mung bean and its sprouts (Vigna radiata). *Chem. Cent. J.* **2014**, *8*, 1–9. [CrossRef] [PubMed]
13. Prakash, D.; Upadhyay, G.; Singh, B.N.; Singh, H.B. Antioxidant and free radical-scavenging activities of seeds and agri-wastes of some varieties of soybean (Glycine max). *Food Chem.* **2007**, *104*, 783–790. [CrossRef]
14. Sahai, V.; Kumar, V. Anti-diabetic, hepatoprotective and antioxidant potential of Brassica oleracea sprouts. *Biocatal. Agric. Biotechnol.* **2020**, *25*, 101623. [CrossRef]
15. Ho, C.Y.; Lin, Y.T.; Labbe, R.G.; Shetty, K. Inhibition of helicobacter pylori by phenolic extracts of sprouted peas (*Pisum sativum* L.). *J. Food Biochem.* **2006**, *30*, 21–34. [CrossRef]
16. Yanaka, A.; Fahey, J.W.; Fukumoto, A.; Nakayama, M.; Inoue, S.; Zhang, S.; Tauchi, M.; Suzuki, H.; Hyodo, I.; Yamamoto, M. Dietary Sulforaphane-Rich Broccoli Sprouts Reduce Colonization and Attenuate Gastritis in Helicobacter pylori–Infected Mice and Humans. *Cancer Prev. Res.* **2009**, *2*, 353–360. [CrossRef]
17. Kim, W.I.; Choi, S.Y.; Han, I.; Cho, S.K.; Lee, Y.; Kim, S.; Kang, B.; Choi, O.; Kim, J. Inhibition of Salmonella enterica growth by competitive exclusion during early alfalfa sprout development using a seed-dwelling Erwinia persicina strain EUS78. *Int. J. Food Microbiol.* **2020**, *312*, 108374. [CrossRef]
18. Teodoro, A.J. Bioactive compounds of food: Their role in the prevention and treatment of diseases. *Oxid. Med. Cell. Longev.* **2019**, *2019*. [CrossRef]
19. Uwineza, P.A.; Waśkiewicz, A. Recent Advances in Supercritical Fluid Extraction of Natural Bioactive Compounds from Natural Plant Materials. *Molecules* **2020**, *25*, 3847. [CrossRef]
20. Alvarez-Jubete, L.; Wijngaard, H.; Arendt, E.K.; Gallagher, E. Polyphenol composition and in vitro antioxidant activity of amaranth, quinoa buckwheat and wheat as affected by sprouting and baking. *Food Chem.* **2010**, *119*, 770–778. [CrossRef]
21. Erba, D.; Angelino, D.; Marti, A.; Manini, F.; Faoro, F.; Morreale, F.; Pellegrini, N.; Casiraghi, M.C. Effect of sprouting on nutritional quality of pulses. *Int. J. Food Sci. Nutr.* **2019**, *70*, 30–40. [CrossRef]
22. Costa, D.C.; Costa, H.S.; Albuquerque, T.G.; Ramos, F.; Castilho, M.C.; Sanches-Silva, A. Advances in phenolic compounds analysis of aromatic plants and their potential applications. *Trends Food Sci. Technol.* **2015**, *45*, 336–354. [CrossRef]
23. Gernand, A.D.; Schulze, K.J.; Stewart, C.P.; West, K.P.; Christian, P. Micronutrient deficiencies in pregnancy worldwide: Health effects and prevention. *Nat. Rev. Endocrinol.* **2016**, *12*, 274–289. [CrossRef]

24. Tucker, K.L. Nutrient intake, nutritional status, and cognitive function with aging. *Ann. N. Y. Acad. Sci.* **2016**, *1367*, 38–49. [CrossRef] [PubMed]
25. Arrigoni, O.; De Tullio, M.C. Ascorbic acid: Much more than just an antioxidant. *Biochim. Biophys. Acta Gen. Subj.* **2002**, *1569*, 1–9. [CrossRef]
26. Poiroux-Gonord, F.; Bidel, L.P.R.; Fanciullino, A.L.; Gautier, H.; Lauri-Lopez, F.; Urban, L. Health Benefits of Vitamins and Secondary Metabolites of Fruits and Vegetables and Prospects To Increase Their Concentrations by Agronomic Approaches. *J. Agric. Food Chem.* **2010**, *58*, 12065–12082. [CrossRef]
27. Godswill, A.G.; Somtochukwu, I.V.; Ikechukwu, A.O.; Kate, E.C. Health Benefits of Micronutrients (Vitamins and Minerals) and their Associated Deficiency Diseases: A Systematic Review. *Int. J. Food Sci.* **2020**, *3*, 1–32. [CrossRef]
28. Matsuo, T.; Asano, T.; Mizuno, Y.; Sato, S.; Fujino, I.; Sadzuka, Y. Water spinach and okra sprouts inhibit cancer cell proliferation. *Vitr. Cell. Dev. Biol. Anim.* **2022**, *58*, 79–84. [CrossRef] [PubMed]
29. Le, L.; Gong, X.; An, Q.; Xiang, D.; Zou, L.; Peng, L.; Wu, X.; Tan, M.; Nie, Z.; Wu, Q.; et al. Quinoa sprouts as potential vegetable source: Nutrient composition and functional contents of different quinoa sprout varieties. *Food Chem.* **2021**, *357*, 129752. [CrossRef]
30. Ebert, A.W. Sprouts and Microgreens—Novel Food Sources for Healthy Diets. *Plants* **2022**, *11*, 571. [CrossRef]
31. Liu, Z.; Shi, J.; Wan, J.; Pham, Q.; Zhang, Z.; Sun, J.; Yu, L.; Luo, Y.; Wang, T.T.Y.; Chen, P. Profiling of Polyphenols and Glucosinolates in Kale and Broccoli Microgreens Grown under Chamber and Windowsill Conditions by Ultrahigh-Performance Liquid Chromatography High-Resolution Mass Spectrometry. *ACS Food Sci. Technol.* **2022**, *2*, 101–113. [CrossRef]
32. Choque-Quispe, D.; Ligarda-Samanez, C.A.; Ramos-Pacheco, B.S.; Leguía-Damiano, S.; Calla-Florez, M.; Zamalloa-Puma, L.M.; Colque-Condeña, L. Phenolic Compounds, Antioxidant Capacity, and Protein Content of Three Varieties of Germinated Quinoa (Chenopodium quinoa Willd). *Ing. Investig.* **2021**, *41*, 1–7. [CrossRef]
33. Wojdyło, A.; Nowicka, P.; Tkacz, K.; Turkiewicz, I.P. Sprouts vs. Microgreens as Novel Functional Foods: Variation of Nutritional and Phytochemical Profiles and Their In vitro Bioactive Properties. *Molecules* **2020**, *25*, 4648. [CrossRef] [PubMed]
34. Abdel-Aty, A.M.; Salama, W.H.; Fahmy, A.S.; Mohamed, S.A. Impact of germination on antioxidant capacity of garden cress: New calculation for determination of total antioxidant activity. *Sci. Hortic.* **2019**, *246*, 155–160. [CrossRef]
35. Majid, I.; Hussain, S.; Nanda, V.; Jabeen, F.; Mehmood Abbasi, A.; Alkahtani, J.; Soliman Elshikh, M.; Azhar Khan, M.; Usmani, S.; Javed Ansari, M. Changes in major flavonols and quercetin glycosides upon sprouting in onion cultivars. *J. King Saud Univ. Sci.* **2021**, *33*, 101222. [CrossRef]
36. Baenas, N.; Gómez-Jodar, I.; Moreno, D.A.; García-Viguera, C.; Periago, P.M. Broccoli and radish sprouts are safe and rich in bioactive phytochemicals. *Postharvest Biol. Technol.* **2017**, *127*, 60–67. [CrossRef]
37. Almuhayawi, S.M.; Almuhayawi, M.S.; Al Jaouni, S.K.; Selim, S.; Hassan, A.H.A. Effect of Laser Light on Growth, Physiology, Accumulation of Phytochemicals, and Biological Activities of Sprouts of Three Brassica Cultivars. *J. Agric. Food Chem.* **2021**, *69*, 6240–6250. [CrossRef]
38. Maoloni, A.; Milanović, V.; Osimani, A.; Cardinali, F.; Garofalo, C.; Belleggia, L.; Foligni, R.; Mannozzi, C.; Mozzon, M.; Cirlini, M.; et al. Exploitation of sea fennel (Crithmum maritimum L.) for manufacturing of novel high-value fermented preserves. *Food Bioprod. Process.* **2021**, *127*, 174–197. [CrossRef]
39. Baenas, N.; Moreno, D.A.; García-Viguera, C. Selecting sprouts of Brassicaceae for optimum phytochemical composition. *J. Agric. Food Chem.* **2012**, *60*, 11409–11420. [CrossRef]
40. Yudiono, K.; Kurniawati, L. Effect of sprouting on anthocyanin, antioxidant activity, color intensity and color attributes in purple sweet potatoes. *Food Res.* **2018**, *2*, 171–176.
41. Miyahira, R.F.; Lopes, J.D.; Antunes, A.E. The Use of Sprouts to Improve the Nutritional Value of Food Products: A Brief Review. *Plant Foods Hum. Nutr.* **2021**, *76*, 143–152. [CrossRef]
42. Azmir, J.; Zaidul, I.S.M.; Rahman, M.M.; Sharif, K.M.; Mohamed, A.; Sahena, F.; Jahurul, M.H.A.; Ghafoor, K.; Norulaini, N.A.N.; Omar, A.K.M. Techniques for extraction of bioactive compounds from plant materials: A review. *J. Food Eng.* **2013**, *117*, 426–436. [CrossRef]
43. Blicharski, T.; Oniszczuk, A. Extraction Methods for the Isolation of Isoflavonoids from Plant Material. *Open Chem.* **2017**, *15*, 34–45. [CrossRef]
44. Lang, Q.; Wai, C.M. Supercritical fluid extraction in herbal and natural product studies—A practical review. *Talanta* **2001**, *53*, 771–782. [CrossRef]
45. Shang, X.; Dou, Y.; Zhang, Y.; Tan, J.N.; Liu, X.; Zhang, Z. Tailor-made natural deep eutectic solvents for green extraction of isoflavones from chickpea (*Cicer arietinum* L.) sprouts. *Ind. Crops Prod.* **2019**, *140*, 111724. [CrossRef]
46. Koraqi, H.; Qazimi, B.; Çesko, C.; Trajkovska Petkoska, A. Environmentally Friendly Extraction of BioactiveCompounds from Rosa canina L. fruits Using DeepEutectic Solvent (DES) as Green Extraction Media. *Acta Chim. Slov.* **2022**, *69*, 3. [CrossRef] [PubMed]
47. Le, T.N.; Chiu, C.H.; Hsieh, P.C. Bioactive Compounds and Bioactivities of Brassica oleracea L. var. Italica Sprouts and Microgreens: An Updated Overview from a Nutraceutical Perspective. *Plants* **2020**, *9*, 946. [CrossRef]
48. Li, T.; Luo, L.; Kim, S.; Moon, S.K.; Moon, B.K. Trans-resveratrol extraction from peanut sprouts cultivated using fermented sawdust medium and its antioxidant activity. *J. Food Sci.* **2020**, *85*, 639–646. [CrossRef]
49. Fernandes, S.R.; Barreiros, L.; Oliveira, R.F.; Cruz, A.; Prudêncio, C.; Oliveira, A.I.; Pinho, C.; Santos, N.; Morgado, J. Chemistry, bioactivities, extraction and analysis of azadirachtin: State-of-the-art. *Fitoterapia* **2019**, *134*, 141–150. [CrossRef]

50. Azwanida, N.N. A review on the extraction methods use in medicinal plants, principle, strength and limitation. *Med. Aromat. Plants* **2015**, *4*, 2167-0412.
51. Kandar, P. Phytochemicals and biopesticides: Development, current challenges and effects on human health and diseases. *J. Biomed. Res.* **2021**, *2*, 3–15.
52. Das, A.; Sharangi, A.B. Postharvest Care of Medicinal and Aromatic Plants: A Reservoir of Many Health Benefiting Constituents. In *Medicinal Plants*; Apple Academic Press: New York, NY, USA, 2022; pp. 387–407.
53. Tzanova, M.; Atanasov, V.; Yaneva, Z.; Ivanova, D.; Dinev, T. Selectivity of current extraction techniques for flavonoids from plant materials. *Processes* **2020**, *8*, 1222. [CrossRef]
54. Moura, P.M.; Prado, G.H.; Meireles, M.A.; Pereira, C.G. Supercritical fluid extraction from guava (*Psidium guajava*) leaves: Global yield, composition and kinetic data. *J. Supercrit. Fluids* **2012**, *62*, 116–122. [CrossRef]
55. Moyo, S.M. *Effects of Cooking and Drying on the Phenolic Compounds, Antioxidant Activity and Antibacterial Activity of Cleome gynandra (Spider Plant)*; University of Johannesburg: Johannesburg, South Africa, 2016.
56. Vongsak, B.; Sithisarn, P.; Mangmool, S.; Thongpraditchote, S.; Wongkrajang, Y.; Gritsanapan, W. Maximizing total phenolics, total flavonoids contents and antioxidant activity of Moringa oleifera leaf extract by the appropriate extraction method. *Ind. Crops Prod.* **2013**, *44*, 566–571. [CrossRef]
57. Ding, J.; Feng, H. Controlled germination for enhancing the nutritional value of sprouted grains. *Sprouted Grains Nutr. Value Prod. Appl.* **2019**, 91–112.
58. Finnie, S.; Brovelli, V.; Nelson, D. Sprouted grains as a food ingredient. *Sprouted Grains Nutr. Value Prod. Appl.* **2019**, 113–142.
59. Omary, M.B.; Fong, C.; Rothschild, J.; Finney, P. REVIEW: Effects of Germination on the Nutritional Profile of Gluten-Free Cereals and Pseudocereals: A Review. *Cereal Chem.* **2012**, *89*, 1–14. [CrossRef]
60. Liu, T.; Hou, G.G.; Cardin, M.; Marquart, L.; Dubat, A. Quality attributes of whole-wheat flour tortillas with sprouted whole-wheat flour substitution. *LWT* **2017**, *77*, 1–7. [CrossRef]
61. Physiological Characteristics and Manufacturing of the Processing Products of Sprout Vegetables-Korean Journal of Food and Cookery Science | Korea Science. Available online: https://koreascience.kr/article/JAKO201025665646714.page (accessed on 24 July 2022).
62. Dal Bosco, A.; Castellini, C.; Martino, M.; Mattioli, S.; Marconi, O.; Sileoni, V.; Ruggeri, S.; Tei, F.; Benincasa, P. The effect of dietary alfalfa and flax sprouts on rabbit meat antioxidant content, lipid oxidation and fatty acid composition. *Meat Sci.* **2015**, *106*, 31–37. [CrossRef]
63. Gawlik-Dziki, U.; Dziki, D.; Pietrzak, W.; Nowak, R. Phenolic acids prolife and antioxidant properties of bread enriched with sprouted wheat flour. *J. Food Biochem.* **2017**, *41*, e12386. [CrossRef]
64. Charoenthaikij, P.; Jangchud, K.; Jangchud, A.; Prinyawiwatkul, W.; Tungtrakul, P. Germination Conditions Affect Selected Quality of Composite Wheat-Germinated Brown Rice Flour and Bread Formulations. *J. Food Sci.* **2010**, *75*, S312–S318. [CrossRef]
65. Bello, C.; Maldini, M.; Baima, S.; Scaccini, C.; Natella, F. Glucoraphanin and sulforaphane evolution during juice preparation from broccoli sprouts. *Food Chem.* **2018**, *268*, 249–256. [CrossRef] [PubMed]
66. Smith, T.K.; Lund, E.K.; Clarke, R.G.; Bennett, R.N.; Johnson, I.T. Effects of Brussels Sprout Juice on the Cell Cycle and Adhesion of Human Colorectal Carcinoma Cells (HT29) In Vitro. *J. Agric. Food Chem.* **2005**, *53*, 3895–3901. [CrossRef] [PubMed]
67. Fahey, J.W.; Zhang, Y.; Talalay, P. Broccoli sprouts: An exceptionally rich source of inducers of enzymes that protect against chemical carcinogens. *Proc. Natl. Acad. Sci. USA* **1997**, *94*, 10367–10372. [CrossRef] [PubMed]
68. Marshall, K.L.; Farah, M. Axonal regeneration and sprouting as a potential therapeutic target for nervous system disorders. *Neural Regen. Res.* **2021**, *16*, 1901–1910.
69. Okumura, K.; Hosoya, T.; Kawarazaki, K.; Izawa, N.; Kumazawa, S. Antioxidant activity of phenolic compounds from fava bean sprouts. *J. Food Sci.* **2016**, *81*, C1394–C1398. [CrossRef]
70. Rebollo-Hernanz, M.; Aguilera, Y.; Herrera, T.; Cayuelas, L.T.; Dueñas, M.; Rodríguez-Rodríguez, P.; Martín-Cabrejas, M.A. Bioavailability of melatonin from lentil sprouts and its role in the plasmatic antioxidant status in rats. *Foods* **2020**, *9*, 330. [CrossRef] [PubMed]
71. Malone, A.; Hamilton, C. The Academy of Nutrition and Dietetics/the American Society for Parenteral and Enteral Nutrition consensus malnutrition characteristics application in practice. *Nutr. Clin. Pract.* **2013**, *28*, 639–650. [CrossRef]
72. Petrov, D.; Mansfield, C.; Moussy, A.; Hermine, O. ALS clinical trials review: 20 years of failure. Are we any closer to registering a new treatment? *Front. Aging Neurosci.* **2017**, *9*, 68. [CrossRef]
73. Masci, A.; Mattioli, R.; Costantino, P.; Baima, S.; Morelli, G.; Punzi, P.; Mosca, L. Neuroprotective effect of brassica oleracea sprouts crude juice in a cellular model of alzheimer's disease. *Oxid. Medi. Cellular Long.* **2015**, *2015*, 781938.
74. Sun, Y.; Yang, T.; Mao, L.; Zhang, F. Sulforaphane protects against brain diseases: Roles of cytoprotective enzymes. *Austin J. Cereb. Dis. Stroke* **2017**, *4*, 1054.
75. Goyal, M.R.; Suleria, H.; Harikrishnan, R. The Role of Herbal Medicines in Female Genital Infections. In *The Role of Phytoconstitutents in Health Care*; Apple Academic Press: New York, NY, USA, 2020; pp. 191–214.
76. Zhao, X.X.; Lin, F.J.; Li, H.; Li, H.B.; Wu, D.T.; Geng, F.; Ma, W.; Wang, Y.; Miao, B.H.; Gan, R.Y. Recent Advances in Bioactive Compounds, Health Functions, and Safety Concerns of Onion (*Allium cepa* L.). *Front. Nutr.* **2021**, *8*, 463. [CrossRef] [PubMed]
77. Koza, L.A.; Winter, A.N.; Holsopple, J.; Baybayon-Grandgeorge, A.N.; Pena, C.; Olson, J.R.; Mazzarino, R.C.; Patterson, D.; Linseman, D.A. Protocatechuic acid extends survival, improves motor function, diminishes gliosis, and sustains neuromuscular junctions in the hSOD1G93A mouse model of amyotrophic lateral sclerosis. *Nutrients* **2020**, *12*, 1824. [CrossRef]

78. Barber, S.C.; Higginbottom, A.; Mead, R.J.; Barber, S.; Shaw, P.J. An in vitro screening cascade to identify neuroprotective antioxidants in ALS. *Free Rad. Bio. Med.* **2009**, *46*, 1127–1138. [CrossRef]
79. Gawlik-Dziki, U.; Świeca, M.; Dziki, D.; Sęczyk, Ł.; Złotek, U.; Rózyło, R.; Kaszuba, K.; Ryszawy, D.; Czyz, J. Anticancer and antioxidant activity of bread enriched with broccoli sprouts. *Biomed. Res. Int.* **2014**, *2014*, 608053. [CrossRef] [PubMed]
80. Khalid, W.; Arshad, M.S.; Jabeen, A.; Muhammad Anjum, F.; Qaisrani, T.B.; Suleria, H.A.R. Fiber-enriched botanicals: A therapeutic tool against certain metabolic ailments. *Food Sci. Nut.* **2022**, *10*. [CrossRef] [PubMed]
81. Appeldoorn, M.M.; Vincken, J.P.; Gruppen, H.; Hollman, P.C.H. Procyanidin Dimers A1, A2, and B2 Are Absorbed without Conjugation or Methylation from the Small Intestine of Rats. *J. Nutr.* **2009**, *139*, 1469–1473. [CrossRef] [PubMed]
82. Cardona, F.; Andres-Lacueva, C.; Tulipani, S.; Tinahones, F.J.; Queipo-Ortuño, M.I. Benefits of polyphenols on gut microbiota and implications in human health. *J. Nutr. Biochem.* **2013**, *24*, 1415–1422. [CrossRef]
83. Kawabata, K.; Yoshioka, Y.; Terao, J. Role of Intestinal Microbiota in the Bioavailability and Physiological Functions of Dietary Polyphenols. *Molecules* **2019**, *24*, 370. [CrossRef]
84. Majid, I.; Dhatt, A.S.; Sharma, S.; Nayik, G.A.; Nanda, V. Effect of sprouting on physicochemical, antioxidant and flavonoid profile of onion varieties. *Int. J. Food Sci. Technol.* **2016**, *51*, 317–324. [CrossRef]
85. Müller, L.; Meyer, M.; Bauer, R.N.; Zhou, H.; Zhang, H.; Jones, S.; Robinette, C.; Noah, T.L.; Jaspers, I. Effect of Broccoli Sprouts and Live Attenuated Influenza Virus on Peripheral Blood Natural Killer Cells: A Randomized, Double-Blind Study. *PLoS ONE* **2016**, *11*, e0147742. [CrossRef]
86. Gill, C.I.R.; Haldar, S.; Porter, S.; Matthews, S.; Sullivan, S.; Coulter, J.; McGlynn, H.; Rowland, I. The Effect of Cruciferous and Leguminous Sprouts on Genotoxicity, In Vitro and In Vivo. *Cancer Epidemiol. Biomark. Prev.* **2004**, *13*, 1199–1205. [CrossRef]
87. Vale, A.P.; Santos, J.; Brito, N.V.; Fernandes, D.; Rosa, E.; Beatriz, M.; Oliveira, P.P. Evaluating the impact of sprouting conditions on the glucosinolate content of Brassica oleracea sprouts. *Phytochemistry* **2015**, *115*, 252–260. [CrossRef]
88. Clarke, J.D.; Hsu, A.; Riedl, K.; Bella, D.; Schwartz, S.J.; Stevens, J.F.; Ho, E. Bioavailability and inter-conversion of sulforaphane and erucin in human subjects consuming broccoli sprouts or broccoli supplement in a cross-over study design. *Pharmacol. Res.* **2011**, *64*, 456–463. [CrossRef] [PubMed]
89. Keum, Y.S.; Oo Khor, T.; Lin, W.; Shen, G.; Han Kwon, K.; Barve, A.; Li, W.; Kong, A.N. Pharmacokinetics and pharmacodynamics of broccoli sprouts on the suppression of prostate cancer in transgenic adenocarcinoma of mouse prostate (TRAMP) mice: Implication of induction of Nrf2, HO-1 and apoptosis and the suppression of Akt-dependent kinase pathway. *Pharm. Res.* **2009**, *26*, 2324–2331. [PubMed]
90. Lee, S.G.; Parks, J.S.; Kang, H.W. Quercetin, a functional compound of onion peel, remodels white adipocytes to brown-like adipocytes. *J. Nutr. Biochem.* **2017**, *42*, 62–71. [CrossRef] [PubMed]
91. Higdon, J.V.; Delage, B.; Williams, D.E.; Dashwood, R.H. Cruciferous vegetables and human cancer risk: Epidemiologic evidence and mechanistic basis. *Pharmacol. Res.* **2007**, *55*, 224–236. [CrossRef] [PubMed]
92. Zhu, C.; Poulsen, H.E.; Loft, S. Inhibition of oxidative DNA damage in vitro by extracts of brussels sprouts. *Free Radic. Res.* **2000**, *33*, 187–196. [CrossRef]
93. Kahlon, T.S.; Chapman, M.H.; Smith, G.E. In vitro binding of bile acids by spinach, kale, brussels sprouts, broccoli, mustard greens, green bell pepper, cabbage and collards. *Food Chem.* **2007**, *100*, 1531–1536. [CrossRef]
94. Zhu, C.Y.; Loft, S. Effects of Brussels sprouts extracts on hydrogen peroxide-induced DNA strand breaks in human lymphocytes. *Food Chem. Toxicol.* **2001**, *39*, 1191–1197. [CrossRef]
95. Chaudhary, A.; Choudhary, S.; Sharma, U.; Vig, A.P.; Arora, S. In vitro Evaluation of Brassica sprouts for its Antioxidant and Antiproliferative Potential. *Indian J. Pharm. Sci.* **2016**, *78*, 615–623. [CrossRef]
96. Hong, Y.H.; Chao, W.W.; Chen, M.L.; Lin, B.F. Ethyl acetate extracts of alfalfa (Medicago sativa L.) sprouts inhibit lipopolysaccharide-induced inflammation in vitro and in vivo. *J. Biomed. Sci.* **2009**, *16*, 1–12. [CrossRef] [PubMed]
97. Abellán, A.; Domínguez-Perles, R.; Moreno, D.A.; García-Viguera, C. Sorting out the Value of Cruciferous Sprouts as Sources of Bioactive Compounds for Nutrition and Health. *Nutrients* **2019**, *11*, 429. [CrossRef]
98. Barba, F.J.; Nikmaram, N.; Roohinejad, S.; Khelfa, A.; Zhu, Z.; Koubaa, M. Bioavailability of Glucosinolates and Their Breakdown Products: Impact of Processing. *Front. Nutr.* **2016**, *3*, 24. [CrossRef] [PubMed]
99. Wagner, A.E.; Terschluesen, A.M.; Rimbach, G. Health promoting effects of brassica-derived phytochemicals: From chemopreventive and anti-inflammatory activities to epigenetic regulation. *Oxid. Med. Cell. Longev.* **2013**, *2013*, 964539. [CrossRef] [PubMed]
100. Xu, Y.; Szép, S.; Lu, Z. The antioxidant role of thiocyanate in the pathogenesis of cystic fibrosis and other inflammation-related diseases. *Proc. Natl. Acad. Sci. USA* **2009**, *106*, 20515. [CrossRef]
101. Hayes, J.D.; Kelleher, M.O.; Eggleston, I.M. The cancer chemopreventive actions of phytochemicals derived from glucosinolates. *Eur. J. Nutr.* **2008**, *47* (Suppl. S2), 73–88. [CrossRef]
102. Li, R.; Zhu, Y. The primary active components, antioxidant properties, and differential metabolite profiles of radish sprouts (Raphanus sativus L.) upon domestic storage: Analysis of nutritional quality. *J. Sci. Food Agric.* **2018**, *98*, 5853–5860. [CrossRef] [PubMed]
103. Jeon, J.; Kim, J.K.; Kim, H.R.; Kim, Y.J.; Park, Y.J.; Kim, S.J.; Kim, C.; Park, S.U. Transcriptome analysis and metabolic profiling of green and red kale (Brassica oleracea var. acephala) seedlings. *Food Chem.* **2018**, *241*, 7–13. [CrossRef]
104. Devi, C.B.; Kushwaha, A.; Kumar, A. Sprouting characteristics and associated changes in nutritional composition of cowpea (Vigna unguiculata). *J. Food Sci. Technol.* **2015**, *52*, 6821. [CrossRef]

105. Gulewicz, P.; Martínez-Villaluenga, C.; Frias, J.; Ciesiołka, D.; Gulewicz, K.; Vidal-Valverde, C. Effect of germination on the protein fraction composition of different lupin seeds. *Food Chem.* **2008**, *107*, 830–844. [CrossRef]
106. Taniguchi, H.; Kobayashi-Hattori, K.; Tenmyo, C.; Kamei, T.; Uda, Y.; Sugita-Konishi, Y.; Takita, T. Effect of Japanese radish (Raphanus sativus) sprout (Kaiware-daikon) on carbohydrate and lipid metabolisms in normal and streptozotocin-induced diabetic rats. *Phytother. Res. Int. J. Devoted Pharmacol. Toxicol. Eval. Natural Product Deri.* **2006**, *20*, 274–278. [CrossRef] [PubMed]
107. Dipnaik, K.; Bathere, D. Effect of soaking and sprouting on protein content and transaminase activity in pulses. *Int. J. Res. Med. Sci.* **2017**, *5*, 4271–4276. [CrossRef]
108. Benincasa, P.; Falcinelli, B.; Lutts, S.; Stagnari, F.; Galieni, A. Sprouted Grains: A Comprehensive Review. *Nutrients* **2019**, *11*, 421. [CrossRef] [PubMed]
109. Bhat, Z.F.; Kumar, S.; Bhat, H.F. In vitro meat: A future animal-free harvest. *Crit. Rev. Food Sci. Nutr.* **2017**, *57*, 782–789. [CrossRef] [PubMed]

molecules

Article

In Vitro Digestion Assessment (Standard vs. Older Adult Model) on Antioxidant Properties and Mineral Bioaccessibility of Fermented Dried Lentils and Quinoa

Janaina Sánchez-García [1], Sara Muñoz-Pina [1,*], Jorge García-Hernández [2,*], Amparo Tárrega [3], Ana Heredia [1] and Ana Andrés [1]

1 Instituto Universitario de Ingeniería de Alimentos (FoodUPV), Universitat Politècnica de València, Camino de Vera s/n, 46022 Valencia, Spain; jasanga7@doctor.upv.es (J.S.-G.); anhegu@tal.upv.es (A.H.); aandres@tal.upv.es (A.A.)
2 Centro Avanzado de Microbiología de Alimentos (CAMA), Universitat Politècnica de València, Camino de Vera s/n, 46022 Valencia, Spain
3 Instituto de Agroquímica y Tecnología de Alimentos (IATA-CSIC), Agustín Escardino 7, 46980 Valencia, Spain; atarrega@iata.csic.es
* Correspondence: samuopi@upvnet.upv.es (S.M.-P.); jorgarhe@btc.upv.es (J.G.-H.)

Abstract: The growing number of older adults necessitates tailored food options that accommodate the specific diseases and nutritional deficiencies linked with ageing. This study aims to investigate the influence of age-related digestive conditions in vitro on the phenolic profile, antioxidant activity, and bioaccessibility of minerals (Ca, Fe, and Mg) in two types of unfermented, fermented, and fermented dried quinoa and lentils. Solid-state fermentation, combined with drying at 70 °C, significantly boosted the total phenolic content in Castellana and Pardina lentils from 5.05 and 6.6 to 10.5 and 7.5 mg gallic acid/g dry weight, respectively, in the bioaccessible fraction following the standard digestion model, compared to the unfermented samples. The phenolic profile post-digestion revealed elevated levels of vanillic and caffeic acids in Castellana lentils, and vanillic acid in Pardina lentils, while caffeic acids in Castellana lentils were not detected in the bioaccessible fraction. The highest antioxidant potency composite index was observed in digested fermented dried Castellana lentils, with white quinoa samples exhibiting potency above 80%. Mineral bioaccessibility was greater in fermented and fermented dried samples compared to unfermented ones. Finally, the digestive changes that occur with ageing did not significantly affect mineral bioaccessibility, but compromised the phenolic profile and antioxidant activity.

Keywords: *Pleurotus ostreatus*; phenolic profile; antioxidant activity; total phenol content; phytic acid

Citation: Sánchez-García, J.; Muñoz-Pina, S.; García-Hernández, J.; Tárrega, A.; Heredia, A.; Andrés, A. In Vitro Digestion Assessment (Standard vs. Older Adult Model) on Antioxidant Properties and Mineral Bioaccessibility of Fermented Dried Lentils and Quinoa. *Molecules* **2023**, *28*, 7298. https://doi.org/10.3390/molecules28217298

Academic Editors: Sascha Rohn and Michał Halagarda

Received: 13 September 2023
Revised: 23 October 2023
Accepted: 26 October 2023
Published: 27 October 2023

1. Introduction

The ageing population is predominantly due to decreased fertility rates and increased life expectancy. It is projected that by 2050, the number of individuals over the age of 60 will reach approximately 2 billion, representing 22% of the global population, with the majority residing in developing nations [1]. Therefore, the forthcoming expansion of older adults' population will cause substantial increased demands for food products that are specifically formulated to meet their preferences and nutritional requirements. Ageing frequently causes digestive disorders related to changes in the oral cavity, including tooth loss and wearing dentures, gingivitis, and reduced saliva production. Furthermore, reduced sense of taste and smell can decrease food palatability and increase inappetence, leading to changes in the type and quantity of food consumed [2]. Gastric emptying slows down, and the gastric lipase and pepsin enzyme secretions are reduced, leading to an alkalisation of the gastric environment. Furthermore, peristalsis in the small intestine decreases, resulting in reduced secretion of pancreatic enzymes and bile salts [3–5]. These gastrointestinal tract alterations may contribute to age-related malnutrition, causing deficiencies in micronutrients, particularly minerals, which can lead to functional decline, fragility, and difficulty in maintaining independent living [6]. Furthermore, phytochemicals such as polyphenols may be considerable in chronic diseases, including cardiovascular disease, type II diabetes, cancer, osteoporosis, and neurodegenerative diseases [4]. Therefore, it is crucial to include antioxidants and minerals in one's diet to maintain brain function, support bone and teeth health, aid cellular and thyroid metabolism, and strengthen the immune system of older individuals [7–10].

In this scenario, assessing the impact of common gastrointestinal conditions in older adults on the digestibility of novel ingredients with enhanced antioxidant properties and improved digestibility is important. This assessment is crucial for creating new highly nutritious foods adapted to older adults. Therefore, grains and seeds, such as lentils or quinoa, could be considered good candidates as raw materials for developing protein-rich functional ingredients. The antioxidant activity of these plant materials is associated with a high content of phenolic compounds. Lentils have a higher reported total phenolic content (7.53 mg GAE/g sample) than other legumes, including peas, chickpeas, soybeans, red kidney, and black beans [11]. Quinoa has a total phenolic content (TPC) of 5.18 mg GAE/g sample [12], possessing antioxidant properties that are more effective than those of other cereals and pseudocereals, such as brown rice, millet, whole wheat, barley, oats, rye, Job's tears, corn, and amaranth [13]. Furthermore, some of the phenolic compounds present in lentils are flavonoids, including kaempferol glycosides, catechin/epicatechin glycosides, and procyanidins [14]. Phenolic acids, namely vanillic acid, ferulic acid, and their derivatives, and flavonoid compounds such as quercetin, kaempferol, and their glycosides, were found in quinoa [12,15]. Nevertheless, consumption of these compounds may not offer full health advantages because of factors such as antinutrients and limited digestibility. Solid-state fermentation (SSF), however, enhances the antioxidant properties and nutritional quality of diverse legumes and cereals. Thus, it is possible for the TPC of fermented plant materials to increase because of the release of phenolic compounds. These compounds are produced due to the structural breakdown of the cell wall after fungal colonisation, the action of ligninolytic and hydrolytic enzymes, or the synthesis of soluble phenolic compounds conducted by the fermentative micro-organism [16]. It is important to note that this increase is not subjective, but based on scientific evidence and observations. The variability in antioxidants altered by SSF relies on the binomial substrate–microorganism and process variables, precluding the generalisation of findings across studies. Furthermore, studies showed that the TPC of fermented black bean, kidney bean, and oat samples increased up to twice as much compared to unfermented digested samples after gastrointestinal digestion, mimicking the healthy adult digestion model [17].

The bioavailability of minerals in plant materials is relatively low due to certain molecules, such as phytates or phenols, forming complexes [18]. However, SSF was discovered to decrease phytates by endogenous phytase action, which is activated during fermentation. This leads to mineral release and increased bioavailability [19].

This study aims to analyse the effect of common age-related digestive conditions on the phenolic profile and antioxidant activity of the bioaccessible fraction together with the bioaccessibility of minerals (Ca, Fe, and Mg) of unfermented, fermented, and fermented dried (hot air drying or lyophilisation) quinoa (white and black) and lentils (Castellana and Pardina). Furthermore, all samples were subjected to in vitro digestion under healthy standard GI conditions for comparison.

2. Results and Discussion

2.1. Impact of GI Conditions on the Release of Phenols and Antioxidant Activity of Unfermented Fermented, and Fermented Dried Lentils and Quinoa

TPC and antioxidant activity changes during digestion were analysed and shown in Figure 1. In undigested samples, SSF and hot air drying at 70 °C resulted in an increase in TPC content in quinoa and lentils, compared to unfermented flours. The reasons behind this increase are elaborated in detail by Sánchez-García et al. [20,21]. During digestion, the gastrointestinal process induced a rise of free phenols in the bioaccessible fraction, regardless of the simulated conditions (standard or older adult) and the processing conducted to obtain flour from the substrates. However, this increase was more significant in lentils, as compared to quinoa, and in the fermented samples (FPL, FCL), compared to unfermented ones (UFPL, UFCL). Furthermore, samples fermented and dried at 70 °C (FPL-70, FCL-70) showed the highest TPC in the bioaccessible fraction. The optimal conditions for extracting phenols were pH, enzymatic activity, temperature, and stirring during the digestion process.

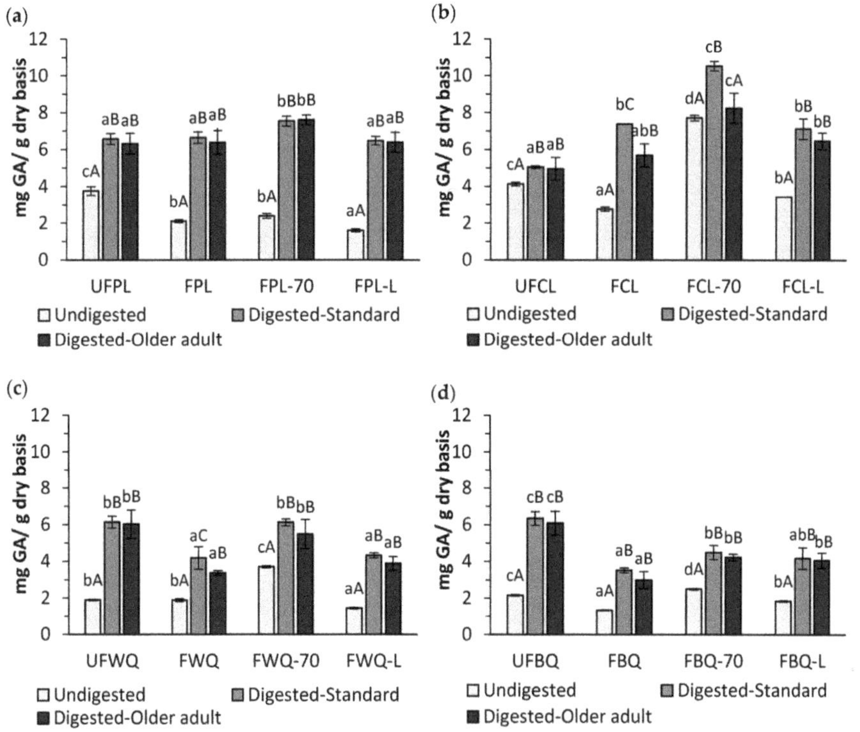

Figure 1. Total phenol content (mg gallic acid/g dry basis) in Pardina (**a**), Castellana lentil (**b**), white (**c**), and black quinoa (**d**) for unfermented flour (UFPL, UFCL, UFWQ, and UFBQ), fermented grain/seed (FPL, FCL, FWQ, and FBQ), fermented dried at 70 °C (FPL-70, FCL-70, FWQ-70, and FBQ-70), and fermented lyophilised (FPL-L, FCL-L, FWQ-L, and FBQ-L) flour obtained with a standard or older adult in vitro digestion model. [a,b,c,d] Different lowercase letters indicate significant differences ($p < 0.05$) between samples. [A,B,C] Different capital letters indicate significant differences ($p < 0.05$) between digestion models.

When TPC in the bioaccessible fractions was measured using two digestive models, the TPC in the fermented dried samples remained largely unaltered. Due to the common altered gastrointestinal conditions that appear with ageing, fermented dried lentils at 70 °C are especially interesting in terms of their bioaccessible TPC.

Tungmunnithum et al. [22] reported similar results in 10 bean varieties consumed in Thailand, with an increase in phenolic and flavonoid content associated with digestion. Phenol and flavonoid content rose with digestion but decreased during bean cooking. After gastrointestinal digestion, the TPC and total flavonoid content increased between 9% and 190%, and 4% and 266%, respectively, across different varieties. Physiological factors, such as digestive enzymes, bile salts, and pH, play a crucial role in the release of these compounds. Certain phenolic compounds are not present in free form in grains or seeds but are bound to the cell wall, creating macromolecular complexes. In addition to gastric digestion, a low pH causes an increase in polyphenols in their undissociated form, facilitating their release from the food matrix into the aqueous phase [23,24]. However, the links between phenolic compounds and carbohydrates are reduced during intestinal digestion by the action of pancreatic enzymes, bile salts, and a neutral pH (6.9) [23,25].

The phenolic fraction's chromatographic analysis presented distinct profiles among the bioaccessible fractions based on both flour variety and the digestive model, as shown in Tables 1–4. Previously published studies [20,21] also performed the same chromatographic analysis on all samples before digestion. Furthermore, the chromatograms of the phenolic profile corresponding to unfermented, fermented, and fermented dried Pardina lentil samples after gastrointestinal digestion, both under the healthy adult (standard) and the older adult digestion model, can be found in the Supplementary Materials (Figures S1–S8). Upon comparison of substrates after in vitro digestion, the bioaccessible fractions of Castellana lentils demonstrated a greater abundance of vanillic and caffeic acids, whereas Pardina exhibited a great abundance of 4-O-caffeoylquinic and vanillic acids. White and black quinoa boasted higher amounts of gallic and vanillic acids, as well as quercitrin, compared to lentils. However, the quantities of these compounds differed depending on the treatment undergone by the flours. The bioaccessible portion of the unfermented flours contained lower levels of these compounds compared to their fermented counterparts, particularly those exposed to hot air drying. Vanillic and caffeic acids demonstrated a greater increase in the Castellana lentil, rising from 6.2 to 20 μg/g dry basis and from 3.4 to 10.8 μg/g dry basis, respectively. In contrast, vanillic acid increased from 7.6 to 20.7 μg/g dry basis in Pardina lentils. Furthermore, an increase in gallic acid was observed in both white and black quinoa samples fermented and dried at 70 °C. The increase in gallic acid was apparent and rose from 20 to 139 μg/g dry basis in white quinoa and from 30 to 42 μg/g dry basis in black quinoa. Consequently, SSF plus drying facilitates the liberation of specific phenolic acids and flavonoids, resulting in their incorporation into the water-soluble bioaccessible fraction during gastrointestinal digestion.

Table 1. Phenolic content (μg/g dry basis) in digested Pardina lentil for unfermented flour (UFPL), fermented grain (FPL), fermented dried at 70 °C (FPL-70), and fermented lyophilised (FPL-L) flour.

	Digested (Standard)				**Digested (Older Adult)**			
	UFPL	**FPL**	**FPL-70**	**FPL-L**	**UFPL**	**FPL**	**FPL-70**	**FPL-L**
Phenolic acids								
Gallic acid	n.d.	n.d.	n.d.	n.d.	n.d.	n.d.	n.d.	n.d.
Caffeic acid	n.d.	n.d.	n.d.	n.d.	n.d.	n.d.	n.d.	n.d.
p-Coumaric acid	4.5 ± 0.5 ^{cB}	2.87 ± 0.08 ^{bB}	3.19 ± 0.19 ^{bA}	2.14 ± 0.07 ^{aA}	2.2 ± 0.2 ^{aA}	2.33 ± 0.04 ^{aA}	3.02 ± 0.18 ^{bA}	2.0 ± 0.3 ^{aA}
Sinapic acid	n.d.	n.d.	n.d.	n.d.	n.d.	n.d.	n.d.	n.d.
4-O-Caffeoylquinic	88 ± 31 ^{bA}	49.5 ± 0.2 ^{aB}	42 ± 3 ^{aB}	60.6 ± 0.3 ^{bB}	86 ± 7 ^{cA}	34 ± 5 ^{aA}	32 ± 3 ^{aA}	46.9 ± 0.3 ^{bA}
4-Hydroxybenzoic acid	4.4 ± 0.2 ^{aA}	4.14 ± 0.15 ^{aB}	4.0 ± 0.5 ^{aB}	3.88 ± 0.13 ^{aB}	3.3 ± 0.3 ^{bA}	3.375 ± 0.010 ^{bA}	2.1 ± 0.3 ^{aA}	2.90 ± 0.17 ^{bA}
Vanillic acid	7.6 ± 0.5 ^{aA}	19.8 ± 0.6 ^{cB}	20.7 ± 0.5 ^{cB}	17.81 ± 0.12 ^{bB}	7.5 ± 0.7 ^{aA}	13.89 ± 0.12 ^{bA}	16.8 ± 0.7 ^{cA}	13.4 ± 1.3 ^{bA}
Ferulic acid	4.6 ± 1.2 ^{bA}	2.29 ± 0.02 ^{aB}	2.1 ± 1.3 ^{aA}	2.58 ± 0.09 ^{aA}	3.0 ± 0.2 ^{cA}	2.03 ± 0.03 ^{aA}	2.06 ± 0.03 ^{aA}	2.47 ± 0.11 ^{bA}
trans-Cinnamic acid	traces	3.02 ± 0.08 ^{aA}	9.5 ± 0.7 ^{bA}	2.19 ± 0.17 ^{aA}	n.d.	2.75 ± 0.02 ^{aA}	9.0 ± 0.6 ^{bA}	2.0 ± 0.3 ^{aA}
Flavonoids								
Rutin	n.d.	n.d.	n.d.	n.d.	n.d.	n.d.	n.d.	n.d.
Epicatechin	n.d.	n.d.	n.d.	n.d.	n.d.	n.d.	n.d.	n.d.
Quercetin 3-glucoside	n.d.	n.d.	n.d.	n.d.	n.d.	n.d.	n.d.	n.d.
Quercitrin	4.7 ± 0.4 ^{B}	n.d.	traces	traces	2.0 ± 0.4 ^{A}	traces	traces	traces
Apigenin-7-glucoside	0.62 ± 0.07 ^{A}	n.d.	traces	traces	2.80 ± 0.06 ^{cB}	2.28 ± 0.03 ^{b}	2.195 ± 0.007 ^{a}	2.67 ± 0.10 ^{c}
Quercetin	5.3 ± 0.4 ^{aB}	5.181 ± 0.006 ^{a}	5.0 ± 0.4 ^{a}	5.776 ± 0.008 ^{a}	0.78 ± 0.05 ^{A}	traces	traces	traces
Naringenin	n.d.	n.d.	n.d.	n.d.	n.d.	n.d.	n.d.	n.d.
Kaempferol	n.d.	n.d.	n.d.	n.d.	n.d.	n.d.	n.d.	n.d.

The results represent the mean of three repetitions with their standard deviation. ^{a,b,c} Different lowercase letters indicate significant differences between flours, and ^{A,B} different capital letters indicate significant differences between digestion models ($p < 0.05$); n.d.: not detected; and traces: not quantifiable.

Table 2. Phenolic content (µg/g dry basis) in digested Castellana lentil for unfermented flour (UFCL), fermented grain (FCL), fermented dried at 70 °C (FCL-70), and fermented lyophilised (FCL-L) flour.

	Digested (Standard)				Digested (Older Adult)			
	UFCL	**FCL**	**FCL-70**	**FCL-L**	**UFCL**	**FCL**	**FCL-70**	**FCL-L**
Phenolic acids								
Gallic acid	n.d.	n.d.	n.d.	n.d.	n.d.	n.d.	n.d.	n.d.
Caffeic acid	3.4 ± 0.4 aB	5.5 ± 0.7 bB	10.8 ± 1.3 cB	5.9 ± 0.4 bA	2.50 ± 0.06 aA	3.3 ± 0.2 abA	4.4 ± 0.2 bA	6.6 ± 0.7 cB
p-Coumaric acid	6.2 ± 1.0 cA	1.91 ± 0.04 aB	3.0 ± 0.2 bB	1.74 ± 0.05 aA	6.4 ± 1.3 bA	1.637 ± 0.007 aA	1.80 ± 0.14 aA	1.86 ± 0.08 aA
Sinapic acid	n.d.	n.d.	n.d.	n.d.	n.d.	n.d.	n.d.	n.d.
4-O-Caffeoylquinic	n.d.	n.d.	n.d.	n.d.	n.d.	n.d.	n.d.	n.d.
4-Hydroxybenzoic acid	n.d.	4.5 ± 1.4 a	5.5 ± 0.9 a	7.4 ± 0.4 a	n.d.	n.d.	n.d.	n.d.
Vanillic acid	6.2 ± 1.2 a	18 ± 3 b	20 ± 2 b	24 ± 2 b	traces	traces	traces	traces
Ferulic acid	traces	n.d.	n.d.	n.d.	n.d.	n.d.	n.d.	n.d.
trans-Cinnamic acid	n.d.	1.6 ± 0.3 aA	8.9 ± 0.9 bB	1.93 ± 0.03 aA	n.d.	2.25 ± 0.06 aB	5.8 ± 0.2 cA	3.3 ± 0.3 bB
Flavonoids								
Rutin	n.d.	n.d.	n.d.	n.d.	n.d.	n.d.	n.d.	n.d.
Epicatechin	n.d.	n.d.	n.d.	n.d.	n.d.	n.d.	n.d.	n.d.
Quercetin 3-glucoside	traces	traces	traces	n.d.	traces	n.d.	n.d.	n.d.
Quercitrin	n.d.	n.d.	n.d.	n.d.	n.d.	n.d.	n.d.	n.d.
Apigenin-7-glucoside	n.d.	2.61 ± 0.09 aB	5.4 ± 0.6 bB	3.23 ± 0.06 aA	n.d.	2.33 ± 0.03 aA	3.8 ± 0.4 cA	3.13 ± 0.15 bA
Quercetin	n.d.	n.d.	n.d.	n.d.	traces	traces	traces	traces
Naringenin	n.d.	traces	traces	traces	n.d.	traces	traces	traces
Kaempferol	n.d.	n.d.	n.d.	n.d.	n.d.	n.d.	n.d.	n.d.

The results represent the mean of three repetitions with their standard deviation. [a,b,c] Different lowercase letters indicate significant differences between flours, and [A,B] different capital letters indicate significant differences between digestion models ($p < 0.05$); n.d.: not detected; and traces: not quantifiable.

Table 3. Phenolic content (µg/g dry basis) in digested white quinoa for unfermented flour (UFWQ), fermented seeds (FWQ), fermented dried at 70 °C (FWQ-70), and fermented lyophilised (FWQ-L) flour.

	Digested (Standard)				Digested (Older Adult)			
	UFWQ	**FWQ**	**FWQ-70**	**FWQ-L**	**UFWQ**	**FWQ**	**FWQ-70**	**FWQ-L**
Phenolic acids								
Gallic acid	20 ± 2 a	77 ± 7 cA	139 ± 13 dB	56 ± 3 bA	traces	68 ± 8 bA	75 ± 9 bA	43 ± 7 aA
Caffeic acid	6.7 ± 0.9 b	2.09 ± 0.09 aB	2.7 ± 0.6 aB	traces	traces	0.83 ± 0.06 aA	0.88 ± 0.02 aA	traces
p-Coumaric acid	2.9 ± 0.9 A	n.d.	traces	traces	6.8 ± 1.0 B	n.d.	traces	traces
Sinapic acid	n.d.	n.d.	n.d.	n.d.	n.d.	n.d.	n.d.	n.d.
4-O-Caffeoylquinic	n.d.	n.d.	n.d.	n.d.	n.d.	n.d.	n.d.	n.d.
4-Hydroxybenzoic acid	n.d.	traces	n.d.	n.d.	n.d.	traces	n.d.	n.d.
Vanillic acid	16 ± 2 bB	3.5 ± 0.4 aA	5.0 ± 0.4 aA	3.1 ± 0.3 aA	8.3 ± 0.9 cA	2.93 ± 0.02 aA	4.1448 ± 0.0014 bA	3.0 ± 0.4 aA
Ferulic acid	9.0 ± 1.0 bA	traces	1.68 ± 0.12 aA	traces	8.6 ± 0.7 bA	traces	1.699 ± 0.004 aA	traces
trans-Cinnamic acid	traces	1.3 ± 0.2 bA	2.95 ± 0.05 cA	0.74 ± 0.05 aA	traces	2.14 ± 0.08 bB	3.69 ± 0.02 cB	1.87 ± 0.10 aB
Flavonoids								
Rutin	n.d.	n.d.	n.d.	n.d.	n.d.	n.d.	n.d.	n.d.
Epicatechin	n.d.	n.d.	n.d.	n.d.	n.d.	n.d.	n.d.	n.d.
Quercetin 3-glucoside	4.5 ± 1.0 a	3.162 ± 0.014 a	n.d.	n.d.	n.d.	n.d.	n.d.	n.d.
Quercitrin	6.2 ± 0.3 abA	6.7 ± 0.9 bB	6.60 ± 0.08 abA	5.3 ± 0.4 aA	6.8 ± 1.7 bA	3.51 ± 0.10 aA	5.9 ± 0.7 abA	4.0 ± 1.1 abA
Apigenin-7-glucoside	2.23 ± 0.04 a	2.62 ± 0.08 aA	9.3 ± 0.4 cB	3.43 ± 0.14 bB	n.d.	2.33 ± 0.03 bA	8.40 ± 0.03 cA	1.99 ± 0.04 aA
Quercetin	traces	traces	traces	traces	traces	traces	traces	traces
Naringenin	n.d.	n.d.	n.d.	n.d.	n.d.	n.d.	n.d.	n.d.
Kaempferol	n.d.	n.d.	n.d.	n.d.	n.d.	n.d.	n.d.	n.d.

The results represent the mean of three repetitions with their standard deviation. [a,b,c,d] Different lowercase letters indicate significant differences between flours, and [A,B] different capital letters indicate significant differences between digestion models ($p < 0.05$); n.d.: not detected; and traces: not quantifiable.

Older adult simulated conditions resulted in a significant reduction in the variety and number of phenolic compounds present in the bioaccessible fraction when compared to standard digestive conditions. However, chromatographic analysis did not detect all compounds in samples digested under the older adult model. In contrast, the same compounds were found in the bioaccessible fraction obtained using the standard model. This applies to vanillic and 4-hydroxybenzoic acids in the Castellana lentil. Phenolic compounds, found in various plant sources, can have preventive health benefits for humans. The extent of these benefits depends on the compounds' structure, such as their degree of glycosylation or acylation, molecular size, solubility, and conjugation with other phenols. These factors ultimately determine their absorption and metabolism [26]. Vanillic acid [27,28], caffeic acid [29,30], and gallic acid [31,32] are widely recognised as common phenolic acids that exhibit diverse chemical and pharmacological properties, such as analgesic, anticancer, anti-inflammatory, antioxidant, antimicrobial, cardioprotective, and neuroprotective activities.

Table 4. Phenolic content (µg/g dry basis) in digested black quinoa for unfermented flour (UFBQ), fermented seed (FBQ), fermented dried at 70 °C (FBQ-70), and fermented lyophilised (FBQ-L) flour.

	Digested (Standard)				Digested (Older Adult)			
	UFBQ	**FBQ**	**FBQ-70**	**FBQ-L**	**UFBQ**	**FBQ**	**FBQ-70**	**FBQ-L**
Phenolic acids								
Gallic acid	30 ± 4^{aB}	21 ± 3^{aB}	42 ± 2^{bB}	39 ± 3^{bB}	15 ± 2^{aA}	16.03 ± 0.18^{aA}	29.81 ± 0.12^{bA}	27.7 ± 2.0^{bA}
Caffeic acid	2.08 ± 0.18^{A}	n.d.	n.d.	n.d.	1.87 ± 0.06^{A}	n.d.	n.d.	n.d.
p-Coumaric acid	traces	traces	traces	traces	1.97 ± 0.06^{b}	traces	traces	1.30 ± 0.07^{a}
Sinapic acid	15 ± 3^{bA}	1.29 ± 0.17^{aA}	traces	2.19 ± 0.03^{aB}	32 ± 3^{bB}	1.34 ± 0.11^{aA}	traces	1.64 ± 0.12^{aA}
4-O-Caffeoylquinic	n.d.	n.d.	n.d.	n.d.	n.d.	n.d.	n.d.	n.d.
4-Hydroxybenzoic acid	n.d.	n.d.	n.d.	n.d.	n.d.	n.d.	n.d.	n.d.
Vanillic acid	n.d.	n.d.	n.d.	n.d.	n.d.	n.d.	n.d.	n.d.
Ferulic acid	4.9 ± 0.4^{A}	traces	traces	traces	8.79 ± 0.04^{B}	traces	traces	traces
trans-Cinnamic acid	traces	traces	traces	traces	traces	traces	traces	traces
Flavonoids								
Rutin	n.d.	n.d.	n.d.	n.d.	n.d.	n.d.	n.d.	n.d.
Epicatechin	traces	n.d.	n.d.	n.d.	traces	traces	traces	traces
Quercetin 3-glucoside	n.d.	n.d.	n.d.	n.d.	n.d.	n.d.	n.d.	n.d.
Quercitrin	10.0 ± 1.0^{bA}	3.7 ± 0.4^{aA}	5 ± 3^{abA}	8.71 ± 0.11^{abB}	11.5 ± 1.3^{bA}	6.95 ± 0.07^{aB}	6.030 ± 0.009^{aA}	5.4 ± 0.3^{aA}
Apigenin-7-glucoside	n.d.	2.23 ± 0.13^{aA}	3.6 ± 0.5^{bA}	3.92 ± 0.02^{bB}	n.d.	3.0 ± 0.3^{aA}	3.2 ± 0.4^{aA}	2.9 ± 0.3^{aA}
Quercetin	n.d.	n.d.	n.d.	n.d.	n.d.	n.d.	n.d.	n.d.
Naringenin	n.d.	1.58 ± 0.18^{aA}	2.6 ± 0.3^{bA}	3.04 ± 0.08^{bB}	n.d.	2.77 ± 0.08^{aB}	2.33 ± 0.15^{aA}	2.26 ± 0.10^{aA}
Kaempferol	n.d.	n.d.	n.d.	n.d.	n.d.	n.d.	n.d.	n.d.

The results represent the mean of three repetitions with their standard deviation. ^{a,b} Different lowercase letters indicate significant differences between flours, and ^{A,B} different capital letters indicate significant differences between digestion models ($p < 0.05$); n.d.: not detected; and traces: not quantifiable.

The antioxidant activity of the bioaccessible fraction was assessed through three assays: ABTS, FRAP, and DPPH. Table 5 displays the indexes for each assay and the antioxidant potency composite index (APCI). The ABTS-antioxidant activity increased following in vitro digestion, whereas no changes or slight decreases were observed in the FRAP and DPPH assays. Significant differences were found in the antioxidant activities of the bioaccessible fraction, with lower values when older adult conditions were used. The ABTS and DPPH antioxidant capabilities were reduced between 1% and 50%. However, the FRAP assay showed a decrease only in lentils, whereas quinoa showed an increase ranging from 10% to 70% for the digesta values of the older adult model compared to those of the standard model. Gallego et al. [33] discovered similar results when evaluating the effect of cooking different legume pastes on antioxidant activity after gastrointestinal digestion using the DPPH, ABTS, and FRAP methods. The study indicated a noteworthy improvement in the antioxidant activity of lentil pastes, reaching 12-fold greater levels than their original undigested content. However, they also found up to a four-fold reduction in pea paste using the DPPH method. The authors explained that these differences were due to the activity of enzymes in the gastrointestinal system. These enzymes promote the breakdown of proteins and peptides, resulting in the release of amino acids and phenolic compounds, and the exposure of internal groups. These factors impact the amount, dimensions, and physicochemical features of these compounds and influence the antioxidant potential. Koehnlein et al. [34] suggested that the high antioxidant capacity of cereals and legumes following gastrointestinal digestion may be due to the partial hydrolysis of total phenols and an increase in their content. Furthermore, the hydroxyl groups on the aromatic rings of the phenolic compounds may be deprotonated. Of all the treatments, flours that were fermented and dried 70 °C displayed the greatest antioxidant activity after gastrointestinal digestion. The fermented flours derived from Castellana lentil (FCL-70) and white quinoa (FWQ-70) demonstrated greater antioxidant capacity with an APCI exceeding 90% and 80%, respectively. Consequently, the results confirm the effectiveness of SSF followed by hot air drying (70 °C) in generating flours that boast an improved functionality of the bioaccessible fraction.

Table 5. Antioxidant activity (mg Trolox/g dry basis) by the ABTS, DPPH, and FRAP methods and total phenol content (mg gallic acid/g dry basis) in undigested and digested Pardina and Castellana lentils and white and black quinoa for unfermented flour (UFPL, UFCL, UFWQ, and UFBQ), fermented grain/seed (FPL, FCL, FWQ, and FBQ), fermented dried at 70 °C (FPL-70, FCL-70, FWQ-70, and FBQ-70), and fermented lyophilised (FPL-L, FCL-L, FWQ-L, and FBQ-L) flour, under standard and older adult simulated gastrointestinal conditions.

	Antioxidant Activity									APCI *		
	ABTS and ABTS Index			DPPH and DPPH Index			FRAP and FRAP Index					
	Undigested	Digested (Standard)	Digested (Older Adult)	Undigested	Digested (Standard)	Digested (Older Adult)	Undigested	Digested (Standard)	Digested (Older Adult)	Undig.	Standard	Older Adult
Pardina Lentil												
UFPL	9.5 ± 0.4 ^{dA} (100.0)	11.5 ± 0.3 ^{aB} (84.0)	11.6 ± 0.5 ^{abB} (92.3)	2.07 ± 0.09 ^{cA} (100.0)	2.29 ± 0.08 ^{cB} (100.0)	2.51 ± 0.09 ^{cC} (100.0)	7.6 ± 0.2 ^{bB} (100.0)	4.1 ± 0.4 ^{cA} (100.0)	3.79 ± 0.18 ^{cA} (100.0)	100.0	94.7	97.4
FPL	5.7 ± 0.5 ^{cA} (60.7)	12.8 ± 0.4 ^{bC} (93.2)	11.4 ± 0.2 ^{abB} (90.9)	0.64 ± 0.04 ^{bC} (30.8)	0.53 ± 0.04 ^{aB} (22.9)	0.44 ± 0.05 ^{aA} (17.5)	0.31 ± 0.02 ^{aA} (4.1)	1.8 ± 0.2 ^{aC} (43.2)	1.35 ± 0.11 ^{aB} (35.6)	31.9	53.1	48.0
FPL-70	3.91 ± 0.16 ^{bA} (41.4)	13.7 ± 0.5 ^{bB} (100.0)	12.5 ± 1.2 ^{bB} (100.0)	0.516 ± 0.010 ^{aA} (25.0)	0.88 ± 0.04 ^{bC} (38.4)	0.76 ± 0.06 ^{bB} (30.2)	0.351 ± 0.007 ^{aA} (4.6)	2.60 ± 0.14 ^{bB} (62.7)	2.38 ± 0.19 ^{bB} (62.8)	23.7	67.0	64.3
FPL-L	3.20 ± 0.04 ^{aA} (33.9)	11.8 ± 0.8 ^{aB} (85.7)	11.0 ± 0.5 ^{aB} (87.9)	0.502 ± 0.014 ^{aA} (24.3)	0.83 ± 0.05 ^{bC} (36.1)	0.70 ± 0.05 ^{bB} (27.8)	0.31 ± 0.02 ^{aA} (4.1)	2.54 ± 0.11 ^{bB} (61.3)	2.3 ± 0.2 ^{bB} (61.3)	20.7	61.0	59.0
Castellana Lentil												
UFCL	8.4 ± 0.4 ^{cA} (100.0)	14.0 ± 1.4 ^{aB} (80.7)	12.2 ± 1.5 ^{aB} (70.5)	1.634 ± 0.015 ^{bA} (72.0)	1.65 ± 0.05 ^{cA} (100.0)	1.78 ± 0.16 ^{cA} (100.0)	8.3 ± 0.2 ^{dB} (100.0)	3.2 ± 0.4 ^{cA} (70.2)	3.02 ± 0.06 ^{cA} (67.9)	90.7	83.6	79.5
FCL	2.50 ± 0.09 ^{aA} (29.9)	16.9 ± 1.3 ^{abC} (97.2)	14.3 ± 1.4 ^{aB} (82.7)	2.27 ± 0.13 ^{cB} (100.0)	0.25 ± 0.04 ^{aA} (15.0)	0.213 ± 0.017 ^{aA} (11.9)	1.10 ± 0.03 ^{aC} (13.4)	0.89 ± 0.02 ^{aB} (19.8)	0.77 ± 0.05 ^{aA} (17.4)	47.7	44.0	37.3
FCL-70	6.2 ± 0.2 ^{bA} (73.9)	17.4 ± 1.8 ^{bB} (100.0)	17.2 ± 1.5 ^{bB} (100.0)	1.71 ± 0.02 ^{bA} (75.1)	1.64 ± 0.10 ^{cA} (99.3)	1.55 ± 0.10 ^{bA} (87.1)	7.0 ± 0.3 ^{cB} (85.3)	4.50 ± 0.15 ^{dA} (100.0)	4.45 ± 0.11 ^{dA} (100.0)	78.1	99.8	95.7
FCL-L	2.32 ± 0.16 ^{aA} (27.7)	14.5 ± 1.7 ^{abB} (83.5)	13.7 ± 1.7 ^{aB} (79.5)	1.093 ± 0.016 ^{aC} (48.2)	0.39 ± 0.04 ^{bB} (23.4)	0.323 ± 0.014 ^{aA} (18.1)	2.14 ± 0.05 ^{bC} (25.9)	1.38 ± 0.07 ^{bB} (30.7)	1.13 ± 0.13 ^{bA} (25.3)	33.9	45.9	41.0
White Quinoa												
UFWQ	1.48 ± 0.08 ^{bA} (64.7)	8.20 ± 1.14 ^{aB} (64.5)	7.1 ± 0.6 ^{aB} (61.5)	1.070 ± 0.005 ^{cC} (80.2)	0.48 ± 0.04 ^{cB} (100.0)	0.23 ± 0.02 ^{bA} (75.0)	1.53 ± 0.05 ^{cB} (81.5)	1.22 ± 0.14 ^{cA} (100.0)	1.70 ± 0.03 ^{cB} (100.0)	75.5	88.2	78.8
FWQ	1.87 ± 0.08 ^{cA} (81.7)	10.1 ± 1.2 ^{abB} (79.4)	9.1 ± 1.2 ^{aB} (78.5)	1.334 ± 0.012 ^{dB} (100.0)	0.127 ± 0.008 ^{aA} (26.4)	0.119 ± 0.011 ^{aA} (38.5)	0.47 ± 0.03 ^{aB} (24.9)	0.34 ± 0.05 ^{aA} (27.8)	0.53 ± 0.07 ^{aB} (31.3)	68.9	44.5	49.5
FWQ-70	2.287 ± 0.006 ^{dA} (100.0)	12.7 ± 1.3 ^{cB} (100.0)	11.6 ± 1.8 ^{bB} (100.0)	0.80 ± 0.04 ^{aB} (60.3)	0.37 ± 0.02 ^{bA} (76.2)	0.31 ± 0.04 ^{cA} (100.0)	1.88 ± 0.09 ^{dB} (100.0)	1.05 ± 0.04 ^{bA} (86.2)	1.15 ± 0.16 ^{bA} (67.7)	86.8	87.5	89.2
FWQ-L	1.12 ± 0.04 ^{aA} (49.1)	10.8 ± 1.0b ^{cC} (84.5)	8.1 ± 0.8 ^{aB} (69.7)	1.00 ± 0.03 ^{bB} (74.8)	0.104 ± 0.009 ^{aA} (21.6)	0.093 ± 0.013 ^{aA} (30.3)	0.63 ± 0.03 ^{bB} (33.3)	0.40 ± 0.02 ^{aA} (32.5)	0.49 ± 0.08 ^{aA} (28.7)	52.4	46.2	42.9
Black Quinoa												
UFBQ	2.48 ± 0.03 ^{cA} (100.0)	10.1 ± 0.7 ^{aC} (99.0)	7.3 ± 0.6 ^{bB} (99.4)	0.88 ± 0.03 ^{bB} (63.6)	0.69 ± 0.04 ^{dA} (100.0)	0.63 ± 0.03 ^{dA} (100.0)	2.74 ± 0.04 ^{dC} (100.0)	1.7 ± 0.2 ^{cA} (100.0)	2.1 ± 0.3 ^{bB} (100.0)	87.9	99.7	99.8
FBQ	1.69 ± 0.10 ^{bA} (68.2)	9.6 ± 0.4 ^{aC} (94.1)	6.7 ± 0.7 ^{abB} (90.8)	1.38 ± 0.02 ^{dB} (100.0)	0.25 ± 0.02 ^{bA} (35.5)	0.26 ± 0.03 ^{cA} (40.9)	0.52 ± 0.02 ^{aA} (19.1)	0.42 ± 0.03 ^{aA} (25.1)	0.71 ± 0.09 ^{aB} (34.3)	62.4	51.6	55.3

Table 5. *Cont.*

	Antioxidant Activity									APCI *		
	ABTS and ABTS Index			DPPH and DPPH Index			FRAP and FRAP Index					
	Undigested	Digested (Standard)	Digested (Older Adult)	Undigested	Digested (Standard)	Digested (Older Adult)	Undigested	Digested (Standard)	Digested (Older Adult)	Undig.	Standard	Older Adult
FBQ-70	1.68 ± 0.06 ^{bA} (67.8)	10.2 ± 0.5 ^{aC} (100.0)	7.4 ± 0.2 ^{bB} (100.0)	0.666 ± 0.014 ^{aC} (48.3)	0.36 ± 0.06 ^{cB} (52.4)	0.20 ± 0.02 ^{bA} (31.0)	1.26 ± 0.04 ^{cB} (46.2)	0.66 ± 0.08 ^{bA} (39.8)	0.74 ± 0.05 ^{aA} (35.5)	54.1	64.1	55.5
FBQ-L	1.33 ± 0.05 ^{aA} (53.7)	9.8 ± 1.2 ^{aC} (96.1)	5.5 ± 1.0 ^{aB} (74.6)	0.96 ± 0.06 ^{cB} (69.6)	0.13 ± 0.02 ^{aA} (19.4)	0.139 ± 0.009 ^{aA} (22.0)	0.82 ± 0.05 ^{bB} (30.0)	0.49 ± 0.08 ^{abA} (29.7)	0.58 ± 0.09 ^{aA} (27.9)	51.1	48.4	41.5

The results represent the mean of three repetitions with their standard deviation. ^{a,b,c,d} Different lowercase letters indicate significant differences ($p < 0.05$) between flours. ^{A,B,C} Different capital letters indicate significant differences ($p < 0.05$) between digestion models. Values in parentheses correspond to ABTS, DPPH, and FRAP indexes, calculated among unfermented, fermented, and fermented dried at 70 °C, and lyophilised samples of each plant food. * APCI: antioxidant potency composite index.

The effect of gastrointestinal conditions shows different trends depending on the pre-treatment of the food and the substrate itself, as well as the methodology used to measure the antioxidant activity. A significant reduction is only observed for DPPH in Pardina lentils and for ABTS in black quinoa after using the older adult model. However, white quinoa and Castellana lentils maintain or even increase the values reported by the control model. These data could help develop new products for this population group, because a high antioxidant capacity is necessary for good health.

2.2. Impact of GI Conditions on the Bioaccessibility of Phytic Acid and Minerals of Unfermented, Fermented, and Fermented Dried Lentils and Quinoa

Minerals are inorganic substances found in all tissues and bodily fluids, which are vital for maintaining specific physicochemical processes essential for life [35,36]. They have structural functions involving the skeleton and soft tissues and regulatory functions including neuromuscular transmission, blood clotting, oxygen transport, and enzyme activity [37]. Legumes and pseudocereals are excellent sources of minerals such as calcium, iron, zinc, potassium, and magnesium [38,39]. The mineral content of unfermented, fermented, and fermented dried samples (Mg, Ca, and Fe) was evaluated pre- and post-gastrointestinal digestion, as shown in Table 6. Undigested samples revealed that SSF and ensuing drying caused an increase in Mg and Ca contents, with increases in Mg ranging from 1% to 20% and Ca from 12% to 59%. Significant differences in Ca content were found in all samples, whereas significant differences in Mg content were found only in Castellana lentil and white quinoa samples dried at 70 °C and lyophilised. In contrast, SSF decreased the Fe content in all samples between 2% and 11%, with a significant difference in Pardina lentil and white quinoa samples. Furthermore, the drying process increased the Fe content between 2% and 23% only in white and black quinoa samples, with significance in the fermented samples dried at 70 °C.

Table 6. Mineral content (µg/g dry basis) in undigested and digested Pardina and Castellana lentils and white and black quinoa for unfermented flour (UFPL, UFCL, UFWQ, and UFBQ), fermented grain/seed (FPL, FCL, FWQ, and FBQ), fermented dried at 70 °C (FPL-70, FCL-70, FWQ-70, and FBQ-70), and fermented lyophilised (FPL-L, FCL-L, FWQ-L, and FBQ-L) flour, under standard and older adult simulated gastrointestinal conditions.

	Magnesium (Mg)			Calcium (Ca)			Iron (Fe)		
	Undigested	Digested (Standard)	Digested (Older Adult)	Undigested	Digested (Standard)	Digested (Older Adult)	Undigested	Digested (Standard)	Digested (Older Adult)
Pardina Lentil									
UFPL	112.6 ± 0.9^{aB}	76 ± 4^{aA}	71 ± 3^{aA}	62 ± 2^{aB}	43 ± 4^{cA}	44.1 ± 0.6^{bA}	11.3 ± 0.3^{cB}	1.60 ± 0.05^{aA}	1.05 ± 0.14^{aA}
FPL	125 ± 3^{bB}	78 ± 2^{aA}	70 ± 3^{aA}	90.5 ± 0.9^{cB}	22 ± 3^{bA}	15 ± 3^{aA}	10.05 ± 0.04^{bC}	3.168 ± 0.004^{cB}	2.45 ± 0.02^{bA}
FPL-70	112 ± 2^{aB}	74 ± 5^{aA}	75 ± 10^{aA}	79.6 ± 0.4^{bB}	12 ± 2^{aA}	12 ± 4^{aA}	8.92 ± 0.05^{aB}	2.6 ± 0.3^{bA}	2.2 ± 0.3^{bA}
FPL-L	109 ± 2^{aB}	71 ± 7^{aA}	76.4 ± 0.6^{aA}	80 ± 2^{bB}	44 ± 5^{cA}	52 ± 2^{cA}	8.9 ± 0.3^{aB}	2.60 ± 0.02^{bA}	2.78 ± 0.08^{cA}
Castellana Lentil									
UFCL	122 ± 2^{aB}	82 ± 7^{aA}	84.8 ± 0.9^{aA}	64.1 ± 0.9^{aB}	48 ± 6^{aA}	43 ± 4^{aA}	9.04 ± 0.07^{aB}	1.4 ± 0.3^{aA}	1.5 ± 0.3^{aA}
FCL	127 ± 4^{abB}	93 ± 5^{abA}	96 ± 2^{cA}	94.0 ± 1.0^{bB}	61 ± 6^{bA}	57 ± 9^{bA}	8.9 ± 0.2^{aC}	5.8 ± 1.0^{bB}	4.43 ± 0.03^{cA}
FCL-70	142 ± 4^{cC}	102 ± 2^{bB}	91 ± 2^{bA}	102 ± 5^{bB}	58 ± 3^{bA}	43 ± 6^{aA}	9.2 ± 0.3^{aC}	5.0 ± 0.2^{bB}	3.8 ± 0.2^{bA}
FCL-L	135 ± 3^{bcC}	107 ± 10^{bB}	92.5 ± 0.4^{bcA}	101 ± 6^{bB}	49 ± 5^{aA}	54 ± 5^{abA}	8.8 ± 0.2^{aB}	6.0 ± 1.4^{bA}	4.56 ± 0.12^{cA}
White Quinoa									
UFWQ	218 ± 2^{aB}	146 ± 3^{aA}	139 ± 6^{abA}	63 ± 2^{aB}	32.1 ± 1.5^{aA}	38 ± 3^{aA}	3.67 ± 0.02^{bB}	3.3 ± 0.2^{bA}	3.1 ± 0.6^{aA}
FWQ	232 ± 10^{aB}	163.7 ± 1.3^{bA}	157 ± 10^{bA}	81 ± 4^{bB}	72 ± 7^{cB}	47 ± 4^{cA}	3.45 ± 0.02^{aB}	2.4 ± 0.2^{aA}	2.0 ± 0.5^{aA}
FWQ-70	261 ± 3^{bB}	163 ± 7^{bA}	156 ± 3^{bA}	79 ± 3^{bC}	65 ± 6^{cB}	45 ± 3^{bA}	4.54 ± 0.07^{cB}	3.50 ± 0.04^{bA}	2.9 ± 0.5^{aA}
FWQ-L	248.3 ± 1.3^{bC}	166 ± 5^{bB}	133 ± 11^{aA}	78.4 ± 0.6^{bB}	49.8 ± 0.7^{bA}	54 ± 3^{cA}	3.78 ± 0.13^{bB}	2.9 ± 0.3^{abA}	2.2 ± 0.6^{aA}
Black Quinoa									
UFBQ	210 ± 8^{aB}	111 ± 10^{aA}	135 ± 6^{aA}	55 ± 2^{aB}	36 ± 4^{aA}	36 ± 5^{aA}	4.3 ± 0.3^{abB}	2.9 ± 0.4^{bA}	3.3 ± 0.2^{bA}
FBQ	212.0 ± 1.4^{aB}	136 ± 2^{abA}	135 ± 11^{aA}	63 ± 2^{bA}	56 ± 5^{bA}	60 ± 10^{cA}	3.86 ± 0.03^{aB}	1.9 ± 0.2^{aA}	1.54 ± 0.09^{aA}
FBQ-70	217 ± 12^{aB}	145 ± 2^{bA}	138.9 ± 0.7^{aA}	62 ± 2^{bA}	57 ± 12^{bA}	41 ± 4^{bA}	4.5 ± 0.2^{bB}	2.2 ± 0.6^{bA}	1.6 ± 0.3^{aA}
FBQ-L	220 ± 7^{aB}	130 ± 16^{abA}	144 ± 2^{aA}	66 ± 4^{bA}	58 ± 10^{bA}	42 ± 5^{bA}	4.4 ± 0.2^{abB}	1.7 ± 0.3^{aA}	1.8 ± 0.3^{aA}

The results represent the mean of three repetitions with their standard deviation. ^{a,b,c} Different lowercase letters indicate significant differences ($p < 0.05$) between flours. ^{A,B,C} Different capital letters indicate significant differences ($p < 0.05$) between digestion models.

When comparing the undigested and digested samples, the findings indicate that the digestive process led to a decrease in mineral content (Mg, Ca, and Fe) in the bioaccessible fraction. Despite the digestion model used, the fermented and fermented dried samples exhibited higher Mg content than the unfermented samples in the 3% to 30% range. However, a notable difference was only observed in Castellana lentils and white quinoa. The fermented and fermented dried samples of Castellana lentil and white and black quinoa exhibited a significant increase in Ca content ranging from 18% to 124%. However, the fermented and fermented dried samples of Pardina lentil exhibited a decrease in Ca content. Regarding Fe content, both Castellana and Pardina lentils experienced a notable increase, ranging from 63% to 329% in their fermented and fermented dried samples. Furthermore, a decrease was observed in the fermented and fermented dried samples of white and black quinoa. Therefore, it can be inferred that mineral bioaccessibility is enhanced through the fermentation process.

When comparing digestion models, it was found that the older adult digestion model demonstrated a decrease in mineral content when evaluated against the standard. Despite individual cases of significant differences, no overall significant differences were observed between the digestion models.

Many legume grains, cereals, and pseudocereal seeds contain varying concentrations of phytic acid. Upon ingestion, it remains undigested in the human digestive system due to the lack of the phytase enzyme. Phytic acid can bind to crucial micronutrients such as iron, calcium, magnesium, and zinc, reducing their absorption during gastrointestinal digestion [18]. Processes such as SSF are used to reduce this anti-nutrient. This study analyses the effects of SSF and drying on the bioaccessibility of phytic acid in unfermented, fermented, and fermented dried samples of Castellana and Pardina lentils and white and black quinoa. The analysis was conducted using standard and older adult in vitro digestion models, as illustrated in Figure 2. A marked reduction in the phytate content of approximately 90% can be observed in undigested fermented and fermented dried samples of Castellana lentil, as well as white and black quinoa as compared to their unfermented counterparts. Furthermore, there is no significant effect on Pardina lentils. These findings indicate that the decrease in this anti-nutritional factor is due to the activation of the endogenous phytase present in each substrate, as discussed previously [20,21]. When the undigested and digested samples were compared, a significant reduction in phytic acid release was observed after gastrointestinal digestion. The reductions ranged from 70% to 80% in unfermented lentil samples (Pardina and Castellana) and quinoa (white and black) regarding their initial content (undigested).

For both digested and fermented dried samples, black quinoa saw a reduction of approximately 40%, whereas Pardina lentils saw a reduction of approximately 80%. In contrast, the reduction in Castellana lentils and white quinoa was 100%, which was due to their minimal phytic acid content in undigested samples rather than the simulated gastrointestinal digestion's physiological conditions. Therefore, a reduction in phytic acid in fermented and dehydrated fermented samples may be associated with increased levels of Mg, Ca, and Fe following gastrointestinal digestion. Chawla et al. [40] evaluated the impact of SSF in black-eyed pea seed flour using an *Aspergillus oryzae* strain on the mineral bioavailability of iron and zinc. They determined that after 96 h of fermentation, iron and zinc increased from 17.3% to 30.2% and from 14.4% to 29.6%, respectively. The authors attributed the improved mineral bioaccessibility to the degradation of anti-nutrient compounds, including phytic acid.

When comparing digestion models, a significant difference was observed between the standard model and the older adult model for Pardina lentils. However, no significant differences were found for Castellana lentils, white quinoa, and black quinoa. Therefore, there is no significant effect of the occurrence of digestive disorders with age. Couzy et al. [41] studied zinc absorption in older and younger subjects (with similar zinc status) using serum concentration curve (SCC). They administered soy milk fortified with 50 mg of zinc containing three levels of phytic acid: 0, 0.13, and 0.26 g/200 mL. They found that phytic acid reduced zinc absorption as the concentration of phytic acid increased in the beverage. Furthermore, they indicated that there were no differences between the older and younger subjects.

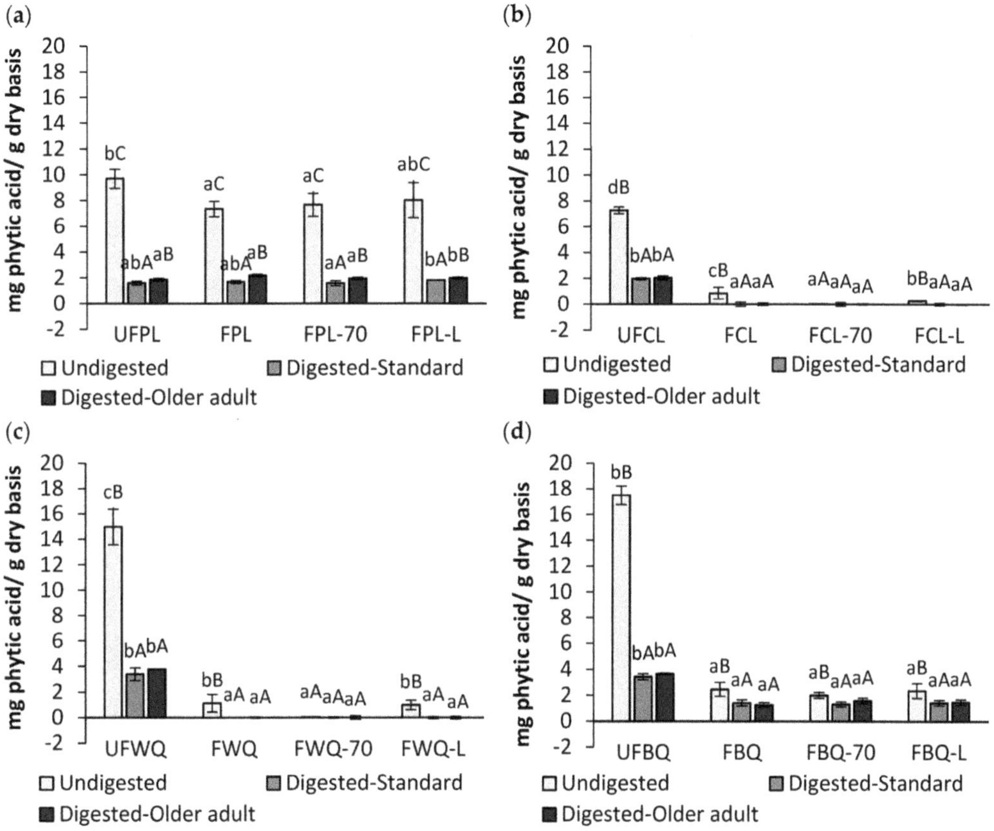

Figure 2. Phytic acid content (mg/g dry basis) in Pardina lentil (**a**) and Castellana lentil (**b**) and white (**c**) and black quinoa (**d**) for unfermented flour (UFPL, UFCL, UFWQ, and UFBQ), fermented grain/seed (FPL, FCL, FWQ, and FBQ), fermented dried at 70 °C (FPL-70, FCL-70, FWQ-70, and FBQ-70), and fermented lyophilised (FPL-L, FCL-L, FWQ-L, and FBQ-L) flour obtained with a standard and older adult in vitro digestion model. ^{a,b,c} Different lowercase letters indicate significant differences ($p < 0.05$) between samples. ^{A,B,C} Different capital letters indicate significant differences ($p < 0.05$) between digestion models.

3. Materials and Methods

3.1. Materials

Lentils (*Lens culinaris*) of the Pardina and Castellana varieties from Hacendado® and quinoa (*Chenopodium quinoa Wild*) of white and black varieties from the Hacendado® and Nut&me brands, respectively, were obtained from local stores in Valencia (Spain). The *Pleurotus ostreatus* strain was acquired from the Spanish Type Culture Collection (CECT20311).

Pepsin from porcine gastric mucosa (≥3200 U/mg), pancreatin from porcine pancreas (8 × USP), bovine bile (dried, unfractionated), p-toluene-sulfonyl-L-arginine methyl ester (TAME, T4626), analytical grade salts (potassium chloride, potassium dihydrogen phosphate, sodium bicarbonate, sodium chloride, magnesium chloride hexahydrate, ammonium carbonate, and calcium chloride), potato starch, sodium phosphate, maltose standard, 3,5-

dinitrosalicylic acid (DNS), potassium sodium tartrate tetrahydrate, sodium hydroxide, thioglycolic acid, phytic acid sodium salt hydrated from rice, 2,2′-bipyridine, formic acid, 2,2-diphenyl-1-picrylhydrazyl (DPPH), 2,2′-azino-bis (3-ethylbenzothiazoline-6-sulphonic acid) (ABTS), Folin–Ciocalteu reagent, 2,4,6-tripyridyl-s-triazine (TPTZ), gallic acid, (±)-6-Hydroxy-2,5,7,8-tetramethylchromane-2-carboxylic acid (Trolox), glucose, and mycopeptone were obtained from Sigma-Aldrich Co. (St. Louis, MO, USA).

For HPLC analysis, vanillic acid, quercetin 3-glucoside, quercetin, quercitrin, 4-hydroxybenzoic acid, rutin, epicatechin, trans-cinnamic acid, ferulic acid, naringenin, caffeic acid, 4-O-caffeoylquinic, p-coumaric acid, apigenin-7-glucoside, kaempferol, and sinapic acid were obtained from Sigma-Aldrich as an analytical standard (HPLC grade). Acetic acid glacial, concentrated hydrochloric acid, ethanol absolute, sodium carbonate, and ammonium iron (III) sulphate dodecahydrate were obtained from Panreac AppliChem (Barcelona, Spain). Acetonitrile (HPLC grade), methanol (HPLC grade), iron (III) chloride hexahydrate, sodium acetate trihydrate, and potassium persulphate were obtained from Honeywell Fluka (Morris Plains, NJ, USA). The malt extract and agar were obtained from Scharlau (Barcelona, Spain).

3.2. Fungal Solid-State Fermentation (SSF) and Flour Production

To perform SSF, a starter culture was first prepared by growing *P. ostreatus* mycelium in a Petri dish containing lentil or quinoa grains/seeds. The fermentation was then conducted by inoculating a portion of the starter culture into a glass jar containing 35 g of lentil or quinoa grains/seeds following the methodology used previously [20]. Fermented lentils and quinoa grains/seeds were dried using hot air drying and freeze drying methods, the latter used as the reference drying method because it was expected to have the best preservation of the sample properties according to literature. Hot air drying was conducted using a convective dryer (Pol-Eko-Aparatura, CLW 750 TOP+, Kokoszycka, Poland) at 70 °C with an air rate of 10.5 ± 0.2 m/s and an air humidity of 8.7 ± 1.2%. The samples were dried for 3.5 to 4 h to a target product moisture of 7% (wet basis). Freeze drying was performed using a freeze dryer (Telstar, Lyoquest-55, Terrassa, Spain) at −45 °C and 0.8 mBar for 48 h. Unfermented and fermented dried samples were then milled using a food processor (Vorwerk, Thermomix® TM6-1, Wuppertal, Germany), applying 10,000 rpm at 15 s intervals for 1 min.

3.3. Simulated In Vitro Gastrointestinal Digestion under Standard and Older Adult Conditions

Unfermented, fermented, and fermented dried samples were digested under two static in vitro digestion models: the older adult model [42] and the healthy adult model (standard; as a control) [43,44] (Table 7). Enzymatic activities were determined before each experiment according to the supplementary information in the protocol published by Brodkorb et al. [43]. Simulated salivary (SSF), gastric (SGF), and intestinal (SIF) fluid were prepared daily for the standard and older adult digestion model considering the concentrations of enzymes, bile salts, and pH of each digestive phase.

To perform the oral stage, 5 g of sample was mixed with 5 mL of SSF containing the enzyme concentration according to the digestion model (Table 7), with adjusted pH, mixed at 25 rpm using an Intelli-Mixer RM-2 (Elmi Ltd., Riga, LV-1006, Latvia), and incubated in a thermostatic chamber (J.P. Selecta SA, Barcelona, Spain) at 37 °C for 2 min.

For the gastric stage, 10 mL of SGF was added to the food bolus according to the conditions simulated in each model (Table 7), the pH, mixed at 55 rpm, and incubated at 37 °C for 2 h. For the intestinal stage, 20 mL of SIF was added to the gastric chyme according to the concentration of enzyme and bile salts (Table 7), adjusted the pH, mixed at 55 rpm, and incubated at 37 °C for 2 h. After gastrointestinal digestion, enzyme activity was inhibited by adjusting the pH to 5 and keeping the samples in an ice bath. Finally, the samples were centrifuged at 8000× *g* for 10 min and aliquots of the bioaccessible fraction were taken for analytical determinations.

Table 7. Gastrointestinal conditions established for an in vitro digestion model for a healthy adult (standard) [43,44] and an older adult [42].

Digestive Stage	Digestion Models	
	Healthy Adult (Standard)	Older Adult
Oral stage	Amylase (75 U/mL) pH 7 2 min	**Amylase (112.5 U/mL)** pH 7 2 min
Gastric stage	Pepsin (2000 U/mL) pH 3 2 h	**Pepsin (1200 U/mL)** **pH 3.7** 2h
Intestinal stage	Pancreatin (100 U/mL) Bile salts (10 mM) pH 7 2 h	**Pancreatin (80 U/mL)** **Bile salts (7 mM)** pH 7 2 h

The alterations made to the model for older adults compared to the standard model are highlighted in bold text.

3.4. Analytical Determinations

3.4.1. Total Phenolic Content (TPC)

TPC of the samples before and after undergoing in vitro digestion was determined using the Folin–Ciocalteu methodology as outlined by Chang et al. [45]. For the samples that were not digested, phenolic compounds were extracted by blending 2.5 g of the sample with 7.5 mL of the extraction solvent (a mixture of double distilled water and ethanol at 70:30) and adjusting the pH to 2 with 2 M HCl. The mixture was then treated to an ultrasonic bath (J.P. Selecta, 3000840) at 30 °C for 2 h. The pH was adjusted to 2 with 2 M HCl. The samples were centrifuged at 8000× *g* for 15 min, and the extraction process was repeated twice, with subsequent mixing of both extracted samples. The bioaccessible fraction determined the digested samples. An aliquot of 125 µL of the extract/digest was taken and mixed with 500 µL of bidistilled water, followed by 125 µL of the Folin–Ciocalteu reagent. This was left to react for 6 min. Then, 1.25 mL of 7% sodium carbonate and 1 mL of bidistilled water were added. The sample was incubated for 30 min at room temperature in darkness. Afterward, the absorbance was measured at 760 nm, and the results were presented in mg gallic acid/g dry basis using a standard curve.

3.4.2. Antioxidant Activity

The antioxidant activity of the samples before and after in vitro digestion was determined by three methods: ABTS, DPPH, and FRAP, following the methodology described by Thaipong et al. [46]. The same extracts used in the TPC section were used for undigested and digested samples.

For the ABTS test, the working solution (7.4 mM ABTS and 2.6 mM potassium persulphate in a 1:1 ratio) was allowed to react for 12 h at room temperature in darkness. The working solution (1 mL) was diluted with methanol to obtain an absorbance close to 1.1 at 734 nm. Extract/digest (150 µL) was reacted with 2.85 mL of ABTS working solution for 2 h in darkness and absorbance was measured at 734 nm.

For the DPPH test, a fresh working solution of 0.039 g/L DPPH was prepared in pure methanol to obtain an absorbance close to 1.1 at 515 nm. Extract/digest (75 µL) reacted with 2.925 mL of DPPH working solution for 30 min in darkness and absorbance was measured at 515 nm.

For the FRAP test, fresh working solution was prepared by mixing 300 mM acetate buffer (3.1 g sodium acetate trihydrate and 16 mL acetic acid glacial in 1 L water, pH 3.6), TPTZ solution (10 mM 2,4,6-tripyridyl-s-triazine dissolved in 40 mM HCl), and 20 mM iron (III) chloride hexahydrate solution in a 10:1:1 ratio, respectively, and incubated at 37 °C before use. Extract/digest (150 µL) reacted with 2.85 mL of FRAP working solution for 30 min in darkness, and the absorbance was measured at 593 nm. The results are expressed as mg Trolox/g dry basis using a standard curve for the three antioxidant determination methods.

The antioxidant index was calculated for each sample for all antioxidant activity assays (ABTS, DPPH, and FRAP). An antioxidant index value of 100 was assigned to the highest sample score in each assay. Then, the antioxidant index was calculated for the entire group of samples in each assay according to Equation (1) [47]:

$$\text{Antioxidant index (\%)} = \left(\frac{\text{sample score}}{\text{highest sample score}}\right) \times 100 \quad (1)$$

The overall APCI was calculated by averaging each sample's antioxidant index (%) of each antioxidant activity assay.

3.4.3. Phenolic Profile by HPLC Analysis

The phenolic profile of the samples after in vitro digestion was determined by filtering the bioaccessible fraction of the digest with a 0.45 μm PTFE filter. The samples were analysed using an HPLC 1200 Series Rapid Resolution coupled to a diode detector Serie (Agilent, Palo Alto, CA, USA) according to the methodology described by Tanleque-Alberto et al. [48]. A Brisa-LC 5 μm C18 column (250 × 4.6 mm) (Teknokroma, Barcelona, Spain) was used. Mobile phase A was 1% formic acid, and mobile phase B was acetonitrile (ACN). The following gradient program was used: 0 min, 90% A; 25 min, 40% A; 26 min, 20% A; held for 30 min; 35 min, 90% A; held for 40 min. Flow rate, injection volume, and working temperature of the column was 0.5 mL/min, 10 μL, and 30 °C, respectively. Unknown compounds were identified by comparing chromatographic retention times with reference standards according to the following wavelengths for each compound: 250 nm for vanillic acid; 260 nm for 4-hydroxybenzoic acid, rutin, quercetin 3-glucoside, and quercitrin; 280 nm for gallic acid, epicatechin, quercetin and trans-cinnamic acid; 290 nm for naringenin; 320 nm for 4-O-caffeoylquinic, caffeic acid, p-coumaric acid, sinapic acid, ferulic acid, and apigenin-7-glucoside; and 380 nm for kaempferol. The results are expressed as μg/g dry basis using a standard curve.

3.4.4. Phytic Acid Content

The phytic acid content was measured before and after in vitro digestion following the protocol described by Haug and Lantzsch [49] and modified by Peng et al. [50]. For undigested samples, the extract was prepared by mixing 50 mg of sample with 10 mL of 0.2 M HCl and left overnight at 4 °C. For digested samples, the determination was performed on the bioaccessible fraction. An aliquot of 500 μL of the extract/digest was taken, and 1 mL of ferric solution (0.2 g of ammonium iron (III) sulphate dodecahydrate dissolved in 100 mL of 2 M hydrochloric acid and made up to 1 L with distilled water) was added. It was incubated in a boiling water bath for 30 min and then cooled to room temperature. The sample was centrifuged for 30 min at 3000× *g*, and 1 mL of the supernatant was taken and mixed with 1.5 mL of 2,2′-bipyridine solution (10 g of 2,2′-bipyridine and 10 mL of thioglycolic acid dissolved in distilled water and made up to 1 L). The results are expressed as mg phytic acid/g dry basis using a standard curve made with a stock solution of 1.3 mg/mL phytic acid concentration and diluted with 0.2 M hydrochloric acid between 0.1 and 1 mL (3.16–31.6 μg/mL phytate phosphorus).

3.4.5. Mineral Quantification

The quantification of minerals (Fe, Ca, and Mg) before and after gastrointestinal digestion was performed by inductively coupled plasma mass spectrometry (ICP-MS). The mineral extract was prepared according to the methodology published by Barrera et al. [51]. A 5 g sample was weighed for undigested food and a 3.5 mL aliquot was taken from the bioaccessible fraction of digested food. The samples were incinerated at 600 °C for 10 h. The ashes were dissolved with 1 mL of 69% nitric acid and re-incinerated until white ashes were obtained. The white ashes were suspended in 1.5 mL of 69% nitric acid and 4 mL of bidistilled water.

The samples were analysed using an ICP-MS equipped with an autosampler (iCAP Q, Thermo, Waltham, MA, USA) according to the methodology proposed by Chen et al. [52]. The conditions of the working equipment were radio frequency power (1550 W), cool gas flow (14 L/min), auxiliary gas flow (0.8 L/min), nebuliser gas flow (1.08 L/min), peristaltic pump speed (40 rpm), sampling depth (5 mm), spray chamber temperature (2.7 °C), and dwell time (20 ms). The results are expressed as μg/g dry basis.

3.5. Statistical Analysis

The experiments were conducted at least in triplicate and results reported as mean ± standard deviation. A one-way ANOVA with a 95% confidence interval ($p < 0.05$) was performed to determine the statistical significance of the variables studied (SSF, drying, and common GI conditions of older adults) on the antioxidant activity, phenolic, phytates, and mineral contents in the bioaccessible fraction in lentil and quinoa samples, employing Statgraphics Centurion versionXV as statistical software.

4. Conclusions

SSF, coupled with drying at 70 °C, had a positive impact on the bioaccessibility of phenolic compounds and antioxidant activity, albeit to various degrees depending on the substrate. The profile of phenolic compounds following gastrointestinal digestion showed an increase in vanillic and caffeic acids in Castellana lentils and in vanillic acid in Pardina lentils, reaching approximately three times the levels of the unfermented samples. There was a significant increase in gallic acid of up to 7 and 1.4 times more than in the unfermented analogue in white and black quinoa, respectively. Regarding antioxidant activity, the Castellana lentil and white quinoa flours fermented and dried at 70 °C showed the highest APCI (>90% and >80%, respectively) after digestion, thus having a higher capacity to neutralise free radicals than the other samples. Fermented and fermented dried samples (at 70 °C and lyophilised) displayed a mineral bioaccessible content that was higher than the unfermented samples. This, together with the low phytic acid content present in fermented dried samples, renders such flours attractive for developing functional products with superior bioaccessibility than unfermented flours. Finally, typical age-related digestive conditions did not appear to affect the mineral bioavailability of Fe, Mg, and Ca in lentils and quinoa flours. However, these conditions reduced the phenolic profile and antioxidant activity of digesta when compared to the results obtained in the standard model.

Fermented Castellana and white quinoa flours are the optimal choice for product development, specifically catering to this population group to maximise health benefits. Furthermore, it is essential to evaluate the techno-functional properties of fermented flours to determine their compatibility with different food applications. Moreover, it is crucial to perform scale-up tests of the fermentation process to facilitate the technological transfer of this process to the food industry.

Supplementary Materials: The following supporting information can be downloaded at: https://www.mdpi.com/article/10.3390/molecules28217298/s1.

Author Contributions: Conceptualization, J.S.-G., S.M.-P., J.G.-H., A.H., A.T. and A.A.; methodology, J.S.-G., S.M.-P., J.G.-H., A.H. and A.A.; formal analysis, J.S.-G. and S.M.-P.; investigation, J.S.-G. and S.M.-P.; writing—original draft preparation, J.S.-G. and S.M.-P.; writing—review and editing, J.S.-G., S.M.-P., J.G.-H., A.T., A.H. and A.A.; supervision, J.G.-H., A.T., A.A. and A.H.; project administration, A.T., A.A. and A.H.; funding acquisition, A.T., A.H. and A.A. All authors have read and agreed to the published version of the manuscript.

Funding: This research was under the project PID2019-107723RB-C22 funded by the Ministry of Science and Innovation MCIN/AEI/10.1309/501100011033. Also, Sara Muñoz Pina was a beneficiary of a postdoctoral grant from the Universitat Politècnica de Valencia (PAID-10-21).

Institutional Review Board Statement: Not applicable.

Informed Consent Statement: Not applicable.

Data Availability Statement: Data is contained within the article.

Conflicts of Interest: The authors declare no conflict of interest.

References

1. WHO Ageing and Health. Available online: https://www.who.int/news-room/fact-sheets/detail/ageing-and-health (accessed on 23 June 2023).
2. Brownie, S. Why Are Elderly Individuals at Risk of Nutritional Deficiency? *Int. J. Nurs. Pract.* **2006**, *12*, 110–118. [CrossRef] [PubMed]
3. Rémond, D.; Shahar, D.R.; Gille, D.; Pinto, P.; Kachal, J.; Peyron, M.-A.; Nunes, C.; Santos, D.; Walther, B.; Bordoni, A.; et al. Understanding the Gastrointestinal Tract of the Elderly to Develop Dietary Solutions That Prevent Malnutrition. *Oncotarget* **2015**, *6*, 13858. [CrossRef] [PubMed]
4. Shang, Y.F.; Miao, J.H.; Zeng, J.; Zhang, T.H.; Zhang, R.M.; Zhang, B.Y.; Wang, C.; Ma, Y.L.; Niu, X.L.; Ni, X.L.; et al. Evaluation of Digestibility Differences for Apple Polyphenolics Using in Vitro Elderly and Adult Digestion Models. *Food Chem.* **2022**, *390*, 133154. [CrossRef] [PubMed]
5. Makran, M.; Miedes, D.; Cilla, A.; Barberá, R.; Garcia-Llatas, G.; Alegría, A. Understanding the Influence of Simulated Elderly Gastrointestinal Conditions on Nutrient Digestibility and Functional Properties. *Trends Food Sci. Technol.* **2022**, *129*, 283–295. [CrossRef]
6. Vural, Z.; Avery, A.; Kalogiros, D.I.; Coneyworth, L.J.; Welham, S.J.M. Trace Mineral Intake and Deficiencies in Older Adults Living in the Community and Institutions: A Systematic Review. *Nutrients* **2020**, *12*, 1072. [CrossRef]
7. Bourre, J.M. Effects of Nutrients (in Food) on the Structure and Function of the Nervous System: Update on Dietary Requirements for Brain. Part 1: Micronutrients. *J. Nutr. Health Aging* **2006**, *10*, 377.
8. Quintaes, K.D.; Diez-Garcia, R.W. The Importance of Minerals in the Human Diet. In *Handbook of Mineral Elements in Food*; Wiley: Hoboken, NJ, USA, 2015; pp. 1–21. ISBN 9781118654316.
9. Lobine, D.; Mahomoodally, M.F. Antioxidants and Cognitive Decline in Elderly. In *Antioxidants Effects in Health*; Elsevier: Amsterdam, The Netherlands, 2022; pp. 651–668.
10. Thangthaeng, N.; Poulose, S.M.; Miller, M.G.; Shukitt-Hale, B. Preserving Brain Function in Aging: The Anti-Glycative Potential of Berry Fruit. *Neuromolecular Med.* **2016**, *18*, 465–473. [CrossRef]
11. Xu, B.J.; Chang, S.K.C. A Comparative Study on Phenolic Profiles and Antioxidant Activities of Legumes as Affected by Extraction Solvents. *J. Food Sci.* **2007**, *72*, S159–S166. [CrossRef]
12. Tang, Y.; Li, X.; Zhang, B.; Chen, P.X.; Liu, R.; Tsao, R. Characterisation of Phenolics, Betanins and Antioxidant Activities in Seeds of Three Chenopodium Quinoa Willd. Genotypes. *Food Chem.* **2015**, *166*, 380–388. [CrossRef]
13. Hirose, Y.; Fujita, T.; Ishii, T.; Ueno, N. Antioxidative Properties and Flavonoid Composition of Chenopodium Quinoa Seeds Cultivated in Japan. *Food Chem.* **2010**, *119*, 1300–1306. [CrossRef]
14. Zhang, B.; Deng, Z.; Ramdath, D.D.; Tang, Y.; Chen, P.X.; Liu, R.; Liu, Q.; Tsao, R. Phenolic Profiles of 20 Canadian Lentil Cultivars and Their Contribution to Antioxidant Activity and Inhibitory Effects on α-Glucosidase and Pancreatic Lipase. *Food Chem.* **2015**, *172*, 862–872. [CrossRef]
15. Safarov, J. Comparative Evaluation of Phenolic and Antioxidant Properties of Red and White Quinoa (*Chenopodium quinoa* Willd.) Seeds. *J. Raw Mater. Process. Foods* **2020**, *1*, 28–33.
16. Bhanja Dey, T.; Chakraborty, S.; Jain, K.K.; Sharma, A.; Kuhad, R.C. Antioxidant Phenolics and Their Microbial Production by Submerged and Solid State Fermentation Process: A Review. *Trends Food Sci. Technol.* **2016**, *53*, 60–74. [CrossRef]
17. Espinosa-Páez, E.; Alanis-Guzmán, M.G.; Hernández-Luna, C.E.; Báez-González, J.G.; Amaya-Guerra, C.A.; Andrés-Grau, A.M. Increasing Antioxidant Activity and Protein Digestibility in Phaseolus Vulgaris and Avena Sativa by Fermentation with the Pleurotus Ostreatus Fungus. *Molecules* **2017**, *22*, 2275. [CrossRef] [PubMed]
18. Ojo, M.A. Phytic Acid in Legumes: A Review of Nutritional Importance and Hydrothermal Processing Effect on Underutilised Species. *Food Res.* **2020**, *5*, 22–28. [CrossRef] [PubMed]
19. Nkhata, S.G.; Ayua, E.; Kamau, E.H.; Shingiro, J.B. Fermentation and Germination Improve Nutritional Value of Cereals and Legumes through Activation of Endogenous Enzymes. *Food Sci. Nutr.* **2018**, *6*, 2446–2458. [CrossRef] [PubMed]
20. Sánchez-García, J.; Muñoz-Pina, S.; García-Hernández, J.; Heredia, A.; Andrés, A. Impact of Air-Drying Temperature on Antioxidant Properties and ACE-Inhibiting Activity of Fungal Fermented Lentil Flour. *Foods* **2023**, *12*, 999. [CrossRef]
21. Sánchez-García, J.; Muñoz-Pina, S.; García-Hernández, J.; Heredia, A.; Andrés, A. Fermented Quinoa Flour: Implications of Fungal Solid-State Bioprocessing and Drying on Nutritional and Antioxidant Properties. *LWT* **2023**, *182*, 114885. [CrossRef]
22. Tungmunnithum, D.; Drouet, S.; Lorenzo, J.M.; Hano, C. Effect of Traditional Cooking and in Vitro Gastrointestinal Digestion of the Ten Most Consumed Beans from the Fabaceae Family in Thailand on Their Phytochemicals, Antioxidant and Anti-Diabetic Potentials. *Plants* **2022**, *11*, 67. [CrossRef]
23. Li, M.; Bai, Q.; Zhou, J.; de Souza, T.S.P.; Suleria, H.A.R. In Vitro Gastrointestinal Bioaccessibility, Bioactivities and Colonic Fermentation of Phenolic Compounds in Different Vigna Beans. *Foods* **2022**, *11*, 3884. [CrossRef]
24. Bohn, T. Dietary Factors Affecting Polyphenol Bioavailability. *Nutr. Rev.* **2014**, *72*, 429–452. [CrossRef] [PubMed]

25. Cárdenas-Castro, A.P.; Pérez-Jiménez, J.; Bello-Pérez, L.A.; Tovar, J.; Sáyago-Ayerdi, S.G. Bioaccessibility of Phenolic Compounds in Common Beans (*Phaseolus vulgaris* L.) after in Vitro Gastrointestinal Digestion: A Comparison of Two Cooking Procedures. *Cereal Chem.* **2020**, *97*, 670–680. [CrossRef]
26. Ozcan, T.; Akpinar-Bayizit, A.; Yilmaz-Ersan, L.; Delikanli, B. Phenolics in Human Health. *Int. J. Chem. Eng. Appl.* **2014**, *5*, 393–396. [CrossRef]
27. Ullah, R.; Ikram, M.; Park, T.J.; Ahmad, R.; Saeed, K.; Alam, S.I.; Rehman, I.U.; Khan, A.; Khan, I.; Jo, M.G.; et al. Vanillic Acid, a Bioactive Phenolic Compound, Counteracts Lps-Induced Neurotoxicity by Regulating c-Jun n-Terminal Kinase in Mouse Brain. *Int. J. Mol. Sci.* **2021**, *22*, 361. [CrossRef]
28. Yalameha, B.; Nejabati, H.R.; Nouri, M. Cardioprotective Potential of Vanillic Acid. *Clin. Exp. Pharmacol. Physiol.* **2023**, *50*, 193–204.
29. Agunloye, O.M.; Oboh, G.; Ademiluyi, A.O.; Ademosun, A.O.; Akindahunsi, A.A.; Oyagbemi, A.A.; Omobowale, T.O.; Ajibade, T.O.; Adedapo, A.A. Cardio-Protective and Antioxidant Properties of Caffeic Acid and Chlorogenic Acid: Mechanistic Role of Angiotensin Converting Enzyme, Cholinesterase and Arginase Activities in Cyclosporine Induced Hypertensive Rats. *Biomed. Pharmacother.* **2019**, *109*, 450–458. [CrossRef]
30. Alam, M.; Ahmed, S.; Elasbali, A.M.; Adnan, M.; Alam, S.; Hassan, M.I.; Pasupuleti, V.R. Therapeutic Implications of Caffeic Acid in Cancer and Neurological Diseases. *Front. Oncol.* **2022**, *12*, 860508.
31. Priscilla, D.H.; Prince, P.S.M. Cardioprotective Effect of Gallic Acid on Cardiac Troponin-T, Cardiac Marker Enzymes, Lipid Peroxidation Products and Antioxidants in Experimentally Induced Myocardial Infarction in Wistar Rats. *Chem.-Biol. Interact.* **2009**, *179*, 118–124. [CrossRef]
32. Wianowska, D.; Olszowy-Tomczyk, M. A Concise Profile of Gallic Acid—From Its Natural Sources through Biological Properties and Chemical Methods of Determination. *Molecules* **2023**, *28*, 1186.
33. Gallego, M.; Arnal, M.; Barat, J.M.; Talens, P. Effect of Cooking on Protein Digestion and Antioxidant Activity of Different Legume Pastes. *Foods* **2021**, *10*, 47. [CrossRef]
34. Koehnlein, E.A.; Koehnlein, É.M.; Corrêa, R.C.G.; Nishida, V.S.; Correa, V.G.; Bracht, A.; Peralta, R.M. Analysis of a Whole Diet in Terms of Phenolic Content and Antioxidant Capacity: Effects of a Simulated Gastrointestinal Digestion. *Int. J. Food Sci. Nutr.* **2016**, *67*, 614–623. [CrossRef]
35. Soetan, K.O.; Olaiya, C.O.; Oyewole, O.E. The Importance of Mineral Elements for Humans, Domestic Animals and Plants: A Review. *Afr. J. Food Sci.* **2010**, *4*, 200–222.
36. Gupta, U.C.; Gupta, S.C. Sources and Deficiency Diseases of Mineral Nutrients in Human Health and Nutrition: A Review. *Pedosphere* **2014**, *24*, 13–38.
37. National Research Council Minerals. *Diet and Health: Implications for Reducing Chronic Disease Risk*; National Academies Press: Washington, DC, USA, 1989; ISBN 0-309-58831-6.
38. Chakraverty, A.; Mujumdar, A.S.; Ramaswamy, H.S. (Eds.) Structure and Composition of Cereal Grains and Legumes. In *Handbook of Postharvest Technology: Cereals, Fruits, Vegetables, Tea, and Spices*; Routledge: London, UK, 2003; Volume 93, pp. 1–16. ISBN 0824705149.
39. Martínez-Villaluenga, C.; Peñas, E.; Hernández-Ledesma, B. Pseudocereal Grains: Nutritional Value, Health Benefits and Current Applications for the Development of Gluten-Free Foods. *Food Chem. Toxicol.* **2020**, *137*, 111178.
40. Chawla, P.; Bhandari, L.; Sadh, P.K.; Kaushik, R. Impact of Solid-state Fermentation (*Aspergillus oryzae*) on Functional Properties and Mineral Bioavailability of Black-eyed Pea (*Vigna unguiculata*) Seed Flour. *Cereal Chem. J.* **2017**, *94*, 437–442. [CrossRef]
41. Couzy, F.; Mansourian, R.; Labate, A.; Guinchard, S.; Montagne, D.H.; Dirren, H. Effect of Dietary Phytic Acid on Zinc Absorption in the Healthy Elderly, as Assessed by Serum Concentration Curve Tests. *Br. J. Nutr.* **1998**, *80*, 177–182. [CrossRef] [PubMed]
42. Menard, O.; Lesmes, U.; Shani-Levi, C.S.; Araiza Calahorra, A.; Lavoisier, A.; Morzel, M.; Rieder, A.; Feron, G.; Nebbia, S.; Mashiah, L.; et al. Static in Vitro Digestion Model Adapted to the General Older Adult Population: An INFOGEST International Consensus. *Food Funct.* **2023**, *14*, 4569–4582. [CrossRef] [PubMed]
43. Brodkorb, A.; Egger, L.; Alminger, M.; Alvito, P.; Assunção, R.; Ballance, S.; Bohn, T.; Bourlieu-Lacanal, C.; Boutrou, R.; Carrière, F.; et al. INFOGEST Static in Vitro Simulation of Gastrointestinal Food Digestion. *Nat. Protoc.* **2019**, *14*, 991–1014. [CrossRef]
44. Minekus, M.; Alminger, M.; Alvito, P.; Ballance, S.; Bohn, T.; Bourlieu, C.; Carrière, F.; Boutrou, R.; Corredig, M.; Dupont, D.; et al. A Standardised Static in Vitro Digestion Method Suitable for Food-an International Consensus. *Food Funct.* **2014**, *5*, 1113–1124. [CrossRef]
45. Chang, C.H.; Lin, H.Y.; Chang, C.Y.; Liu, Y.C. Comparisons on the Antioxidant Properties of Fresh, Freeze-Dried and Hot-Air-Dried Tomatoes. *J. Food Eng.* **2006**, *77*, 478–485. [CrossRef]
46. Thaipong, K.; Boonprakob, U.; Crosby, K.; Cisneros-Zevallos, L.; Hawkins Byrne, D. Comparison of ABTS, DPPH, FRAP, and ORAC Assays for Estimating Antioxidant Activity from Guava Fruit Extracts. *J. Food Compos. Anal.* **2006**, *19*, 669–675. [CrossRef]
47. Sharma, S.; Kataria, A.; Singh, B. Effect of Thermal Processing on the Bioactive Compounds, Antioxidative, Antinutritional and Functional Characteristics of Quinoa (*Chenopodium quinoa*). *LWT* **2022**, *160*, 113256. [CrossRef]
48. Tanleque-Alberto, F.; Juan-Borrás, M.; Escriche, I. Antioxidant Characteristics of Honey from Mozambique Based on Specific Flavonoids and Phenolic Acid Compounds. *J. Food Compos. Anal.* **2020**, *86*, 103377. [CrossRef]
49. Haug, W.; Lantzsch, H.-J. Sensitive Method for the Rapid Determination of Phytate in Cereals and Cereal Products. *J. Sci. Food Agric.* **1983**, *34*, 1423–1426. [CrossRef]

50. Peng, W.; Tao, Z.; Ji Chun, T. Phytic Acid Contents of Wheat Flours from Different Mill Streams. *Agric. Sci. China* **2010**, *9*, 1684–1688. [CrossRef]
51. Barrera, C.; Betoret, N.; Corell, P.; Fito, P. Effect of Osmotic Dehydration on the Stabilization of Calcium-Fortified Apple Slices (Var. Granny Smith): Influence of Operating Variables on Process Kinetics and Compositional Changes. *J. Food Eng.* **2009**, *92*, 416–424. [CrossRef]
52. Chen, L.; Li, X.; Li, Z.; Deng, L. Analysis of 17 Elements in Cow, Goat, Buffalo, Yak, and Camel Milk by Inductively Coupled Plasma Mass Spectrometry (ICP-MS). *RSC Adv.* **2020**, *10*, 6736–6742. [CrossRef]

molecules

MDPI

Article

The Content of Bioactive Compounds and Technological Properties of Matcha Green Tea and Its Application in the Design of Functional Beverages

Katarzyna Najman [1], Anna Sadowska [1,*], Monika Wolińska [1], Katarzyna Starczewska [1] and Krzysztof Buczak [2]

[1] Department of Functional and Organic Food, Institute of Human Nutrition Sciences, Warsaw University of Life Sciences, Nowoursynowska 159c, 02-776 Warsaw, Poland; katarzyna_najman@sggw.edu.pl (K.N.)

[2] Department of Surgery, Faculty of Veterinary Medicine, Wroclaw University of Environmental and Life Science, Pl. Grunwadzki 51, 50-366 Wroclaw, Poland; krzysztof.buczak@upwr.edu.pl

* Correspondence: anna_sadowska@sggw.edu.pl

Abstract: Matcha is a powdered green tea obtained from the *Camellia sinensis* L. plant intended for both "hot" and "cold" consumption. It is a rich source of bioactive ingredients, thanks to which it has strong antioxidant properties. In this research, an organoleptic evaluation was carried out, and the physical characteristics (i.e., instrumental color measurement ($L^*a^*b^*$), water activity, water solubility index (WSI), water holding capacity (WHC) of 10 powdered Matcha green teas, and in the 2.5% Matcha water solutions, pH, °Brix and osmolality were tested. Also, the content of phenolic ingredients, i.e., selected phenolic acids, flavonoids and total polyphenols, was assessed. The content of chlorophyll, vitamin C and antioxidant potential were also examined. Matcha M-4 was used to design two functional model beverages, in the form of ready-to-use powdered drinks, consisting of Matcha green tea, protein preparations, inulin, maltodextrin and sugar. The obtained powdered drink, when dissolved in the preferred liquid (water, milk, juice), is regenerative, high-protein and rich in bioactive ingredients from the Matcha drink, with prebiotic properties derived from the added inulin. The beverage is also characterized by low osmolality. It can be recommended as a regenerating beverage for a wide group of consumers, athletes and people with deficiencies, among others protein, and elderly people, as well as in the prevention and supportive treatment of bone and joint tissue diseases.

Keywords: Matcha; *Camelia sinensis*; green tea; polyphenols; antioxidant activity; $L^*a^*b^*$; WHC; WSI; °Brix; pH; osmolality; functional drinks

Citation: Najman, K.; Sadowska, A.; Wolińska, M.; Starczewska, K.; Buczak, K. The Content of Bioactive Compounds and Technological Properties of Matcha Green Tea and Its Application in the Design of Functional Beverages. *Molecules* **2023**, *28*, 7018. https://doi.org/10.3390/molecules28207018

Academic Editors: Sascha Rohn, Jason Tze Cheng Tzen and Michał Halagarda

Received: 29 August 2023
Revised: 21 September 2023
Accepted: 6 October 2023
Published: 10 October 2023

1. Introduction

In recent years, there has been an increasing interest in functional products and low-processed food naturally rich in antioxidants that can protect against the harmful effects of free radicals and oxidative stress. Such products, rich in bioactive substances, include Matcha green tea, which is gaining more and more popularity around the world [1,2].

Matcha is obtained from the *Camellia sinensis* L. plant (*Thea sinensis*), which in the botanical classification belongs to the plant kingdom (*Plantae*), the vascular group (*Tracheophyta*), the tea family (*Theaceae*), the genus (*Camellia*). There are two types of Matcha tea: Matcha-Koicha and Matcha-Usucha, intended for making a decoction or infusion, both "hot" and "cold". Koicha has the form of thick, viscous tea with an intense, bitter aroma, while Usucha has a diluted form due to the fact that it is brewed in more water than Matcha-Usucha [3–5].

Matcha comes from the regions of Nishio and Uji Tawara in Aichi Prefecture (Japan). Its history dates back to the 13th century when Japanese Zen monks used it as a means of relaxing and maintaining concentration during long hours of meditation. It is commonly consumed during the Japanese Cha-no-you tea ceremony. It is part of the Japanese culinary culture and is gaining more and more popularity among consumers of other cultures and regions of the world [1–3,6].

Three weeks before harvest, tea bushes are covered with reed mats to protect the young leaves from direct sunlight [7]. Thanks to this, the content of amino acids and bioactive compounds (especially chlorophyll and theanine) increases; this is responsible for the characteristic color and taste of Matcha [1]. The tea harvest begins at the end of April; the first one lasts until the end of May, the next ones fall at the turn of June and July, while the third harvest falls in August. The leaves collected from the top of the shoot are of the best quality [6,8].

Black and Oolong teas are produced as a result of total or partial fermentation of *Camelia sinensis* L. leaves, leading to their typical black or brown color, which is a result of enzymatic oxidation of catechins present in leaves [9,10]. Unlike them, green tea (from which Matcha is made) is produced without the fermentation process [8,11]. After the leaves are harvested, the first stage in the production of green teas is high-temperature heat treatment (steaming, roasting or steaming at 80–90 °C), preventing oxidation [12,13]. At this stage, enzymes (polyphenol oxidase and peroxidase), catalyzing the reactions of catechins contained in the leaves with oxygen, are inactivated [9,10,14]. This prevents fermentation processes and inhibits the decomposition of color pigments contained inside the leaves, which allows the tea to retain its intense green color [11,14]. The next step is drying at a moderate temperature, thanks to which the leaves shrink, and finally grinding them into a powder in granite mills, i.e., slowly rotating stones [8,12]. The final stage is roasting at a high temperature, which allows for an intense tea flavor [8,12,13].

Due to the unique composition of bioactive compounds, Matcha has a wide range of health benefits. It is characterized by a high content of antioxidant compounds, which results from the method of shade cultivation and leads to the production of larger amounts of amino acids and bioactive compounds [1]. Matcha contains high concentrations of phenolic acids], rutin, quercetin [2,13,15–22], theanine, chlorophyll and other carotenoid [12,23] amounts exceeding traditional green teas. Due to the richness of bioactive ingredients, especially those with antioxidant properties (polyphenols, mainly flavonoids) [12,15,24–27], it is more and more often used in the prevention and treatment of many civilization diseases, e.g., heart disease, diabetes and hypertension [1,13,22,24,28,29]. Its infusions can also be used in the prevention of inflammatory [1,30,31] or viral diseases [20,32]. Matcha may have a positive effect on weight loss by reducing the level of triglycerides, fat and glucose in the blood and additionally may contribute to increasing muscle mass [22,28,29]. Matcha is also known for its anti-aging properties [33].

In recent years, Matcha has become an increasingly popular product among consumers [6]. On the food market, Matcha is most often found in powdered form [16,24,26], intended for both "hot" and "cold" consumption [4,34], but it is also increasingly used as an additive to various products, such as bars, jellies, cakes, cookies, chocolates, candies, puddings, drinks, cocktails or ice cream [4,34], becoming a promising ingredient in the functional food industry [22,35–37].

Despite the ever-growing popularity of this product, knowledge about Matcha green tea is insufficient and constantly expanded with scientific research, which is why this research was carried out in two stages. In the first stage of this research, the physicochemical and bioactive properties of 10 types of Matcha green tea products available on the Polish market were assessed. Physicochemical tests (solubility, degree of water binding, color, osmolality, soluble solids content) were carried out in powdered teas. The content of bioactive components, i.e., selected phenolic acids and flavonoids, total polyphenols, chlorophyll, vitamin C and antioxidant activity, were also assessed. In the second stage of the research, the composition of powdered beverages containing Matcha green tea, selected from among those tested in the first stage of the research, was designed as a source of bioactive substances as well as to shape the taste profile of the designed beverages. Therefore, in this study, research was conducted on designing innovative functional beverages in powder form, containing bioactive ingredients from Matcha and easily digestible proteins from milk or enzymatically hydrolyzed collagen proteins and additionally containing a prebiotic (inulin). Such composition and convenient form (powder) allow direct consumption of the drink after dissolving it in water, milk or juice according to consumer preference. Designed functional beverages can be intended for a wide group of consumers, especially physically active people, as a recovery drink after exercise due to its rich composition and low osmolality (hypotonic and isotonic drink), as well as for the elderly in the prevention and treatment of diseases of joint and bone tissue.

2. Results

2.1. Physicochemical and Bioactive Properties of Matcha Green Tea

2.1.1. Physicochemical Properties of Matcha Green Tea

Organoleptic Evaluation of Matcha Green Tea Powders

Figure 1 presents the general appearance of the tested samples of 10 market products—Matcha green tea powders.

Figure 1. The general appearance of the tested Matcha green tea powders. Symbols from M-1 to M-10 indicate the 10 tested types of green Matcha tea.

As results from the organoleptic evaluation of Matcha teas, the tested samples were characterized by a diversified general appearance and color. Among the evaluated products, samples M-6 and M-10 were characterized by a lower degree of fragmentation and a darker color compared to the remaining teas, showing a higher degree of fragmentation and a saturated, light green color. All evaluated samples were characterized by an intense tea aroma and a very bitter taste. No visible lumps were found in the tested products; the powders were loose and dry, with the consistency of a light, aerated powder. No solid impurities were found in the tested products.

Instrumental Measurement of the Color Parameters of Matcha Green Tea Powders

The results of the instrumental color measurement of the tested Matcha samples in the $L^*a^*b^*$ color space are summarized in Table 1. As results from the conducted research, the tested Matcha samples differed significantly ($p \leq 0.05$) in color parameters.

In terms of the L^* parameter, defining the brightness, significant differences were found in the vast majority of the tested samples. This parameter ranged from 45.90 ± 0.58 to 68.76 ± 2.51, with the highest value in sample M-8 (68.76 ± 2.51) and the lowest in M-10, which means that these products showed the greatest differences in terms of brightness.

The color parameter a^*, referring to red ($+a^*$) and green ($-a^*$) in most of the tested samples had negative values, which means that shades of green prevailed in these Matcha samples. The lowest ($p \leq 0.05$) value of the a^* parameter was found in the M-4 (-4.21 ± 0.15), which means that the color of this Matcha tea was shifted towards green shades to the greatest extent. On the other hand, positive values for the a^* parameter were found in samples M-3, M-6 and M-10, which means that they were characterized by the largest color shift towards red, with the highest values in M-6 and M-10 (average 0.96 ± 0.04).

In terms of the b^* parameter, defining the yellow tone, the highest values were found in samples M-2 and M-7 (average 34.19 ± 0.49), were slightly lower in M-4 and M-8 (average 31.75 ± 0.61) and were the lowest in M-10 and M-6 (average 19.53 ± 0.72), which meant that these products showed the fewest shades of yellow.

Table 1. Color parameters (CIE $L^*a^*b^*$) in the tested Matcha green tea powders.

Sample	L^*	a^*	b^*
M-1	63.15 ± 0.58 [d]	-1.10 ± 0.02 [c]	30.17 ± 0.74 [d]
M-2	59.82 ± 0.31 [c]	-2.29 ± 0.11 [b]	34.04 ± 0.36 [g]
M-3	55.25 ± 1.45 [b]	0.71 ± 0.09 [g]	26.69 ± 0.58 [b]
M-4	56.28 ± 2.10 [b]	-4.21 ± 0.15 [a]	31.51 ± 0.69 [ef]
M-5	56.44 ± 0.31 [b]	-0.55 ± 0.04 [e]	30.40 ± 1.11 [de]
M-6	46.35 ± 1.45 [a]	0.96 ± 0.06 [h]	19.78 ± 0.73 [a]
M-7	64.10 ± 0.72 [d]	-0.74 ± 0.04 [d]	34.35 ± 0.62 [g]
M-8	68.76 ± 2.51 [e]	-1.07 ± 0.05 [c]	31.99 ± 0.39 [f]
M-9	57.02 ± 0.35 [b]	-0.33 ± 0.07 [f]	28.80 ± 0.60 [c]
M-10	45.90 ± 0.58 [a]	0.94 ± 0.04 [h]	19.28 ± 0.45 [a]

Mean values ± standard deviation with different letters ([a–h]) in the same column differ significantly (Duncan's test, $p \leq 0.05$).

Water Activity, Water Solubility Index (WSI) and Water Holding Capacity (WHC) of Matcha Green Tea Powders

The results of water activity (a_w), water solubility index (WSI) and water holding capacity (WHC) in the tested Matcha green tea powders are presented in Table 2.

Table 2. Water activity (a_w), water solubility index (WSI) and water holding capacity (WHC) in the tested Matcha green tea powders.

Sample	Water Activity (a_w)	WSI (%)	WHC (g/g)
M-1	0.2655 ± 0.001 [c]	25.45 ± 0.56 [f]	2.04 ± 0.01 [b]
M-2	0.3000 ± 0.000 [d]	22.14 ± 0.48 [d]	2.04 ± 0.16 [b]
M-3	0.3024 ± 0.000 [d]	23.30 ± 0.45 [e]	3.06 ± 0.18 [e]
M-4	0.4082 ± 0.003 [h]	23.33 ± 0.20 [e]	2.19 ± 0.21 [bc]
M-5	0.2238 ± 0.001 [b]	21.05 ± 0.37 [c]	2.44 ± 0.16 [cd]
M-6	0.3926 ± 0.003 [g]	17.43 ± 0.76 [a]	3.71 ± 0.02 [f]
M-7	0.3743 ± 0.001 [f]	25.80 ± 0.22 [f]	1.48 ± 0.09 [a]
M-8	0.3375 ± 0.001 [e]	22.72 ± 0.18 [de]	2.01 ± 0.14 [b]
M-9	0.1546 ± 0.002 [a]	20.89 ± 0.23 [e]	2.47 ± 0.14 [d]
M-10	0.3956 ± 0.004 [g]	18.80 ± 0.21 [b]	3.56 ± 0.26 [f]

Mean values ± standard deviation with different letters ([a–h]) in the same column differ significantly (Duncan's test, $p \leq 0.05$).

As the conducted research shows, Matcha teas significantly ($p \leq 0.05$) differed in terms of water activity (a_w), reaching values ranging from 0.1546 ± 0.002 to 0.4082 ± 0.003. The highest water activity (0.4082 ± 0.003) was shown by Matcha M-4 (0.4082 ± 0.003); a lower water activity was shown by M-10 and M-6 (which reached a_w values at the average level 0.3941 ± 0.003), while the lowest a_w ($p \leq 0.05$) was distinguished by M-9 (0.1546 ± 0.002). All the tested products were characterized by a relatively low water activity, which proves their high durability and microbiological safety, due to the fact that most microorganisms do not develop and grow at a water activity level of $a_w < 0.60$.

According to the conducted tests, the water solubility index (WSI) of the tested Matcha powders was 22.09 ± 2.59%, with significant differences in this parameter for the tested products. The highest WSI was found in M-4 and M-1 (average 25.63 ± 0.43%); it was lower in Matcha M-7, M-3, M-8 and M-9, with an average WSI of 22.56 ± 1.07%. The lowest WSI was found in M-6, which dissolved only to about 17% in water at room temperature.

In the case of water holding capacity (WHC), the tested Matcha samples bound an average of 2.50 ± 0.51 g of water per g of powder, with significant differences between the products tested in the study. The highest WHC was shown by Matcha M-6 and M-10 (3.64 ± 0.19 g/g on average); a significantly lower WHC was shown by Matcha M-3 (3.06 ± 0.18 g/g) and M-9 and M-5 (on average 2.45 ± 0.14 g/g). Four of the tested Matcha products, i.e., M-7, M-1, M-2 and M-8, did not differ statistically in this parameter and bound an average of 2.07 ± 0.15 g of water per one gram of powder, while the lowest WHC was in M-4 (1.48 ± 0.09 g/g).

pH, Soluble Solids Content and Osmolality in 2.5% Aqueous Solutions of Matcha Green Tea

In the tested market products of Matcha, basic physicochemical parameters such as pH, soluble solids content (°Brix) and osmolality were assessed, and the results are presented in Table 3.

Table 3. pH, °Brix and osmolality of 2.5% aqueous Matcha green tea solutions.

Sample	pH	°Brix (%)	Osmolality (mOsm/kg·H_2O)
M-1	5.84 ± 0.01 ^{c}	1.23 ± 0.06 ^{abc}	1.00 ± 0.00 ^{a}
M-2	5.83 ± 0.02 ^{c}	1.30 ± 0.10 ^{cde}	2.33 ± 0.58 ^{bc}
M-3	5.59 ± 0.02 ^{a}	1.13 ± 0.06 ^{a}	1.00 ± 0.00 ^{a}
M-4	5.94 ± 0.03 ^{d}	1.40 ± 0.10 ^{e}	5.67 ± 0.58 ^{d}
M-5	5.61± 0.02 ^{a}	1.27 ± 0.06 ^{bcd}	3.33 ± 0.58 ^{c}
M-6	5.63 ± 0.09 ^{a}	1.17 ± 0.06 ^{ab}	1.67 ± 0.58 ^{ab}
M-7	5.72 ± 0.05 ^{b}	1.37 ± 0.06 ^{de}	5.33 ± 0.58 ^{d}
M-8	5.75 ± 0.02 ^{b}	1.27 ± 0.06 ^{bcd}	3.00 ± 0.00 ^{c}
M-9	5.61 ± 0.04 ^{a}	1.13 ± 0.06 ^{a}	0.67 ± 0.58 ^{a}
M-10	5.58 ± 0.04 ^{a}	1.17 ± 0.06 ^{ab}	1.67 ± 0.58 ^{ab}

Mean values ± standard deviation with different letters ($^{a–e}$) in the same column differ significantly (Duncan's test, $p \leq 0.05$).

The average pH of aqueous solutions (2.5%) of all tested Matcha teas was 5.71 ± 0.13, with the products significantly ($p \leq 0.05$) differing in this parameter. The highest pH was found in M-4 (5.94 ± 0.03); it was significantly lower in M-1 and M-2 (average 5.84 ± 0.01) and M-8 and M-7, with an average pH value of 5.74 ± 0.04. The lowest ($p \leq 0.05$) pH values were found in M-6, M-5, M-9 and M-10 teas (average 5.60 ± 0.04), with no significant differences between these Matcha teas. The average soluble solids content (°Brix) in 2.5% water solutions of Matcha green tea powders was 1.24 ± 0.11%, with the highest °Brix in the aqueous solution of M-4 tea (1.40 ± 0.10%) and the lowest in M-3 and M-10 (average 1.13 ± 0.05%). As in the case of soluble solids content, the tested Matcha water solutions were characterized by very low osmolality (on average for all samples 2.57 ± 1.77 mOsm/kg·H_2O), similar to the osmolality of water, although significant ($p \leq 0.05$) differences in this parameter were found. The lowest osmolality was found in M-9, M-1, M-3, M-6 and M-10, with an average value of 1.20 ± 0.56 mOsm/kg·H_2O; it was slightly higher in M-2, M-8 and M-5 (2.89 ± 0.78 mOsm/kg·H_2O), while the highest was shown by two water solutions of the tested Matcha teas, i.e., M-7 and M-4 (5.50 ± 0.55 mOsm/kg·H_2O).

2.1.2. Bioactive Properties of Matcha Green Tea

Phenolic Ingredients in Matcha Green Tea

The research carried out in the study showed a high content of phenolic compounds, both determined by the HPLC method (Table 4, Figure 2a) and by the spectrophotometric method using the Folin–Ciocalteu reagent (Figure 2b), in all tested Matcha green teas.

According to the conducted research (Table 4), Matcha green tea was characterized by varied contents of determined phenolic acids, i.e., from 2.96 ± 0.03 to 25.10 ± 0.47 mg/g d.m. (gallic acid) and from 5.78 ± 0.17 to 41.16 ± 0.33 mg/g d.m. (*p*-coumaric acid), with the significantly ($p \leq 0.05$) highest content of these bioactive ingredients found in M-4 (total 66.26 ± 0.79 mg/g d.m.) and the lowest in the M-2 sample (total 8.74 ± 0.17 mg/g d.m.).

Table 4. Selected phenolic compounds identified by the HPLC method in the tested Matcha green tea powders.

Sample	Gallic Acid	P-Coumaric Acid	Catechin	Epigallocatechin	Gallate Epigallocatechin	Quercetin	Rutoside-3-*O*-quercetin
M-1	3.75 ± 0.19 ^{b}	20.64 ± 0.53 ^{f}	3.41 ± 0.10 ^{e}	3.43 ± 0.33 ^{d}	22.67 ± 1.03 ^{b}	0.45 ± 0.01 ^{d}	3.46 ± 0.18 ^{a}
M-2	2.96 ± 0.03 ^{a}	5.78 ± 0.17 ^{ab}	2.49 ± 0.15 ^{d}	5.51 ± 0.11 ^{f}	10.41 ± 0.15 ^{a}	0.75 ± 0.02 ^{e}	5.22 ± 0.37 ^{b}
M-3	2.86 ± 0.06 ^{a}	15.94 ± 0.77 ^{e}	1.68 ± 0.05 ^{c}	4.43 ± 0.21 ^{e}	26.13 ± 1.35 ^{c}	0.95 ± 0.01 ^{g}	13.00 ± 3.08 ^{c}
M-4	25.10 ± 0.47 ^{h}	41.16 ± 0.33 ^{g}	0.80 ± 0.01 ^{a}	6.61 ± 0.08 ^{g}	25.66 ± 0.70 ^{c}	0.29 ± 0.00 ^{bc}	2.31 ± 0.34 ^{a}
M-5	6.58 ± 0.08 ^{c}	7.32 ± 1.31 ^{cd}	1.14 ± 0.05 ^{b}	1.39 ± 0.24 ^{b}	27.39 ± 0.42 ^{c}	0.31 ± 0.01 ^{c}	2.66 ± 0.13 ^{a}
M-6	22.65 ± 0.05 ^{f}	6.27 ± 0.37 ^{bc}	0.69 ± 0.01 ^{a}	6.68 ± 0.03 ^{g}	38.45 ± 0.26 ^{d}	0.28 ± 0.01 ^{b}	2.30 ± 0.12 ^{a}
M-7	24.16 ± 0.50 ^{g}	7.54 ± 0.61 ^{d}	4.70 ± 0.05 ^{f}	4.25 ± 0.40 ^{e}	36.79 ± 3.63 ^{d}	0.23 ± 0.01 ^{a}	2.23 ± 0.34 ^{a}
M-8	7.33 ± 0.39 ^{d}	6.48 ± 0.67 ^{bcd}	3.52 ± 0.22 ^{e}	1.00 ± 0.02 ^{a}	25.49 ± 0.11 ^{c}	0.44 ± 0.03 ^{d}	2.51 ± 0.09 ^{a}
M-9	3.63 ± 0.19 ^{b}	7.15 ± 0.43 ^{cd}	1.03 ± 0.06 ^{b}	2.38 ± 0.09 ^{c}	11.62 ± 0.10 ^{a}	0.29 ± 0.00 ^{bc}	2.52 ± 0.28 ^{a}
M-10	18.79 ± 0.46 ^{e}	4.95 ± 0.08 ^{a}	2.51 ± 0.06 ^{d}	4.19 ± 0.12 ^{e}	22.46 ± 2.08 ^{b}	0.84 ± 0.03 ^{f}	2.29 ± 0.07 ^{a}

Mean values ± standard deviation with different letters ($^{a-h}$) in the same column differ significantly (Duncan's test, $p \leq 0.05$).

The tested Matcha green teas were characterized by a varied content of flavonoids, and the sum of these compounds ranged from 17.84 ± 0.17 mg/g d.m. (M-9) up to 48.40 ± 0.36 mg/g d.m. (M-7). Among the determined flavonoids, flavanols dominated, including catechins, and the main flavanol in all tested Matcha green teas was gallate epigallocatechin, with an average content of 24.71 ± 8.78 mg/g d.m. The highest ($p \leq 0.05$) content of this flavonoid was found in samples M-7 and M-6 (average 37.62 ± 2.48 mg/g d.m.), and the lowest was found in M-9 and M-2 (average 11.02 ± 0, 68 mg/g d.m.). The remaining flavanols were present in much lower concentrations, with the epigallocatechin content being over six times lower and the catechin content over eleven times lower, for all analyzed Matcha green tea samples, reaching an average value of 3.99 ± 1.92 mg/g d.m. and 2.20 ± 1.31 mg/g d.m., respectively.

Regarding the flavanols determined in the tested Matcha samples, the content of rutoside-3-*O*-quercetin was over eight times higher compared to the content of quercetin; the significantly ($p \leq 0.05$) highest concentration of these flavonoids was found in the M-3 sample (13.00 ± 3.08 and 0.95 ± 0.01 mg/g d.m., respectively), and the lowest ($p \leq 0.05$) was found in the M-7 sample (2.23 ± 0.34 and 0.23 ± 0.01 mg/g d.m., respectively) (Table 4). Figure 2a shows the sum of identified phenolic compounds determined by the HPLC method. The average content of phenolic bioactive ingredients identified by the chromatographic method was 59.33 ± 21.74 mg/g d.m., and there were statistically significant differences between the tested Matcha green tea samples. The highest ($p \leq 0.05$) content of these ingredients was found in M-4 (101.91 ± 1.12 mg/g d.m.); it was significantly ($p \leq 0.05$) lower in samples M-7 and M-6 (average 78.60 ± 2.62 mg/g d.m.), M-3 (64.99 ± 2.44 mg/g d.m.) or M-1 and M-10 (average 56.92 ± 2.08 mg/g d.m.), while the lowest ($p \leq 0.05$) was found in M-9 (28.62 ± 0.50 mg/g d.m.).

Similar trends were also noted for the content of total polyphenolic compounds, determined by the spectrophotometric method using the Folin–Ciocalteu reagent (Figure 2b).

As can be seen from the data in Figure 2b, the samples were characterized by an average content of total polyphenols at the level of 136.76 ± 28.47 mg GAE/g d.m., with significant ($p \leq 0.05$) differences between individual Matcha products. The highest polyphenol content was found in Matcha M-4 (190.96 ± 2.14 mg GAE/g d.m.); it was lower in M-7 (169.42 ± 0.43 mg GAE/g d.m.) and M-6 (155.28 ± 0.51 mg GAE/g d.m.). The lowest content of total polyphenols among all tested Matcha samples was recorded in M-9

(89.91 ± 3.17 mg GAE/g d.m.) and it was more than two times lower than in Matcha M-4, with the highest concentration of these bioactive ingredients.

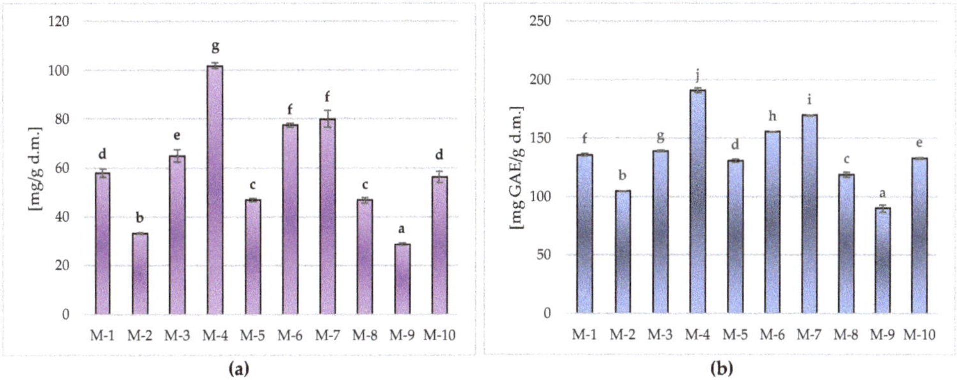

Figure 2. Phenolic ingredients in the tested Matcha green tea samples: the sum of identified phenolic compounds determined by the HPLC method (**a**) and the total polyphenols content measured by the spectrophotometric method (**b**). Mean values marked in bars by different letters ([a–j]) differ significantly (Duncan's test, $p \leq 0.05$).

Vitamin C and Chlorophylls in Matcha Green Tea

As can be seen from the data presented in Figure 3a, the tested samples of powdered Matcha green tea differed significantly ($p \leq 0.05$) in terms of vitamin C content. The highest ($p \leq 0.05$) concentration of this ingredient was found in sample M-4 (2.03 ± 0.12 mg/g d.m.); it was significantly ($p \leq 0.05$) lower in M-7 (1.54 ± 0.03 mg/g d.m.), M-3 and M-6 (average 1.36 ± 0.07 mg/g d.m.), M-10, M-1, M-5 and M-8 (average 1.08 ± 0.05 mg/g d.m.) or in sample M-2 (0.81 ± 0.02 mg/g d.m.). The lowest ($p \leq 0.05$) vitamin C concentration was recorded in sample M-9, in which the amount of this bioactive ingredient was almost six times lower than in sample M-4, reaching only 0.36 ± 0.02 mg/g s.m.

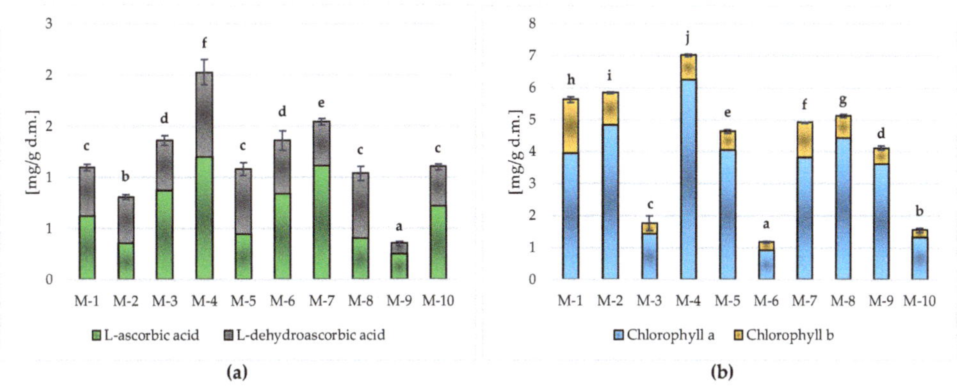

Figure 3. Vitamin C (**a**) and chlorophylls (**b**) content in the tested Matcha green tea samples. Mean values marked in bars by different letters ([a–j]) differ significantly (Duncan's test, $p \leq 0.05$).

The average content of total chlorophylls in the tested Matcha green tea samples (Figure 3b) was within a very wide range, i.e., from 1.16 ± 0.03 mg/g d.m. (M-6) up to 7.01 ± 0.05 mg/g d.m. (M-4), and all tested market products differed significantly ($p \leq 0.05$) in terms of the content of these bioactive ingredients. In all tested Matcha green tea samples, the main chlorophyll was chlorophyll a, whose share was on average 82.68 ± 5.59%.

Antioxidant Activity in Matcha Green Tea

All teas tested in the study were characterized by high antioxidant activity in a range from 1443.56 ± 3.61 µM TEAC/g d.m. (in M-9) to 2076.35 ± 59.11 µM TEAC/g d.m. (in M-4) (Figure 4), with significant ($p \leq 0.05$) differences between the tested products. Slightly lower antioxidant potential compared to M-4 was found in M-7 (1814.88 ± 14.45 µM TEAC/g d.m.), M-3 and M-6 (average 1705.39 ± 8.11 µM TEAC/g d.m.) or M-10, for which the antioxidant activity reached the values of 1624.94 ± 7.82 µM TEAC/g d.m. Among the Matcha samples tested in the study, M-1 and M-5 showed significantly ($p \leq 0.05$) lower abilities to deactivate synthetic cation radicals $ABTS^{+\bullet}$, with an average value of 1579.00 ± 4.63 µM TEAC/g d.m., or M-2 and M-8 (1533.53 ± 11.65 µM TEAC/g d.m.).

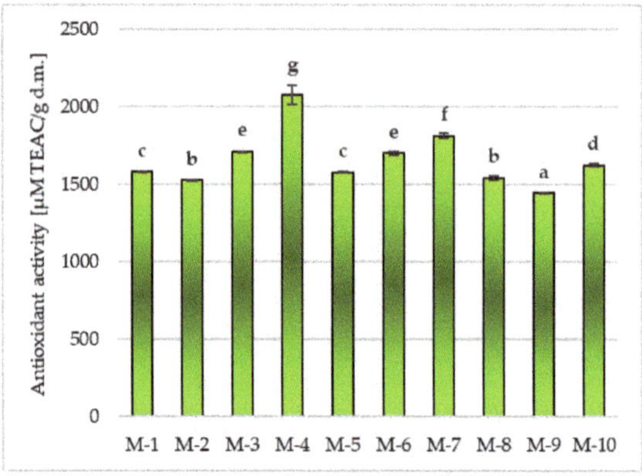

Figure 4. Antioxidant activity in the tested Matcha green tea samples. Mean values marked in bars by different letters ($^{a-g}$) differ significantly (Duncan's test, $p \leq 0.05$).

2.2. The Possibility of Using Matcha Green Tea in the Design of Functional Protein Drinks

In the second stage of the research, it was decided to design a model functional beverage with the addition of Matcha green tea. For this purpose, a sample M-4 tea was selected. All tested samples can be a valuable source of polyphenolic compounds in the daily diet, but sample M-4 had the highest antioxidant properties and the highest polyphenol content. Taking into account the obtained results of physicochemical properties, it can be assumed that for the design of a model functional beverage, it would be possible to use any of the tested samples and obtain beverages with similar properties.

The recipe for the composition of two drinks was developed in the form of a ready-to-use powdered mix, which should be mixed with water, milk or another selected liquid product, e.g., a vegetable drink. The recipe composition and appearance of the designed beverages are presented in Table 5.

In both drinks, the addition of maltodextrin as a source of complex carbohydrates and inulin as a source of dietary fiber was used. The addition of inulin and maltodextrin resulted in a slight thickening of the drinks. The drinks also contained the addition of protein preparations—hydrolyzed collagen protein and/or whey isolate. Collagen after dissolving is transparent and tasteless and does not affect the color and taste of the drink; therefore, its addition was used in both drinks. In the case of a drink prepared on the basis of milk, the addition of whey isolate was included in its composition, which also did not affect the color of the drink. The addition of collagen enriched the drink with protein without affecting the taste, consistency and color, while the addition of whey isolate deepened the milky taste of the drink. A slight sediment was visible in the beverages due to the presence of insoluble tea powder fractions.

The first designed drink was to resemble the so-called "Matcha Latte", which is tea with an addition of milk. To prepare it, milk and tea brewed or mixed with cold water were mixed. One of the advantages of this drink was its characteristic, vivid green color. Matcha is so intense in taste that a small amount is enough to give the drink the right taste. The drink gained a taste characteristic of tea, green, slightly earthy and bitter, balanced by the sweet taste of milk and the addition of sugar.

The second drink designed was a water-based drink. This drink was supposed to be similar to the so-called "Cold Brew Matcha" or "Cold Matcha Tea", which is mixed with cold water and optionally sweetened as desired.

Table 5. Recipe composition of functional drinks with Matcha flavor based on milk and water.

Component	**Matcha Drink Based on Milk**		**Matcha Drink Based on Water**	
General appearance of Matcha drinks				
	[g/100 g drink]	[g/100 g powder]	[g/100 g of drink]	[g/100 g powder]
Water	43	-	90	-
Milk 1.5% fat	43	-	-	-
Inulin	3	21	2.5	25
Collagen hydrolyzate	3	21	3	30
Whey isolate	3.5	25	-	-
Maltodextrin	2.5	18	2.5	25
Sugar	1.5	11	1.5	15
Matcha M-4	0.5	4	0.5	5

2.2.1. Physicochemical Evaluation of Designed Functional Protein Drinks

The designed mixes of ready-to-use powders for the drinks preparation and the prepared Matcha functional drinks were evaluated in terms of basic physicochemical characteristics, including a_w (powder mix) and pH, °Brix and osmolality (liquid beverages, ready to drink), and the obtained results are presented in Table 6.

Table 6. Water activity (powder mix), pH, soluble solids content (°Brix) and osmolality of designed Matcha functional drinks.

Parameter	**Matcha Drink Based on Milk**	**Matcha Drink Based on Water**
Water activity (a_w)	0.19 ± 0.00^{a}	0.27 ± 0.01^{b}
pH	6.46 ± 0.05^{a}	6.58 ± 0.02^{b}
°Brix	16.33 ± 1.03^{b}	8.83 ± 1.24^{a}
Osmolality (mOsm/kg·H_2O)	313.00 ± 5.65^{b}	147.00 ± 1.63^{a}

Mean values ± standard deviation with different letters (^{a,b}) in the same line differ significantly (Duncan's test, $p \leq 0.05$).

The water activity (a_w) of the ready-to-use powders was 0.186 for the milk-based drink and 0.272 for the water-based Matcha drink. The beverages had pH values ranging from 6.46 for the milk-based drink to 6.58 for the water-based drink. The designed drinks also differed significantly in terms of soluble solids content and osmolality, with significantly ($p \leq 0.05$) higher °Brix and osmolality for the Matcha drink based on milk (16.33 ± 1.03% and 313 ± 5.65 mOsm/kg·H_2O). In the Matcha drink based on water, these parameters were about two times lower, 8.83 ± 1.24% (°Brix) and 147 ± 1.63 mOsm/kg·H_2O (osmolality).

2.2.2. Sensory Evaluation of Designed Functional Protein Drinks

The designed Matcha functional drinks were assessed using the descriptive and point method, paying attention to their features such as their appearance, smell, taste, consistency and overall assessment, and the results of the assessment are presented in Table 7.

Table 7. Evaluation of organoleptic features of designed Matcha functional drinks by descriptive and point method.

Sensory Features	Matcha Drink Based on Milk	Matcha Drink Based on Water
Appearance	light green color with a delicate sediment and foam on the surface	dark green color, cloudy drink
Points (0–5)	4.8	4.0
Smell	milk, tea, vegetable, herbal, sweet	herbal, tea, spicy, sweet, vegetable, grassy
Points (0–5)	4.5	4.2
Taste	milk, tea, herbal, slightly sweet, slightly bitter, vegetable	vegetable, tea, herbal, almond, slightly sweet, bitter, burning, astringent, raw, grassy
Points (0–5)	4.8	4.0
Consistency	liquid, causing astringency on the tongue, perceptible fine particles suspended in the drink	fluid, slightly viscous, causing tightening on the tongue
Points (0–5)	4.7	4.0
Overall score (0–5)	4.8	4.0

Matcha functional drinks differed depending on the selected liquid component (milk vs. water). The aroma of the drinks, described as herbal, tea, vegetable or sweet, was at a similar level in both variants. Both beverages had a fluid consistency, smooth, free from lumps and large particles, with a marked astringent sensation on the tongue. The taste of the drinks was tea, herbal, slightly sweet and slightly bitter and milky (in the case of a milk-based drink). The overall scores, being a harmonization of all assessed attributes, were high for both designed Matcha drinks. However, it is worth emphasizing that the higher score, i.e., 4.8/5 points, was scored by a milk-based Matcha drink compared to a water-based Matcha drink, which scored 4.0/5 points.

2.2.3. Nutritional Value of Designed Functional Protein Drinks

Table 8 lists the nutritional value of the designed Matcha functional drinks. The energy value of the drinks was about 33–63 kcal/100 mL. The Matcha drink prepared on the basis of water was characterized by a lower energy value; it did not contain fat and salt, and it was characterized by a lower content of protein, sugar and fiber. This was mainly due to the preparation of this water-based drink, without using a lot of ingredients (Table 5). The addition of milk caused a significant increase in energy value, fat, carbohydrate and protein content. Nevertheless, these were not high values. Taking into account the nutritional value of drinks, they can be labeled with nutrition claims in accordance with Regulation (EC) 1924/2006 [38]. The list of claims for individual beverages is also given in Table 8.

Table 8. Nutritional value of designed functional Matcha drinks in 100 mL of beverage.

Nutritional Value/ Nutrition Claim	Matcha Drink Based on Milk	Matcha Drink Based on Water
Energy value (kJ/kcal)	265.19/63.34	139.29/33.27
Fat (g)	0.7	<0.1
Saturated fatty acids (g)	0.47	<0.01
Carbohydrates (g)	8.8	6.25
Sugars (g)	4.28	2.18
Proteins (g)	6.7	2.7
Fiber (g)	2.7	2.25
Salt (g)	0.05	<0.01
Nutrition claims according to Regulation (EC) 1924/2006	High protein content Low fat Low salt content High content of dietary fiber	High protein content Does not contain fat It does not contain salt High content of dietary fiber

3. Discussion

3.1. Physicochemical and Bioactive Properties of Matcha Green Tea

The conducted organoleptic evaluation of Matcha green tea powders concerning general appearance, degree of fragmentation, granulation, flowability, as well as color, aroma and taste allowed us to conclude that the tested market products were of good quality. In the organoleptic assessment, powdered Matcha green teas were characterized by an intense green color, which proved the correct course of the technological process of its production and the preservation of a large amount of chlorophyll giving the green color [39]. In the case of M-6 and M-10 Matcha samples, which showed a much lower degree of fragmentation (Figure 1), these products were characterized by a less intense color, reminiscent of the typical color of green leaf tea, which made them less attractive than the others. All the tested Matcha samples had an intense tea aroma and were characterized by a distinctly bitter aftertaste, resulting, among others, from the presence of tannins [11].

One of the most important attributes of the appearance of food products, which strongly influences consumer acceptance, is color, and undesirable colors may lead consumers to reject products [40,41]. Considering that the basic selection criterion for tea consumers is precisely this attribute [42], and from the point of view of organoleptic assessment and overall quality of products, it is important to determine the color. A detailed analysis of the color of powdered Matcha green teas with the instrumental method is carried out in this paper and significant ($p \leq 0.05$) differences between the tested products for individual $L^*a^*b^*$ chromatic coordinates in the CIE color space are demonstrated (Table 1).

The available literature lacks studies on the instrumental measurement of the color of powdered Matcha green teas. The intense, green color of the tested products resulted from the high content of pigments, mainly chlorophylls [23]. Chlorophylls a and b, which occur in all photosynthetic plants, are among the most common in nature, and their significant content is responsible for the green color of plants [23]. According to the literature, the content of chlorophylls and their derivatives in the leaves of Tencha tea, intended for the production of Matcha, is significantly higher (5.65 mg/g d.m.) than in traditional green tea (4.33 mg/g d.m.) [12]. The research carried out in this study was consistent with the literature data [12]. The results obtained for the total chlorophyll content in the tested samples of powdered Matcha green tea ranged from 1.16 ± 0.03 mg/g d.m. to 7.01 ± 0.05 mg/g d.m. and indicated a high content of these pigments in the 10 tested Matcha products.

Taking into account the instrumental color measurement (CIE $L^*a^*b^*$) (Table 1) and the obtained results of high total chlorophyll content (Figure 3b) in the tested Matcha samples, a significant ($p \leq 0.05$) relationship between the content of total chlorophylls and the chromatic coordinate a^*, which determined the intensity of the green color ($-a^*$) in the tested Matcha green tea powders, was found. This relationship is presented in Figure 5.

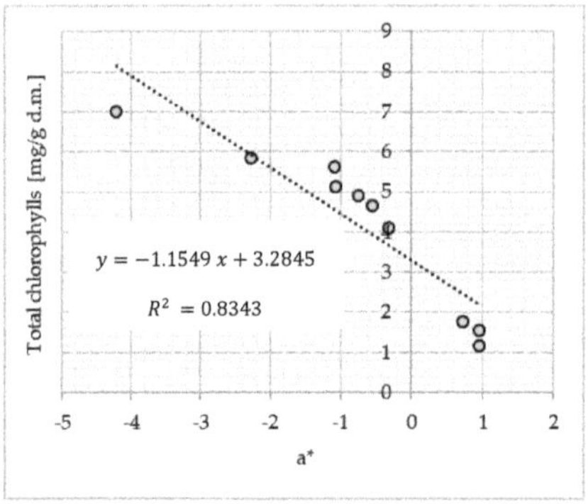

Figure 5. Relation between the total chlorophylls (mg/g d.m.) content and the chromatic coordinate a^* (CIE $L^*a^*b^*$) in the tested Matcha green tea samples.

The high content of bioactive ingredients, including specific dyes in Matcha green tea, is largely influenced by the technological processes used during its production [11]. Black teas are the result of the complete fermentation of tea leaves, leading to the typical brown-black color, resulting from the enzymatic oxidation of the catechins present in the tea leaves. In green tea production, the phenol oxidase (catalyzing oxidation reactions during the leaf fermentation stage) is deactivated by high-temperature treatment, which allows the green color to be preserved [11]. Differences in the obtained shades of color in the conducted research could result not only from the different impact of technological processes used to obtain powdered Matcha tea [11,16] but also from the degree of fragmentation of the product [26] or agro-climatic conditions of tea cultivation, such as the number and distribution of sunny and rainy days, possible fertilization and plant protection measures during the growth period [25].

The powdered Matcha teas tested in this study were characterized by low water activity (a_w). Many physical, chemical and microbiological changes can occur in food products, and the a_w values can affect the speed and intensity of these changes [43]. Each food product has a specific water activity that affects the adsorption or desorption of water from the environment. Powdered products have a low water activity of 0.15–0.40, e.g., in granulated tea, it is about 0.35 [44]. The average value of this parameter for all products tested in the study was 0.32 ± 0.08, which proves the high durability and microbiological safety of the products, due to the fact that most microorganisms do not develop at water activity at the level of $a_w < 0.60$ [43].

According to the conducted research, 2.5% water solutions of the tested Matcha teas showed an acid reaction (pH < 7), which also increases the stability of the tested powders by inhibiting the development of microorganisms (e.g., bacteria or yeast) and protection against rotting during storage [43]. In addition, the tested market products had a stable color and intense taste, and, according to the literature, the acidity of teas affects the stability of dyes and enhances the flavor and aroma characteristics [43]. The soluble solids content (°Brix) and osmolality in 2.5% aqueous solutions of Matcha green tea were very low. The content of soluble solids (°Brix) indicates, among others, the content of carbohydrates in the tested solutions, and compared to sweetened beverages, tea has a lower content, which makes it suitable for diabetics or people on a slimming diet [45,46]. The low osmolality of the tested teas allows them to be classified as hypotonic beverages, such as spring or mineral waters [46]. Such drinks are very well absorbed into the body's cells and can quickly quench thirst [45].

Matcha is characterized by a high content of antioxidant compounds, in particular polyphenols [12,15,24–27], including, e.g., flavonoids [2,13,16,17,19,21], phenolic acids [13, 17,18], carotenoids [12,23] or vitamins, thanks to which it has a strong antioxidant potential [13–18,26]. The conducted research showed a high content of phenolic compounds, determined by both the HPLC (Table 4, Figure 2a) and the spectrophotometric (Figure 2b) methods, in tested Matcha green tea samples.

Chromatographic analysis of selected phenolic compounds in the tested Matcha green tea samples (Table 4) showed the dominant share of catechins in the identified flavonoids, which was confirmed by other authors [1,30,31]. According to the literature, the dominant catechins in green tea are epicatechin, epicatechin-3-gallate, epigallocatechin and epigallocatechin-3-gallate [1], with the highest concentration of epigallocatechin-3-gallate [30,31]. The research results obtained in this study confirmed the tendency described in the literature, because gallate epigallocatechin was present in the highest concentration, while the content of epigallocatechin and catechin was over six times and over eleven times lower, respectively, in the tested Matcha green tea samples (Table 4).

The studies of Matcha green tea showed a high content of total polyphenols, ranging for the tested Matcha market products from 89.91 ± 3.17 to 190.96 ± 2.14 mg GAE/g d.m. (Figure 2b), which was confirmed by the test results of other authors. In the studies of Koláčková et al. (2019) [17], the total content of polyphenols ranged from 64.4 to 93.9 mg GAE/g d.m. in aqueous solutions and from 169 to 273 mg GAE/g d.m. in 80% methanol solutions. Analyzing the research results obtained in this work and the other authors results, it can be seen that methanolic solutions (usually at a concentration of 80%) showed a higher content of total polyphenols, which may be justified by a more effective extraction process of polyphenols from tea [17].

In the study by Jakubczyk et al. (2020) [13], the content of total polyphenols in infusions of two types of Matcha green tea was determined: traditional Matcha (from the first and second harvest) and daily Matcha (from the second and third harvest). The content of polyphenols in both infusions was high, but it differed depending on the temperature of the water used to prepare the infusion. The content of flavonoids was also determined in the same study. Infusions were characterized by a high content of these compounds, but higher concentrations were noted in teas from the second and third harvests. The results for traditional Matcha ranged from 1222.60 to 1514.28 mg/L, and for daily Matcha, they ranged from 1379.82 to 1968.79 mg/L.

In other studies [26], it was shown that powdered teas compared to leaf teas were characterized by a higher content of polyphenols using the same amount of leaves and powder during extraction, which may suggest that the grinding process may also affect the increasing the efficiency of extraction of polyphenolic compounds from tea [16,47].

Based on the research on the content of total polyphenols and the analysis of the literature, it can be concluded that Matcha contains large amounts of bioactive substances, in particular those with antioxidant activity [1]. According to the research, Matcha green teas available on the Polish market were characterized by a high antioxidant potential, ranging from 1443.56 ± 3.61 to 2076.35 ± 59.11 μM TEAC/g d.m. (Figure 4). The obtained results were confirmed in the available literature [17,18,22]. In the studies of Koláčková et al. (2020) [17], the antioxidant activity in Matcha teas from different producers was determined using $ABTS^{+\bullet}$. The authors showed differences in the Matcha antioxidant activity in 80% methanol extracts (from 306 to 368 mg TE/g) and in aqueous extracts (from 246 to 382 mg TE/g). The highest antioxidant activity was found in Matcha methanol extracts. In other studies conducted by Jakubczyk et al. (2020) [13], the antioxidant potential of the tested Matcha tea infusions ranged from 5767.30 to 6129.53 μM Fe^{2+}/L. In these studies, higher values were noted in teas brewed at higher temperatures (90 °C), which was most likely related to better release of biologically active compounds and higher kinetic energy at higher temperatures [13]. In other studies on various green teas, for all the analyzed products, an increase in antioxidant potential was shown with increasing water temperature during brewing, and it was found that the highest antioxidant activity was shown by infusions brewed at 100 °C for 3 min [27]. In addition, these authors showed that the antioxidant potential correlates with the content of polyphenolic compounds in the samples they studied, which was also found in the studies obtained in this work (Figure 6a).

The conducted research showed the existence of a significant ($p \leq 0.05$), positive relationship between the content of total polyphenols and antioxidant activity ($R^2 = 0.8937$), as well as between the content of vitamin C and antioxidant activity ($R^2 = 0.9061$), measured by the ability to deactivate synthetic $ABTS^{+\bullet}$ cation radicals, as presented in Figure 6. This study is consistent with the results achieved by other authors, who confirmed the high antioxidant properties of Matcha green tea, depending on the high concentration of phenolic compounds and the high vitamin C content [13,16,25,26,47].

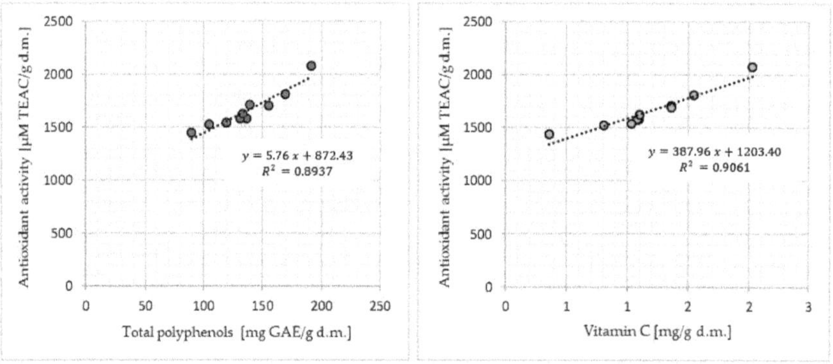

(**a**) (**b**)

Figure 6. Relation between the total polyphenols (mg GAE/g d.m.) (**a**) or vitamin C content (mg/g d.m.) (**b**) and antioxidant activity (μM TEAC/g d.m.) in the tested Matcha green tea samples.

In the literature studies, the content of bioactive components and antioxidant activity in Matcha fall within very wide ranges of values. These differences can be explained by the influence of various factors, determining not only the content and profile of phenolic components but also the antioxidant potential in the raw material itself. These factors include species and type of tea [2,3], production processes [13,17,25] and agro-climatic conditions of cultivation [1,25], including the degree of shading and insolation of the plant during its growth [8,12,36] or the conditions of tea harvesting [13,36]. The process of green tea and Matcha production also plays a very important role [25,26]: the form of the final product (leaves, bags, powder) [16,24,26], degree of fragmentation (milling) [26,47] or packaging and storage conditions [48]. The key role in shaping the bioactive properties of consumed products is also played by the methods and conditions of preparing Matcha drinks, including brewing time and temperature [24,47,49].

Due to the richness of bioactive ingredients (mainly polyphenols), especially those with antioxidant properties, Matcha green tea is increasingly used in the prevention and treatment of many diseases, including inflammatory diseases [30,31], viral diseases [20,32] and many civilization diseases, including diabetes [22,50,51], obesity [28], cancer [2], hypertension and heart disease [1,13,22,24,29,52], or for slowing down the aging process [33]

In addition, thanks to its health-promoting properties, as well as specific physicochemical properties (finely ground powder with an intense green color intended for consumption both "hot" and "cold"), Matcha is more and more often used as an addition to various products, including bars, jellies, pastries, drinks, cocktails or ice cream [4,34], becoming a promising ingredient in the functional food industry [22,35–37].

3.2. *The Possibility of Using Matcha Green Tea in the Design of Functional Protein Drinks*

The above-presented results of the conducted research on the assessment of physicochemical and bioactive features of various types of Matcha green teas available on the Polish market were the basis for the selection of a product with the most appropriate features for the design of functional beverages. Based on the obtained results, a sample of Matcha M-4 green tea, characterized by the most desirable bioactive properties, was selected for further (design) research. This sample was used to develop the composition of the model beverages that can be recommended to a wide range of consumers, depending on their individual expectations or needs. Thanks to the high content of protein in the drinks, in an easily accessible form, an important group of recipients may be athletes or people with an increased need for protein. Moreover, the addition of maltodextrin, which is an easily available source of carbohydrates, indicates that the drinks can be especially recommended to athletes or physically active people. Maltodextrin is quickly released in the body, thanks to which muscle glycogen stores are quickly replenished and energy is supplied. This ingredient is often used in food products for athletes [53]. It has been shown that the addition of protein to a carbohydrate drink results in better glycogen storage after exercise compared to drinks of the same energy value consisting of carbohydrates or carbohydrates and fat [54]. For the elderly or malnourished, these drinks can also be a helpful supplement to the diet due to the high protein content with low energy value [55]. In addition, the fiber in the form of inulin added to the recipe increases the nutritional value of the drink and is a source of prebiotics. Inulin has prebiotic properties, it can contribute to lowering the concentration of triglycerides and cholesterol in the blood, as well as improving the absorption of minerals such as calcium and magnesium [56]. These ingredients may be present in the liquid component with which the beverage is prepared and thus better absorbed, e.g., calcium in milk or magnesium in water. The bioactive compounds from Matcha present in the recipe composition increase the health-promoting qualities of the prepared drinks.

The available literature contains conflicting data on the effect of adding milk to tea on the antioxidant activity of the beverage so obtained. The differences in the obtained results are mainly due to the use of different types of tea and milk for the preparation of beverages, the composition of the beverage, and the method of their preparation. Interactions between polyphenolic compounds present in tea and proteins present in milk (mainly catechins and caseins) can lead to a decrease in the antioxidant activity of the prepared beverage. Few data on the explanation of this phenomenon can be found in the literature [57]. In addition, milk contains fat globules, which may also interact with tea catechins. Although the negative effect of milk addition on antioxidant activity in tea has been widely described, there is a lack of information on the effect of milk addition on the activity or bioavailability of caffeine in tea infusions [58]. In this study, in the context of sensory characteristics, it was confirmed that the addition of milk to tea infusion improves the taste characteristics of the prepared drink by counteracting the occurrence of astringent and bitter taste. These flavors are characteristic of tea infusions, which is due to the presence of tannins in them. Additionally, the presence of milk in the tea infusion increases the nutritional value of this drink.

Despite the rich and growing market of functional drinks, there is still little information and research on their formulation and design [59]. There is also little research on the interaction of polyphenols with compounds present in food, i.e., carbohydrates, lipids or proteins [60–62]. Therefore, the presented research may be the starting point for the design of functional drinks using protein preparations with the addition of powdered Matcha green tea and, in the future, also for testing the bioavailability of polyphenolic components contained in various food products and food matrices.

4. Materials and Methods

4.1. Materials

The research material consisted of 10 powdered Matcha green teas, which were market products and the most frequently chosen brands in Poland. The products were purchased in specialist tea shops in 100g individual packages. The research was carried out in two stages.

In the first stage of the research, the physicochemical and bioactive properties of 10 selected Matcha products were assessed. An organoleptic evaluation was carried out, the physical properties of samples in the form of powders and 2.5% aqueous solutions of

Matcha were assessed, and the bioactive properties were also tested. On the basis of the research results obtained in the first stage, one of ten tested products was selected; it was characterized by the most desirable organoleptic, physicochemical and bioactive properties, which was used to design the composition of functional drinks with the Matcha addition.

In the second stage of the research, the composition of the ready-to-use powdered mix was designed, containing the addition of Matcha green tea as a source of bioactive substances and shaping the taste profile of the functional drinks designed later. The composition was designed, and 2 types of Matcha functional drinks were obtained, i.e., milk-based and water-based. The detailed composition of the beverage formulations is presented in Section 3. Results (Table 5).

4.2. Methods

In Matcha powder samples, organoleptic evaluation, measurements of water activity (a_w), instrumental color measurements ($L^*a^*b^*$), water solubility index (WSI) and water holding capacity (WHC) as well as total polyphenol content and antioxidant activity of the tested 10 market products of Matcha green tea were determined.

4.2.1. Instrumental Color Measurement (CIE $L^*a^*b^*$) of Powders

Color parameters of Matcha green tea powders were measured at room temperature (20 °C) using a colorimeter (Konica Minolta CR-400, Konica Minolta, BSP, Warsaw, Poland) in the CIE Lab color space ($L^*a^*b^*$) (L-brightness; +*a*-red; −*a*-green; +*b*-yellow; −*b*-blue). After calibration using a white CR-A43 reflective plate and placing the samples in a glass dish (Ø 60 mm) on the measuring head, the results were read.

4.2.2. Water Activity (a_w) of Powders

Water activity (a_w) of Matcha green tea powders and designed ready-to-use drink powdered mixes were measured using a handheld AquaLab Water Activity Meter (Decagon Devices. Inc., Pullman, WA, USA) with a temperature stabilizer

4.2.3. Water Solubility Index (WSI) of Powders

The water solubility index (WSI) was determined by the gravimetric method according to Yousf et al. (2017) [63]. In plastic Falcone tubes (with a capacity of 50 mL), 2.5 g (with an accuracy of 0.001 g) of Matcha green tea powders were weighed on an analytical balance (AS 220/X, Radwag, Radom, Poland); 30 mL of distilled water was added, mixed in a centrifuge tube, incubated in a heated shaker (IKA KS 4000i Control, IKA® Ltd., Warsaw, Poland) (37 °C ± 1 °C, 30 min) and then centrifuged (MPW-380 R, MPW Med. Instruments, Warsaw, Poland) for 20 min at 10,000 rpm. The supernatant was collected into pre-weighed weighing bottles and dried in an incubator (SUP 200W, Wamed, Warsaw, Poland) at 103 °C ± 2 °C. After drying, it was weighed and the WSI was expressed as % dissolved substance in water.

4.2.4. Water Holding Capacity (WHC) of Powders

The water holding capacity (WHC) was determined by the gravimetric method according to the methodology described by Sudha et al. (2007) [64]. In plastic Falcone tubes (with a capacity of 50 mL), 1.0 g (with an accuracy of 0.001 g) of Matcha green tea powders were weighed on an analytical balance (AS 220/X, Radwag, Radom, Poland); 50 mL of distilled water was added, mixed in a centrifuge tube and then centrifuged (MPW-380 R, MPW Med. Instruments, Warsaw, Poland) for 15 min at 10,000 rpm. Excess water was poured off, and the water-absorbed powder was reweighed and the WHC was expressed as g/g (grams of water per gram of powder).

4.2.5. pH

The pH of 2.5% aqueous Matcha green tea solutions and designed functional drinks was measured by the potentiometric method using a laboratory pH-meter probe (Elmetron CP-511, Elmetron Rp., Zabrze, Poland) at room temperature (20 °C).

4.2.6. Soluble Solids Content (°Brix)

The soluble solids content (°Brix) in 2.5% aqueous Matcha green tea solutions and designed functional drinks was measured using an Abbe refractometer (ORT-1, Kern & Sohn GmbH, Balingren-Frommern, Balingen, Germany), using the refractometric method, according to the Polish Standard (PN-EN 12143:2000) [65]. The results (expressed in %) were read from the sugar scale at room temperature (20 °C).

4.2.7. Osmolality

The osmolality of 2.5% aqueous Matcha green tea solutions and designed functional drinks was measured using an osmometer (Osmometr Krioskop 800CL, Trident Med. Clp, Warsaw, Poland) by measuring the crystallization temperature of a supercooled solution with a one-point calibration. After measuring 100.0 μL of the sample into the osmometric tube and placing it in the cooling chamber, as a result of supercooling the sample and initiating crystallization, the heat of crystallization was measured with a thermistor and automatically converted to mOsm/kg·H_2O.

4.2.8. Selected Phenolic Acids and Flavonoids Identified by the HPLC Method

In plastic sterile test tubes (10 mL capacity), 0.1 g (accuracy to 0.001 g) of powdered Matcha green tea was weighed on an analytical balance (AS 220/X, Radwag, Radom, Poland); 5.0 mL of 80% methanol (Sigma-Aldrich, Poznań, Poland) was added and vortexed (Wizard Advanced IR Vortex Mixer, VELP Scientifica Srl, Usmate, Italy) (30 s, 2000 rpm) to mix thoroughly, and then incubated in an ultrasonic bath (PolSonic, Warsaw, Poland) for 10 min (5.5 kHz, 30 °C). After the incubation, the samples were centrifuged in a refrigerated centrifuge (MPW-380 R, MPW Med. Instruments, Poland, Warsaw) for 15 min (6000 rpm, 2 °C), and the obtained supernatants (1.0 mL) were collected into the chromatographic vials and analyzed by HPLC.

The contents of selected phenolic acids and flavonoids in Matcha green tea were determined by the HPLC method, according to Hallmann et al. (2017) [66], using the Shimadzu HPLC kit (USA Manufacturing Inc., Waltham, MA, USA), consisting of two LC-20AD pumps, a CMB-20A system controller, a CTD-20AC controller, a SIL-20AC autosampler and a UV/VIS SPD-20AV detector. Identification and separation of selected phenolic compounds were performed on a Synergi Fusion-RP 80i chromatographic column (250 × 4.60 mm) (Phenomenex, Shimpol, Warsaw, Poland), using a flow gradient with two phases: acetonitrile/deionized water (55% and 10%) at pH 3.00. The analysis time was 38 min, with a flow rate of 1.0 mL/min and detection at λ = 250–370 nm. External standards, i.e., gallic acid, *p*-coumaric acid, catechin, epigallocatechin, gallate epigallocatechin, quercetin and rutoside-3-*O*-quercetin, with a purity of 99.98% (Fluka and Sigma-Aldrich Poznań, Poland), were used to identify the substances. Five injections of phenolic standard solutions were made in order to prepare standard curves. The content of selected phenolic compounds was calculated based on the standard curves, and the results were expressed as mg/g d.m. (d.m.—dry matter).

4.2.9. Vitamin C Content

In plastic sterile test tubes (10 mL capacity), 0.1 g (accuracy to 0.001 g) of powdered Matcha green tea was weighed on an analytical balance (AS 220/X, Radwag, Radom, Poland); 5.0 mL of 5% metha-phosphoric acid (Sigma-Aldrich, Poznań, Poland) was added and vortexed (Wizard Advanced IR Vortex Mixer, VELP Scientifica Srl, Usmate, Italy) (30 s, 2000 rpm) to mix thoroughly, and then incubated in an ultrasonic bath (PolSonic, Warsaw, Poland) for 10 min (5.5 kHz, 30 °C). After the incubation, the samples were centrifuged in a refrigerated centrifuge (MPW-380 R, MPW Med. Instruments, Poland, Warsaw) for 10 min (6000 rpm, 0 °C), and the obtained supernatants (1.0 mL) were collected into the chromatographic vials and analyzed by HPLC.

The vitamin C content in Matcha green tea was determined by the HPLC method, according to Ponder and Hallmann (2020) [67], using the Shimadzu HPLC kit (USA Manufacturing Inc., Waltham, MA, USA), consisting of two LC-20AD pumps, a CMB-20A system controller, a SIL-20AC autosampler and a UV/VIS SPD-20AV detector. Identification and separation of L-ascorbic and L-dehydroascorbic acids were performed on a Hydro 80-A RP column (250 × 4.6 mm) (Phenomenex, Shimpol, Warsaw, Poland), using a mobile phase: 0.1 M acetic acid (glacial, 99.9% purity) (Sigma-Aldrich, Poznań, Poland) and 0.1 M sodium acetate (Sigma-Aldrich, Poznań, Poland) (in volume proportions of 63:37 *v*/*v*) at pH 4.4. The analysis time was 18 min, with an isocratic flow of 1.0 mL/min and detection at λ = 255–260 nm. External standards, i.e., L-ascorbic acid (L-ASC) and L-dehydroascorbic acid (L-DHA) (Sigma-Aldrich, Poznań, Poland), with a purity of 99.5%, were used to identify the substances. Five injections of L-ASC and L-DHA standards solutions were made in order to prepare standard curves. The vitamin C content was calculated based on standard curves, and the results were expressed (as the sum of L-ASC and L-DHA acids) in mg/g d.m. (d.m.—dry matter).

4.2.10. Chlorophylls Content

In plastic sterile test tubes (10 mL capacity), 0.1 g (accuracy to 0.001 g) of powdered Matcha green tea was weighed on an analytical balance (AS 220/X, Rad-wag, Radom, Poland); 5.0 mL of cold acetone (Sigma-Aldrich, Poznań, Poland) and 10.0 mg of magnesium carbonate $MgCO_3$ (Sigma-Aldrich, Poznań, Poland) were added and vortexed (Wizard Advanced IR Vortex Mixer, VELP Scientifica Srl, Usmate, Italy) (30 s, 2000 rpm) to mix thoroughly, and then incubated in an ultrasonic bath (PolSonic, Warsaw, Poland) for 10 min (5.5 kHz, 0 °C). After the incubation, the samples were centrifuged in a refrigerated centrifuge (MPW-380 R, MPW Med. Instruments, Poland, Warsaw) for 10 min (6000 rpm, 0 °C), and the obtained supernatants (1.0 mL) were collected into the chromatographic vials and analyzed by HPLC.

The chlorophyll content in Matcha green tea was determined by the HPLC method, according to Hallmann et al. (2017) [66], using the Shimadzu HPLC kit (USA Manufacturing Inc., Waltham, MA, USA), consisting of two LC-20AD pumps, a CMB-20A system controller, a SIL-20AC autosampler and a UV/VIS SPD-20AV detector. Identification and separation of chlorophyll a and chlorophyll b were performed on a Synergi Max-RP 80A column (250 mm × 4.60 mm) (Phenomenex, Shimpol, Warsaw, Poland), using two mobile phases: acetonitrile/methanol (Sigma-Aldrich, Poznań) (90:10 *v*/*v*) and methanol/ethyl acetate (Sigma-Aldrich, Poznań) (64:36 *v*/*v*). The analysis time was 28 min, with a flow rate of 1.0 mL/min and detection at λ = 450 nm. External standards, i.e., chlorophyll a and chlorophyll b, with a purity of 99.98% (Sigma-Aldrich Poznań, Poland) were used to identify the substances. Five injections of chlorophyll a and chlorophyll b standard solutions were made in order to prepare standard curves. The total chlorophyll content was calculated based on the standard curves, and the results were expressed (as the sum of chlorophyll a and b) in mg/g d.m. (d.m.—dry matter).

4.2.11. Preparation of Water Extracts for Total Polyphenol Content and Antioxidant Activity Assay

In plastic sterile Falcone tubes with a cap (50 mL capacity), 0.5 g (accuracy to 0.001 g) of powdered Matcha green tea was weighed on an analytical balance (AS 220/X, Radwag, Radom, Poland); 40 mL of distilled water was added and vortexed (Wizard Advanced IR Vortex Mixer, VELP Scientifica Srl, Usmate, Italy) (60 s, 2000 rpm) to mix thoroughly, and then incubated in a shaking incubator (IKA KS 4000i Control, IKA® Ltd., Warsaw, Poland) for 60 min (60 °C, 200 rpm). After incubation, the samples were again vortexed for 60 s for thorough mixing and then centrifuged in a refrigerated centrifuge (MPW-380 R, MPW Med. Instruments, Warsaw, Poland) for 15 min (4 °C, 10,000 rpm). The supernatant obtained in this way was used to determine the total polyphenol content and antioxidant activity in Matcha green tea.

4.2.12. Total Polyphenol Content

Total polyphenol content in Matcha green tea was determined using the Folin–Ciocalteu reagent (Sigma-Aldrich, Poznań, Poland) according to the Singleton and Rossi (1965) [68] method. The solution (1.0 mL of diluted supernatant, 2.5 mL of Folin–Ciocalteu reagent, 5.0 mL of 20% Na_2CO_3 (sodium carbonate) in 41.5 mL of distilled water) was incubated for 60 min (20 °C, no access to light) and the absorbance was measured in a spectrophotometer (UV/Vis UV-6100A, Metash Instruments Co., Ltd., Shanghai, China) at $\lambda = 750$ nm. After taking into account the dilutions used, the obtained results of absorbance measurements were recalculated based on the standard curve ($y = 2.127x + 0.1314$, $R^2 = 0.9994$) for gallic acid (Sigma-Aldrich, Poznań, Poland) as a standard substance and the total polyphenol content was expressed as mg GAE/g d.m. (GAE—gallic acid equivalent, d.m.—dry matter).

4.2.13. Antioxidant Activity

Antioxidant activity in Matcha green tea was determined using the cation radical $ABTS^{+\bullet}$ (2,2′-azino-bis 3-ethylbenzothiazolin-6-sulfonic acid) (Sigma-Aldrich, Poznań, Poland) according to the Re et al. (1999) [69] method. The solution (1.5 mL of diluted supernatant, 3.0 mL radical cations $ABTS^{+\bullet}$ in PBS solution (PBS—phosphate buffer solution)) was vortexed (Wizard Advanced IR Vortex Mixer, VELP Scientifica Srl, Usmate, Italy) and incubated for 6 min (20 °C), and the absorbance was measured in a spectrophotometer (UV/Vis UV-6100A, Metash Instruments Co., Ltd., Shanghai, China) at $\lambda = 734$ nm. After taking into account the dilutions used, the obtained results of absorbance measurements were recalculated based on the standard curve ($y = -5.6067x + 0.7139$, $R^2 = 0.9998$) for Trolox (Sigma-Aldrich, Poznań, Poland) as a standard substance and the antioxidant activity was expressed as μM TEAC/g d.m. (TEAC—Trolox equivalent antioxidant capacity, dm.—dry matter).

4.2.14. Sensory Evaluation of Functional Matcha Drinks

Sensory evaluation of the designed functional Matcha drinks was carried out by six panelists with extensive experience in assessing the sensory characteristics of various products using the descriptive and point method. The drinks were assessed in the following categories: taste, smell, color, consistency and overall rating on a five-point scale (5—very good, 4—good, 3—sufficient, 2—insufficient, 1—bad).

4.2.15. Nutritional Value of Functional Matcha Drinks

The nutritional value of the designed functional Matcha drinks was calculated on the basis of the information provided on the labels of the beverage components regarding their energy value and the content of individual nutrients and salt. The study calculated the energy value, fat content (including saturated fatty acids), carbohydrate content (including sugars), fiber, protein and salt content of the prepared drinks.

4.2.16. Statistical Analysis

The differences between the samples were considered statistically significant at $p \leq 0.05$. Statistical analysis of the obtained results was performed using the Statistica 13.0 software (Tibco Software Inc., Palo Alto, CA, USA). The significance of differences between the obtained results was determined by performing a one-way analysis of variance (ANOVA). In order to show the differences between the individual groups, the Duncan test was used at the assumed significance level of $p \leq 0.05$. The results were presented in tables and graphs in which the division into homogeneous groups was marked and the mean values and standard deviations (SD) were presented.

5. Conclusions

The Matcha green teas tested in the study, in the organoleptic assessment regarding the general appearance, degree of fragmentation, granulation, flowability, as well as color, aroma and taste, were characterized by good quality and their solubility in water, giving mostly homogeneous mixtures with a pleasant, typical tea aroma and intense green color. The soluble solids content of 2.5% aqueous solutions of Matcha green tea powders was close to the °Brix of pure water, and their osmolality was within the range of hypotonic beverages. Knowledge of such properties is crucial due to the possibility of using Matcha green tea powder as a source of bioactive ingredients in the design of functional food and various types of food products with its participation.

The tests carried out showed a high content of bioactive ingredients, including selected phenolic acids and flavonoids identified by the HPLC method, as well as total polyphenols, determined by the spectrophotometric method using the Folin–Ciocalteu reagent. Moreover, the tested Matcha green tea samples were characterized by a high content of chlorophylls, vitamin C and high antioxidant activity.

Functional food is a constantly developing market segment. It affects health and can prevent the development of civilization diseases. Many active or nutritional ingredients are used in its production. Currently, most products are enriched with proteins, pro and prebiotics, polyunsaturated fatty acids and antioxidants, including polyphenols. Beverages are the most extensive category of functional foods. Despite the great interest in the group of functional products, which include the designed drinks, there is little research on their formulation, the interaction of the ingredients of such food as well as the sensory characteristics of protein drinks with the addition of various protein preparations and additional inulin preparation. Due to the wide range of health-promoting properties of Matcha green tea, it is worth including this product in daily diet, both in the form of functional drinks or as an addition to other food products, such as cocktails, desserts or snacks. In the present study, a formulation of an innovative ready-to-drink protein drink was developed, containing both Matcha bioactive ingredients and digestible whey proteins, amino acids and peptides of enzymatically hydrolyzed collagen proteins rich in bone-joint tissue hydroxyamino acids (hydroxyproline and hydroxylysine) and the prebiotic inulin. This preparation, when dissolved in the consumer's preferred liquid (water, milk, juice), constitutes a ready-to-drink regenerative drink of low osmolality (an iso- and hypotonic drink). This beverage can be recommended to a wide range of consumers, including physically active people and malnourished people with protein deficiency, and as a functional beverage, it can be used in the prevention and treatment of connective tissue diseases.

Author Contributions: Conceptualization, K.N., A.S., M.W. and K.S.; methodology, K.N., A.S. and K.B.; formal analysis, K.N., A.S., M.W., K.S. and K.B.; investigation, M.W. and K.S.; resources, K.N. and A.S.; data curation, K.N. and A.S.; writing—original draft preparation, K.N., A.S., M.W. and K.S.; writing—review and editing, K.N., A.S. and K.B.; visualization, K.N., A.S. and K.B. All authors have read and agreed to the published version of the manuscript.

Funding: This research was prepared as part of the statutory activity of the Department of Functional and Ecological Food, Warsaw University of Life Sciences, Warsaw, Poland.

Institutional Review Board Statement: Not applicable.

Informed Consent Statement: Not applicable.

Data Availability Statement: Not applicable.

Conflicts of Interest: The authors declare no conflict of interest.

Sample Availability: Not applicable.

References

1. Kochman, J.; Jakubczyk, K.; Antoniewicz, J.; Mruk, H.; Janda, K. Health benefits and chemical composition of Matcha green tea: A review. *Molecules* **2021**, *26*, 85. [CrossRef] [PubMed]
2. Schröder, L.; Marahrens, P.; Koch, J.G.; Heidegger, H.; Vilsmeier, T.; Phan-Brehm, T.; Hofmann, S.; Mahner, S.; Jeschke, U.; Richter, D.U. Effects of green tea, matcha tea and their components epigallocatechin gallate and quercetin on MCF-7 and MDA-MB-231 breast carcinoma cells. *Oncol. Rep.* **2019**, *41*, 387–396. [CrossRef]

3. Horie, H.; Kaori, E.; Sumikawa, O. Chemical components of Matcha and powdered green tea. *J. Cook. Sci. Jpn.* **2017**, *50*, 182–188. [CrossRef]
4. Yüksel, K.A.; Yüksel, M.İ.; Şat, G.İ. Determination of certain physicochemical characteristics and sensory properties of green tea powder (Matcha) added ice creams and detection of their organic acid and mineral contents. *GIDA—J. Food* **2017**, *42*, 116–126. [CrossRef]
5. Sivanesan, I.; Gopal, J.; Muthu, M.; Chun, S.; Oh, J.W. Retrospecting the antioxidant activity of Japanese Matcha green tea Lack of enthusiasm? *Appl. Sci.* **2021**, *11*, 5087. [CrossRef]
6. Ahmed, S.; Stepp, J. Chapter 2—Green Tea: The Plants, Processing, Manufacturing and Production. In *Tea in Health and Disease Prevention*; Preedy, V.R., Ed.; Academic Press: Cambridge, MA, USA; Elsevier Inc.: Amsterdam, The Netherlands, 2013; pp. 19–31. [CrossRef]
7. Unno, K.; Furushima, D.; Hamamoto, S.; Iguchi, K.; Yamada, H.; Morita, A.; Horie, H.; Nakamura, Y. Stress-reducing function of Matcha green tea in animal experiments and clinical trials. *Nutrients* **2018**, *10*, 1468. [CrossRef] [PubMed]
8. Sano, T.; Horie, H.; Matsunaga, A.; Hirono, Y. Effect of shading intensity on morphological and color traits and on chemical components of new tea (*Camellia sinensis* L.) shoots under direct covering cultivation. *J. Sci. Food Agric.* **2018**, *98*, 5666–5676. [CrossRef]
9. Queiroz, C.; Lopes, M.L.M.; Fialho, E.L.; Valente-Mesquita, V.L. Polyphenol oxidase: Characteristics and mechanisms of browning control. *Food Rev. Int.* **2008**, *24*, 361–375. [CrossRef]
10. Chen, C.; Yang, C.Y.; Tzen, J.T.C. Molecular characterization of polyphenol oxidase between small and large leaf tea cultivars. *Sci. Rep.* **2022**, *12*, 12870. [CrossRef]
11. Jyotismita, K.; Chandan, R.; Arindam, R. Determination of tannin content by titrimetric method from different types of tea. *J. Chem. Pharm. Res.* **2015**, *7*, 238–241.
12. Ku, K.M.; Choi, J.N.; Kim, J.; Kim, J.K.; Yoo, L.G.; Lee, S.J.; Hong, Y.-S.; Lee, C.H. Metabolomics analysis reveals the compositional differences of shade grown tea (*Camellia sinensis* L.). *J. Agric. Food Chem.* **2010**, *58*, 418–426. [CrossRef] [PubMed]
13. Jakubczyk, K.; Kochman, J.; Kwiatkowska, A.; Kałduńska, J.; Dec, K.; Kawczuga, D.; Janda, K. Antioxidant properties and nutritional composition of Matcha green tea. *Foods* **2020**, *9*, 483. [CrossRef]
14. Tang, M.-G.; Zhang, S.; Xiong, L.-G.; Zhou, J.-H.; Huang, J.-A.; Zhao, A.-Q.; Liu, Z.-H.; Liu, A.-L. A comprehensive review of polyphenol oxidase in tea (*Camellia sinensis*): Physiological characteristics, oxidation manufacturing, and biosynthesis of functional constituents. *Compr. Rev. Food Sci. Food Saf.* **2023**, *22*, 2267–2291. [CrossRef] [PubMed]
15. Yamamoto, T.; Juneja, L.R.; Chu, D.; Kim, M. *Chemistry and Applications of Green Tea*; CRC Press: New York, NY, USA, 1997.
16. Farooq, S.; Sehgal, A. Antioxidant activity of different forms of green tea: Loose leaf, bagged and Matcha. *Curr. Res. Nutr. Food Sci.* **2018**, *6*, 35–40. [CrossRef]
17. Koláčková, T.; Kolofiková, K.; Sytařová, I.; Snopek, L.; Sumczynski, D.; Orsavová, J. Matcha tea: Analysis of nutritional composition, phenolics and antioxidant activity. *Plant Foods Hum. Nutr.* **2020**, *75*, 48–53. [CrossRef] [PubMed]
18. Cabrera, C.; Giménez, R.; López, M.C. Determination of tea components with antioxidant activity. *J. Agric. Food Chem.* **2003**, *51*, 4427–4435. [CrossRef] [PubMed]
19. Pastoriza, S.; Mesías, M.; Cabrera, C.; Rufián-Henares, J.A. Healthy properties of green and white teas: An update. *Food Funct.* **2017**, *8*, 2650–2662. [CrossRef]
20. Xu, J.; Xu, Z.; Zheng, W. A review of the antiviral role of green tea catechins. *Molecules* **2017**, *22*, 1337. [CrossRef] [PubMed]
21. Xu, D.; Hu, M.J.; Wang, Y.Q.; Cui, Y.L. Antioxidant activities of quercetin and its complexes for medicinal application. *Molecules* **2019**, *24*, 1123. [CrossRef]
22. Xu, P.; Ying, L.; Hong, G.; Wang, Y. The effects of the aqueous extract and residue of Matcha on the antioxidant status and lipid and glucose levels in mice fed a high-fat diet. *Food Funct.* **2016**, *7*, 294–300. [CrossRef]
23. Suzuki, Y.; Shioi, Y. Identification of chlorophylls and carotenoids in major teas by high-performance liquid chromatography with photodiode array detection. *J. Agric. Food Chem.* **2003**, *51*, 5307–5314. [CrossRef] [PubMed]
24. Komes, D.; Horžić, D.; Belščak, A.; Ganić, K.K.; Vulić, I. Green tea preparation and its influence on the content of bioactive compounds. *Food Res. Int.* **2010**, *43*, 167–176. [CrossRef]
25. Jeszka-Skowron, M.; Krawczyk, M.; Zgoła-Grześkowiak, A. Determination of antioxidant activity, rutin, quercetin, phenolic acids and trace elements in tea infusions: Influence of citric acid addition on extraction of metals. *J. Food Compos. Anal.* **2015**, *40*, 70–77. [CrossRef]
26. Fujioka, K.; Iwamoto, T.; Shima, H.; Tomaru, K.; Saito, H.; Ohtsuka, M.; Yodhidome, A.; Kawamura, Y.; Monome, Y. The powdering process with a set of ceramic mills for green tea promoted catechin extraction and the ROS inhibition effect. *Molecules* **2016**, *21*, 474. [CrossRef] [PubMed]
27. Liu, S.; Ai, Z.; Qu, F.; Chen, Y.; Ni, D. Effect of steeping temperature on antioxidant and inhibitory activities of green tea extracts against α-amylase, α-glucosidase and intestinal glucose uptake. *Food Chem.* **2017**, *234*, 168–173. [CrossRef]
28. Cardoso, G.A.; Salgado, J.M.; Cesar, M.C.; Pestana-Donado, C.M. The effects of green tea consumption and resistance training on body composition and resting metabolic rate in overweight or obese women. *J. Med. Food* **2013**, *16*, 120–127. [CrossRef] [PubMed]
29. Zhou, J.; Yu, Y.; Wang, Y. Matcha green tea alleviates non-alcoholic fatty liver disease in high-fat diet-induced obese mice by regulating lipid metabolism and inflammatory responses. *Nutrients* **2021**, *13*, 1950. [CrossRef] [PubMed]

30. Reygaert, W.C. Green tea catechins: Their use in treating and preventing infectious diseases. *BioMed Res. Int.* **2018**, *2018*, 9105261. [CrossRef]
31. Ohishi, T.; Goto, S.; Monira, P.; Isemura, M.; Nakamura, Y. Anti-inflammatory action of green tea. *Anti-Inflamm. Anti-Allergy Agents Med. Chem.* **2016**, *15*, 74–90. [CrossRef]
32. Mahmood, M.S.; Mártinez, J.L.; Aslam, A.; Rafique, A.; Vinet, R.; Laurido, C.; Hussain, I.; Abbas, R.Z.; Khan, A.; Ali, S. Antiviral effects of green tea (*Camellia sinensis*) against pathogenic viruses in human and animals (a Mini-Review). *Afr. J. Tradit. Complement. Altern. Med.* **2016**, *13*, 176. [CrossRef]
33. Prasanth, M.I.; Sivamaruthi, B.S.; Chaiyasut, C.; Tencomnao, T. A review of the role of green tea (*Camellia sinensis*) in antiphotoaging, stress resistance, neuroprotection, and autophagy. *Nutrients* **2019**, *11*, 474. [CrossRef]
34. Dietz, C.; Dekker, M.; Piqueras-Fiszman, B. An intervention study on the effect of matcha tea, in drink and snack bar formats, on mood and cognitive performance. *Food Res. Int.* **2017**, *99*, 72–83. [CrossRef] [PubMed]
35. Čížková, H.; Voldřich, M.; Mlejnecká, J.; Kvasnička, F. Authenticity evaluation of tea-based products. *Czech. J. Food Sci.* **2008**, *26*, 259–267. [CrossRef]
36. Topuz, A.; Dinçer, G.; Torun, M.; Tontul, İ.; Şahin-Nadeem, H.; Haznedar, A.; Ozdemir, F. Physicochemical properties of Turkish green tea powder: Effects of shooting period, shading, and clone. *Turk. J. Agric. For.* **2014**, *38*, 233–241. [CrossRef]
37. Ujihara, T.; Hayashi, N.; Ikezaki, H. Objective evaluation of astringent and umami taste intensities of matcha using a taste sensor system. *Food Sci. Technol. Res.* **2013**, *19*, 1099–1105. [CrossRef]
38. European Union. Regulation (EC) No 1924/2006 of the European Parliament and of the Council of 20 December 2006 on nutrition and health claims made on foods. The European Parliament and Council of the European Union, 30 December 2006, L 404/9-25. *Off. J. Eur. Union.* **2006**. Available online: http://data.europa.eu/eli/reg/2006/1924/oj (accessed on 30 December 2006).
39. Szwedziak, K.; Polańczyk, E.; Dąbrowska-Molenda, M.; Płuciennik, K. The impact of brewing time and the fineness of the color of black tea. *Postępy Tech. Przetwórstwa Spożywczego* **2016**, *2*, 76–79. (In Polish)
40. Maskan, M. Kinetics of colour change of kiwifruits during hot air and microwave drying. *J. Food Eng.* **2001**, *48*, 169–175. [CrossRef]
41. Maskan, M. Production of pomegranate (*Punica granatum* L.) juice concentrate by various heating methods: Colour degradation and kinetics. *J. Food Eng.* **2006**, *72*, 218–224. [CrossRef]
42. Ceni, G.C.; Baldissera, E.M.; Primo, M.S.; Antunes, O.A.C.; Dariva, C.; Oliveira, J.V.; Oliveira, D. Influence of application of microwave energy on quality parameters of mate tea leaves (*Ilex paraguariensis* St. Hill.). *Food Technol. Biotechnol.* **2009**, *47*, 221–226.
43. Erkmen, O.; Bozoglu, T.F. (Eds.) Factors affecting microbial growth in foods. In *Food Microbiology: Principles into Practice*; John Wiley & Sons, Ltd.: Hoboken, NJ, USA, 2016; Volume 1, Section 2; pp. 91–106. [CrossRef]
44. Kowalska, J.; Majewska, E.; Lenart, A. Water activity of powdered cocoa beverage with a modified composition of raw materials. *Zywnosc Nauka Technol. Jakosc.* **2011**, *4*, 57–65. (In Polish) [CrossRef]
45. Gisolfi, C.V.; Summers, R.W.; Lambert, G.P.; Xia, T. Effect of beverage osmolality on intestinal fluid absorption during exercise. *J. Appl. Physiol.* **1998**, *85*, 1595–2000. [CrossRef] [PubMed]
46. Bonetti, D.L.; Hopkins, W.G. Effects of hypotonic and isotonic sports drinks on endurance performance and physiology. *Sportscience* **2010**, *14*, 63–70.
47. Shishikura, Y.; Khokhar, S. Factors affecting the levels of catechins and caffeine in tea beverage: Estimated daily intakes and antioxidant activity. *J. Sci. Food Agric.* **2005**, *85*, 2125–2133. [CrossRef]
48. Park, J.H.; Back, C.N.; Kim, J.K. Recommendation of packing method to delay the quality decline of green tea powder stored at room temperature. *Korean J. Hortic. Sci.* **2005**, *23*, 499–506. [CrossRef]
49. Koch, W.; Kukula-Koch, W.; Głowniak, K. Catechin composition and antioxidant activity of black teas in relation to brewing time. *J. AOAC Int.* **2017**, *100*, 1694–1699. [CrossRef]
50. Yamabe, N.; Kang, K.S.; Hur, J.M.; Yokozawa, T. Matcha, a powdered green tea, ameliorates the progression of renal and hepatic damage in type 2 diabetic OLETF rats. *J. Med. Food* **2009**, *12*, 714–721. [CrossRef] [PubMed]
51. Zhang, H.H.; Liu, J.; Lv, Y.J.; Jiang, Y.L.; Pan, J.X.; Zhu, Y.J.; Huang, M.G.; Zhang, S.K. Changes in intestinal microbiota of type 2 diabetes in mice in response to dietary supplementation with instant tea or Matcha. *Can. J. Diabetes* **2020**, *44*, 44–52. [CrossRef] [PubMed]
52. Miura, Y.; Chiba, T.; Tomita, I.; Koizumi, H.; Miura, S.; Umegaki, K.; Hara, Y.; Ikeda, M.; Tomita, T. Tea catechins prevent the development of atherosclerosis in apoprotein E—Deficient mice. *J. Nutr.* **2001**, *131*, 27–32. [CrossRef]
53. Orrù, S.; Imperlini, E.; Nigro, E.; Alfieri, A.; Cevenini, A.; Polito, R.; Daniele, A.; Buono, P.; Mancini, A. Role of functional beverages on sport performance and recovery. *Nutrients* **2018**, *10*, 1470. [CrossRef]
54. Spaccarotella, K.J.; Andzel, W.D. Building a beverage for recovery from endurance activity: A review. *J. Strength Cond. Res.* **2011**, *25*, 3198–3204. [CrossRef]
55. Leidy, H.J. Consumption of protein beverages as a strategy to promote increased energy intake in older adults. *Am. J. Clin. Nutr.* **2017**, *106*, 715–716. [CrossRef] [PubMed]
56. Kulczyński, B.; Gramza-Michałowska, A. Health benefits of inulin-type fructans. *Med. Rodz.* **2016**, *19*, 86–90. (In Polish)
57. Rashidinejad, A.; Birch, J.; Sun-Waterhouse, D.; Everett, D.W. Addition of milk to tea infusions: Helpful or harmful? Evidence from in vitro and in vivo studies on antioxidant properties. *Crit. Food Sci. Nutr.* **2017**, *57*, 3188–3196. [CrossRef] [PubMed]
58. Rashidinejad, A.; Birch, E.J.; Everett, D.W. Interactions between milk fat globules and green tea catechins. *Food Chem.* **2016**, *199*, 347–355. [CrossRef] [PubMed]

59. Tireki, S. A review on packed non-alcoholic beverages: Ingredients, production, trends and future opportunities for functional product development. *Trends Food Sci.* **2021**, *112*, 442–454. [CrossRef]
60. Serra, A.; Macià, A.; Romero, M.P.; Valls, J.; Bladé, C.; Arola, L.; Motilva, M. Bioavailability of procyanidin dimers and trimers and matrix food effects in in vitro and in vivo models. *Br. J. Nutr.* **2010**, *103*, 944–952. [CrossRef] [PubMed]
61. Schramm, L. Going green: The role of the green tea component EGCG in chemoprevention. *J. Carcinog. Mutagen.* **2013**, *4*, 1000142. [CrossRef] [PubMed]
62. Jakobek, L. Interactions of polyphenols with carbohydrates, lipids and proteins. *Food Chem.* **2015**, *175*, 556–567. [CrossRef]
63. Yousf, N.; Nazir, F.; Salim, R.; Ahsan, H.; Adnan Sirwal, A. Water solubility index and water absorption index of extruded product from rice and carrot blend. *J. Pharmacogn. Phytochem.* **2017**, *6*, 2165–2168.
64. Sudha, M.L.; Baskaran, V.; Leelavathi, K. Apple pomace as a source of dietary fiber and polyphenols and its effect on the rheological characteristics and cake making. *Food Chem.* **2007**, *104*, 686–692. [CrossRef]
65. *PN-EN 12143:2000*; Determination of the Content of Soluble Substances—Refractometric Method. Department of Standardization Publishing, Polish Committee for Standardization: Warsaw, Poland, 2000.
66. Hallmann, E.; Kazimierczak, R.; Marszałek, K.; Drela, N.; Kiernozek, E.; Toomik, P.; Matt, D.; Luik, A.; Rembiałkowska, E. The nutritive value of organic and conventional white cabbage (*Brassica oleracea* L. var. *Capitata*) and anti-apoptotic activity in gastric adenocarcinoma cells of sauerkraut juice produced therof. *J. Agric. Food Chem.* **2017**, *65*, 8171–8183. [CrossRef] [PubMed]
67. Ponder, A.; Hallmann, E. The nutritional value and vitamin C content of different raspberry cultivars from organic and conventional production. *J. Food Compos. Anal.* **2020**, *87*, 103429. [CrossRef]
68. Singleton, V.L.; Rossi, J.A., Jr. Colorimetry of total phenolics with phosphomolybdic acid reagents. *Am. J. Enol. Vitic.* **1965**, *16*, 144–158. [CrossRef]
69. Re, R.; Pellegrini, N.; Proteggente, A.; Pannala, A.; Yang, M.; Rice-Evans, C. Antioxidant activity applying an improved ABTS radical cation decolorization assay. *Free Radic. Biol. Med.* **1999**, *26*, 1231–1237. [CrossRef] [PubMed]

molecules

MDPI

Article

Reduction of Nitrite in Canned Pork through the Application of Black Currant (*Ribes nigrum* L.) Leaves Extract

Karolina M. Wójciak [1], Karolina Ferysiuk [1], Paulina Kęska [1,*], Małgorzata Materska [2], Barbara Chilczuk [2], Monika Trząskowska [3], Marcin Kruk [3], Danuta Kołożyn-Krajewska [3] and Rubén Domínguez [4]

[1] Department of Animal Raw Materials Technology, Faculty of Food Science and Biotechnology, University of Life Sciences in Lublin, 20-704 Lublin, Poland
[2] Department of Chemistry, Faculty of Food Science and Biotechnology, University of Life Sciences in Lublin, 20-950 Lublin, Poland
[3] Department of Food Gastronomy and Food Hygiene, Institute of Human Nutrition Sciences, Warsaw University of Life Sciences SGGW, 02-776 Warsaw, Poland
[4] Centro Tecnológico de la Carne de Galicia, Avd. Galicia n° 4, Parque Tecnológico de Galicia, 32900 San Cibrao das Viñas, Ourense, Spain
* Correspondence: paulina.keska@up.lublin.pl; Tel.: +48-81-4623340

Abstract: Sodium nitrite is a multifunctional additive commonly used in the meat industry. However, this compound has carcinogenic potential, and its use should be limited. Therefore, in this study the possibility of reducing the amount of sodium(III) nitrite added to canned meat from 100 to 50 mg/kg, while enriching it with freeze-dried blackcurrant leaf extract, was analyzed. The possibility of fortification of canned meat with blackcurrant leaf extract was confirmed. It contained significant amounts of phenolic acids and flavonoid derivatives. These compounds contributed to their antioxidant activity and their ability to inhibit the growth of selected Gram-positive bacteria. In addition, it was observed that among the three different tested doses (50, 100, and 150 mg/kg) of the blackcurrant leaf extract, the addition of the highest dose allowed the preservation of the antioxidant properties of canned meat during 180 days of storage (4 °C). At the end of the storage period, this variant was characterized by antiradical activity against ABTS (at the level of 4.04 mgTrolox/mL) and the highest reducing capacity. The addition of 150 mg/kg of blackcurrant leaf extract caused a reduction in oxidative transformations of fat in meat products during the entire storage period, reaching a level of TBARS almost two times less than in the control sample. In addition, these products were generally characterized by stability (or slight fluctuations) of color parameters and good microbiological quality and did not contain N-nitrosamines.

Keywords: canned meat; nitrite-free; antioxidants; lyophilization; *R. nigrum* L.; black currant leaves; TBARS

Citation: Wójciak, K.M.; Ferysiuk, K.; Kęska, P.; Materska, M.; Chilczuk, B.; Trząskowska, M.; Kruk, M.; Kołożyn-Krajewska, D.; Domínguez, R. Reduction of Nitrite in Canned Pork through the Application of Black Currant (*Ribes nigrum* L.) Leaves Extract. *Molecules* **2023**, *28*, 1749. https://doi.org/10.3390/molecules28041749

Academic Editor: Adele Papetti

Received: 11 January 2023
Revised: 6 February 2023
Accepted: 8 February 2023
Published: 12 February 2023

1. Introduction

Oxidation of meat and meat products leads to the deterioration of their nutritional value, color, and sensory properties, as well as the formation of harmful compounds (e.g., aldehydes and ketones) [1,2]. The mechanism of oxidation of lipids, proteins, and pigments has been very well explained by various authors [1–5]. In general, free radicals are considered the main elements responsible for the oxidation of lipids and proteins. Therefore, in the last decades, several strategies to improve the oxidative stability of meat and meat products were studied, including the addition of natural antioxidants [6–8], the use of active packaging [9,10], or a combination of both. These substances can inhibit the activity of free radicals and improve the shelf life of the meat products [2,10]. Sodium(III) nitrite is a strong antioxidant that not only protects these products against the negative effects of oxygen, but also renders them with the characteristic color, taste, and flavor [11,12]. In addition, this salt exhibits antimicrobial action against various foodborne pathogens (*Escherichia coli,*

Clostridium botulinum, Clostridium perfringens, and *Bacillus cereus*) [11]. For these reasons, sodium(III) nitrite can be considered as a multifunctional additive, and replacing it in meat products' processing is a major challenge for researchers and professionals in the meat industry [13].

Unfortunately, the latest research points out that nitrite salts pose harmful effects on human health, which was confirmed by the detection of *N*-nitrosocompounds [14]. *N*-nitrosamines (NAs) are mutagenic, genotoxic, and cancerogenic compounds that are formed in cured products during the process of production or storage [15,16]. In addition, NAs could also be produced during the complex digestive process in the human body [15,16]. Hence, there is a search for new methods to protect the safety of meat products. To balance the benefits of sodium(III) nitrite and its adverse effects, it was suggested that the amount of nitrite added to meat products can be reduced. However, to maintain food safety, some additional methods of protection should also be applied [16].

In this sense, some studies have proposed the use of high-nitrate plant extracts or powders, such as those obtained from *Beta vulgaris* [13], radish [17], or cruciferous vegetables [18], to reduce the addition of synthetic nitrite to meat products. Additionally, because secondary plant metabolites are known to have various positive effects on human health, scavenge free radicals, and act as reactive metal chelators, their addition to meat products was also proposed [19,20]. For this purpose, the extracts of various plant materials have been tested: black barberry (*Berberis crataegina* L.) [21]; pomegranate (*Punica granatum* L.) and pistachio (*Pistacia vera* L.) green hull [22]; grape seed and chestnut together with olive pomace hydroxytyrosol [23]; *Caesalpinia sappan* L. [24]; tomato (*Solanum lycopersicum* L.), pomace extract, and peppermint (*Mentha piperita* L.) essential oil [25]; red dragon fruit (*Hylocereus polyrhizus*) peel extract [26]; and bilberry (*Vaccinium myrtillus*) and sea buckthorn (*Hippophae rhamnoides*) leaves [27]. It should be noted that most of the research works focused on cooked pork or beef sausages, while canned meat has not been tested so far.

Black currant (*Ribes nigrum* L.) belongs to the *Grossulariaceae* family, which contains over 150 species. This plant does not require specific conditions for cultivation and can grow in various types of soils [28,29]. Its fruits are rich in different polyphenolic compounds, especially anthocyanins [28,29], while the leaves contain mainly quercetin-3-O-derivatives, myricetin, and kaempferol derivatives, as well as neochlorogenic acid, caffeic acid, gallic acid, chlorogenic acid, and catechins [30–32]. However, the content of polyphenols is higher in leaves than in fruits [29–31]. The soil type and harvest time determine the antioxidant activity and chemical composition of the extracts obtained from these raw materials [29,30]. Nevertheless, both the leaves and fruits of black currant exhibit various biological effects, such as antimicrobial, antihypertensive, antimutagenic, antidiabetic, analgesic, anti-inflammatory, and anticancer properties, and can inhibit cell proliferation [29,31]. Leaves of black currant plants were used in natural medicine in the form of extracts to treat inflammatory disorders and as a diuretic, while in the form of infusions, they were used to regulate the function of kidneys [30].

Despite the high antioxidant potential of black currant leaves and their possible use as an antioxidant in meat products, studies focusing on this subject are limited. Among them, only in the study of Nowak et al. [33], extracts obtained from the leaves of cherry (*Prunus cerasus* L.) and black currant (*R. nigrum* L.) were analyzed to determine their antimicrobial properties in pork sausages with no nitrite addition.

Our previous research [34] proved that the amount of sodium(III) nitrite added to canned pork can be safely reduced without adverse effects, such as the presence of NA, exceeding the amount of malondialdehyde (MDA), and formation of pathogenic bacteria. However, it was noted that the antioxidant properties of the product containing only 50 mg/kg of sodium(III) nitrite were reduced during storage.

Therefore, the present study aimed to evaluate the possibility of producing canned pork, in which the amount of sodium(III) nitrite added was reduced from 100 mg/kg (following Regulation No. 1333/2008 [35] for canned meat products) to 50 mg/kg, along with simultaneous fortification with water-lyophilized leaf extract of black currant during six months (180 days) of storage. The assessment of the quality of the products was based on the assessment of color-forming properties (L*, a*, b*, and nitrosochemochrome), antioxidant properties (in the ABTS, DPPH, and FRAP tests as well as the TBARS test) and antimicrobial properties, important from the point of view of consumers. N-nitrosamines (NAs) content as a critical point in assessing product safety was also assessed in these studies.

2. Results and Discussion

2.1. Characteristics of Black Currant Leaf Extract

2.1.1. Chemical Analysis and Antioxidant Capacity

The lyophilized aqueous extract of black currant leaves (Figure 1) was characterized by a slightly acicular consistency, mild golden color, and delicate fruity aroma. The extract was easily soluble in water.

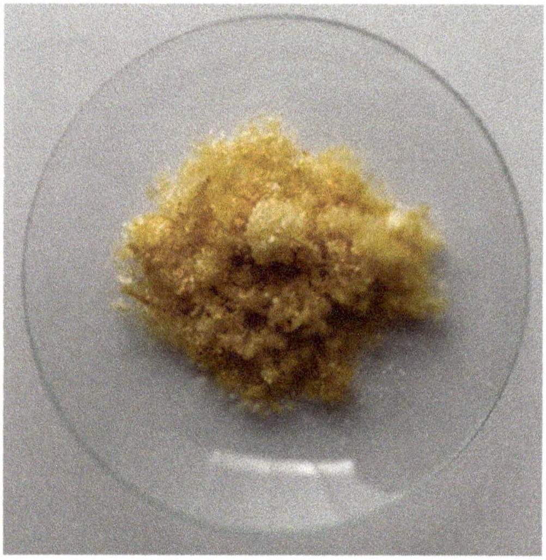

Figure 1. Water extract of black currant leaves after lyophilization.

The extraction yield was satisfactory and amounted to over 18%. Chemical analysis revealed the complex chemical composition of the prepared extract (Table 1). The antiradical activity of the tested extract, determined in the ABTS radical model system and expressed as the EC50 value (11.1 $\pm$ 0.06 μg/mL), was significant ($p < 0.05$). The EC50 value was even lower than of DPPH (32.5 $\pm$ 0.21 μg/mL). The active compounds present in the prepared extract were phenolic acids and derivatives of flavonoids with the concentration of acids found to be dominant over that of flavonoids. This was evidenced by the results of both spectrophotometric analyses and the detailed analysis of the phenolic compound profile using the HPLC method (Table 1 and Figure 2). Of the ten compounds identified in the extract, half were phenolic acids and the other half were glycosidic derivatives of quercetin, luteolin, and apigenin (Figure 2). However, in terms of quantity, the content of phenolic acids, calculated in accordance with the HPLC results, exceeded 99% (Table 1). The chemical content of the investigated dry black currant leaf extract differed from that reported by other authors both quantitatively and qualitatively. The highest concentration of chlorogenic acid was recorded in the tested extract, while in other studies the highest concentration was noted as quercetin-3-*O*-glucoside [30]. This may be due to differences in the methods used for extract preparation as well as analysis.

Table 1. Chemical content and antioxidant activity of water extract obtained from black currant leaves.

Analyzed Parameters	Content
Extraction yield (%)	18.05 ± 0.15
Vitamin C (mg L-ascorbic acid/100 g)	3.11 ± 0.03
DPPH (EC50, μg/mL)	32.5 ± 0.21
ABTS (EC50, μg/mL)	11.1 ± 0.06
Total phenolic content (mg gallic acid/g)	100.5 ± 9.1
Flavonoid content (mg quercetin/g)	10.02 ± 0.1
Dihydroxycinnamic acid content (mg chlorogenic acid/g)	53.54 ± 2.85
Main phenolic compounds (HPLC) (mg/g):	
1. Caffeoyl malic acid	0.335 ± 0.005
2. Chlorogenic acid	5.273 ± 0.015
3. Caffeic acid	3.391 ± 0.021
4. Coumaric acid	0.264 ± 0.004
5. Ferulic acid	0.297 ± 0.025
6. Rutin	0.006 ± 0.001
7. Quercetin 3-*O*-glucoside	0.009 ± 0.001
8. Luteolin-7-*O*-rhamnoside	0.024 ± 0.002
9. Apigenin-7-*O*-glucoside	0.017 ± 0.001
10. Luteolin-7-*O*-glucoside	0.018 ± 0.001

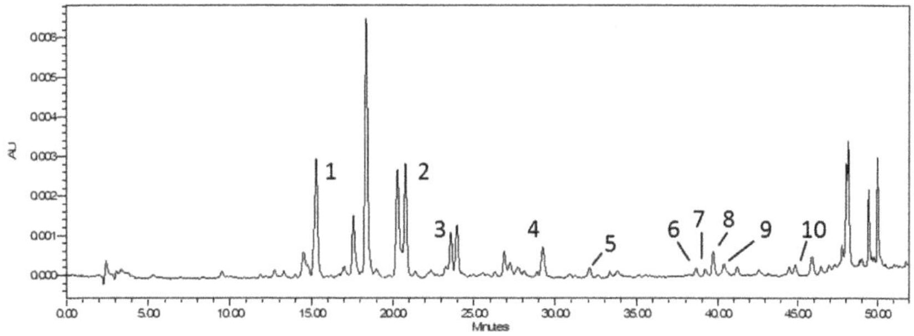

Figure 2. Chromatogram of black currant leaf extract with marked peaks of the identified compounds. The numbers 1–10 refer to the names of the compounds listed in Table 1.

2.1.2. Antimicrobial Activity of Black Currant Leaf Extract

MIC refers to the lowest concentration of the antimicrobial agent that completely inhibits the visible growth of the tested microorganisms [36]. As shown in Table 2, depending on the species, the tested strain displayed different levels of susceptibility to the extract. The black currant leaf extract mostly inhibited the growth of Gram-positive bacteria. However, no growth inhibitory effect toward LAB, *E. coli*, *Enterococcus faecalis*, *Clostridium sporogenes*, and *Salmonella Hofit* IFM 2318 was observed. This lack of suppressive effect on LAB is advantageous as it may allow the use of the extract in fermented meat products. For *C. sporogenes*, a reduction in the size of bacterial colonies was noted with an increase in the amount of extract in the culture medium. This suggests that a higher concentration of black currant leaf extract may be needed to inhibit the growth of this species. The results of our study agree with those reported in the works of Staszowska-Karkut et al. [30] and Paunović et al. [29], which demonstrated the ability of black currant leaf extract to limit the growth of microorganisms. However, the likelihood of bacterial growth inhibition was influenced by the strain properties and the method used for MIC determination as well as for the extraction of black currant leaves.

Table 2. Minimum inhibitory concentration (mg/mL) of black currant leaf extract.

Bacterial Strain	MIC	Bacterial Strain	MIC
Bacillus cereus ATCC 11778	5	*Listeria monocytogenes* ATCC 15313	-
Bacillus subtilis ATCC 6633	5	*L. monocytogenes* ATCC 19111	5
Clostridium sporogenes ATCC 11437	-	*L. monocytogenes* ATCC 7644	5
Enterococcus faecalis ATCC 51229	-	*L. monocytogenes* IFM 1011	5
Escherichia coli ATCC 10536	-	*Salmonella enterica* ATCC 29631	5
E. coli ATCC 25922	-	*Salmonella* Hofit IFM 2318	-
Starter culture for meat fermentation M892	-	*Staphylococcus aureus* 4.4	2.5
Lactobacillus plantarum 299v	-	*S. aureus* 4538	1.5
Lacticaseibacillus rhamnosus ŁOCK 0900	-	*S. aureus* ATCC 25923	2.5
Listeria innocua ATCC 33090	-	*S. aureus* ATCC 6535	5.0

2.2. *Characterization of Canned Pork Containing Black Currant Leaf Extract*

2.2.1. Color Parameters and Nitrosohemochrome

The color properties of canned pork samples with leaf extract added are shown in Table 3. In general, the effect of time on the lightness of the products was not confirmed. All products were characterized by the stability of the L* parameter during refrigerated storage, except for sample B100 for which a significant increase in lightness ($p < 0.05$) was noted on the 180th day of the analysis. For this sample, the L* parameter was on average higher by almost 16 units compared to the other test variants at that time. Taking into account the effect of the dose of lyophilized extract on the lightness of meat products, no significant differences were found between the samples with a lower dose of the extract (B50) and the control sample (C). On the other hand, for sample B100 and B150, a significant ($p < 0.05$) increase in the lightness value was noted after 180 and 90 days of storage, respectively. Moreover, during two months of storage, the a* parameter (redness) was stable for samples B100 and B150 and the values were similar between the samples. Changes were observed after the next 60 days in the samples containing 100 mg/kg of leaf extract, with a slight but statistically significant ($p < 0.05$) increase in a* value. An opposite trend was noted for changes in the a* value in the B150 sample. A gradual increase in the value of the b* (yellowness) parameter was noted in samples B50 and B150 (from 7.39 to 9.18 and from 8.08 to 8.89, respectively). However, a significant ($p < 0.05$) increase in the value of this parameter was noted in sample B100 (from 8.37 to 13.31, difference: approximately 4.94). Moreover, it was observed that this particular sample was highly unstable during the six months of storage in comparison to other samples with respect to both redness and yellowness parameters.

During 180 days of storage, samples with lyophilized black currant leaf extract at the level of 50 mg/kg (B50) and the batches with the highest amount of extract (B150) contained a stable amount of nitrosohemochrome (NO-Mb), and only a slight but significant ($p < 0.05$) decrease in content was observed during the storage time, compared to sample B100, in which sharp, alternating increases and decreases in the amount of NO-Mb were noted. However, at the end of the storage period, all the samples had a similar amount of this pigment (about 22.38 mg/kg). The protein responsible for the color of meat is myoglobin (Mb). Mb is influenced by the state of its heme group, which contains an iron ion. Based on the content of Mb, meat products can appear dull brown or purplish red. The addition of nitrite leads to the formation of nitrosylmyoglobin, which is later stabilized to nitrosyl hemochrome upon thermal treatment, and as a result meat products acquire the characteristic pink color [12,37]. It should be mentioned that the iron ion from the heme group acts as a double agent—it contributes to color formation and also acts as a pro-oxidant due to its ability to donate electrons with ease [38]. Suman and Joseph [37] reported that lipid oxidation products (e.g., aldehydes) can also affect the color of meat products by destabilizing the heme group. The addition of antioxidant substances (natural or synthetic) causes the chelation of Fe^{2+} ions [38]. Riviera et al. [12] pointed out that for producing meat products with stable and acceptable color for commercial purposes, about 10–15 ppm of $NaNO_2$ should be added, while Food Chain Evaluation Consortium [16] suggests an amount of 55–70 mg/kg (for nontraditional products).

Table 3. The color properties (CIE L*a*b* system) and amount of meat pigment (nitrosohemochrome) in canned meat samples with a reduced amount of sodium nitrite and fortified with the extract of black currant leaves.

Parameter	Sample	Storage Time (Days)			
		1	60	90	180
L*	C	63.50 ± 3.01 Aa	62.35 ± 2.98 Aa	62.15 ± 1.60 Aa	62.80 ± 2.58 Aa
	B50	63.55 ± 3.29 Aa	62.23 ± 3.02 Aa	62.12 ± 1.64 Aa	62.89 ± 2.63 Aa
	B100	62.37 ± 2.94 Aa	64.02 ± 2.57 Aa	62.34 ± 2.48 Aa	78.84 ± 2.91 Bb
	B150	62.01 ± 3.06 Aa	63.91 ± 3.30 Aa	66.01 ± 2.36 Ba	63.32 ± 2.24 Aa
a*	C	9.68 ± 1.15 Ba	10.59 ± 1.05 Bb	10.71 ± 0.50 ABb	9.90 ± 1.15 Ba
	B50	9.72 ± 1.17 Ba	10.68 ± 1.04 Bb	10.78 ± 0.55 ABb	9.95 ± 1.00 Ba
	B100	10.13 ± 1.17 Aa	10.01 ± 0.93 Aa	10.73 ± 1.21 Aa	12.38 ± 0.97 Ab
	B150	10.29 ± 0,97 Ab	10.11 ± 1.07 Bb	9.45 ± 0.91 Bab	9.83 ± 0.77 Ba
b*	C	7.60 ± 0.70 Ab	8.10 ± 0.57 Aab	8.79 ± 0.62 Aa	8.77 ± 0.73 Bb
	B50	7.59 ± 0.68 Ab	8.17 ± 0.65 Ab	8.86 ± 0.44 Aa	8.77 ± 0.84 Ba
	B100	8.36 ± 0.65 Ab	8.70 ± 0.56 Ab	9.23 ± 0.64 Ab	13.23 ± 0.54 Aa
	B150	8.17 ± 0.76 Aa	8.31 ± 0.58 Aa	8.69 ± 0.58 Bab	8.87 ± 0.73 Bb
Nitrosohemochrome	C	21.95 ± 0.77 Ba	24.37 ± 1.76 Ba	21.98 ± 1.49 Ba	22.51 ± 2.39 Aa
	B50	22.73 ± 0.97 Ba	25.59 ± 1.83 Ba	22.51 ± 1.43 Ba	22.48 ± 2.72 Aa
	B100	20.26 ± 1.37 Aa	29.16 ± 1.62 Cc	19.40 ± 1.55 Aa	22.67 ± 1.26 Ab
	B150	27.84 ± 2.4 Cb	23.54 ± 1.72 Aa	24.24 ± 1.38 Cb	22.00 ± 2.85 Aa

C—control; B50—black currant leaves 50 mg/kg; B100—black currant leaves 100 mg/kg; B150—black currant leaves 150 mg/kg. Means with different capital letters are significantly different ($p < 0.05$) in the same column. Means with different small letters are significantly different ($p < 0.05$) in the same row. Results are presented as mean ± SD.

Bae et al. [39] investigated the effect of alternating cured process on, among others, the number of pigments in pork products. They observed no significant differences between the control (0.1% of $NaNO_2$) and test samples with radish or celery powder (0.15% and 0.30%). In general, the redness of all samples varied from 8.30 to 8.55 (control), but the values of b* parameter were higher in products containing celery powder at an amount of 0.30% (b* = 8.66). Bae et al. [39] also noted that the addition of radish powder and a long incubation time resulted in higher yellowness in the samples. The authors suspected that pigments presented in celery could affect the value of the b* parameter. In addition, they observed that incubation time affected the amount of nitrosyl hemochrome to a greater extent than the quantity of the added powder. This is explained by the fact that a higher volume of nitrite is converted by bacteria from a vegetable source. In our study, low-nitrate plant material was used, which may explain the low value of NO-Mb (<30 mg/kg in comparison to the value reported by Bae et al. [39]). It could also be assumed that 50 mg/kg of sodium(III) nitrite was not enough to support a higher amount of nitrosyl hemochrome but was high enough to result in an appropriate value of the a* parameter (approximately 10 throughout the storage period in comparison to the value of about 8 after production in Bae et al.'s study). Data obtained from our study show the influence of black currant leaf extract on the yellowness of canned meat—the value of the b* parameter is higher than that determined by Bae et al. [39]. Moreover, Nowak et al. [33] observed an increase of the b* parameter during storage in pork sausages containing cherry and black currant leaf extract. The authors concluded that this could be related to the color of plant extracts. Similarly, Sun and Xiong [40] noted that the amount of NO-Mb was not directly proportional to the redness of beef patties (samples containing pea protein isolate and pea protein hydrolysate had a lower value of nitrosyl hemochrome compared to their a* values). In addition, the authors assumed that other pigments may also influence the color of beef patties. Moreover, it could be possible that nitrogen groups from pea proteins may bind heme iron and, therefore, decrease the level of nitrosyl hemochrome.

2.2.2. Antioxidant Abilities

Antiradical Activity and FRAP

The results obtained in the analysis of the antioxidant abilities ($ABTS^+$, DPPH, FRAP) of canned pork samples are presented in Table 4. There were no differences in the ability to neutralize synthetic radicals ($ABTS^+$ or DPPH) as well as in the FRAP between the control sample (C) and the B50 sample during refrigerated storage. However, referring to the ABTS test, after 1 day of storage, the free radical scavenging capacity was high in samples B50 and C (3.95 and 3.90, respectively, $p < 0.05$). On the other hand, immediately after production (1 day) there was no effect of the extract dose on the antiradical effect against DPPH. Although the present study confirmed the antioxidant effect of extracts in model systems, immediately after the production of canned meat, no effect of the amount of plant extract addition on the ability to neutralize free radicals was observed, both in the ABTS and DPPH tests. The effect of the antioxidant compound in the food matrix may differ significantly in activity from the purified extract. This is influenced by many factors, including matrix properties (pH, water activity), antioxidant structure, and the resulting structure–activity relationship (SAR). In addition, it is well known that antioxidants can be significantly altered by processing. In particular, exposure to high temperatures, such as pasteurization [41] or sterilization [42], may be the main cause of the reduction of the natural amounts/properties of plant antioxidants. Many plant secondary metabolites act as antioxidants and pro-oxidants and can affect the concentration of reactive oxygen species depending on the reaction conditions. Under certain conditions, flavonoids and phenolic acids have the ability to act as pro-oxidants, and their activity depends on the chemical structure and concentration of the compound and the environment of their reaction, such as pH, the presence of metals (Cu and Fe), and temperature. This may explain the lower antioxidant activity of products containing the addition of plant extract (containing, among others, polyphenols) immediately after production, in that the greater the addition of the extract, the lower the effectiveness against ABTS radicals (Table 4). In addition, the heating process also disrupts muscle cell structure, inactivates antioxidant enzymes, and produces myoglobin catalytic iron, leading to an intense pro-oxidative environment, which may have affected the antioxidant activity of canned meat in the early post-production period. With the increase in storage time, the antiradical effect of extracts from meat products increased and was higher in samples with a higher content of extracts from currant leaves. Notably, this increase was more intense in the test with ABTS than with DPPH, which was probably related to the low antiradical activity of hydrophobic compounds present in the products. The antioxidant properties of sample B50 were lower than sample B100 and B150 with the ABTS test. At the end of the storage period, samples with the lowest amount of extract addition showed a similar value, while canned pork with 150 mg/kg of black currant leaves extract presented the strongest antioxidant properties (4.04 mgTrolox/mL). However, statistical analysis did not show any significant difference in the values. Conversely, on the 60th day of the analysis, variants with higher doses of lyophilized extract obtained from black currant leaves (B100 and B150) were characterized by a higher antiradical capacity than the remaining variants, although it decreased over the time of the analysis (Table 4). Khaleghi et al. [21] observed an increase in antioxidant properties (DPPH method) in beef sausages with an increase in the amount of black barberry extract added; however, when sodium(III) nitrite was also added at an amount of 90 mg/kg, the DPPH value was lower. Seo et al. [24] noted higher DPPH scavenging ability in the sample with 0.1% *C. sappan* extract than the sample with 0.05% of extract and 0.004% of nitrite. In addition, it can be postulated that in addition to the added antioxidants present in the extracts, other substances, such as proteins present in the meat samples, may affect the antioxidant activity of the samples. On the one hand, the presence of myoglobin, which is easily oxidized, may be associated with increased antioxidant activity in the DPPH test [43]. This can be explained by the fact that the oxidation of meat leads to the formation of compounds that act as scavengers of free radicals that could not be detected by the TBA test, which measures the outcome of fat oxidation [44]. On the other hand, meat whites may be a source of biologically active peptides, and their release from the protein chain may increase the antiradical effect of canned meat extracts. In this context, the addition of natural polyphenolic compounds may promote this bioactive role of peptides [45].

Table 4. Antioxidant abilities of canned meat fortified with the extract from black currant leaves during 180 days of storage (4 °C).

Parameter	Sample	Storage Time (Days)			
		1	60	90	180
ABTS•+ [mg$_{TROLOX}$/mL]	C	3.90 ± 0.42 ^{Aa}	3.20 ± 0.29 ^{Bb}	2.71 ± 0.31 ^{Bc}	3.65 ± 0.20 ^{ABa}
	B50	3.95 ± 0.61 ^{Aa}	3.22 ± 0.35 ^{Bb}	2.67 ± 0.21 ^{Bc}	3.71 ± 0.22 ^{ABa}
	B100	2.37 ± 0.54 ^{Bc}	4.42 ± 0.44 ^{Aa}	4.02 ± 0.39 ^{Aa}	3.61 ± 0.43 ^{Bab}
	B150	3.49 ± 0.31 ^{Ab}	3.91 ± 0.63 ^{Aab}	3.76 ± 0.23 ^{Aab}	4.04 ± 0.26 ^{ABa}
DPPH [mg$_{TROLOX}$/mL]	C	0.019 ± 0.001 ^{Aa}	0.015 ± 0.001 ^{Bb}	0.015 ± 0.001 ^{Bb}	0.013 ± 0.001 ^{Bc}
	B50	0.018 ± 0.001 ^{Aa}	0.015 ± 0.001 ^{Bb}	0.015 ± 0.001 ^{Bb}	0.013 ± 0.001 ^{Bc}
	B100	0.018 ± 0.001 ^{Ab}	0.02 ± 0.000 ^{Aa}	0.015 ± 0.001 ^{Bc}	0.014 ± 0.002 ^{Ad}
	B150	0.018 ± 0.001 ^{Ab}	0.02 ± 0.001 ^{Aa}	0.017 ± 0.001 ^{Ab}	0.013 ± 0.001 ^{Bc}
FRAP [$A_{700 nm}$]	C	1.67 ± 0.08 ^{Ba}	1.40 ± 0.04 ^{Bc}	1.49 ± 0.10 ^{Cb}	1.70 ± 0.04 ^{Ba}
	B50	1.68 ± 0.09 ^{Ba}	1.38 ± 0.05 ^{Bc}	1.52 ± 0.11 ^{Cb}	1.74 ± 0.04 ^{Ba}
	B100	1.64 ± 0.16 ^{Bb}	1.91 ± 0.08 ^{Aa}	1.71 ± 0.03 ^{Bb}	1.73 ± 0.02 ^{Bb}
	B150	1.84 ± 0.07 ^{Ab}	1.87 ± 0.07 ^{Ab}	1.91 ± 0.11 ^{Aab}	2.00 ± 0.05 ^{Aa}

C—control; B50—black currant leaves—50 mg/kg; B100—black currant leaves—100 mg/kg; B150—black currant leaves—150 mg/kg. Means with different capital letters are significantly different ($p < 0.05$) in the same column. Means with different small letters are significantly different ($p < 0.05$) in the same row. Results are presented as mean ± SD.

Metals, such as copper and iron, can act as pro-oxidants, which is related to the Fenton reaction and change of their oxidation form. Therefore, it was assumed that metals can determine the oxidation stability of meat products [2]. The ability of polyphenols to donate electrons demonstrates the reducing power of these biomolecules and is also representative of their antioxidant activity. As observed in this study, a systematic increase (from 1.84 to 2.00) in iron ion-reducing power was noted in sample B150, with the highest amount of black currant leaf extract (150 mg/kg) (Table 4). Samples C (control), B50, and B100 presented similar values of this parameter (1.67, 1.68, and 1.64, respectively) at the beginning of the storage period. After two months, the values were reduced in C and B50 samples compared to samples B100. Subsequently, a gradual increase in the values of the FRAP parameter was noted in C and B50 samples, while the sample with 100 mg/kg of black currant leaves extract was stable. However, at the end of the storage period, both samples showed similar values (1.74 and 1.73, respectively). Mira et al. [46] investigated the ability of selected flavonoids to reduce copper and iron and concluded that they depend both on the standard metal redox potential (E0) as well as on the flavonoid structure. Changes in both of these parameters may have contributed to the observed results in the FRAP antioxidant test.

Secondary Lipid Oxidation Products

The results obtained in the analysis of the secondary lipid oxidation products were detected (Figure 3). Determination of the MDA content is one of the most commonly used methods for evaluating the degree of lipid rancidity. During lipid oxidation, various hazardous products are produced, among which aldehydes are highly toxic as they are characterized by mutagenic, cytotoxic, and pro-inflammatory properties [2]. As presented in Figure 3, from the first day of storage (4 °C) there was a significantly ($p < 0.05$) higher

MDA content in the sample with only 50 mg/kg of extract compared to other variants, although the B50 sample had the same TBARS level as the control sample throughout the whole period up to 180 days. After 90 days of storage, a rapid increase in the amount of MDA was observed in samples B50 and B100 (approximately 0.034 and 0.021, respectively) compared to sample B150. The addition of 100 and 150 mg/kg black currant leaf extract resulted in lower ($p < 0.05$) values of the TBARS parameter compared with the C and B50 sample. However, the TBARS values were stable during six months of storage only for sample B150. In fact, similar values were described in the B150 samples between 0 and 180 days (0.033 and 0.029 mg MDA/kg in 0 and 180 days, respectively). Additionally, the significant decrease of the MDA content between 90 and 180 days in the B50 and B100 samples is expected since the secondary lipid oxidation products could be further degraded to other more stable compounds [2]. The acceptable amount of MDA in meat products is around 2–2.5 mg/kg [2]. Other authors conclude that in some meat products, this limit could be 0.6 mg MDA/kg [47,48]. In our study, none of the canned pork samples showed MDA levels exceeding this limit since very low values were observed in all samples (between 0.018 and 0.047 mg MDA/kg) but sample B50 after 90 days of storage (0.07 mg MDA/kg). A possible explanation for these low TBARS values could be related to the limited exposition of canned meat to oxygen [47]. It is well known that oxygen plays a vital role in lipid oxidation, both in the initiation and propagation phases [2]. Thus, canned meat in oxygen-free packaging determines the low lipid oxidation.

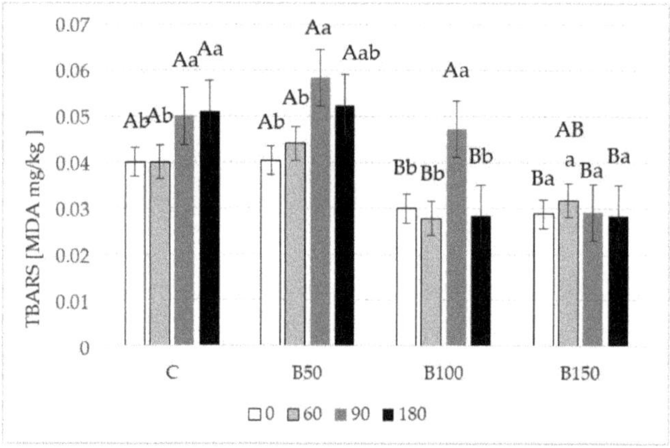

Figure 3. Amount of secondary lipid oxidation products (TBARS) in canned meat samples fortified with the extract of black currant leaves during 180 days of storage (4 °C). C—control; B50—black currant leaves 50 mg/kg; B100—black currant leaves 100 mg/kg; B150—black currant leaves 150 mg/kg. Means with different capital letters are significantly different ($p < 0.05$) on the day of the study. Means with different small letters are significantly different ($p < 0.05$) for a single variant during the storage time. Results are presented as mean ± SD.

Natural substances with antioxidant properties are added to meat products to prevent oxidation. In this sense, the antioxidant capacity of the extract from black currant leaves could be another explanation for the limited lipid oxidation observed in our samples. However, the use of antioxidants in combination does not always give positive results. During 30 days of storage (4 °C) of beef sausages with the addition of nitrite and black barberry extract (30, 60, and 90 mg/kg), a negative interaction was observed between the two additives at the highest concentrations. The synergistic effect was observed when 90 mg/kg extract was used in combination with 30 mg/kg sodium(III) nitrite [21]. However, different results were observed in the study of Šojić et al. [25], in which a combination of sodium nitrite, tomato pomace extract, and peppermint essential oil allowed for the obtaining of much lower levels of MDA in cooked pork sausages (60 days of storage at 4 °C) than samples containing nitrite and a single plant additive. A similar observation was made by Seo et al. [24]. They observed that samples with *C. sappan* (separately and with nitrite) had a similar amount of MDA as samples containing only sodium(III) nitrite (0.007%) during one month of storage (4 °C). However, the authors noted that a combination of two antioxidants allowed for the obtaining of slightly better results. As mentioned earlier, the sample containing both antioxidants showed greater antioxidant capacity (DPPH method) than other samples, but this did not translate into better MDA inhibition [24]. A similar finding was also noted in our study (Table 1). Although, Mäkinen et al. [27] found higher antioxidant properties in pork sausages containing bilberry and sea buckthorn leaf extract in a higher concentration than the control sample (nitrite and ascorbic acid) during 20 days of storage. The addition of 2.0% and 1.0% of bilberry leaves and 0.2% of sea buckthorn leaves resulted in the inhibition against lipid oxidation.

2.2.3. Consumer Safety

NAs (i.e., NDBA, NDMA, NDEA, NDPA, NMOR, NPIP, and NPYR) were not detected after 1 and 180 days of storage in the samples (data not presented). These compounds are formed from nitrosating derivatives (e.g., NO) and secondary amines—a combination that allows the formation of the most stable NA [49]. During the reaction of nitrite with the heme group or free thiol groups, nitric oxide is released [50]. NA can be formed in an acidic environment (stomach) or in the presence of microorganisms (large intestine). They are also formed in the product itself during production or while a meal is prepared at home (when the applied temperature is higher than 100 °C and in the presence of fat or during reheating) [49,50]. Endogenous NA formation in the colon is associated with the nitrosylation of heme (resulting from nonabsorbed red meat residues) and products formed from the metabolism of amino acids by microorganisms [51].

De Mey et al. [52] pointed out that following good manufacturing practices are necessary to prevent the formation of NA—storage of meat for a longer period may increase the chance of NA formation, while spices containing pyrrolidine or pyrroperine may act as NA precursors. Moreover, Park et al. [51] conducted extensive research on the presence of NA in various types of food (including milk and milk products, seafood, alcoholic beverages, oils, meat, and processed meat). The authors detected *N*-nitrosodimethylamine (NDMA) in all the examined processed meat products and *N*-nitrosodiethylamine (NDEA) and N-nitrosopiperidine in most of the tested products. According to Özbay and Şireli [53], NDMA can lead to diarrhea, abdominal cramps, vomiting, headache, as well as carcinogenic and mutagenic effects. NDEA also has carcinogenic effect on the esophagus and liver, while *N*-nitrosodipropylamine exhibits a carcinogenic effect on the lungs and esophagus. The addition of antioxidant substances (e.g., ascorbic acid or tocopherol) may inhibit the formation of NA [53].

After the first and last days of storage, no pathogenic microorganisms (i.e., *Clostridium perfringens, Listeria monocytogenes*, or *Salmonella*) were detected in the tested samples (data not presented). Paunović et al. [29] tested the effect of extracts from fruits and leaves of black currant obtained from various types of soils on selected microorganisms (such as *Staphylococcus aureus, Micrococcus lysodeikticus, Bacillus mycoides, Klebsiella pneumoniae, Pseudomonas glycinea, E. coli, Candida albicans, Fusarium oxysporum, Penicillium canescens*, and *Aspergillus glaucus*).

The MIC determined for the extract from fruits was 38.2–170.1 μg/mL, while for the extract from leaves it was 123.0–389.2 μg/mL. Raudsepp et al. [32] compared the ethanolic (20% and 96%) extracts obtained from the fruits and leaves of rhubarb, blue honeysuckle, and black currant for, among others, antimicrobial properties. They found that black currant extract (in all dilutions) inhibited the growth of *Campylobacter jejuni, Salmonellae Enteritidis, E. coli, L. monocytogenes* (extraction with 96% of ethanol), and *B. cereus* (20% and 96% ethanol). The extract from fruits showed slightly lower antimicrobial effects on Gram-positive bacteria; in general, less diluted extracts presented stronger inhibitory properties against microorganisms. However, the extracts from fruits and leaves of black currant did not present strong antimicrobial properties in comparison to rhubarb extracts. Furthermore, Majou and Christieans [54] pointed out that the antimicrobial properties of nitrite depend on various factors (nitrite concentration, pH value, temperature, and presence of curing accelerator). They showed that chemical transformations lead to the formation of peroxynitrite from sodium(III) nitrite, which is suspected to be responsible for antimicrobial properties.

3. Materials and Methods

3.1. Black Currant Leaf Extract Preparation

The plant material consisted of black currant (*R. nigrum* L.) leaves, which were collected in May before the flowering of the bushes. The extraction conditions were established in our previous studies [55–57]. The leaves were initially dried in a shaded, airy place and then in a dryer at 60 °C. Dry leaves were used for preparing water extract by ultrasound-assisted extraction (10 min) with hot distilled water (90 °C) at a plant-to-solvent ratio of 1:10 (m/v). The infusions were placed in an ultrasonic bath for extraction using Sonic 6D equipment (Polsonic Palczynki Sp. J., Warsaw, Poland). The ultrasound frequency was set as 40 kHz and sound intensity as 320 W/cm^2, temperature as 30 °C. The obtained infusions were filtered after 30 min, cooled, frozen (−18 °C), and then lyophilized. Freeze drying was carried out for 72 h using a freeze dryer (Free Zone 12 lyophilizer, Labconco Corporation, Kansas City, MO, USA) at −80 °C and 0.04 mbar. The lyophilized dry extracts were stored in airtight plastic containers, protected from light, at room temperature until analysis.

3.2. Chemical Analysis and Antioxidant Capacity of Black Currant Leaf Extract

3.2.1. Total Phenolic Content

The total phenolic content of the extracts was analyzed by the Folin–Ciocalteu (FC) method using gallic acid (25–500 mg/L) as standard [58]. Briefly, 0.06 mL of extract was mixed with 0.54 mL of distilled water, 1.5 mL of FC reagent (diluted 1:10 with distilled water), and 1.2 mL of sodium bicarbonate solution (7.5%). The samples were incubated in the dark at room temperature for 30 min and then the absorbance was measured (750 nm) on a Cary 50 spectrophotometer (Varian, Palo Alto, CA, USA). The results expressed as mg gallic acid equivalent/g dry extract.

3.2.2. Total Flavonoids

Total flavonoids of the extracts were determined using a spectrophotometric method based on the formation of a colored complex between the flavonoid and $AlCl_3$ [59]. Briefly, the lyophilized extract was dissolved in water to prepare a solution with a starting concentration of 1 mg/mL. Then, 0.25 mL of the prepared solution was mixed with 0.75 mL of ethanol (96%), 0.05 mL of $AlCl_3$ (10%), 0.05 mL of sodium acetate (1 M), and 1.4 mL of distilled water. The absorbance of the resulting solution was measured at 415 nm. Total flavonoids were expressed as quercetin equivalent based on a calibration curve previously prepared for this compound and were presented in mg quercetin/g dry extract.

3.2.3. Dihydroxycinnamic Acids

Samples of dissolved extracts were prepared as described for the analysis of polyphenolic compounds. The content of total hydroxycinnamic acid was expressed as chlorogenic acid equivalent, as described by Nicolle et al. [60]. Briefly, 1 mL of the extract was added to 2 mL of 0.5 M HCl, 2 mL of Arnow's reagent (10 g of sodium nitrite and 10 g of sodium molybdate, made up to 100 mL with distilled water), 2 mL of NaOH (at a concentration of 2.125 M), and 3 mL of water. Each of the prepared solutions was compared with the corresponding mixture that did not contain Arnow's reagent. The absorbance was read at 525 nm.

3.2.4. HPLC Analysis

The high-performance liquid chromatography (HPLC) method was used for a detailed investigation of changes in the lyophilized water extracts obtained from the leaves of black currant. The analysis was performed using an Empower-Pro chromatograph (Waters), equipped with a quaternary pump (M2998 Waters) with a degasser and a UV–VIS diode array detection system. Separation was performed on a column filled with modified silica gel RP-18 (Atlantis T3, Waters; 3 μm, 4.6 × 150 mm). The mobile phase consisted of A (1% acetic acid) and B (acetonitrile), in which the concentration of solvent B changed as follows: until 0–8th min, 8–12%; in the 10th min, 20%; and in the 25th min, 25%; the flow speed was set at 1 mL min^{-1}. Detection was carried out at 330 nm. The identified compounds (marked by peak numbers on the chromatogram) were quantified according to the calibration curves prepared for each compound, and their content was expressed as mg/g dry extract.

3.2.5. Vitamin C Content

The content of vitamin C was determined with the spectrofluorimetric method described by Wu [61] with necessary modifications. Compounds including 3% metaphosphoric acid, 7 M HCl, 0.1 M $Na_2S_2O_3$, and 0.005 M H_2SO_4 were used in the analysis. An oxidizing solution was prepared by dissolving 1.3 g I_2 in 10 mL of 40% KI, in which 0.1 mL of 7 M HCl and distilled water were added to a volume of 100 mL. The 0.1 M $Na_2S_2O_3$ solution was made by dissolving 1.25 g of the reagent and 0.01 g Na_2CO_3 in 50 mL of water. The derivatization reagent was prepared using 10 mg o-phenylenediamine (OPDA) dissolved in 10 mL of 0.005 M H_2SO_4.

The lyophilized extracts of black currant leaves (0.2 g) were diluted in 10 mL of 3% metaphosphoric acid. To the sample extract (2 mL) or standard (L-ascorbic acid, 50–500 μg/L), 0.3 mL portions of a 0.005 M solution of iodine in potassium iodide were added. The solution was vortexed for 1 min and supplemented with 0.3 mL portions of 0.01 M $Na_2S_2O_3$. The pH of the sample was adjusted to approximately 6.0 by adding 0.3 mL of 2 M NaOH, and derivatization was carried out by adding 0.3 mL portions of the OPDA solution. The solution was stirred for 30 min at the maximum stirring force. Then, it was diluted to 50 mL with distilled deionized water. The determinations were conducted on a Cary Eclipse spectrofluorometer (Varian, Palo Alto, CA, USA) at an excitation wavelength of $\lambda = 365$ nm and an emission wavelength of $\lambda = 425$ nm. The content of vitamin C was calculated based on a standard curve prepared using an aqueous solution of L-ascorbic acid standard and expressed as L-ascorbic acid (μg)/100 g sample (dry basis).

3.2.6. Antioxidant Capacity

The lyophilized extract was dissolved in water to obtain a solution with a starting concentration of 1 mg/mL. The solution was then diluted to produce a series of samples with concentrations ranging from 0.1 to 1 mg/mL. The ability of the samples to scavenge ABTS radical cation ($ABTS^{\bullet+}$) [2,2-azinobis(3-ethyl-benzothiazoline-6-sulfonate)] was determined as described by Re et al. [62]. For this purpose, the working solution of $ABTS^{\bullet+}$ radical was prepared by mixing ABTS (7 mM) with potassium persulfate (2.45 mM). The solution was kept in the dark at room temperature for 18 h and then diluted with 95% ethanol until an absorbance of 0.70 (±0.02) was reached at 734 nm. Next, the sample (20 μL) and the $ABTS^{\bullet+}$ solution (3 mL) were mixed and incubated at room temperature for 10 min. The absorbance of the mixture was measured at 734 nm on a Cary 50 spectrophotometer (Varian, Palo Alto, CA, USA). The control was prepared by adding 20 μL of methanol to $ABTS^{\bullet+}$ solution instead of the sample. In addition to ABTS's radical scavenging ability, the antiradical capacity of the samples against the DPPH (2,2-diphenyl-1-picrylhydrazyl) radical was measured [63]. Briefly, 0.1 mL of the extract solution was mixed with a freshly prepared methanolic solution of DPPH (0.1 mM, 4 mL). The resulting mixture was vortexed and incubated in the dark at room temperature for 30 min. The absorbance was measured at 517 nm on a Cary 50 spectrophotometer (Varian, Palo Alto, CA, USA). A control was prepared by adding 4 mL of 0.1 mM DPPH to 0.1 mL of methanol (analytical grade) instead of the sample.

The percentage of absorbance reduction in comparison to the initial value was determined to calculate the inhibition percentage, according to the formula:

$$\%AA = \left[1 - \left(A_p - A_b\right)\right] \times 100\% \quad (1)$$

where AA is the antioxidant activity of the analyzed sample, A_p is the absorbance of the analyzed sample, and A_b is the absorbance of the blank sample.

Based on the dependence between the antiradical activity of the samples and their concentration (f(c) = $\%AA$), the EC50 values were calculated.

3.3. Antimicrobial Activity of Black Currant Leaf Extract

The minimum inhibitory concentration (MIC) of black currant leaf extract was determined according to the recommendations of EUCAST [36]. Mueller–Hinton (MH) agar (Bio-Rad, Hercules, CA, USA) was used as a nonselective medium for the growth of tested microorganisms (Table 2). For lactic acid bacteria (LAB), 10 g/L glucose (Sigma-Aldrich, Poznań, Poland) was added to the MH medium. The bacterial suspensions were prepared using overnight cultures. Aqueous black currant leaf extract was diluted with a molten MH medium to prepare tested concentrations (1–5 mg/mL). After solidification of the MH agar, the tested strain with the adjusted density of 10^4 colony-forming units (cfu)/mL was spread on the medium. Then, the samples were incubated at 37 °C for 24 h. The positive control consisted of MH agar inoculated with the test bacteria without the extract, while uninoculated plates containing black currant leaf extract served as the negative control. When the visible growth inhibition was observed (judged by the naked eye), regardless of the appearance of a single colony or a thin haze, the MIC of the extract was determined.

3.4. Canned Meat Preparation

Canned meat was prepared using pork shoulder and pork dewlap (80%:20%), salt (2%), water (5%), a reduced amount of sodium(III) nitrite (50 mg/kg of meat), and lyophilized black currant leaf extract (0 for control (C)) and 50, 100, and 150 mg/kg of meat for tested batches (B50, B100, and B150, respectively). Meat was purchased from an organic farm (Zakład Mięsny Wasąg SP. J., Hedwiżyn, Poland, organic certificate no. PL-EKO-093027/18). After initial and final grinding (universal machine KU2-3E, Mesko-AGD, Skarżysko-Kamienna, Poland), the material was divided into three variants and subjected to mixing (4–5 min/variant; universal machine KU2-3E, Mesko-AGD, Poland). Then, the meat stuffing was transferred to metal cans (meat filling: 250 g) and sterilized on a vertical steam sterilizer (TYP-AS2, Poland) at 121 °C. After sterilization, the cans were cooled in water and stored (at 4 °C) for further analysis. The experimental canned pork samples were heated at 121 °C, assuming that their degree of heating was achieved as measured with the sterilization value ($F \approx 4$ min) determined from the formula:

$$F = \int_0^1 L dt \quad (2)$$

$$L = 10\frac{T - T_0}{z} \quad (3)$$

where F is the sterilization value, L is the lethality degree, T_0 is the reference temperature (121 °C), and z is the sterilization effect factor (10 °C).

The sterilization values were calculated by determining the degree of lethality by measuring temperature every minute. The limits of integration were assumed from 90 °C during the growth phase to 90 °C during the decrease (i.e., cooling). The degree of heating was determined for the cans in their critical zone using an electric thermometer equipped with a thermoelectric sensor. After sterilization, the products were cooled in water and stored. Then, they were divided into four groups, and each group was tested (according to the analysis listed below) immediately after production (day 1), and after 60, 90, and 180 days of storage. The experiment included one-time preparation of 48 (+5 inventory) canned pork samples (12 cans from each test variant: C, B50, B100, and B150). In each study period (1, 60, 90, and 180 days) 3 cans of each variant were tested. The experiment thus planned was repeated three times at about two-week time intervals.

3.5. *Canned Meat Analysis*

3.5.1. Color Parameters (CIE L*a*b*) and No-Mb

The color of the canned meat samples was measured using an X-Rite Color 8200 spectrophotometer (X-Rite Inc., Grand Rapids, MI, USA; port size: 13 mm; standard observer: 10°; illuminant: D65). The samples were cut into cuboids and analyzed at three points [64]. For the determination of nitrosohemochrome, 5 g of samples were homogenized for 1 min in acid acetone. After 30 min of storage in the dark, the samples were centrifuged (4000 rpm for 10 min), and the absorbance was measured (540 nm) using a UV–VIS spectrophotometer (HITACHI U-5100). The results were obtained by multiplying absorbance by 290 and expressed as mg/kg [65].

3.5.2. Antioxidant Properties

Antiradical Properties and FRAP

Sample Preparation

For the analysis of antioxidant properties, samples were prepared as described by Jung et al. [66] with some modifications. Briefly, 10 mL of ethanol was mixed with 5 g of minced canned meat and homogenized (1000 rpm). Then, the samples were centrifuged (10,000 rpm for 20 min) and filtered. Absorbance was measured using a UV–VIS spectrophotometer (HITACHI U-5100).

$ABTS^{\bullet+}$ and DPPH

Free radical scavenging activity was determined as described by Jung et al. [66] with some modifications. In the $ABTS^{\bullet+}$ method, 12 μL of supernatant was added to 1.8 mL of diluted $ABTS^{\bullet+}$ solution. Absorbance was measured at 734 nm after 3 min. In the DPPH method, absorbance was measured after 3 min at 517 nm using a diluted DPPH solution with an absorbance of 0.9 ± 0.02. In both these methods, the antioxidant capacity of the samples was calculated from a standard curve (concentration: 0.025–0 mg/mL for DPPH; 15–0 mg/mL for ABTS) and expressed as mg Trolox equivalent/mL.

FRAP

FRAP (ferric ion-reducing antioxidant power) of the samples was measured at 700 nm as described by Oyaizu [67]. The results were expressed as absorbance.

Secondary Lipid Oxidation Products

TBARS (thiobarbituric acid reactive substance) parameter was determined as described by Fan et al. [68]. Briefly, the meat samples (10 g) were homogenized using a mixture of ethylenediaminetetraacetic acid (0.1%) and trichloroacetic acid (7.5%). After shaking (30 min) and filtration, the samples were boiled at 100 °C for 40 min, cooled, and centrifuged. Then, chloroform was added, and the samples were manually shaken. After the solutions became clear, measurement was carried out at two wavelengths (532 and 600 nm). The results were expressed as malondialdehyde (MDA) mg/kg according to the formula:

$$\text{TBARS} = \frac{(A_{532} - A_{600})}{\left(155 \times \frac{1}{10} \times 72.6\right)} \times 1000 \quad (4)$$

where A refers to the absorbance measured at 532 and 600 nm.

3.6. Consumer Safety

3.6.1. *N*-Nitrosoamines (NAs) Content

The volatile NAs (VNAs) in the samples were analyzed using the method of Drabik and Markiewicz [69] and DeMey [52]. Briefly, meat samples (50 g) were mixed with 200 mL of 3 N KOH and then VNAs were extracted by vacuum distillation (Heidolph Laborota 4010-digital, Schwabach, Germany). After distillation, 4 mL of 37% HCl was added, and the distillate was extracted three times with 50 mL of dichloromethane. Subsequently, the obtained extract was concentrated. The detection and quantification of selected NAs (*N*-nitrosodibutylamine (NDBA), *N*-nitrosodimethylamine (NDMA), *N*-nitrosodimethylamine (NDEA), *N*-nitrosodipropylamine (NDPA), *N*-nitrosomorpholine (NMOR), *N*-nitrosopiperidine (NPIP), and *N*-nitrosopyrrolidine (NPYR), μg/kg) was performed using a gas chromatograph coupled to a thermal energy analyzer (TEA; Thermo Electron Cooperation). For this, the extracts (5 μL) were injected on a packed column, and chromatographic separation was carried out using argon as carrier gas (25 mL/min). The injection port was set at 175 °C, and the oven temperature was increased from 110 °C to 180 °C at 5 °C/min. The interface and pyrolizer of the TEA were set at 250 °C and 500 °C, respectively. The content of NA in the samples was estimated after 1 and 180 days of storage.

3.6.2. Number of Selected Microorganisms

The microbial count was calculated after 1 and 180 days of storage. The C. perfringens were measured according to [70], Listeria monocytogenes according to [71], and Salmonella according to [72]. For preparing the appropriable dilutions, 180 mL of peptone water was homogenized with 20 g of sample (Stomacher Lab-Blender 400, Seward Medical, London, UK). The results are expressed as cfu/g products.

3.7. Statistical Analysis

A two-way analysis-of-variance (ANOVA) model was used for analysis. It included the main effects of the level of extract (0, 50, 100, and 150 mg/kg) and the storage period (1, 60, 90, and 180 days) as well as their interactions. All measurements were carried out in triplicate. The results were analyzed statistically using STATISTICA® 13.1 statistical package (StatSoft) and presented as mean ± standard error using T-Tukey's range test.

4. Conclusions

The present study evaluated the possibility of reducing sodium(III) nitrite in canned meat, together with simultaneous fortification with lyophilized leaf extract of black currant leaves and its influence on the antioxidant stability of the product during 180 days of storage. The analyses of free radical scavenging ability ($ABTS^{\bullet+}$, DPPH) and iron ion-reducing potential allowed us to confirm the high antioxidant properties of meat products with black currant leaf extract. These results were further confirmed by the analysis of secondary lipid oxidation products, which showed low amounts of MDA in the tested products. In addition, no negative, pro-oxidative interactions were noted between sodium(III) nitrite and black currant leaves extract, even in the sample with the highest (150 mg/kg) amount of added extract. Moreover, no NAs were detected after production in any of the samples. The addition of black currant leaf extract at an amount of 150 mg/kg of meat stuffing can be recommended.

Our research confirms the possibility of reducing the amount of sodium(III) nitrite in canned meat and at the same time maintaining the safety and quality of the product.

Author Contributions: Conceptualization, K.M.W., K.F.; methodology, K.F., K.M.W., M.T. and M.M.; validation and formal analysis, K.F., M.T., M.M., D.K.-K., R.D., B.C. and M.K.; investigation, K.F., M.T., M.M. and D.K.-K.; writing—original draft preparation, K.F., M.T. and M.M.; writing—review and editing, K.M.W., R.D. and P.K.; supervision, K.M.W. and D.K.-K.; project administration, K.M.W.; funding acquisition, K.M.W. All authors have read and agreed to the published version of the manuscript.

Funding: Project financed under the program of the Minister of Education and Science under the name "Regional Initiative of Excellence" in 2019–2023, project number 029/RID/2018/19, funding amount 11,927,330.00 PLN.

Institutional Review Board Statement: Not applicable.

Informed Consent Statement: Not applicable.

Data Availability Statement: Not applicable.

Acknowledgments: Thanks to GAIN (Axencia Galega de Innovación) for supporting this study (grant number IN607A2019/01). Rubén Domínguez is a member of the Healthy Meat network funded by CYTED (ref. 119RT0568).

Conflicts of Interest: The authors declare no conflict of interest.

Sample Availability: Samples of the compounds are not available from the authors.

References

1. Lorenzo, J.M.; Pateiro, M.; Domínguez, R.; Barba, F.J.; Putnik, P.; Kovačević, D.B.; Shpigelmand, A.; Granatoe, D.; Franco, D. Berries extracts as natural antioxidants in meat products: A review. *Food Res. Int.* **2018**, *106*, 1095–1104. [CrossRef] [PubMed]
2. Domínguez, R.; Pateiro, M.; Gagaoua, M.; Barba, F.J.; Zhang, W.; Lorenzo, J.M. A Comprehensive Review on Lipid Oxidation in Meat and Meat Products. *Antioxidants* **2019**, *8*, 429. [CrossRef] [PubMed]
3. Kumar, Y.; Yadav, D.N.; Ahmad, T.; Narsaiah, K. Recent Trends in the Use of Natural Antioxidants for Meat and Meat Products. *Compr. Rev. Food Sci. Food Saf.* **2015**, *14*, 796–812. [CrossRef]
4. Ribeiro, J.S.; Santos, M.J.M.C.; Silvac, L.K.R.; Pereira, L.C.L.; Santos, I.A.; da Silva Lannesd, S.C.; da Silva, M.V. Natural antioxidants used in meat products: A brief review. *Meat Sci.* **2019**, *148*, 181–188. [CrossRef]
5. Falowo, A.B.; Fayemi, P.O.; Muchenje, V. Natural antioxidants against lipid–protein oxidative deterioration in meat and meat products: A review. *Food Res. Int.* **2014**, *64*, 171–181. [CrossRef] [PubMed]
6. Domínguez, R.; Munekata, P.E.; Pateiro, M.; Maggiolino, A.; Bohrer, B.; Lorenzo, J.M. Red beetroot. A potential source of natural additives for the meat industry. *Appl. Sci.* **2020**, *10*, 8340. [CrossRef]
7. Munekata, P.E.; Gullón, B.; Pateiro, M.; Tomasevic, I.; Domínguez, R.; Lorenzo, J.M. Natural antioxidants from seeds and their application in meat products. *Antioxidants* **2020**, *9*, 815. [CrossRef] [PubMed]
8. Domínguez, R.; Gullón, P.; Pateiro, M.; Munekata, P.E.; Zhang, W.; Lorenzo, J.M. Tomato as potential source of natural additives for meat industry. A review. *Antioxidants* **2020**, *9*, 73. [CrossRef] [PubMed]
9. Domínguez, R.; Barba, F.J.; Gómez, B.; Putnik, P.; Kovačević, D.B.; Pateiro, M.; Lorenzo, J.M. Active packaging films with natural antioxidants to be used in meat industry: A review. *Food Res. Int.* **2018**, *113*, 93–101. [CrossRef]
10. Lorenzo, J.M.; Domínguez, R.; Carballo, J. Control of Lipid Oxidation in Muscle Food by Active Packaging Technology. In *Natural Antioxidants: Applications in Foods of Animal Origin*; Banerjee, R., Verma, A.K., Siddiqui, M.W., Eds.; Apple Academic Press: Cambridge, MA, USA, 2017; pp. 343–382. Available online: https://www.taylorfrancis.com/chapters/edit/10.1201/9781315365916-10/potential-applications-natural-antioxidants-meat-meat-productsCHAPTERBOOK (accessed on 15 February 2021).
11. Alahakoon, A.U.; Jayasenab, D.D.; Ramachandrac, S.; Jo, C. Alternatives to nitrite in processed meat: Up to date. *Trends Food Sci. Technol.* **2015**, *45*, 37–49. [CrossRef]
12. Rivera, N.; Bunning, M.M. Uncured-Labeled Meat Products Produced Using Plant-Derived Nitrates and Nitrites: Chemistry, Safety, and Regulatory Considerations. *J. Agric. Food Che.* **2019**, *67*, 8074–8084. [CrossRef] [PubMed]
13. Munekata, P.E.; Pateiro, M.; Domínguez, R.; Pollonio, M.A.; Sepúlveda, N.; Andres, S.C.; Lorenzo, J.M. Beta vulgaris as a Natural Nitrate Source for Meat Products: A Review. *Foods* **2021**, *10*, 2094. [CrossRef] [PubMed]
14. Honikel, K.O. The use and control of nitrate and nitrite for the processing of meat products. *Meat Sci.* **2008**, *78*, 68–76. [CrossRef] [PubMed]
15. EFSA. *Scientific Opinion. Re-Evaluation of Potassium Nitrite (E 249) and Sodium Nitrite (E 250)*; EFSA: Parma, Italy, 2017.
16. FECIS. Food Chain Evaluation Consortium. Study on the Monitoring of the Implementation of Directive 2006/52/EC as Regards the Use of Nitrites by Industry in Different Categories of Meat Products: Final Report. 2016. Available online: http://www.fecic.es/img/galeria/file/BUTLLETi%20INTERNACIONAL/ARXIUS%20BUTLLETI%20INTERNACIONAL/setmana%205/05.pdf (accessed on 15 February 2021).
17. Ozaki, M.M.; Dos Santos, M.; Ribeiro, W.O.; de Azambuja Ferreira, N.C.; Picone, C.S.F.; Domínguez, R.; Pollonio, M.A.R. Radish powder and oregano essential oil as nitrite substitutes in fermented cooked sausages. *Food Res. Int.* **2021**, *140*, 109855. [CrossRef] [PubMed]
18. Munekata, P.E.; Pateiro, M.; Domínguez, R.; Santos, E.M.; Lorenzo, J.M. Cruciferous vegetables as sources of nitrate in meat products. *Curr. Opin. Food Sci.* **2021**, *38*, 1–7. [CrossRef]
19. Fraga, C.G.; Croft, K.D.; Kennedy, D.O.; Tomás-Barberán, F.A. The effects of polyphenols and other bioactives on human health. *Food Funct.* **2019**, *10*, 514–528. [CrossRef]

20. Nikmaram, N.; Budaraju, S.; Barba, F.J.; Lorenzo, J.M.; Cox, R.B.; Mallikarjunan, K.; Roohineja, S. Application of plant extracts to improve the shelf-life. nutritional and health-related properties of ready-to-eat meat products. *Meat Sci.* **2018**, *145*, 245–255. [CrossRef]
21. Khaleghi, A.; Kasaai, R.; Khosravi-Darani, K.; Rezaei, K. Combined Use of Black Barberry (*Berberis crataegina* L.) Extract and Nitrite in Cooked Beef Sausages during the Refrigerated Storage. *J. Agr. Sci. Technol.* **2016**, *18*, 601–614.
22. Aliyari, P.; Kazaj, F.B.; Barzegar, M.; Gavlighi, H.A. Production of Functional Sausage Using Pomegranate Peel and Pistachio Green Hull Extracts as Natural Preservatives. *J. Agr. Sci. Technol.* **2020**, *22*, 159–172.
23. Aquilania, C.; Sirtoria, F.; Floresb, M.; Bozzia, R.; Lebretc, B.; Pugliese, C. Effect of natural antioxidants from grape seed and chestnut in combination with hydroxytyrosol. as sodium nitrite substitutes in *Cinta Senese* dry-fermented sausages. *Meat Sci.* **2018**, *145*, 389–398. [CrossRef]
24. Seo, J.-K.; Parvin, R.; Yim, D.-G.; Zahid, A.; Yang, H.-S. Effects on quality properties of cooked pork sausages with *Caesalpinia sappan* L. extract during cold storage. *J. Food Sci. Technol.* **2019**, *56*, 4946–4955. [CrossRef]
25. Šojić, B.; Pavlić, B.; Tomović, V.; Kocić-Tanackov, S.; Đurović, S.; Zeković, Z.; Belović, M.; Torbica, A.; Jokanović, M.; Urumović, N.; et al. Tomato pomace extract and organic peppermint essential oil as effective sodium nitrite replacement in cooked pork sausages. *Food Chem.* **2020**, *330*, 127202. [CrossRef] [PubMed]
26. Manihuruka, F.M.; Suryatib, T.; Arief, I.I. Effectiveness of the red dragon fruit (*Hylocereus polyrhizus*) peel extract as the colorant, antioxidant, and antimicrobial on beef sausage. *Media Peternak* **2017**, *40*, 47–54. [CrossRef]
27. Mäkinen, S.; Hellström, J.; Mäki, M.; Korpinen, R.; Mattila, P.H. Bilberry and sea buckthorn leaves and their suB50ritical water extracts prevent lipid oxidation in meat products. *Foods* **2020**, *9*, 265. [CrossRef] [PubMed]
28. Djordjević, B.; Rakonjac, V.; Akšić, F.M.; Katarina Šavikin Vulić, T. Pomological and biochemical characterization of European currant berry (*Ribes* sp.) cultivars. *Sci. Hortic.* **2014**, *165*, 156–162. [CrossRef]
29. Paunović, S.M.; Mašković, P.; Nikolić, M.; Miletić, R. Bioactive compounds and antimicrobial activity of black currant (*Ribes nigrum* L.) berries and leaves extract obtained by different soil management system. *Sci. Hortic.* **2017**, *222*, 69–75. [CrossRef]
30. Staszowska-Karkut, M.; Materska, M. Phenolic Composition, Mineral Content. and Beneficial Bioactivities of Leaf Extracts from Black Currant (*Ribes nigrum* L.), Raspberry (*Rubus idaeus*), and Aronia (*Aronia melanocarpa*). *Nutrients* **2020**, *12*, 463. [CrossRef]
31. Ferlemi, A.-V.; Lamari, F.N. Berry Leaves: An Alternative Source of Bioactive Natural Products of Nutritional and Medicinal Value. *Antioxidants* **2016**, *5*, 17. [CrossRef]
32. Raudsepp, P.; Koskar, J.; Anton, D.; Meremäe, K.; Kapp, K.; Laurson, P.; Believe, U.; Kaldmäe, H.; Roasto, M.; Püssa, T. Antibacterial and antioxidative properties of different parts of garden rhubarb, blackcurrant, chokeberry and blue honeysuckle. *J. Sci. Food Agric.* **2019**, *99*, 2311–2320. [CrossRef] [PubMed]
33. Nowak, A.; Czyzowska, A.; Efenberger, M.; Krala, L. Polyphenolic extracts of cherry (*Prunus cerasus* L.) and blackcurrant (*Ribes nigrum* L.) leaves as natural preservatives in meat products. *Food Microbiol.* **2016**, *59*, 142–149. [CrossRef] [PubMed]
34. Ferysiuk, K.; Wójciak, K.M. The possibility of reduction of synthetic preservative E 250 in canned pork. *Foods* **2020**, *9*, 1–20. [CrossRef]
35. Regulation (EC) No 1333/2008 of the European Parliament and of the Council of 16 December 2008 on Food Additives. Available online: https://eur-lex.europa.eu/eli/reg/2008/1333/2016-05-25 (accessed on 15 February 2021).
36. EUCAST. Determination of Minimum Inhibitory Concentrations (MICs) of Antibacterial Agents by Agar Dilution. *Clin. Microbiol. Infect.* **2000**, *6*, 509–515. [CrossRef] [PubMed]
37. Suman, S.P.; Joseph, P. Myoglobin Chemistry and Meat Color. *Annu. Rev. Food Sci. Technol.* **2013**, *4*, 79–99. [CrossRef]
38. Amaral, A.B.; da Solva, M.V.; da Silva Lannes, S.C. Lipid oxidation in meat: Mechanisms and protective factors—A review. *Food Sci. Technol.* **2018**, *38* (Suppl. S1), 1–15. [CrossRef]
39. Bae, S.M.; Choi, J.H.; Jeong, J.Y. Effects of radish powder concentration and incubation time on the physicochemical characteristics of alternatively cured pork products. *J. Anim. Sci. Technol.* **2020**, *62*, 922–932. [CrossRef] [PubMed]
40. Sun, W.; Xiong, Y.L. Stabilization of cooked cured beef color by radical-scavenging pea protein and its hydrolysate. *LWT-Food Sci. Technol.* **2015**, *61*, 352–358. [CrossRef]
41. Anese, M.; Manzocco, L.; Nicoli, M.C.; Lerici, C.R. Antioxidant properties of tomato juice as affected by heating. *J. Sci. Food Agric.* **1999**, *79*, 750–754. [CrossRef]
42. Wolosiak, R.; Druzynska, B.; Piecyk, M.; Worobiej, E.; Majewska, E.; Lewicki, P.P. Influence of industrial sterilisation, freezing and steam cooking on antioxidant properties of green peas and string beans. *Int. J. Food Sci. Technol.* **2011**, *46*, 93–100. [CrossRef]
43. Tepe, B.; Sokmen, M.; Akpulat, H.A.; Daferera, D.; Polissiou, M.; Sokmen, A. Antioxidative activity of the essential oils of *Thymus sipyleus* subsp. sipyleus var. sipyleus and *Thymus sipyleus* subsp. sipyleus var. rosulans. *J. Food Eng.* **2005**, *66*, 447–454. [CrossRef]
44. Guillen-Sans, R.; Guzman-Chozas, M. The thiobarbituric acid (TBA) reaction in foods: A review. *Crit. Rev. Food Sci. Nutr.* **1998**, *38*, 315–350. [CrossRef]
45. Ferysiuk, K.; Wójciak, K.M.; Kęska, P. Effect of willow herb (*Epilobium angustifolium* L.) extract addition to canned meat with reduced amount of nitrite on the antioxidant and other activities of peptides. *Food Funct.* **2022**, *13*, 3526–3539. [CrossRef]
46. Mira, L.; Tereza Fernandez, M.; Santos, M.; Rocha, R.; Helena Florêncio, M.; Jennings, K.R. Interactions of flavonoids with iron and copper ions: A mechanism for their antioxidant activity. *Free. Radic. Res.* **2002**, *36*, 1199–1208. [CrossRef] [PubMed]
47. Pateiro, M.; Domínguez, R.; Bermúdez, R.; Munekata, P.E.; Zhang, W.; Gagaoua, M.; Lorenzo, J.M. Antioxidant active packaging systems to extend the shelf life of sliced cooked ham. *Curr. Res. Food Sci.* **2019**, *1*, 24–30. [CrossRef]

48. de Carvalho, F.A.L.; Munekata, P.E.; Pateiro, M.; Campagnol, P.C.; Domínguez, R.; Trindade, M.A.; Lorenzo, J.M. Effect of replacing backfat with vegetable oils during the shelf-life of cooked lamb sausages. *LWT* **2020**, *122*, 109052. [CrossRef]
49. Bonifacie, A.; Promeyrat, A.; Nassy, G.; Gatellier, P.; Santé-Lhoutellier, V.; Théron, L. Chemical reactivity of nitrite and ascorbate in a cured and cooked meat model implication in nitrosation, nitrosylation and oxidation. *Food Chem.* **2021**, *348*, 129073. [CrossRef] [PubMed]
50. Johnson, I.T. The cancer risk related to meat and meat products. *Br. Med. Bull.* **2017**, *121*, 73–81. [CrossRef] [PubMed]
51. Park, J.; Seo, J.; Lee, J.; Kwon, H. Distribution of Seven N-Nitrosamines in Food. *Toxicol. Res.* **2015**, *31*, 279–288. [CrossRef] [PubMed]
52. De Mey, E.; De Maere, H.; Paelinck, H.; Fraeye, I. Volatile N-nitrosamines in meat products: Potential precursors, influence of processing, and mitigation strategies. *Crit. Rev. Food Sci. Nutr.* **2017**, *57*, 2909–2923. [CrossRef]
53. Özbay, S.; Şireli, U.T. Volatile N-nitrosamines in processed meat products and salami from Turkey. *Food Addit. Contam. Part B* **2021**, *14*, 110–114. [CrossRef]
54. Majou, D.; Christieans, S. Mechanisms of the bactericidal effects of nitrate and nitrite in cured meats. *Meat Sci.* **2018**, *145*, 273–284. [CrossRef]
55. Staszowska-Karkut, M.; Materska, M.; Kulik, B.; Waraczewski, R. Określenie wpływu rodzaju wody na potencjał antyoksydacyjny naparów z wybranych roślin ziołowych. Rośliny- przegląd wybranych zagadnień. Wydawnictwo Naukowe TYGIEL sp. Z o. o. *Lublin* **2016**, 175–184. (In Polish)
56. Staszowska-Karkut, M.; Materska, M.; Chilczuk, B.; Pabich, M.; Sachadyn-Król, M. Wpływ rodzaju rozpuszczalnika na aktywność przeciwrodnikową ekstraktów ziołowych. Żywność dla przyszłości, XLIII Sesja Naukowa Nauk o Żywności i Biotechnologii. *Wrocław* **2017**, 221–228. (In Polish)
57. Chilczuk, B.; Materska, M.; Staszowska-Karkut, M.; Kulik, B.; Stępnikowska, A. Porównanie metod ekstrakcyjnych przy przygotowywaniu preparatów roślinnych o korzystnych właściwościach prozdrowotnych. Żywność–tradycja i nowoczesność-Prozdrowotne właściwości żywności, aspekty żywieniowe i technologiczne. Red. M. Karwowska, I. Jackowska. *Lublin* **2018**, 17–28. (In Polish)
58. Singleton, V.L.; Rossi, J.A. Colorimetry of total phenolics with phosphomolybdic-phosphotungstic acid reagents. *Am. J. Enol. Viticult* **1965**, *16*, 144–158.
59. Chang, C.C.; Yang, M.H.; Wen, H.M.; Chern, J.C. Estimation of total flavonoid content in propolis by two complementary colorimetric methods. *J. Food Drug Anal.* **2002**, *10*, 178–182.
60. Nicolle, C.; Carnat, A.; Fraisse, D.; Lamaison, J.L.; Rock, E.; Michel, H.; Amouroux, P.; Remesy, C. Characterisation and variation of antioxidant micronutrients in lettuce (*Lactuca sativa* folium). *J. Sci. Food Agric.* **2004**, *84*, 2061–2069. [CrossRef]
61. Wu, X.; Diao, Y.; Sun, C.; Yang, J.; Wang, Y.; Sun, S. Fluorimetric determination of ascorbic acid with o-phenylenediamine. *Talanta* **2003**, *59*, 95–99. [CrossRef] [PubMed]
62. Re, R.; Pellegrini, N.; Proteggente, A.; Pannala, A.; Yang, M.; Rice-Evans, C.A. Antioxidant activity applying an improved ABTS radical cation decolorization assay. *Free Rad. Bio. Med.* **1999**, *26*, 1231–1237. [CrossRef]
63. Brand-Williams, W.; Cuvelier, M.; Berset, C. Use of a free radical method to evaluate antioxidant activity. *LWT-Food Sci. Technol.* **1995**, *28*, 25–30. [CrossRef]
64. Hunt, R.W.G. A model of colour vision for predicting colour appearance in various viewing conditions. *Color Res. Appl.* **1987**, *12*, 297–314. [CrossRef]
65. Hornsey, H.C. The colour of cooked cured pork. I.—Estimation of the Nitric oxide Haem Pigment. *J. Sci. Food Agric.* **1956**, *7*, 534–540. [CrossRef]
66. Jung, S.; Choe, J.C.; Kim, B.; Yun, H.; Kruk, Z.A.; Jo, C. Effect of dietary mixture of gallic acid and linoleic acid on antioxidative potential and quality of breast meat from broilers. *Meat Sci.* **2010**, *86*, 520–526. [CrossRef] [PubMed]
67. Oyaizu, M. Studies on the product of browning reaction prepared from glucose amine. *Jpn. J. Nutr.* **1986**, *44*, 307–315. [CrossRef]
68. Fan, X.J.; Liua, S.; Lia, H.H.; Hea, J.; Fenga, J.T.; Hanga, X.; Yana, H. Effects of *Portulaca oleracea* L. extract on lipid oxidation and color of pork meat during refrigerated storage. *Meat Sci.* **2019**, *147*, 82–90. [CrossRef] [PubMed]
69. Drabik-Markiewicz, G.; Dejaegher, B.; De Mey, E.; Kowalska, T.; Paelinck, H.; Vander Heyden, Y. Influence of putrescine, cadaverine, spermidine or spermine on the formation of N-nitrosamine in heated cured pork meat. *Food Chem.* **2011**, *126*, 1539–1545. [CrossRef]
70. *PN-EN ISO 7937*; Horizontal Method for the Determination of Clostridium Perfringens. Polish Committee for Standardization: Warsaw, Poland, 2005.
71. *PN-EN ISO 11290-2*; Food Chain Microbiology—Horizontal Method for Determining the Presence of Hepatitis A Virus and Norovirus Using Real-Time RT-PCR—Part 2: Detection Method. Polish Committee for Standardization: Warsaw, Poland, 2017.
72. *PN-EN ISO 6579-1*; Food Chain Microbiology—Horizontal Method for the Detection, Determination and Serotyping of Salmonella—Part 1: Detection. Polish Committee for Standardization: Warsaw, Poland, 2017.

molecules

MDPI

Article

The Influence of Pulsed Electric Field and Air Temperature on the Course of Hot-Air Drying and the Bioactive Compounds of Apple Tissue

Agnieszka Ciurzynska, Magdalena Trusinska, Katarzyna Rybak, Artur Wiktor and Malgorzata Nowacka *

Department of Food Engineering and Process Management, Institute of Food Sciences, Warsaw University of Life Sciences (WULS-SGGW), 159c Nowoursynowska St., 02-776 Warsaw, Poland
* Correspondence: malgorzata_nowacka@sggw.edu.pl; Tel.: +48-22-593-75-79

Abstract: Drying is one of the oldest methods of obtaining a product with a long shelf-life. Recently, this process has been modified and accelerated by the application of pulsed electric field (PEF); however, PEF pretreatment has an effect on different properties—physical as well as chemical. Thus, the aim of this study was to investigate the effect of pulsed electric field pretreatment and air temperature on the course of hot air drying and selected chemical properties of the apple tissue of Gloster variety apples. The dried apple tissue samples were obtained using a combination of PEF pretreatment with electric field intensity levels of 1, 3.5, and 6 kJ/kg and subsequent hot air drying at 60, 70, and 80 °C. It was found that a higher pulsed electric field intensity facilitated the removal of water from the apple tissue while reducing the drying time. The study results showed that PEF pretreatment influenced the degradation of bioactive compounds such as polyphenols, flavonoids, and ascorbic acid. The degradation of vitamin C was higher with an increase in PEF pretreatment intensity level. PEF pretreatment did not influence the total sugar and sorbitol contents of the dried apple tissue as well as the FTIR spectra. According to the optimization process and statistical profiles of approximated values, the optimal parameters to achieve high-quality dried apple tissue in a short drying time are PEF pretreatment application with an intensity of 3.5 kJ/kg and hot air drying at a temperature of 70 °C.

Keywords: pulsed electric field; plant tissue; hot air drying; chemical properties; pretreatment

Citation: Ciurzynska, A.; Trusinska, M.; Rybak, K.; Wiktor, A.; Nowacka, M. The Influence of Pulsed Electric Field and Air Temperature on the Course of Hot-Air Drying and the Bioactive Compounds of Apple Tissue. *Molecules* **2023**, *28*, 2970. https://doi.org/10.3390/molecules28072970

Academic Editors: Michał Halagarda and Sascha Rohn

Received: 24 February 2023
Revised: 23 March 2023
Accepted: 25 March 2023
Published: 27 March 2023

1. Introduction

Drying is one of the oldest and most commonly used food processing techniques. It consists of the exchange of heat and mass between a dried product and a drying agent, which is associated with a phase change, i.e., evaporation. Thanks to this process, the course of chemical reactions and the development of microorganisms in products can be inhibited, which extends the shelf-life of food products [1]. Although there are other advantages of drying food, there are also disadvantages, such as changes in sensory characteristics and color, as well as the degradation of nutrients susceptible to high temperatures, which are all associated with deterioration of the quality of the final product [2]. Undesirable changes that occur during the drying process can be minimized by selecting appropriate drying parameters, for example, temperature, humidity, and speed of air drying [3]. To improve the quality of the final product, new, more efficient drying methods have been developed. Hybrid drying using ultrasonics or microwaves is increasingly used [1,4]. The wide range of possibilities for combining various drying techniques have resulted in many benefits, and therefore, have contributed to the continuous development of science in this field and the improvement of existing technologies [5].

Microwave-convective drying uses the action of microwaves, i.e., electromagnetic waves, which have a frequency in the range from 300 MHz to 300 GHz. During application, microwaves are absorbed by a material, where they are converted into heat, which increases the temperature of the material. Inside the material, a higher temperature is reached in relation to the temperature outside, which accelerates heat and mass exchange [6].

Microwave-convective drying (hot air drying) is characterized by better heat transfer, which results in faster drying of a material, and the product is characterized by better sensory properties in relation to the dried material obtained using convective drying. These advantages were proven by, among others, Szadzińska and Mierzwa [7] who studied the effect of microwave-convective drying on the kinetics of drying and the quality of white mushrooms.

The goal of pretreatment is to shorten the drying time, which is associated with a reduction in energy costs during drying. In addition, applying pretreatments with properly selected parameters can effectively maintain the quality of the final material, for example, with high contents of certain nutritive compounds (anthocyanin, total phenolics, vitamin C, and antioxidant activity). Pulsed electric field (PEF) pretreatment is a type of non-thermal technology that is becoming more and more popular. Under the influence of an applied electric field of appropriate intensity, pores are formed that increase membrane permeability and facilitate the removal of water from the material during drying [8]. Due to the holes created, it is easier to transfer various components, for example, ions, to the inside of the cell [9]. PEF sets in motion ions that are located on the inside and outside of material particles. The mechanism of electroporation is related to transmembrane ion transfer. The movement of ions takes place analogously to the direction of the applied electric field. As a result, oppositely charged electric charges accumulate on each side of the cell membrane, which interact with each other, and therefore, lead to an increase in the pressure in the cell, resulting in modification of the cell membrane thickness, and then its rupture [10].

The pulsed electric field method, through the phenomenon of electroporation, significantly affects the drying process of plant origin materials [11]. Rahaman et al. [12] studied the effect of pulsed electric field as a pretreatment on the kinetics of plum drying. They found that as the intensity of the pulsed electric field increased, the drying rate of the plum and the amount of water removed increased. Mirzaei-Baktash et al. [13] studied the effect of PEF on the kinetics of convective drying of mushrooms. They showed that, in mushrooms treated with PEF, from 20 to 32% shorter drying time was needed to obtain a constant moisture content compared to a control sample. The use of PEF pretreatment resulted in obtaining dried material characterized by high contents of L-ascorbic acid and total polyphenol content, as well as high antioxidant capacity compared to the samples obtained without this treatment [14]. The selection of appropriate conditions for PEF application and the type of drying used have a significant impact on the quality of the dried material subjected to PEF pretreatment [15].

Pulsed electric field pretreatment with appropriate parameter values has a positive effect on the bioactive components of the dried material, while PEF application that is too intense (3 kJ/kg used in presented investigations) may have a negative impact on the final quality of the dried material [15]. Studies have confirmed the beneficial effect of PEF on preserving the total polyphenol content and ascorbic acid content, as well as on the antioxidant capacity of the dried material [9]. Spinach dried using hot air drying and treated with a pulsed electric field had a higher level of L-ascorbic acid and better color compared to untreated samples [14]. Mango dried by convective and vacuum with PEF pretreatment was characterized by high levels of polyphenols and flavonoids; however, the level of carotenoids was slightly reduced during drying. In addition, the application of pulsed electric field at a lower intensity was more beneficial, which may have been due to oxidation of carotenoids that were sensitive to the presence of air or free radicals formed during pulsed electric field treatment [15].

The aim of the work was to investigate the impact of pulsed electric field and air temperature on the course of hot air drying, and the selected properties of apple tissue. Apple tissue was subjected to different intensity levels for the parameter pulsed electric field intensity (1, 3.5, and 6 kJ/kg), and then hot air drying at three different temperatures (60, 70, and 80 °C). The following properties were determined: water activity, total polyphenol content, flavonoid content, antioxidant activity based on the degree of quenching of the synthetic DPPH radical and reducing power in the obtained dried apple tissue, vitamin C content, sugar content, and characteristic bonds between molecules according to Fourier transform infrared spectroscopy (FTIR).

2. Results

2.1. The Influence of PEF on the Kinetics of Drying Apple Tissue

Figure 1 shows the drying time of apple tissue subjected to different PEF pretreatment intensity levels and dried at different drying air temperatures in relation to the relative water content (humidity ratio, MR). The shortest drying time was characterized by the sample marked as PEF 6, 70 °C, with an average drying time of 120 min, which was about 37% shorter than the control sample marked as Control 70 °C. The longest drying time was obtained for the sample marked as Control 60 °C. In this case, the time needed for the complete evaporation of water was equal to 220 min. The air temperature of 80 °C resulted in faster drying of the apple tissue compared to other temperatures. The response surface results of the drying time showed a good fit for the model. According to Figure 2, it can be seen that the dried material is obtained in a shorter drying time when a higher temperature is used, while the PEF pretreatment does not affect it significantly. Whereas a longer drying time is obtained when the lower temperature is applied.

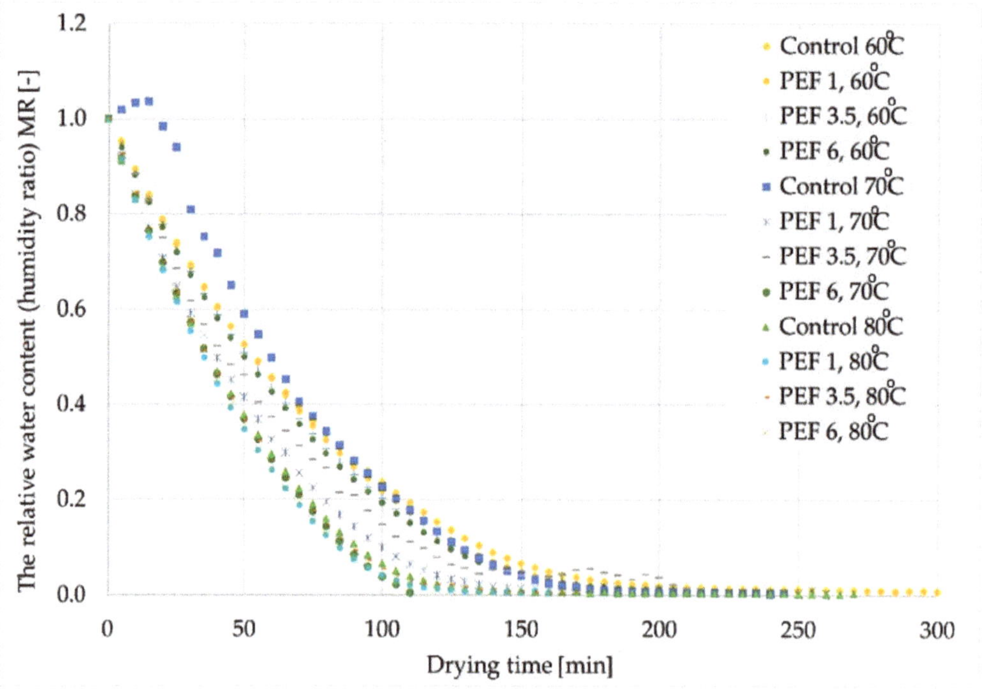

Figure 1. Kinetics of drying apple tissue at temperatures of 60, 70, and 80 °C that has been pretreated with PEF at electric field intensity levels of 1, 3.5, and 6 kJ/kg.

2.2. Influence of PEF on Water Activity of Dried Apple Tissue

The fresh apples were characterized by high water activity of 0.986 (Table 1). The drying resulted in a high reduction in water activity.

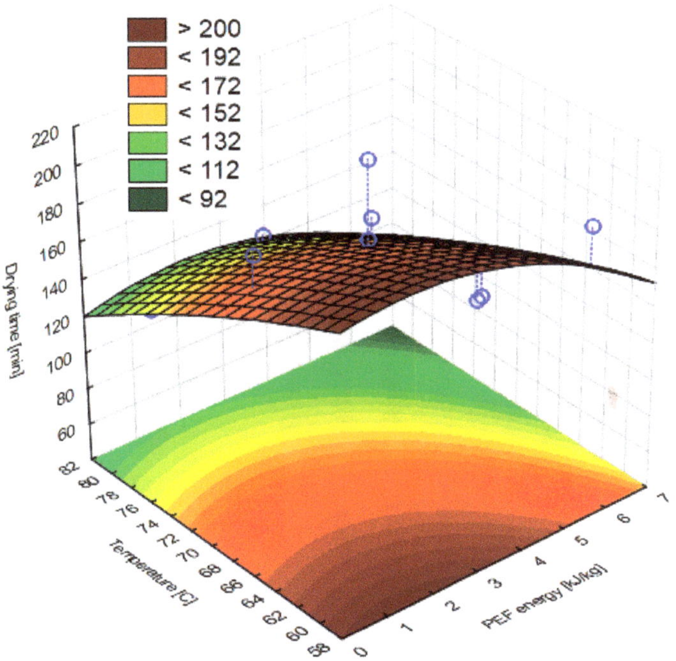

Figure 2. Response surface results for drying time, depending on the pulsed electric field intensity and the temperature obtained in the case of apple tissue subjected to PEF pretreatments with different electric field intensities (1, 3.5, 6 kJ/kg) and hot air drying at different temperatures (60, 70 and 80 °C). Experimental values are marked with points.

Table 1. Properties of the fresh apple sample.

Material	Fresh Apple
Water content (%)	16 ± 0.01
Water activity (-)	0.986 ± 0.04
Bioactive compounds	
TPC (mg ChlA/100 g d.m.)	1683 ± 2
TFC (mg QE/100 g d.m.)	715 ± 8
Vitamin C (mg/100 g d.m.)	143.2 ± 5.0
Antioxidant capacity	
AC (mg TE/100 g d.m.)	4.33 ± 0.06
RP (mg TE/100 g d.m.)	20.48 ± 0.72
Sugar content (g/100 g d.m.)	36.71 ± 2.99
Sucrose (g/100 g d.m.)	4.94 ± 0.24
Glucose (g/100 g d.m.)	6.11 ± 0.68
Fructose (g/100 g d.m.)	24.8 ± 2.02
Sorbitol (g/100 g d.m.)	0.86 ± 0.06

The dried apple tissue was characterized by water activity values ranging from 0.211 to 0.331. Figure 3 shows the water activity of the dried apple tissue obtained using the different values of the selected parameters. The water activity values were affected by temperature and the interaction of PEF pretreatment with temperature, with a greater effect of temperature (η^2 = 0.93), as shown by the two-factor analysis of variance. The lowest water activity, equal to 0.211, was found in the dried apple tissue obtained after PEF pretreatment with an electric field intensity level of 6 kJ/kg and dried using the hot air drying temperature of 70 °C. The highest water activity was found in the the dried apple tissue subjected to PEF pretreatment with an electric field intensity level of 6 kJ/kg but dried at the temperature of 60 °C. Based on the homogeneous groups, it can be concluded that, in the most cases, there were significant differences between the water activities in the obtained dried samples. Since the response surface results of the water activity were not a good fit for the model, this figure was not shown.

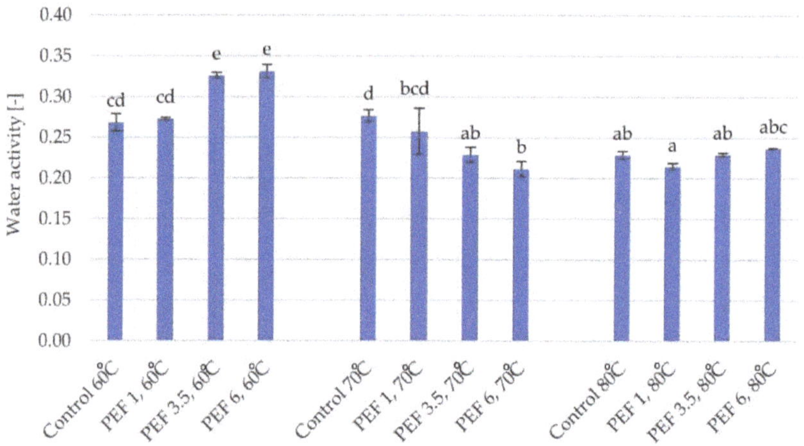

Figure 3. Water activity of dried apple tissue obtained using different values for the parameters of pulsed electric field intensity (1, 3.5, and 6 kJ/kg) and hot air drying temperature (60, 70, and 80 °C). The columns with different letters are significantly different ($p < 0.05$).

2.3. *Influence of PEF on Total Polyphenol Content and Flavonoid Content in Dried Apple Tissue*

Based on the results of the conducted research, it can be observed that the total polyphenol content in the dried apple tissue was affected by both the electric field intensity level ($\eta^2 = 0.83$) and the drying temperature ($\eta^2 = 0.78$) as well as the interaction between these two parameters ($\eta^2 = 0.93$). Figure 4 shows changes in the total polyphenol content of dried apple tissue depending on the applied electric field intensity level and the temperature. The total polyphenol content in the fresh material was 1682.9 mg of chlorogenic acid per 100 g of dry substance (ChlA/100 g d.m.). At a temperature of 60 °C, the total polyphenol content was equal to 1277.38 mg ChlA/100 g d.m. PEF pretreatment at each intensity level resulted in a decrease in the total polyphenol content compared to that of the dried material obtained without the use of pretreatment. With an increase in the intensity level of PEF, degradation of polyphenols in the apple tissue was observed. For PEF pretreatment with an electric field intensity of 1 kJ/kg, the total polyphenol content was the highest (849 mg ChlA/100 g d.m) compared to those of other dried fruits obtained with prior pretreatment, which was about 33% lower than the value of apples not treated with PEF. Increasing the PEF pretreatment electric field intensity levels to 3.5 and 6 kJ/kg made it possible to obtain dried fruit with total polyphenol contents of 844.9 and 790.4 mg ChlA/100 g d.m., respectively. This means that the result was about 34% and 38% lower as compared to the dried fruit obtained with a PEF pretreatment electric field intensity of 1 kJ/kg.

When drying at 70 °C, it can be concluded that, as in the case of drying at 60 °C, the use of the pretreatment reduced the number of polyphenols compared to the control samples (dried apple tissue obtained without pretreatment). In the case of the pretreatment, the highest total polyphenol content was equal to 1201.1 mg ChlA/100 g d.m., which was found in the dried material obtained as a result of applying a pulsed electric field intensity of 1 kJ/kg. This result is about 19% lower than the material that was not treated with

PEF. Increasing the pulsed electric field intensity resulted in a further decrease in total polyphenol content. For PEF pretreatment with an electric field intensity of 3.5 kJ/kg, the total polyphenol content is about 34% lower than the result at the lowest PEF pretreatment electric field intensity, and at 6 kJ/kg, this value is 40% lower than the result at a PEF pretreatment electric field intensity of 1 kJ/kg.

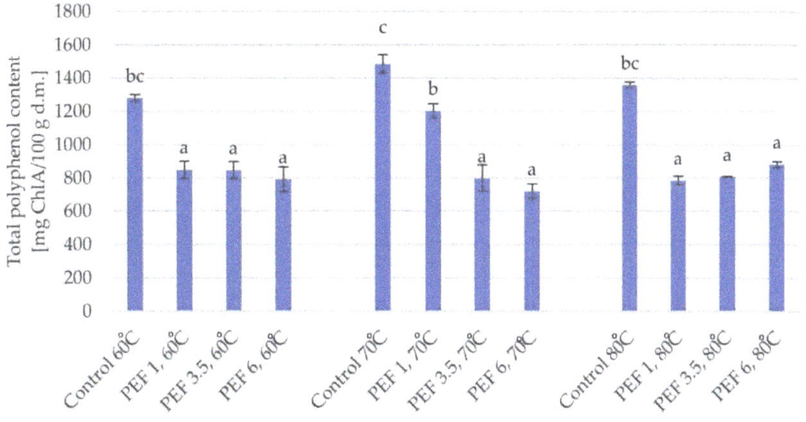

Figure 4. Total polyphenol content of dried apple tissue obtained using different values for the parameters of pulsed electric field intensity (1, 3.5, and 6 kJ/kg) and hot air drying temperature (60, 70, and 80 °C) ($n = 2$). The columns with different letters are significantly different ($p < 0.05$).

Analyzing the results obtained when the drying temperature was at 80 °C, it can be seen that an increase in the PEF pretreatment intensity resulted in a slight increase in the total polyphenol content in the apple tissue; however, it should be highlighted that the highest value for total polyphenol content was found in the dried apple tissue obtained without pulsed electric field pretreatment. For example, apple tissue pretreated with a PEF pretreatment electric field intensity of 1 kJ/kg was characterized by a total polyphenol content that was 42% lower than the reference sample. Increasing the PEF pretreatment electric field intensity to 3.5 kJ/kg resulted in an increase in the retention of phenolic compounds by approximately 22.3 mg ChlA/100 g d.s. compared to that of the dried apple tissue obtained at an electric field intensity of 1 kJ/kg. In the case of the drying temperature of 80 °C, a pulsed electric field pretreatment with an electric field intensity of 6 kJ/kg was the most advantageous procedure, since the dried apple tissue reached the highest total polyphenol content among the pretreated variants, i.e., equal to 882.7 mg ChlA/100 g d.m. However, this value was about 35% lower than the result obtained in the dried apple tissue obtained without pretreatment.

Figure 5 shows the response surface results of the total polyphenol content of dried apple tissue after applying pulsed electric field at different intensity levels and then hot air drying at different temperatures. The statistical analysis showed a good fit for the model. According to the model, the highest value of the total polyphenol content was obtained by applying the PEF pretreatment with the lowest electric field intensity (1 kJ/kg) and a drying temperature of 70 °C.

Similar dependencies of the effect of PEF pretreatment on the total flavonoid content in dried apple tissue were shown as in the case of the analysis of total polyphenol content (Figure 6). The two-factor analysis showed that both parameters, i.e., PEF pretreatment intensity ($\eta^2 = 0.94$) and the temperature of drying ($\eta^2 = 0.73$), as well as the interaction between the parameters ($\eta^2 = 0.96$) had a significant impact on the total flavonoid content. It was found the PEF pretreatment caused a decrease in the total flavonoid content, and that increasing the PEF energy input, in most cases, caused a slight decrease in the analyzed results. While the drying temperature had no significant effect on the total flavonoid content.

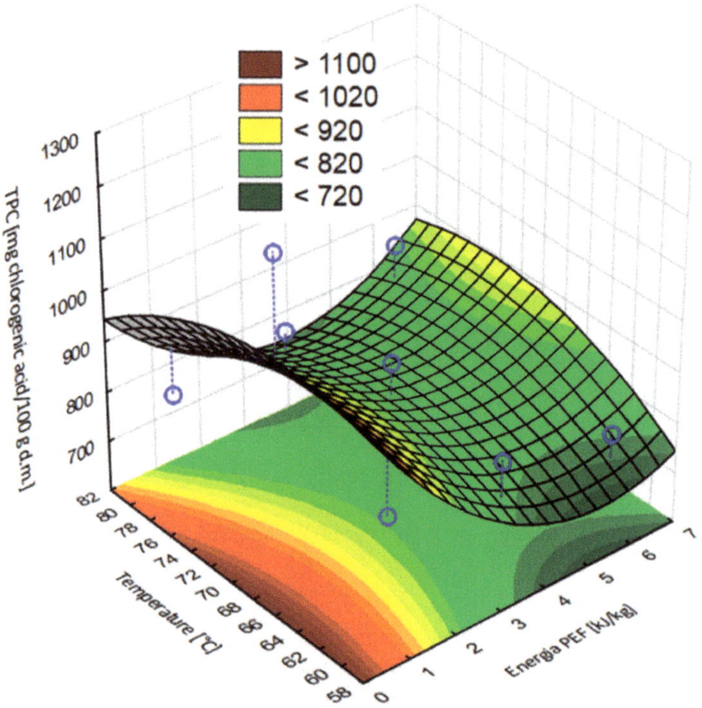

Figure 5. Response surface results of the total polyphenol content, depending on the pulsed electric field intensity and the temperature, obtained in apple tissue subjected to different PEF pretreatment intensities (1, 3.5, and 6 kJ/kg) and hot air drying at different temperatures (60, 70, and 80 °C). Experimental values are marked with points.

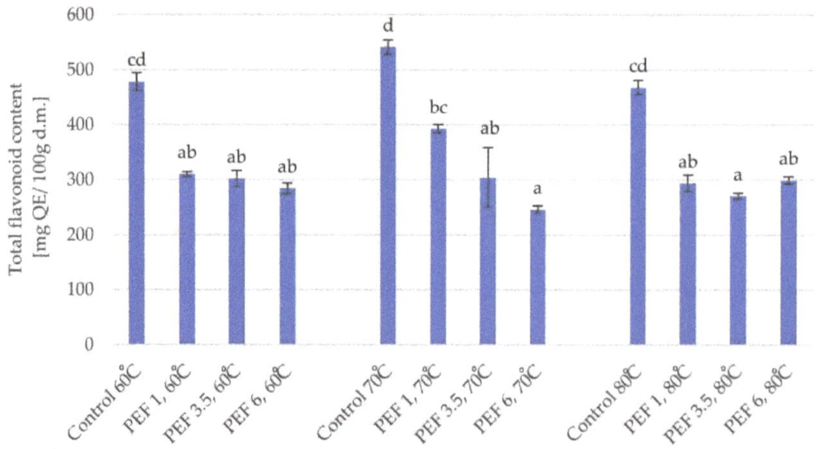

Figure 6. Total flavonoid content of dried apple tissue obtained using different values for the parameters of pulsed electric field intensity (1, 3.5, and 6 kJ/kg) and hot air drying temperature (60, 70, and 80 °C). The columns with different letters are significantly different ($p < 0.05$).

Figure 7 shows the response surface results of the total flavonoid content of dried apple tissue obtained as a result of pulsed electric field pretreatment at different intensity levels and hot air drying in a convective dryer at different temperatures. The statistical analysis showed a good fit for the model. In addition, it showed that the highest value of the total flavonoid content was obtained for apple tissue subjected to a PEF pretreatment with an electric field intensity of 1 kJ/kg and dried at 70 °C.

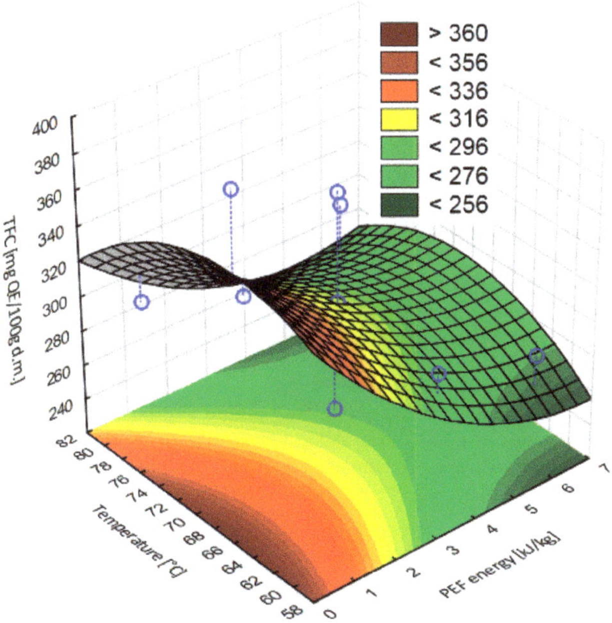

Figure 7. Response surface results of the total flavonoid content, depending on the pulsed electric field intensity and the temperature, obtained in the case of apple tissue subjected to different PEF pretreatment intensities (1, 3.5, and 6 kJ/kg) and hot air drying at different temperatures (60, 70, and 80 °C). Experimental values are marked with points.

2.4. The Influence of PEF on the Vitamin C Content in Dried Apple Tissue

Figure 8 shows the change in vitamin C content in dried apple tissue. The statistical analysis showed that the change in vitamin C content in the dried apple tissue was affected by both the pulsed electric field intensity level (η^2 = 0.97) and the drying temperature (η^2 = 0.90) as well as the interaction between both of these parameters (η^2 = 0.91). The two-factor analysis showed that the PEF pretreatment had a more significant effect on the content of vitamin C. The pretreatments caused a decrease in the amount of ascorbic acid compared to the fresh apple sample with a vitamin C content equal to 143.2 mg/100 g d.m. The content of ascorbic acid ranged from 130.6 to 6.5 mg/100 g d.m. In each of the variants, there was a tendency of decreasing ascorbic acid content caused by the application of pulsed electric field pretreatment. Taking into account all drying temperatures and PEF pretreatment intensities, the highest vitamin C content (19.7 mg/100 g d.m.) in dried apple tissue treated with an electric field was found in Sample PEF 1, 70 °C and the lowest content (6.5 mg/100 g d.m.) in dried apple tissue was found in Sample PEF 6, 60 °C. The conducted one-factor analysis of variance showed that the tested samples differed significantly based on the examined parameter.

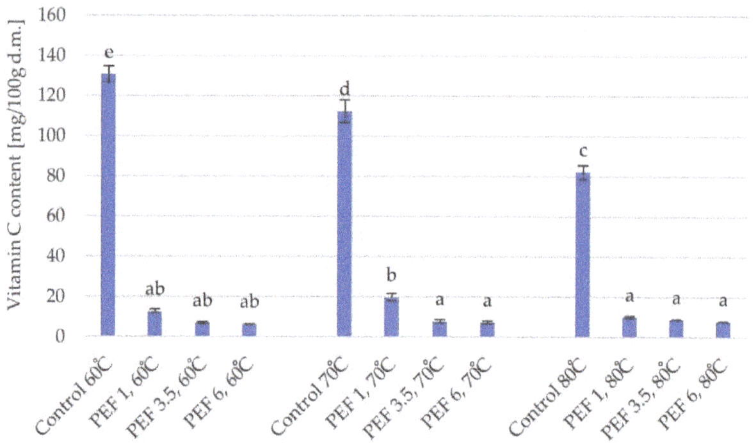

Figure 8. Vitamin C content of dried apple tissue obtained using different values for the parameters of pulsed electric field intensity (1, 3.5, and 6 kJ/kg) and hot air drying temperature (60, 70, and 80 °C). The columns with different letters are significantly different ($p < 0.05$).

Figure 9 shows the response surface results of vitamin C content in dried apple tissue obtained as a result of applying pulsed electric field at different intensity levels and hot air drying at different temperatures. The statistical analysis showed a lack of fit of the model in statistical terms, thus, this figure was not shown.

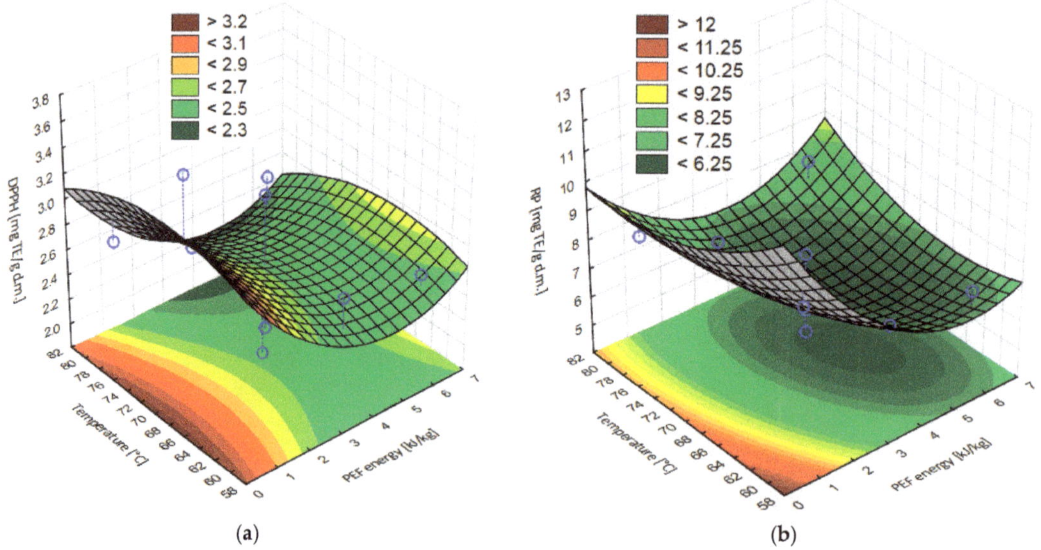

Figure 9. Response surface results of (**a**) the antioxidant capacity with DPPH radical and (**b**) reducing power RP, depending on the pulsed electric field intensity and the hot air drying temperature, obtained in the case of apple tissue subjected to different PEF pretreatment intensities (1, 3.5, and 6 kJ/kg) and hot air drying at different temperatures (60, 70, and 80 °C). Experimental values are marked with points.

2.5. The Influence of PEF on the Antioxidant Activity of Dried Apple Tissue

Table 2 shows the change in the antioxidant capacity of dried apple tissue obtained by applying PEF pretreatment at different intensities and hot air drying at different drying temperatures. The antioxidant capacity was determined taking into account the degree of quenching of the synthetic DPPH radical, presenting the results as the content of mg of trolox per 1 g of dry substance (mg TE/g d.m.) as well as the iron ion reducing power expressed in mg of trolox per 1 g of dry substance (mg TE/g d.m).

Table 2. Antioxidant capacity (with DPPH radicals and iron ion reducing power RP) of dried apple tissue obtained using different values for the parameters of pulsed electric field intensity (1, 3.5, and 6 kJ/kg) and hot air drying temperature (60, 70, and 80 °C).

Symbol	Process Parameters		Antioxidant Capacity (AC)	
	Temperature (°C)	PEF (kJ/kg)	DPPH (mg TE/g d.m.)	RP (mg TE/g d.m.)
Control 60 °C	60	-	3.31 ± 0.12 [abcd*]	14.34 ± 0.99 [c]
PEF 1, 60 °C	60	1	2.61 ± 0.04 [abc]	9.42 ± 0.44 [b]
PEF 3.5, 60 °C	60	3.5	2.73 ± 0.13 [abcd]	7.40 ± 0.26 [ab]
PEF 6, 60 °C	60	6	2.61 ± 0.28 [abc]	7.19 ± 0.17 [ab]
Control 70 °C	70	-	3.80 ± 0.11 [d]	17.72 ± 0.14 [d]
PEF 1, 70 °C	70	1	3.56 ± 0.28 [cd]	9.64 ± 0.20 [b]
PEF 3.5, 70 °C	70	3.5	2.51 ± 0.55 [ab]	6.36 ± 1.40 [a]
PEF 6, 70 °C	70	6	2.26 ± 0.17[a]	5.35 ± 0.19 [a]
Control 80 °C	80	-	3.46 ± 0.09 [bcd]	15.23 ± 0.71 [cd]
PEF 1, 80 °C	80	1	2.63 ± 0.03 [abcd]	7.96 ± 0.28 [ab]
PEF 3.5, 80 °C	80	3.5	2.29 ± 0.15 [ab]	6.67 ± 0.17 [ab]
PEF 6, 80 °C	80	6	2.59 ± 0.14 [abc]	8.06 ± 0.27 [ab]

SD—standard deviation. *,a,b,c,d—the values in the same column with different letters are significantly different ($p < 0.05$).

The parameters used for processing reduced the antioxidant capacity of the dried apple tissue compared to fresh apple tissue, whose antioxidant capacity was 4.33 mg TE/g d.m. (see Table 2). The antioxidant activity evaluated with DPPH radicals for the obtained dried apple tissue ranged from 2.26 to 3.8 mg TE/g d.m. The highest antioxidant capacity was shown by the dried apple tissue Sample Control 70 °C, and the lowest antioxidant capacity was shown by Sample PEF 6, 70 °C. Taking into account all the drying temperatures, the application of each of the selected electric field intensities resulted in a decrease in the content of mg TE/g d.m. compared to the control samples (dried samples obtained without pretreatment). The smallest loss of antioxidant activity compared to the control samples was obtained in the case of apple tissue dried at 70 °C and PEF pretreatment with an electric field intensity of 1 kJ/kg, where the antioxidant capacity only decreased by 6.3% and this change was not statistically significant. The conducted two-factor analysis of variance showed that the PEF pretreatment intensity ($\eta^2 = 0.74$) and the interaction between PEF pretreatment and temperature ($\eta^2 = 0.84$) had a significant impact on the antioxidant capacity of the tested material.

Similar results were obtained for reducing power (RP) determination (Table 1). Samples PEF pretreatment applied obtained significantly lower reducing power than apple tissue dried without pretreatment. An increase in the PEF pretreatment intensity, in most cases, had a decreasing effect but generally it was statistical insignificant. However, in the case of both parameters: PEF pretreatment intensity ($\eta^2 = 0.97$), temperature ($\eta^2 = 0.75$), and the interaction between the parameters (96) had significant effects on the RP. Furthermore, Figure 9a,b show the response surface results of the antioxidant capacity with DPPH radicals and reducing power for dried apple tissue obtained with different parameter process values. The statistical analysis showed a good fit for the model for both parameters.

2.6. The Influence of PEF on the Content of Sugars in Dried Apple Tissue

Figure 10 shows changes in the total sugar content including sucrose, glucose, fructose, and sorbitol in apple samples that were hot air dried at different temperatures after PEF pretreatment at various intensity levels. The total sugar content in dried apple tissue varied from 24.29 g /100 g d.m. (for PEF 3.5, 60 °C) to 37.34 g d.m. (for PEF 1, 60 °C), whereas the total sugar content in the fresh apples was equal to 36.71 g/100 g d.m. For almost all tested samples (except for apple tissue dried at 60 °C and PEF pretreatment with an electric field intensity of 1 kJ/kg) the total sugar content decreased as a result of the applied treatment. When the electric field intensity was 1 kJ/kg, the total sugar content was higher than in the case of the PEF pretreatments with electric field intensities of 3.5 or 6 kJ/kg, regardless of the drying temperature. For all samples, on the one hand, fructose accounted for the highest share of the total sugar content, ranging from 62.87 to 69.71%. On the other hand, sorbitol accounted for the lowest share of the total sugar content, ranging from 1.49 to 2.67%. Furthermore, glucose and sucrose accounted for between 14.48 and 21.99% and between 6.93 and 19.98% of the total sugar content, respectively. Additionally, increasing fructose and glucose content was noted with an increase in the drying temperature for samples not subjected to PEF pretreatment. The fructose content for apple tissue dried at 60, 70, and 80 °C was 16.54, 19.85, and 22.23 g/100 g d.m., respectively, and the glucose content was 4.29, 4.73, and 6.07 g/100 g d.m., respectively. On the contrary, with the higher drying temperature, sucrose content was lower or constant and it was equal to 4.01, 3.54, and 3.54 for apple tissue dried at 60, 70, and 80 °C, respectively. Sorbitol content was the lowest among the no PEF pretreated samples when convective drying was carried out at a temperature of 70 °C (0.44 g/100 g d.m.); both an increase and a decrease in drying temperature caused an increase in sorbitol content in the analyzed samples. The statistical analysis showed that both sucrose content and sorbitol content did not differ significantly among the tested samples but some significant differences in glucose content and fructose content were observed among the samples.

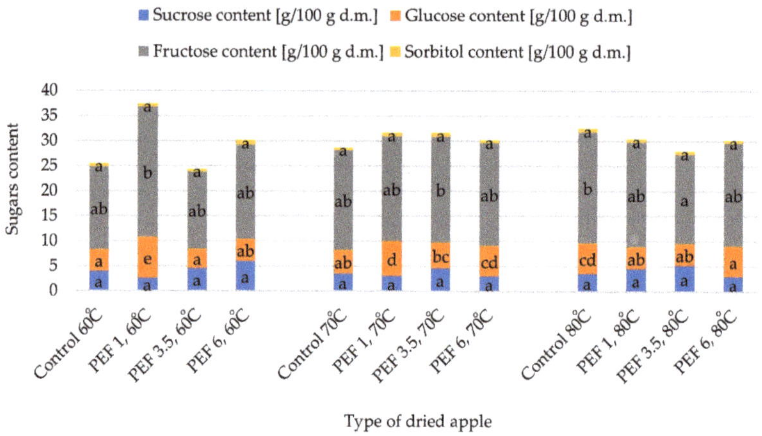

Figure 10. Total sugar content (sucrose, glucose, fructose, and sorbitol) of dried apple tissue obtained using different values for the parameters of pulsed electric field intensity (1, 3.5, and 6 kJ/kg) and hot air drying temperature (60, 70, and 80 °C). The different letters for the one type of sugars shows the significant differences between the samples ($p < 0.05$).

The two-factor analysis showed that both parameters (PEF pretreatment intensity and drying temperature) and the interaction between them significantly influenced the total sugar content as well as sucrose, fructose, glucose, and sorbitol contents.

2.7. The Influence of PEF on the Chemical Properties of Dried Apple Tissue (FTIR Method)

Figure 11 shows the FTIR spectra of a dried apple tissue. The spectra reflect the correlations between the absorbance intensity of the radiation and its energy described as the wave number (cm^{-1}). Six peaks of the IR spectrum are listed. Each of the obtained dried samples showed a similar pattern of spectra.

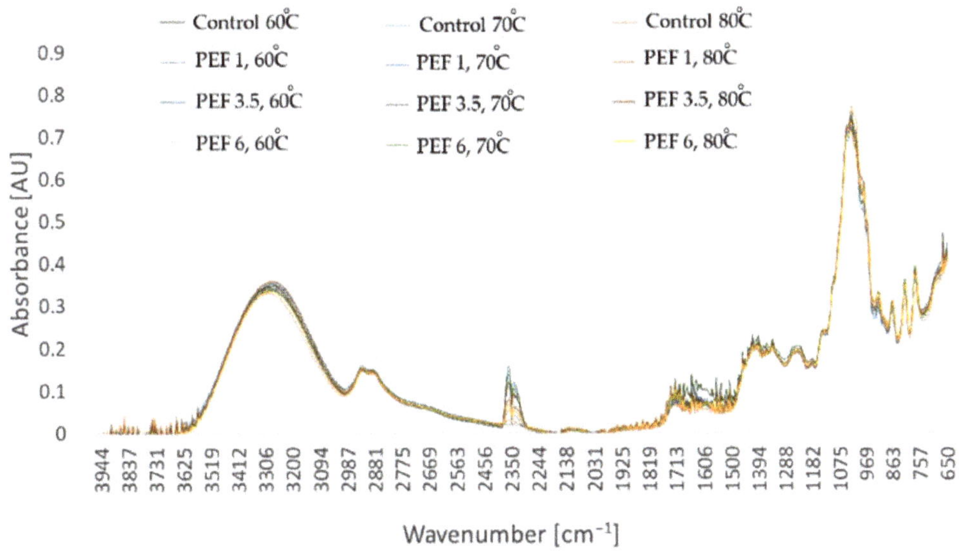

Figure 11. FTIR spectra of dried apple tissue obtained using different values for the parameters of pulsed electric field intensity (1, 3.5, and 6 kJ/kg) and hot air drying temperature (60, 70, and 80 °C). The columns with different letters are significantly different ($p < 0.05$).

3. Discussion

PEF pretreatment is usually used to improve the drying process. In our study, we found that, in most cases, after applying PEF pretreatment, the time needed for complete evaporation of water from apple tissue was shortened along with an increase in the pulsed electric field intensity. Such results can be explained by electroporation during PEF pretreatment. Electroporation is the formation of pores in tissue, which facilitates the process of water evaporation from a material [8]. Wiktor et al. [16] argued that the cell disintegration index increased with both an increase in electric field intensity and in pulse numbers for apple tissue air dried with PEF pretreatment. As the intensity of the electric field increases, the number of pores increases, and therefore, the average drying time of the product is shortened. Ostremeier et al. [17] dried onion tissue at three drying air temperatures, i.e., 65, 75, and 85 °C, and studied the impact of PEF pretreatment (PEF pretreatment with an electric field intensity of 4 kJ/kg) on the time of convective drying of plant tissue. Their research showed that the use of PEF pretreatment shortened the drying time by an average of approx. 25% compared to the control samples (without PEF application). Chauhan et al. [18] studied the effects of PEF pretreatment intensity, duration of the pulsed electric field, and the water temperature during pretreatment on the drying time of apple slices. The experiment showed that an increase in the electric field intensity accelerated the apple drying process due to the increasing degree of disintegration of plant tissue cells. A statistical analysis of the obtained results proved that, in the case of the combined application of the mentioned process parameters, the number of pulses used and the water temperature during the pretreatment had more significant impacts on the time of the apple drying process. This was most likely due to the fact that the applied electric field intensity was relatively low and did not cause sufficient disintegration of apple tissue cells, which led to the acceleration of mass exchange. The increasing number of delivered impulses meant that the investigated material was subjected to pretreatment for a longer period of time, and thus, stayed in hot water longer, which also strongly influenced the breakdown of apple tissue cells, and therefore, resulted in a noticeable shortening of the duration of the apple drying process.

Removing water from food extends its shelf-life, which, at the same time, extends the possible storage time. It is necessary to select the appropriate parameters of the drying process in order to obtain dried material of the best quality in terms of, for example, the content of nutrients and chemical ingredients [19]. Water availability in plant material is characterized by the concept of water activity, which affects the chemical processes taking place, as well as the limited development of microflora in food products. It has been reported that water activity below 0.6 prevented the development of microorganisms [20]; each of the dried materials had a water activity below 0.6, which allowed for inhibition of the growth of microorganisms in the material. It has been shown that temperature has an important effect on water activity, and the use of PEF pretreatment improves this effect. Ostremeier et al. [17] showed that samples, in which PEF pretreatment was applied, were obtained that had lower residual moisture even if a lower temperature of drying was used. The authors supposed that PEF pretreatment not only facilitated mass transfer in the first drying stage but also improved the second drying stage, which was characterized by the migration of moisture from the inner interstices of the sample to the outer surface.

Polyphenols are compounds found in food, which belong to the group of bioactive ingredients that are important for the human body. The content of these compounds affects the quality and antioxidant properties of food products [21]. The degradation of bioactive compounds can be influenced by numerous technological processes, for example, drying, extrusion, pasteurization, sterilization, and the use of preliminary treatments, including pulsed electric field pretreatment [22]. In the conducted research, we showed that the content of polyphenols and flavonoids in the dried apple tissue was affected by both the supplied electric field intensity and the drying temperature. The use of PEF pretreatment caused a decrease in total polyphenol content and total flavonoid content compared to the control samples (dried without PEF application). An increase in PEF pretreatment intensity was accompanied by a decrease in the total polyphenol content in the dried apple tissue. Changes in total polyphenol content depend on the efficiency of electroporation, measured, for example, by determining the degree of disintegration. A PEF pretreatment intensity that is too high can lead to the degradation of bioactive compounds, and properly selected values for the parameters can improve the extractability of these compounds. Similar relationships were observed by Wang et al. [23], who studied the impact of pulsed electric field treatment on the extraction of bioactive ingredients from apple skins. Mello et al. [9] indicated that pulsed electric field pretreatment before the drying process positively affected preservation of the total polyphenol content, ascorbic acid, and the antioxidant capacity of the dried material. Additional benefits can be achieved by the combination of pulsed electric field pretreatment with hybrid drying. Orange peel dried with PEF pretreatment has been characterized with a higher total polyphenol content in relation to orange peel dried in a hybrid way without pretreatment. In addition, Lammerkitten et al. [15] showed that PEF pretreatment had a positive effect on the contents of phenols and flavonoids in dried mango, which was associated with the phenomenon of electroporation that increased the permeability of the cell membrane and consequently improved the extractability of phenolic compounds. They showed that the use of PEF pretreatment with a lower specific intensity was the most beneficial. Lammerskitten et al. [24] showed that the total phenolic content was increased by up to 47% for PEF pretreated freeze-dried apple tissue compared to untreated freeze-dried apple tissue.

Vitamins are a large group of organic chemical compounds that are characterized by diverse structures. Their task is to support the proper functioning of the body. They belong to the group of exogenous compounds, for example, compounds that the human body is unable to produce, and therefore, it is necessary to supply them through food intake.

Vitamin C is a bioactive compound and has strong antioxidant properties, which help to prevent heart and capillary diseases. Ascorbic acid can also be used as a food additive. Due to its good antioxidant properties, it slows down oxidation reactions, which allows products to extend their shelf-life [25,26]. Yamakage et al. [14] studied the effect of pulsed electric field pretreatment on changes in the quality of spinach during hot air drying and found that the use of PEF pretreatment obtained a dried product with a high content of L-ascorbic acid compared to the dried product obtained without the PEF pretreatment. The intensity of the applied PEF pretreatment had a greater influence on the content of vitamin C. A decrease in the amount of ascorbic acid was found in relation to the control sample. Ascorbic acid is very susceptible to degradation under the influence of the drying process. The use of PEF pretreatment reduced the acid content of the orange peel compared to the control sample. This relationship was most likely due to the fact that the greatest degradation of ascorbic acid was caused by drying itself. Pulsed electric field pretreatment leads to the electroporation process, which increases the exposure of nutrients to hot air during drying, and therefore, contributes to the degradation of ascorbic acid [9]. The effect of pulsed electric field pretreatment was used by Morales-de la Peña et al. [27], who studied the effect of high-voltage PEF pretreatment on changes in the content of vitamin C in stored fruit juices. The results were compared to the content of ascorbic acid in juices not subjected to PEF pretreatment and to the fruit juices subjected to thermal treatment (90 °C). On the basis of vitamin C content, studies have shown that a shorter duration of PEF pretreatment is more beneficial. However, regardless of the duration time of PEF pretreatment, samples subjected to PEF pretreatment retained a higher amount of vitamin C compared to juices treated at a high temperature, but each treatment resulted in a greater decrease in the amount of vitamin C versus the control (untreated). Since PEF pretreatment is a non-thermal treatment, the higher content of vitamin C in juices with electric field obstruction may result from the low resistance of ascorbic acid to the high temperature used during thermal treatment. With PEF pretreatment, greater degradation of vitamin C in the juice obtained after a longer duration pretreatment may be due to the increase in the temperature of the material; the juice was more intensively subjected to high-voltage electricity, which caused the Joule effect, resulting in an increase in the temperature of the material, which resulted in a higher degree of degradation of ascorbic acid. When comparing the results of these studies and those presented in this paper, one should bear in mind the differences between the compared processes and the form ("state of aggregation") of the tested matrices.

Antioxidants are chemical compounds that belong to the group of bioactive ingredients present mainly in fruits and vegetables. One of the characteristics of biologically active compounds is that living organisms are not able to produce them, and therefore, they should be supplied through food intake. The role of antioxidant substances is to protect cells and tissues against the adverse effects of emerging free radicals (unpaired oxygen atoms), which cause undesirable changes in the human body. Antioxidants inhibit the action of free radicals by "connecting" with unpaired electrons, slowing down the oxidation reactions taking place in the body. The antioxidant compounds include polyphenols, flavonoids, vitamin C, carotenoids, and xanthophylls [28,29]. The conducted analysis showed that the applied PEF pretretment intensity, in most cases, had an insignificant impact on the antioxidant capacity of dried apple tissue and that PEF pretreatment caused loss of antioxidant activity compared to the control samples (dried without PEF application). Lammerskitten et al. [24] also showed that, for freeze-dried apple tissue, PEF pretreatment decreased antioxidant activity up to ~60% compared to a reference sample. A different effect of PEF pretreatment on the antioxidant properties of dried material was shown by Huang et al. [30], who applied PEF PREtreatment to fresh apricots. The use of a higher electric field intensity improved the antioxidant capacity of the dried material. The differences in the results could have been influenced by the sodium sulfite used, which has an oxidizing capacity. The combination of these two factors enhanced the ability to reduce free radicals in dried apricots. Mello et al. [9] showed that the dried material obtained with PEF pretreatment had a higher antioxidant capacity for a shorter application time and this activity was similar to the dried material obtained using hybrid drying without pretreatment. Reducing the antioxidant capacity when using higher intensity values of the PEF pretreatment parameter may be caused by greater exposure of bioactive cells to high temperature during drying, which leads to a decrease in the value of bioactive compounds and the antioxidant capacity of the dried material. Therefore, PEF pretreatment with appropriate parameter values positively affects the quality of the dried material, while too intensive PEF pretreatment may have a negative impact on the final quality of the dried material.

Reducing power has also been used to evaluate the ability of natural antioxidants in dried apple tissue pretreated using PEF and the obtained results correlated with the polyphenol and flavonoid contents. In addition, Yakubu et al. [31] showed a significant decrease in the reducing power and the DPPH scavenging activity of blanched bitter leaves. The increase or decrease in antioxidant activity was explained by Adefegha and Oboh [32] by the fact that heat treatments can soften the matrix and can improve or degrade the extractability of phytochemicals involved with antioxidant activity.

Sugars are essential in the human diet for proper body functioning. Fructose and glucose are important monosaccharides. Chemically combined, these sugars form a disaccharide, i.e., sucrose. Such sugars as glucose, fructose, and sucrose are present, among others, in fruit and vegetables. Fructose supplies the human body with quick energy, whereas glucose is necessary as one of the primary sources of fuel for cellular metabolism. Furthermore, sorbitol belongs to polyols, which are saccharide derivatives. Compared with sugars, polyols, including sorbitol, are poorly absorbed, and therefore provide fewer calories and lower glycemic responses [33,34]. Regarding sugars, apple tissue contains mostly fructose, but certain amounts of glucose and sucrose are usually also present in these fruits, whereas sorbitol is present in apple tissue in a much smaller quantity [35–37]. This was also confirmed by our study (Table 2). The total sugar content in apple tissue as well as specific sugar contents (i.e., fructose, glucose, and sucrose) vary depending on the variety, weather conditions, culture technology, and also the position and exposition of the apple tissue in the crown. The specific sugar content during storage may both decrease or increase, which is also determined by the apple variety [38]. Wojdyło et al. [39] studied the influence of different drying methods and conditions on the quality parameters of red-fleshed apple fruit snacks, including the total sugar content. The authors observed that sugar content was strongly determined by the drying method. The samples obtained by freeze-drying were characterized by the highest sugar content which was 34.10 g/100 g d.m., whereas the sugar contents of convective dried samples were 24.3 $\pm$ 2.5, 26.7 $\pm$ 3.1, and 21.7 $\pm$ 1.8 g/100 g d.m. depending on the drying temperatures which were 50, 60, and 70 °C, respectively. These results were similar to the results reported in this research. When analyzing the apple tissue samples without PEF pretreatment, the increasing contents of both fructose and glucose together with the decreasing or constant content of sucrose was observed with an increase in the drying temperature. Similarly, Delgado et al. [40] reported the phenomenon of increased fructose content during convective drying of chestnuts, while Mitrović et al. [41] reported a decrease in sucrose content and, simultaneously, an increase in inverted sugars content after convective drying of plums at 70 or 90 °C. Furthermore, Macedo et al. [42] studied the effect of convective drying temperature (40, 60, and 80 °C) on the drying kinetics and the physicochemical properties of dried bananas, and also observed that the reduced sugar content of the dried fruits was higher as the drying air temperature increased. The authors explained that the results were probably related to sucrose hydrolysis to glucose and fructose during the convective drying [40–42]. In the present research, the effect of PEF pretreatment was not clear and it varied depending on the PEF pretreatment intensity applied and also the temperature of the following hot air drying. The content of sucrose did not differ significantly but statistically significant differences in both glucose and fructose content between the samples were noted, which was in accordance with the research of Rybak et al. [43], in which the influences of various treatments, including PEF pretreatment with an electric field intensity of 1 or 3 kJ/kg, on the selected properties of red bell pepper were analyzed. Additionally, the authors suggested that the increase in the glucose content of PEF-treated samples might be caused by the decomposition of carbohydrates in such a way that it can affect other carbohydrates, and also improve the extractability of sugars [43]. Furthermore, PEF pretreatment of chokeberry juice sources from six different farms resulted in an increase in the total sugar content in all cases [44]. In turn, Rybak et al. [45] recorded that PEF pretreatment with an electric field intensity of 1 or 3 kJ/kg conducted before convective or microwave-convective drying affected the significant decrease in total sugar content in red bell pepper samples. It was explained by the fact that the electroporation occurring as a result of the PEF pretreatment that caused a rupture in the cell membrane which resulted in sugar leakage.

Infrared spectroscopy (FTIR) is a method used to determine the structure of particles as well as the composition of molecular mixtures. Infrared radiation is absorbed in frequencies related to the vibration energy of the bonds between atoms in a molecule [46]. In this study, it was shown that the obtained dried apple tissue showed a similar pattern of spectra, which may have resulted from the same structure and chemical composition of the tested samples, and at the same time from the presence of the same functional groups in the tested material. According to the literature, the range of peaks from 1200 to 1500 cm^{-1} refers to the vibrations of COH, CCH, and COH bonds; additionally, the range from 1630 to 1680 cm^{-1} corresponds to the amide functional groups associated with the CO carbonyl group. The peaks at the wave values between 1149 cm^{-1} and 1336 cm^{-1} indicate the presence of COC glycosidic bonds for pectin. The IR spectra in the ranges of 3200–3500 cm^{-1} and 2900–2920 cm^{-1} are related to the stretching vibrations of the OH, NH, and CH bonds. The range associated with the CH group mainly relates to the presence of these bonds in the cellulose and hemicellulose present. The absorbed light at the value of 1710 cm^{-1} is related to the absorption of the carbonyl group of fatty acids that are present in the fibers of the raw material. In addition, the CO bonds present in lignin are absorbed at the peaks of 1634 cm^{-1} and 1374 cm^{-1}. Bioactive ingredients such as phenolic compounds are identified at 1560 cm^{-1} and 1630 cm^{-1}. With regard to the presence of sugars, the peak near 922 cm^{-1} is related to the α-anomeric linkage between glucose and fructose in sucrose; additionally, the glycosidic linkage in sucrose COC is about 990 cm^{-1}. In fructose, the vibrations corresponding to the stretching of CO and CC bonds are at the wave of 1046 cm^{-1}, in the case of sucrose, the stretching of CO bonds takes place at 1048 cm^{-1}, and the value of 1102 cm^{-1} is the stretching of CO and CC bonds in glucose, as well as bending of COH bonds. The wavelength range from 896 cm^{-1} to 900 cm^{-1} is associated with β-glucosidic bonds in hemicellulose and cellulose [47–49]. The influence of PEF p retreatment on the chemical composition of plant tissue was studied by Ahmed et al. [50]. The research was carried out using the juice of wheat plants. PEF pretreatment with an intensity of 9 kV/cm was applied. In addition, sonication and the combination of PEF + US were used to determine the difference in the effect of pretreatment on the material properties. Similar IR spectra with similar wave values were obtained for all samples. In addition, Niu et al. [51] obtained similar results for freeze-dried naringin in pomelo tissue. Several PEF pretreatment intensities were used. Most of the peaks obtained, as well as those obtained for the control samples not subjected to PEF, coincided, and therefore, it was concluding that PEF pretreatment had no significant effect on the absorption of infrared light.

4. Materials and Methods

4.1. Materials

For this study, the the Gloster apple variety was selected from an organic farm (Warsaw, Poland). The apples were cut into 5 mm thick slices, and then they were cut into four parts, using a sharp knife. Before the apples were cut and dried, a pulsed electric field (PEF) pretreatment was applied. The properties of the fresh apple tissue are presented in Table 2.

4.2. *Technological Methods*

4.2.1. Pulsed Electric Field Application

Before the drying process, the tested material was pretreated by applying a pulsed electric field (PEF) in an ELEA Pi-lot-Dual pulsed reactor (Elea Vertriebs- und Vermarktungsgesellschaft mbHVer, Quakenbrück, Germany). The PEF device has stainless steel electrodes, where the gap between them is 28 cm. The device delivers a 2 Hz frequency of exponential decay pulses with a monopolar signal and a width of 40 ms.

The treatment consisted of placing whole apples (approx. 150 g) in the chamber, and then adding tap water at room temperature in such an amount that the mass of the entire system was approx. 1000 g. The electric field intensity during the PEF application was 1 kV/cm, and the energy values delivered to the system during pretreatment were 1, 3.5, and 6 kJ/kg.

4.2.2. Hot Air (Convective) Drying

The material was dried using the convective method in a prototype laboratory dryer (Promis-Tech, Wrocław, Poland) at the drying air temperature of 60, 70, and 80 °C. The air flowed through the material perpendicularly to the screen and its velocity was 1.2 m/s. Before putting the raw material into the dryer, the material was placed on a sieve in the amount of about 150 g, and then the parameters were set and the drying time was recorded every 5 min using a scale that automatically registered weight changes.

4.3. *Analytical Methods*

4.3.1. Determination of the Kinetics of the Drying Process

The kinetics of drying are presented in the form of drying curves, which show the relationship between the relative water content (MR) and the progress of the process [52]. The relative water content (humidity ratio, MR) was calculated based on the following relationship:

$$MR = \frac{u_\tau}{u_0}, \quad (1)$$

where u_0 is the water content in raw material (kg H_2O/kg d.m.) and u_τ if the water content of the material at the time of drying (kg H_2O/kg d.m.).

4.3.2. Determination of the Water Content

About 0.2 g (with an accuracy of 0.0001 g) of dried material, homogenized using an analytical mill, was weighed in previously dried weighing bottles, with an accuracy of 0.0001 g. Then, the vials were dried in a laboratory dryer at 70 °C for 24 h [53]. After this time, the samples were placed in a desiccator and kept until they reached room temperature. Then, the cups were weighed with an accuracy of 0.0001 g and the content of dry matter in apple tisssue was calculated. The assay was performed in duplicate.

4.3.3. Determination of Water Activity

The water activity measurement was carried out in two repetitions for each type of dried material. The Aqualab 4 TE (Decagon, Pullman, WA, USA) water activity meter was used for the measurement. The measurement was made at a temperature of 25 °C [54].

4.3.4. Determination of Total Polyphenol Content (TPC)

The determination was carried out using the spectroscopic method based on a color reaction of the analyte with the Folin Ciocalteau (F-C) reagent [55]. The material was ground using an analytical mill (IKA A11 basic, IKA-Werke GmbH & Co., Staufen, Germany).Then, 0.3 g of dried material was weighed into the test tube with an accuracy of ±0.0001 g and diluted with 10 mL of 80% ethyl alcohol. The solution was shaken on an orbital shaker (Multi Reax, Heidolph Instruments, Schwabach, Germany) at 20 °C and after 12 h it was centrifuged (MegaStar 600, VWR, Leuven, Belgium) at 4350 rpm for 2 min. Such extract was used for the further analysis.

In a well of a 96-well plate, 10 μL of the supernatant was placed, which was diluted twice with distilled water. Then, 40 μL of F-C reagent (5 times diluted with distilled water) was added to the solution, and after 3 min, 250 μL of 7% sodium carbonate solution was added. The samples were incubated for 60 min at 20 °C and protected from light. Absorbance was measured at a wavelength of 750 nm on a plate reader (against reagent blank). The result was expressed in mg/100 g of dry matter of chlorogenic acid using a calibration curve for chlorogenic acid standard at a concentration of 0–100 μL. The assays were performed in duplicate for each extract.

4.3.5. Determination of Total Flavonoid Content

The method with aluminium (III) chloride was used to determine the total flavonoid content [56]. First, 20 μL of the extract was diluted with 80 μL of distilled water and mixed with 10 μL of $NaNO_2$ (5% *w*/*v*). After 5 min, 10 μL of $AlCl_3$ (10% *w*/*v*) was added and mixed, and after 6 min, 40 μL of 1 M NaOH solution was added and mixed. After 20 min, the absorbance of the solutions was measured at 510 nm. The quantitative content of flavonoids was calculated based on a calibration curve for quercetin in the range of 0–500 μg/mL. Measurements were made in duplicate.

4.3.6. Determination of Antioxidant Capacity (AC) on the Basis of Free Radical Scavenging DPPH and Ferric Antioxidant Reducing Power (RP)

The antioxidant capacity was determined using a spectrophotometric method to determine the degree of quenching of the synthetic 2,2-diphenyl-1-picrylhydrazyl (DPPH) radical [57]. In order to prepare the initial DPPH solution, 0.025 g of 2,2-diphenyl-1-picrylhydrazyl was diluted to 100 mL with 99% methanol. The solution was stored for a minimum of 24 h protected from light at 4 °C to generate the radical. The working solution was prepared immediately before analysis. First, 9 mL of the starting solution was diluted with 80% ethanol solution, in which absorbance measured at 515 nm was in the range of 0.700 ± 0.020. Measurements were made in 96-well plates. Then, 250 μL of the radical solution was added to the well with 10 μL of the 5 times diluted sample extract. After 30 min storage at room temperature, without access to light, the absorbance was measured using a plate reader (Multiskan Sky, Thermo Electron Co., St. Louis, MO, USA) at a wavelength of 515 nm. The assays were performed in duplicate for each extract. The antioxidant activity was determined on the basis of a decrease in the absorbance of the radical solution in the presence of the antioxidant and expressed as the Trolox equivalent antioxidant capacity (TEAC) coefficient, corresponding to the concentration of Trolox with the same antioxidant capacity as the tested sample (mg Trolox/g d.m.).

To determine the reduction power (RP) of iron ions by the analyte, 25 μL of the extract, 50 μL of a 1% aqueous solution of potassium ferricyanide, and 75 μL of distilled water were pipetted into the well. The whole was mixed, and then placed in an incubator (INCU-Line ILS 10; VWR, Radnor, PA, USA) at 50 °C. After 20 min, 50 μL of 10% trichloroacetic acid was added. Then, 100 μL of the reaction mixture were taken into an empty well, 100 μL of distilled water and 20 μL of 0.1% iron (III) chloride solution were added, and the whole was mixed. After 10 min, the absorbance at 700 nm was measured against a blank [58]. The RP value is expressed as mg of Trolox.

4.3.7. Determination of Vitamin C Content

The UPLC-PDA method (WATERS Acquity H-Class, Milford, MA, USA) was used to determine the content of L-ascorbic acid [59]. First, 10 mL of cooled extraction reagent (3% metaphosphoric acid and 8% acetic acid) was added to 0.3 g of ground dried material, the solution was stirred with vortex for 5 min, and then centrifuged (2 min, 4350 rpm, 5 °C). The assay was carried out with limited access to light. The solution was filtered using 0.2 μm GHP syringe filters (Acrodisc, Pall Corporation, Port Washington, NY, USA). Next, 1 mL of the solution was added to 1 mL of the eluent, and then injected into the column. A WATERS Acquity UPLC HSS T3 column (2.1 × 100 mm, 1.8 μm, Waters, Ireland) with a BEH C18 pre-column (2.1 × 5 mm, 1.7 μm, Waters, Ireland) was used for separation. The mobile phase flow (Milli-Q water with 0.1% formic acid) was 0.25 mL/min. The column thermostat temperature was 25 °C and the autosampler tempurature was 4 °C. The spectral analysis was performed at a wavelength of 245 nm. The vitamin C content was calculated against a calibration curve prepared for an analytical standard of L-ascorbic acid (0.005–0.100 mg/mL). The analysis was performed in duplicate.

4.3.8. Total Sugar Content (Sucrose, Glucose, Fructose, and Sorbitol)

The sugar content was determined by liquid chromatography [60]. The system consisted of a quadruple pump (Waters 515, Milford, MA, USA), an autosampler (Waters 717, Milford, MA, USA), a column thermostat, and an RI detector (Waters 2414, Milford, MA, USA). First, 0.3 g of the material was poured with MilliQ water at a temperature of 80 °C. The samples were placed in a circular-vibrating shaker and sugars were extracted for 4 h. The solution was centrifuged (5 min, 4350 rpm), filtered through a 0.22 μm hydrophobic PTFE syringe filter (Millex-FG, Millipore, Milford, MA, USA), and 1 μL was injected into the column. Separation was carried out using a 300 × 6.5 mm Waters Sugar Pak I column with a Sugar-Pak pre-column. The assay was carried out with a constant composition of the mobile phase (Milli-Q redistilled water), the flow rate of which was 0.6 mL/min, the detector temperature was 50 °C, and the column temperature was 90 °C. The quantitative analysis was performed on the basis of prepared calibration curves for glucose, fructose, sucrose, and sorbitol standards. The assays were carried out in duplicate.

4.3.9. Fourier Transform Infrared Spectroscopy (FTIR)

FTIR spectra of dried apple tissue were performed in infrared using Cary 630 (Agilent Technologies Inc., Santa Clara, CA, USA) with a single reflection diamond attenuated total reflection ATR [58]. The analysis was carried out at a wavelength in the range of 650–4000 cm^{-1} with a resolution of 4 cm^{-1}, with 32 scans of the spectrum. The dried sample was pressed against the crystal with a pressure clamp. Each material was scanned five times. The analysis data was recorded using the MicroLab FTIR software.

4.4. Statistical Methods

The experiment was organized using response surface methodology (RSM) with experimental planning of two factors: drying temperature and energy consumption during PEF application, at three levels (Table 3).

Table 3. List of dryings, taking into account the temperature and energy supplied during application of the PEF pretreatment—experiment plan.

		Drying Parameters	
Drying Number	**Symbol**	**Temperature (°C)**	**Energy PEF (kJ/kg)**
1	PEF 1, 60 °C	1	60
2	PEF 1, 80 °C	1	80
3	PEF 6, 60 °C	6	60
4	PEF 6, 80 °C	6	80
5	PEF 1, 70 °C	1	70
6	PEF 6, 70 °C	6	70
7	PEF 3.5, 60 °C	3.5	60
8	PEF 3.5, 80 °C	3.5	80
9 (A)	PEF 3.5, 70 °C	3.5	70
10 (B)	PEF 3.5, 70 °C	3.5	70
11 (C)	PEF 3.5, 70 °C	3.5	70
12	Control 60 °C	-	60
13	Control 70 °C	-	70
14	Control 80 °C	-	80

The levels of energy consumption during PEF pretreatment applications were selected based on the efficiency of electroporation, determined by measuring the specific electrical conductivity (preliminary studies), while the temperature levels were selected based on the literature. The analysis of the impact of the pulsed electric field intensity and the applied temperature on the quality of the dried samples was assessed on the basis of a one-way ANOVA analysis of variance, and homogeneous groups were determined on the basis of Tukey's test. Furthermore, in order to determine the impact of PEF pretreatment intensity and hot air drying temperature as well as the interaction of these two factors on the obtained results, a two-factor analysis of variance was performed at the significance level $\alpha = 0.05$. To conduct the statistical analysis, the Statictica program ver. 13 of the TIBCO company software (Palo Alto, CA, USA) was used.

On the basis of the obtained results, statistical profiles of approximated values of the PEF pretreatment intensity levels and hot air drying temperatures as well as the usability for the drying time and chosen properties (water activity, antioxidant capacity, and total sugars content) were made and presented in Figure 12.

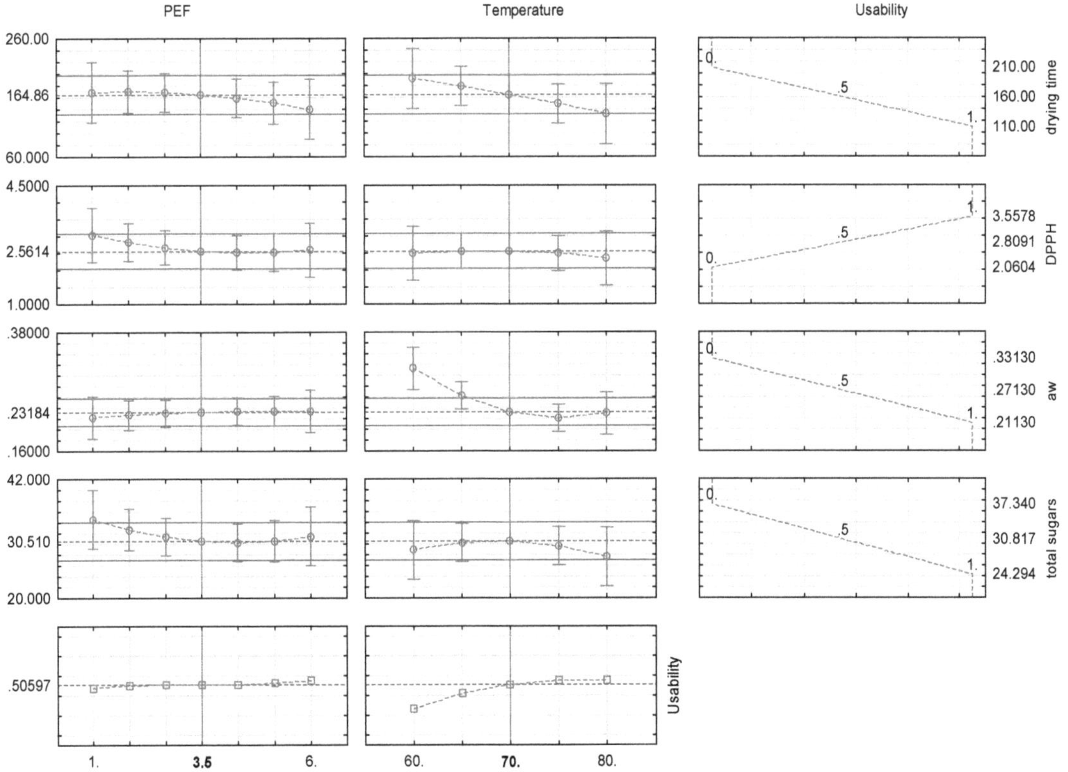

Figure 12. Approximation profiles and usability for dried apple tissue.

Optimal properties were chosen as the general and most important for obtaining good quality dried material in a short time period. For approximation profiles, the drying time, water activity, and total sugar content were set to obtain lower values, while antioxidant capacity was set to gain the highest values. The projection showed that the best parameters to obtain such dried material would be application of a PEF pretreatment intensity of 3.5 kJ/kg and drying at 70 °C.

5. Conclusions

Our results showed that pulsed electric field applied as a pretreatment before the hot air drying process shortened the drying time of the material due to the pores formed that facilitated the evaporation of water from the material. The shortest drying time was obtained when the higher temperature was used. In the case of samples with PEF pretreatment, the shortest drying time was characterized by the material dried at 70 °C and treated using pulsed electric field pretreatment with an intensity of 6 kJ/kg.

Dried materials were characterized by water activity not exceeding 0.6, and the highest value did not exceed 0.331, which was obtained by PEF pretreatment of apple tissue with an electric field intensity of 6 kJ/kg and dried at 60 °C. In addition, electroporation caused a change in the content of chemical compounds in the apple tissue. The application of each electric field intensity caused the degradation of polyphenols from apple tissue compared to samples not subjected to PEF pretreatment. The smallest loss of these substances was noticed in the dried material obtained by applying PEF pretreatment with an electric field intensity of 1 kJ/kg, and then dried at 70 °C. The same trend was observed for flavonoid content and antioxidant capacity. Furthermore, the reducing power investigations confirmed the antioxidant capacity results. PEF pretreatment decreases the reducing power of dried apple tissue compared to the untreated samples. Additionally, ascorbic acid was degraded by pulsed electric field at any level of applied intensity. An increase in energy supplied during PEF application led to an increase in the degradation of vitamin °C in dried apple tissue.

The application of pretreatment did not influence changes in the total sugar content and sorbitol content of dried apple tissue. However, increased fructose and glucose contents were noted with an increase in the drying temperature and the PEF pretreatment electric field intensity. PEF pretreatment did not effect the samples' FTIR spectra, which were similar for all samples.

The optimization process, as well as statistical profiles of approximated values, showed that the best parameter values for obtaining a high-quality product with a short drying time were the application of PEF pretreatment with an intensity of 3.5 kJ/kg and hot air drying at a temperature of 70 °C.

Author Contributions: Conceptualization, M.N., A.W. and A.C.; methodology, K.R., A.W. and M.N.; software, A.C., A.W. and M.N.; formal analysis, A.C., K.R., A.W. and M.N.; investigation, K.R., A.W. and M.N.; resources, M.N.; data curation, A.C. and M.N.; writing—original draft preparation, A.C., K.R., A.W., M.T. and M.N.; writing—review and editing, A.C. and M.N.; visualization, A.C., M.T. and M.N.; supervision, M.N.; project administration, M.N.; funding acquisition, M.N. All authors have read and agreed to the published version of the manuscript.

Funding: This project received funding from transnational funding bodies, partners of the H2020 ERA-NETs SUSFOOD2 and CORE Organic Cofunds, under the Joint SUSFOOD2/CORE Organic Call 2019 (MILDSUSFRUIT), as well as the National Centre for Research and Development (POLAND, decision DWM/SF-CO/31/2021). The research for this publication was carried out with the use of research equipment purchased as part of the "Food and Nutrition Centre—modernisation of the WULS campus to create a Food and Nutrition Research and Development Centre (CŻiŻ)" co-financed by the European Union from the European Regional Development Fund under the Regional Operational Programme of the Mazowieckie Voivodeship for 2014–2020 (project no. RPMA.01.01.00-14-8276/17).

Institutional Review Board Statement: Not applicable.

Informed Consent Statement: Not applicable.

Data Availability Statement: Not applicable.

Conflicts of Interest: The authors declare no conflict of interest.

Sample Availability: Samples of the dried material are available from the authors.

References

1. Vargas, L.; Kapoor, R.; Nemzer, B.; Feng, H. Application of different drying methods for evaluation of phytochemical content and physical properties of broccoli, kale, and spinach. *LWT* **2022**, *155*, 112892. [CrossRef]
2. An, N.-N.; Sun, W.-H.; Li, B.-Z.; Wang, Y.; Shang, N.; Lv, W.-Q.; Li, D.; Wang, L.-J. Effect of different drying techniques on drying kinetics, nutritional components, antioxidant capacity, physical properties and microstructure of edamame. *Food Chem.* **2022**, *373*, 131412. [CrossRef] [PubMed]
3. Janowicz, M.; Domian, E.; Lenart, A.; Pomarańska-Łazuka, W. Charakterystyka suszenia konwekcyjnego jabłek odwadnianych osmotycznie w roztworze sacharozy. *Żywn. Nauka Technol. Jakość* **2008**, *4*, 190–198.
4. Kaur, M.; Modi, V.K.; Sharma, H.K. Effect of carbonation and ultrasonication assisted hybrid drying techniques on physical properties, sorption isotherms and glass transition temperature of banana (Musa) peel powder. *Powder Technol.* **2021**, *396*, 519–534. [CrossRef]
5. Sagar, V.R.; Kumar, S.P. Recent advances in drying and dehydration of fruits and vegetables: A review. *J. Food Sci. Technol.* **2010**, *47*, 15–26. [CrossRef] [PubMed]
6. Kouchakzadeh, A.; Shafeei, S. Modeling of microwave-convective drying of pistachios. *Energy Convers. Manag.* **2010**, *51*, 2012–2015. [CrossRef]
7. Szadzińska, J.; Mierzwa, D. The influence of hybrid drying (microwave-convective) on drying kinetics and quality of white mushrooms. *Chem. Eng. Process.-Process Intensif.* **2010**, *167*, 108532. [CrossRef]
8. Yu, Y.; Jin, T.Z.; Xiao, G. Effects of pulsed electric fields pretreatment and drying method on drying characteristics and nutritive quality of blueberries. *J. Food Process. Preserv.* **2017**, *41*, 13303. [CrossRef]
9. Mello, R.E.; Fontana, A.; Mulet, A.; Correa, J.L.G.; Caler, J.A. PEF as pretreatment to ultrasound-assisted convective drying: Influence on quality parameters of orange peel. *Innov. Food Sci. Emerg. Technol.* **2021**, *72*, 102753. [CrossRef]
10. Wiktor, A.; Witrowa-Rajchert, D. Zastosowanie pulsacyjnego pola elektrycznego do wspomagania procesów usuwania wody z tkanek roślinnych. *Żywn. Nauka Technol. Jakość* **2012**, *2*, 22–32.
11. Thamkaew, G.; Gómez Galindo, F. Influence of pulsed and moderate electric field protocol on the reversible permeabilization and drying of Thai basil leaves. *Innov. Food Sci. Emerg. Technol.* **2020**, *64*, 102430. [CrossRef]
12. Rahamana, A.; Siddeeg, A.; Manzoor, M.F.; Zeng, X.-A.; Ali, S.; Baloch, Z.; Li, J.; Wen, Q.-H. Impact of pulsed eletric field treatment on drying kinetics, mass transfer, colour parameters and microstructure of pulm. *J. Food Sci. Technol.* **2019**, *56*, 2670–2678. [CrossRef] [PubMed]
13. Mirzaei-Baktash, H.; Hamdami, N.; Torabi, P.; Fallah-Joshaqani, S.; Dalvi-Isfahan, M. Impact of different pretreatments on drying kinetics and quality of buttom mushroom slices dried by hot-air or electrohydrodynamic drying. *Food Sci. Technol.* **2022**, *155*, 112894.
14. Yamakage, K.; Yamada, T.; Takahashi, K.; Takaki, K.; Komuro, M.; Sasaki, K.; Aoki, H.; Kamagata, J.; Koide, S.; Orikasa, T. Impact of pre-treatment with pulsed electric field on drying rate and changes in spinach quality during hot air drying. *Innov. Food Sci. Emerg. Technol.* **2021**, *68*, 102615. [CrossRef]
15. Lammerskitten, A.; Shrstkii, I.; Parniakov, O.; Mykhailyk, V.; Toepfl, S.; Rybak, K.; Dadan, M.; Nowacka, M.; Wiktor, A. The effect of different methods of mango drying assisted by a pulsed electric field on chemical and physical properties. *J. Food Process. Preserv.* **2020**, *44*, 14973. [CrossRef]
16. Wiktor, A.; Iwaniuk, M.; Śledź, M.; Nowacka, M.; Chudoba, T.; Witrowa-Rajchert, D. Drying kinetics of apple tissue treated by pulsed electric field. *Dry. Technol.* **2013**, *31*, 112–119. [CrossRef]
17. Ostermeier, R.; Parniakov, O.; Töpfl, S.; Jäger, H. Applicability of Pulsed Electric Field (PEF) Pre-Treatment for a Convective Two-Step Drying Process. *Foods* **2020**, *9*, 512. [CrossRef]
18. Chauhan, O.P.; Sayanfar, S.; Teopfl, S. Effect of pulsed electric field on texture and drying time of apple slices. *J. Food Sci. Technol.* **2018**, *55*, 2251–2258. [CrossRef]
19. Skorupska, E. Badanie procesu suszenia konwekcyjnego pietruszki korzeniowej. *Inż. Rol.* **2005**, *9*, 313–320.
20. Pałacha, Z. Aktywność wody ważny parametr trwałości żywności. *Przem. Spoż.* **2008**, *4*, 22–26.
21. Sadowska, A.; Świderski, F.; Kromołowska, R. Polifenole-źródło naturalnych przeciwutleniaczy. *Postępy Tech. Przetw. Spoż.* **2011**, *1*, 108–111.
22. Nowacka, M.; Witrowa-Rajchert, D.; Ruła, J. Wpływ procesów technologicznych na aktywność przeciwutleniającą i zawartość polifenoli w tkance jabłka. *Postępy Tech. Przetw. Spoż.* **2011**, *2*, 12–15.
23. Lammerskitten, A.; Wiktor, A.; Siemer, C.; Toepfl, S.; Mykhailyk, V.; Gondek, E.; Rybak, K.; Witrowa-Rajchert, D.; Parniakov, O. The effects of pulsed electric fields on the quality parameters of freeze-dried apples. *J. Food Eng.* **2019**, *252*, 36–43. [CrossRef]
24. Wang, L.; Boussetta, N.; Lebovka, N.; Vorobiev, E. Cell disintegration of apple peels induced by pulsed electric field and efficiency of bio-compound extraction. *Food Bioprod. Process.* **2020**, *122*, 13–21. [CrossRef]
25. Janda, K.; Kasprzak, M.; Wolska, J. Witamina C—Budowa, właściwości, funkcje i występowanie. *Pom. J. Life Sci.* **2015**, *61*, 419–425. [CrossRef]
26. Zawada, K. Znaczenie witaminy C dla organizmu człowieka. *Herbalism* **2016**, *1*, 22–34. [CrossRef]
27. Morales-de la Peña, M.; Salvia-Trujillo, L.; Rojas-Graü, M.A.; Martín-Belloso, O. Impact of high intensity pulsed electric field on antioxidant properties and quality parameters of a fruit juice–soymilk beverage in chilled storage. *LWT* **2010**, *43*, 872–881. [CrossRef]

28. Olędzki, R. Potencjał antyoksydacyjny owoców i warzyw oraz jego wpływ na zdrowie człowieka. *Nauk. Inż. Technol.* **2012**, *1*, 44–54.
29. Zych, I.; Krzepiłko, A. Pomiar całkowitej zdolności antyoksydacyjnej wybranych antyoksydantów i naparów metodą redukcji rodnika DPPH. *Chem. Dydakt. Ekol. Metrol.* **2010**, *15*, 51–54.
30. Huang, W.; Feng, Z.; Aila, R.; Hou, Y.; Carne, A.; Ahmed Bekhit, A.E.-D. Effect of pulsed electric fields (PEF) on physico-chemical properties, β-carotene and antioxidant activity of air-dried apricots. *Food Chem.* **2019**, *291*, 253–262. [CrossRef]
31. Yakubu, N.; Amuzat, A.O.; Hamza, R.U. Effect of processing methods on the nutritional contents of bitter leaf (Vernonia amygdalina). *Am. J. Food Nutr.* **2012**, *2*, 26–30. [CrossRef]
32. Adefegha, S.A.; Oboh, G. Enhancement of total phenolics and antioxidant properties of some tropical green leafy vegetables by steam cooking. *J. Food Process. Preserv.* **2011**, *35*, 615–622. [CrossRef]
33. Edwards, C.H.; Rossi, M.; Corpe, C.P.; Butterworth, P.J.; Ellis, P.R. The role of sugars and sweeteners in food, diet and health: Alternatives for the future. *Trends Food Sci. Technol.* **2016**, *56*, 158–166. [CrossRef]
34. Misra, V.; Shrivastava, A.K.; Shukla, S.P.; Ansari, M.I. Effect of sugar intake towards human health. *Saudi J. Med.* **2016**, *1*, 29–36.
35. Begić-Akagić, A.; Spaho, N.; Gaši, F.; Drkenda, P.; Vranac, A.; Meland, M.; Salkić, B. Sugar and organic acid profiles of the traditional and international apple cultivars for processing. *J. Hyg. Eng. Des.* **2014**, *7*, 190–196. [CrossRef]
36. Ticha, A.; Salejda, A.M.; Hyšpler, R.; Matejicek, A.; Paprstein, F.; Zadak, Z. Sugar composition of apple cultivars and its relationship to sensory evaluation. *Żywn. Nauka Technol. Jakość* **2015**, *22*, 137–150. [CrossRef]
37. Fang, T.; Cai, Y.; Yang, Q.; Ogutu, C.O.; Liao, L.; Han, Y. Analysis of sorbitol content variation in wild and cultivated apples. *J. Sci. Food Agric.* **2020**, *100*, 139–144. [CrossRef]
38. Cichowska, J.; Woźniak, Ł.; Figiel, A.; Witrowa-Rajchert, D. The influence of osmotic dehydration in polyols solutions on sugar profiles and color changes of apple tissue. *Period. Polytech. Chem. Eng.* **2020**, *64*, 530–538. [CrossRef]
39. Wojdyło, A.; Lech, K.; Nowicka, P. Effects of Different Drying Methods on the Retention of Bioactive Compounds, On-Line Antioxidant Capacity and Color of the Novel Snack from Red-Fleshed Apples. *Molecules* **2020**, *25*, 5521. [CrossRef]
40. Delgado, T.; Pereira, J.A.; Ramalhosa, E.; Casal, S. Effect of hot air convective drying on sugar composition of chestnut (Castanea sativa Mill.) slices. *J. Food Process. Preserv.* **2018**, *42*, e13567. [CrossRef]
41. Mitrović, O.; Popović, B.; Miletić, N.; Leposavić, A.; Korićanac, A. Effect of drying on the change of sugar content in plum fruits. In Proceedings of the X International Scientific Agriculture Symposium AGROSYM, Jahorina, Bosnia and Herzegovina, 3–6 October 2019; pp. 372–378.
42. Macedo, L.L.; Vimercati, W.C.; da Silva Araújo, C.; Saraiva, S.H.; Teixeira, L.J.Q. Effect of drying air temperature on drying kinetics and physicochemical characteristics of dried banana. *J. Food Process Eng.* **2020**, *43*, e13451. [CrossRef]
43. Rybak, K.; Wiktor, A.; Witrowa-Rajchert, D.; Parniakov, O.; Nowacka, M. The effect of traditional and non-thermal treatments on the bioactive compounds and sugars content of red bell pepper. *Molecules* **2020**, *25*, 4287. [CrossRef] [PubMed]
44. Oziembłowski, M.; Trenka, M.; Czaplicka, M.; Maksimowski, D.; Nawirska-Olszańska, A. Selected Properties of Juices from Black Chokeberry (*Aronia melanocarpa* L.) Fruits Preserved Using the PEF Method. *Appl. Sci.* **2022**, *12*, 7008. [CrossRef]
45. Rybak, K.; Wiktor, A.; Kaveh, M.; Dadan, M.; Witrowa-Rajchert, D.; Nowacka, M. Effect of Thermal and Non-Thermal Technologies on Kinetics and the Main Quality Parameters of Red Bell Pepper Dried with Convective and Microwave–Convective Methods. *Molecules* **2022**, *27*, 2164. [CrossRef]
46. Bureau, S.; Cozzolino, D.; Clark, C.J. Contributions of Fourier-transform mid infrared (FT-MIR) spectroscopy to the study of fruit and vegetables: A review. *Postharvest Biol. Technol.* **2019**, *148*, 1–14. [CrossRef]
47. Lan, W.; Renard, C.M.G.C.; Jaillais, B.; Leca, A.; Bureau, S. Fresh, freeze-dried or cell wall samples: Which is the most appropriate to determine chemical, structural and rheological variations during apple processing using ATR-FTIR spectroscopy. *Food Chem.* **2020**, *330*, 127357. [CrossRef]
48. Nizamlioglu, N.M.; Yasar, S.; Bulut, Y. Chemical versus infrared spectroscopic measurements of quality attributes of sun or oven dried fruit leathers from apple, plum and apple-plum mixture. *LWT-Food Sci. Technol.* **2022**, *153*, 112420. [CrossRef]
49. Iskandar, W.M.E.; Ong, H.R.; Khan, M.R.; Ramli, R. Effect of ultrasonication on alkaline treatment of empty fruit bunch fibre: Fourier Transform Infrared Spectroscopy (FTIR) and morphology study. *Mater. Today Proc.* **2022**, *1092*, 2840–2843. [CrossRef]
50. Ahmed, Z.; Manzoor, M.F.; Hussain, A.; Hanif, M.; Din, Z.u.; Zeng, X.-A. -A. Study the impact of ultra-sonication and pulsed electric field on the quality of wheat plantlet juice through FTIR and SERS. *Ultrason. Sonochem.* **2021**, *76*, 105648. [CrossRef]
51. Niu, D.; Ren, E.-F.; Li, J.; Zeng, X.-A.; Li, S.-L. Effects of pulsed electric field-assisted treatment on the extraction, antioxidant activity and structure of naringin. *Sep. Purif. Technol.* **2021**, *265*, 118480. [CrossRef]
52. Seremet, L.; Botez, E.; Nistor, O.-V.; Andronoiu, D.G.; Mocanu, G.-D. Effect of different drying methods on moisture ratio and rehydration of pumpkin slices. *Food Chem.* **2016**, *195*, 104–109. [CrossRef]
53. AOAC International. *Official Methods of Analysis of AOAC International*, 17th ed.; The Association of Official Analytical Chemists: Gaithersburg, MD, USA, 2002.
54. Kahraman, O.; Malvandi, A.; Vargas, L.; Feng, H. Drying characteristics and quality attributes of apple slices dried by a non-thermal ultrasonic contact drying method. *Ultrason. Sonochem.* **2021**, *73*, 105510. [CrossRef]
55. Vidal-Gutiérrez, M.; Robles-Zepeda, R.E.; Vilegas, W.; Gonzalez-Aguilar, G.A.; Torres-Moreno, H.; López-Romero, J.C. Phenolic composition and antioxidant activity of Bursera microphylla A. Gray. *Ind. Crops Prod.* **2020**, *152*, 112412. [CrossRef]

56. Tian, W.; Chen, G.; Tilley, M.; Li, Y. Changes in phenolic profiles and antioxidant activities during the whole wheat bread-making process. *Food Chem.* **2021**, *345*, 128851. [CrossRef] [PubMed]
57. Zhao, H.; Avena-Bustillos, R.J.; Wang, S.C. Extraction, purification and in vitro antioxidant activity evaluation of phenolic compounds in California Olive Pomace. *Foods* **2022**, *11*, 174. [CrossRef]
58. Simonovska, J.; Škerget, M.; Knez, Ž.; Srbinoska, M.; Kavrakovski, Z.; Grozdanov, A.; Rafajlovsk, V. Physicochemical characterization and bioactive compounds of stalk from hot fruits of *Capsicum annuum* L. Maced. *J. Chem. Chem. Eng.* **2016**, *35*, 199–208. [CrossRef]
59. Spínola, V.; Mendes, B.; Câmara, J.S.; Castilho, P.C. Effect of time and temperature on vitamin C stability in horticultural extracts. UHPLC-PDA vs iodometric titration as analytical methods. *LWT-Food Sci. Technol.* **2013**, *50*, 489–495. [CrossRef]
60. El Kossori, R.L.; Villaume, C.; El Boustani, E.; Sauvaire, Y.; Méjean, L. Composition of pulp, skin and seeds of prickly pears fruit *(Opuntia ficus indica* sp.). *Plant Foods Hum. Nutr.* **1998**, *52*, 263–270. [CrossRef]

molecules

MDPI

Article

Influence of Post-Harvest Processing on Functional Properties of Coffee (*Coffea arabica* L.)

Michał Halagarda * and Paweł Obrok

Department of Food Product Quality, Krakow University of Economics, Ul. Sienkiewicza 5, 30-033 Krakow, Poland
* Correspondence: michal.halagarda@uek.krakow.pl

Abstract: Coffee is one of the most popular beverages worldwide, valued for its sensory properties as well as for its psychoactive effects that are associated with caffeine content. Nevertheless, coffee also contains antioxidant substances. Therefore, it can be considered a functional beverage. The aim of this study is to evaluate the influence of four selected post-harvest coffee fruit treatments (natural, full washed, washed–extended fermentation, and anaerobic) on the antioxidant and psychoactive properties of Arabica coffee. Additionally, the impact of coffee processing on the selected quality parameters was checked. For this purpose, results for caffeine content, total phenolic content (TPC), DPPH assay, pH, titratable acidity, and water content were determined. The results show that natural and anaerobic processing allow the highest caffeine concentration to be retained. The selection of the processing method does not have a significant influence on the TPC or antiradical activity of coffee. The identified differences concerning water content and pH along with lack of significant discrepancies in titratable acidity may have an influence on the sensory profile of coffee.

Keywords: coffee processing; antioxidants; polyphenols; caffeine; full washed; natural; anaerobic; washed–extended fermentation

Citation: Halagarda, M.; Obrok, P. Influence of Post-Harvest Processing on Functional Properties of Coffee (*Coffea arabica* L.). *Molecules* **2023**, *28*, 7386. https://doi.org/10.3390/molecules28217386

Academic Editor: Tristan Richard

Received: 8 September 2023
Revised: 25 October 2023
Accepted: 30 October 2023
Published: 1 November 2023

1. Introduction

Coffee is one of the most popular beverages worldwide. It is highly valued for its sensory properties as well as for its psychoactive effects that are associated with caffeine content. However, coffee is also a good source of antioxidant compounds, mainly phenolics, but also Maillard reaction products that are generated in the roasting process [1–4]. This makes coffee a functional beverage [5]. The roasting process itself has been thoroughly studied in terms of its impact on the antioxidant properties of coffee [2,6–10]. The impact of brewing time and method have also been verified [1,11]. However, post-harvest (pre-roasting) coffee bean preparation steps have an influence on their exact chemical composition [1,9,12] and thus may affect the content of compounds responsible for antioxidant activity and caffeine concentration.

The essence of coffee plant cultivation is extraction of its beans. To obtain them the fruit must be processed. Each of the layers covering the coffee bean when processed may affect the chemical composition of the bean itself. Therefore, coffee makers use different methods of coffee fruit treatment to achieve certain flavors. However, fruit processing not only has an impact on flavor precursors but may also influence the content of functional compounds. The fruits are processed immediately after harvesting to limit the occurrence of unwanted fermentation and reduce contamination. The most common methods include natural and full washed processing [13,14].

The dry method of coffee processing, also known as the natural method, is one of the oldest techniques for processing coffee cherries. In this method, fruits are spread in thin layers and dried in the sun. Depending on the specific region or place of cultivation, the drying stations may look different; some plantations will use the simplest brick terraces for this purpose, while others will use special beds that allow air to flow freely between the fruits, thanks to which drying takes place more evenly. The fruit is turned regularly to avoid mold, rotting, or fermentation. On larger plantations, mechanical drying devices are sometimes used to speed up the process [15]. However, this may have an influence on the coffee quality [16]. The sun-drying process itself takes about 3–4 weeks, until the cherries become hard to the touch, shrink, and take on a dark brown color [17]. When the fruit reaches a moisture content close to 11%, it is considered dry [18]. Then, to achieve higher quality, beans can be stored for some months in special silos. There they rest and the flavor of the beans matures fully [17]. This method ends when the skin and pulp are mechanically removed from the fruit, and the coffee beans are sorted, bagged, and exported to customers [19]. The dry method is used for approximately 90% of the Arabica coffee produced in Brazil, most of the coffee produced in Ethiopia, Haiti and Paraguay, and some Arabica produced in India and Ecuador. Almost all Robusta coffee is processed using this method [17,20,21]. The natural process is common primarily in places where there is no access to water. However, it is not practical in very rainy regions where the humidity is too high or where it often rains during the harvest months. Regardless of the variety and region of cultivation, the dry process primarily gives the coffee a fruity aroma and sweetness, as the drying process enhances the sugar profiles [21].

The full-washed coffee cherry processing method is by far the fastest, probably the most efficient, and therefore the most commonly used method. The first stage of the process is placing freshly picked cherries in a flotation tank filled with water, in which the ripe cherries sink and the unripe cherries—undesirable in harvesting—float to the surface, making it possible to remove them from further stages of processing [17,22]. The next step is to transfer the fruit to the depulper—a device that is responsible for splitting and squeezing the coffee cherry to separate the beans from the outer skin and pulp. After depulping, the coffee beans are still covered with a thin and sticky layer of mucus. Their resistance to pressure is due to the combination of sugars and pectins, and the best way to remove it is a fermentation process [13,15,17]. For this purpose, coffee beans are placed in fermentation tanks filled with water for 6 to 72 h [17,22,23]. During this time, thanks to the activity of enzymes, the pectins contained in the mucus are broken down. The duration of fermentation depends on many factors, including: altitude above sea level, ambient temperature, volume of coffee, and type of beans. The fermentation period has a significant impact on the flavor of the coffee, so knowing when to stop fermentation is a key factor in this process. If fermentation takes too long, undesirable flavors may occur. However, when properly carried out, washed coffee can acquire a characteristic, clean acidity [22]. Furthermore, the presence of bacteria and fungi that are specific for different areas and altitudes may affect the sensory profile of fermented coffee. The most common microorganisms that could be associated with coffee and its processing include: *Debaryomyces*, *Pichia*, *Candida*, *Saccharomyces kluyveri*, *S. Ceverisiae*, *Aspergillus*, *Penicillium*, *Fusarium*, *Trichoderma*, *Lactobacillus*, *Bacillus*, *Arthrobacter*, *Acinetobacter*, and *Klebsiella* [14]. After the fermentation process, the beans are washed again with clean water and left to dry in the sun. As with the dry method, beans can be dried on concrete terraces, tables, or beds. Depending on the prevailing weather conditions, the drying period may last up to 21 days, but is mostly completed between 2 and 15 days [15,22]. Already dried beans covered with parchment have a light beige color. To remove this thin layer, the beans are transferred to a dry mill where the parchment is rubbed off their surface. The final stage is the sorting and packaging of green coffee beans [15]. The taste and aroma of washed coffee can be described as a clean, light-bodied profile with pronounced acidity [24].

The washed–extended fermentation method is used when a given batch of coffee is harvested over several days. Each day, extracted beans are added to the fermentation tanks containing the previous days' harvest. In this method, each subsequent batch of beans increases the pH level in the tank. This inhibits the growth of bacteria present in an acid environment, the activity of which may lead to excessive fermentation of the bed. Such a processing results in more sophisticated flavor profile of the coffee beans [25].

For some time now, in the world of specialty coffees, new coffees produced using new coffee cherry treatment techniques can be seen on the shelves. Experimental trials of new coffee processing methods began when Sasa Sestic won the 2015 World Barista Championship with carbon maceration coffee. Since then, many coffee producers, wanting to increase the cupping score of their coffees, have been trying their hand at producing perfect beans using innovative methods. Of these, anaerobic fermentation has attracted the most interest. The first stage is like that in the wet method. The fresh fruits are placed in a depulper, and then the separated beans are placed together with part of the pulp and outer skin in vacuum-sealed tanks equipped with a non-return valve to stop air from entering the tank. During the process, microorganisms begin to break down glucose molecules, resulting in the release of heat and carbon dioxide, which, being a heavier gas, displaces oxygen from the tank. In an anaerobic environment, bacteria naturally found in coffee cherries produce enzymes that break down sugars into less complex compounds such as organic acids or alcohols. Anaerobic fermentation allows for better control of the process by measuring the pH, sugar content, and temperature inside the tank. Controlling and prolonging the fermentation of coffee causes a change in its chemical composition, something which is also associated with changes in the flavor profile [23,26]. Coffees from this method are characterized by a silky, creamy texture and complex acidity [27].

Coffee studies considering post-harvest (pre-roasting) processing focus mainly on sensory characteristics, e.g., [14,26,28,29]. However, research concerning the impact of coffee fruit processing methods on the functional properties of coffee is scarce. Therefore, the aim of this study was to evaluate the influence of post-harvest coffee fruit treatments on the antioxidant and psychoactive properties of coffee. Furthermore, to the best of the authors' knowledge, the manuscript presents the results of the first research directly comparing four different coffee fruit processing methods (natural, full washed, washed with extended fermentation, and anaerobic). Implementation of the processing on a coffee plantation allowed the quality of coffee cherries and the influence of the actual process conditions to be maintained, along with that of the site-specific microbiota to be reflected in the final characteristics of coffee beans.

2. Results and Discussion

The results of the study (Table 1) show that the water content was significantly higher in coffee beans from full-washed (2.68 ± 0.03 g/100 g) and anaerobic (2.63 ± 0.02 g/100 g) processing than in beans processed with the natural method (2.24 ± 0.03 g/100 g). These differences may be a consequence of the strong dehydration resulting from the sun drying used in natural processing. Other methods involve immersing coffee in water. The level of the retained water in natural processed coffee is consistent with the results obtained by Baggenstoss et al. [30] indicating that natural processed coffee after the roasting process contains 2.3 g/100 g of water.

The pH value was significantly higher in washed–extended fermentation coffees (5.08 ± 0.03) than in coffee beans processed using the anaerobic method (4.98 ± 0.02). Although both methods rely on fermentation, in the case of anaerobic coffees the process is carried out to a specific pH value of fermenting mass. Still, due to the continuous addition of successive portions of fresh beans to the fermentation tank, coffee beans from prolonged fermentation treatment tend to have a higher pH. This is connected with additions of extra amounts of sugars when dosing subsequent portions of beans. They are a product of the degradation of organic compounds and subsequently act as a nutrient medium for microorganisms to produce acids and alcohols [25]. The average titratable acidity of the tested coffees range from 18.5 to 18.83 with no significant differences. In view of this fact, it can be confirmed that the processing conditions influence the acidic profile of the coffee, which in turn has an influence on the sensory parameters of the beverage.

Table 1. Selected functional and quality parameters of tested coffee samples representing different post-harvest (pre-roasting) processing methods.

Parameter		Sample (n = 3)				p
		Washed–Extended Fermentation—A	Full Washed—B	Natural—C	Anaerobic—D	
Water (g/100 g)	mean ± SD	2.46 ± 0.02	2.68 ± 0.03	2.24 ± 0.03	2.63 ± 0.02	p = 0.019 * B,D > C
pH	mean ± SD	5.08 ± 0.03	5.04 ± 0.02	5.04 ± 0.01	4.98 ± 0.02	p = 0.019 * A > D
Titratable acidity (mol L^{-1} NaOH per 100 g)	mean ± SD	18.5 ± 0.5	18.83 ± 0.76	18.83 ± 0.76	18.83 ± 0.29	p = 0.811
Caffeine (g/100 g)	mean ± SD	1.672 ± 0.010	1.666 ± 0.009	1.758 ± 0.008	1.758 ± 0.014	p = 0.04 * D > A,B,C > B
TPC (mg GAE/g)	mean ± SD	38.81 ± 1.88	37.51 ± 0.78	37.93 ± 2.21	40.12 ± 1.81	p = 0.273
DPPH IC_{50} (μg mL^{-1})	mean ± SD	21.59 ± 1.8	19.37 ± 0.18	20.16 ± 2.57	23.54 ± 1.3	p = 0.129

*—indicates statistical significance.

The presence of caffeine, which has a centrally excitatory effect resulting from its structure (Figure 1), makes coffee a functional beverage. The more caffeine is in the bean, the more will diffuse into the beverage and the higher the stimulating effect for a coffee consumer. In this study, each of the coffees contained the amount of caffeine typical for the *C. arabica* (0.7–1.7 g/100) [31]. Lower values were noted by Eshetu et al. [32] for full- washed sun-dried Ethiopian Arabica (1.06–1.28 g/100 g, depending on variety). The results of this study indicated that the fruit-processing-influenced caffeine content in the roasted beans. The caffeine concentration in anaerobic coffees (1.758 ± 0.014 g/100 g) was significantly higher than in washed–extended fermentation coffees (1.672 ± 0.010 g/100 g) and full-washed coffees (1.666 ± 0.009 g/100 g). Moreover, coffee beans processed with the use of the natural method (1.758 ± 0.008 g/100 g) contained significantly more of this functional compound than full-washed coffees. This is in agreement with the findings of Guyot et al. [33], who detected small losses of caffeine (3%) as a result of the soaking phase in the wet process in comparison to the natural process. In contrast, Mintesnot and Dechassa [34] did not report any difference between caffeine content in coffees processed with the dry and wet method.

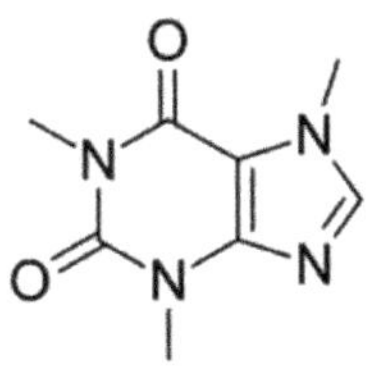

Figure 1. Chemical structure of caffeine.

The functional properties of coffee are also connected with the content of antioxidant compounds, mostly polyphenols. The total phenolic content (TPC) ranges between 37.51 and 40.12 mg GAE/g. It seems to be typical for coffees from Indonesia, as those values are close to that determined by Jeszka-Skowron et al. [35]—38.5 mg GAE/g. At the same time, the determined TPC was higher than the values noted by Odžaković et al. [2] for Brazilian Arabica of three roasting degrees (23.66–32.78 mg GAE/g), as well as those for Chinese (36.17 mg GAE/g) and Thai (33.76 mg GAE/g) coffees studied by Cheong et al. [5] and was lower than the TPC determined by Bobková et al. [6] for light roasted coffees of

Colombian, Indian, and Ethiopian origin (38.34–59.79 mg GAE/g) as well as by Cheong et al. [5] for Indonesian coffee (48.51 mg GAE/g).

The fruit processing itself did not have significant influence on TPC. Correspondingly, there were no significant differences in antiradical activity among the tested coffees. Nevertheless, Haile Bae and Kang [36] showed that, when considering light roasted coffees, wet processed coffee exhibits better antiradical activity against DPPH and higher TPC than the dry processed equivalent. However, surprisingly, in their research, there were no differences in DPPH inhibition for medium and dark roasted coffees, whereas discrepancies in TPC were noted for medium roasted beans.

The values of DPPH IC_{50} obtained in this study seem to be typical and close to those measured by Vignoli, Bassoli, and Benassi [3] for light, medium, and dark roasted Arabica extract (16.11–24.92 $\mu g\ mL^{-1}$).

3. Materials and Methods

3.1. Coffee Samples

Research samples of Arabica (*Coffea arabica* L.) S-795 cultivar beans were acquired from a plantation in the village of Beiposo, located in the Indonesian island of Flores. Coffee from Flores is of high quality thanks to the know-how of the local women who cultivate it. Only fully ripe cherries are harvested. Furthermore, they are grown in the fertile lands of the Bajawa Plateau, located between two volcanoes, at altitudes from 1300 up to 1600 m above sea level. Due to the favorable growing conditions and care for the quality of harvested fruit, the coffee from Beiposo is mild and has a balanced acidity and bitterness. Therefore, it is considered to be a specialty coffee [37].

The coffee cherries used for samples were processed using four different methods:

- Natural: coffee cherries were dried under the sun for 30 days.
- Full washed: coffee cherries were washed and then fermented for 24 h inside the tank with filled with water. The external temperature during fermentation was kept between 11 °C and 20 °C. The fermentation process was finished when the pH level reached 4.4.
- Washed–extended fermentation: Coffee cherries were washed and fermented 5 times for 24 h. Each fermentation process was followed by washing. The external temperature during fermentation was maintained between 12 and 18 °C.
- Anaerobic: The coffee cherries were fermented inside vacuum-sealed containers for 7 days, until the pH level reached 4.2.

The bean roasting process was performed using an SR3 Coffee Roaster (Coffed, Piła, Poland) equipped with temperature sensors placed in the roasted coffee and exhaust fumes. The process parameters were monitored using Artisan 2.4.6 software. The coffee beans were roasted to the same, light degree. For the purpose of the analyses the coffees were ground with the use of a Retsch GM 200 (Haan, Germany) mill.

3.2. Water

Water content was determined with the use of the oven-drying method according to the PN-A-76100:2009 Standard [38]. Samples of 5 g were dried at 103 ± 2 °C for 2 h. Afterwards they were cooled to room temperature in a desiccator and weighed. The procedure was repeated until a constant weight of dried sample was reached.

3.3. Titratable Acidity and pH

Titratable acidity and pH were measured according to the AOAC methodology [39]. The samples of coffee beans were milled. The samples of 2 g of coffee were homogenized with 100 mL water, kept in a water bath (60 °C) for 30 min, and cooled to room temperature. The pH values were measured using a SevenCompact digital pH meter (Mettler Toledo, Greifensee, Switzerland) equipped with an InLAb Expert Pro-ISM (Mettler Toledo, Greifensee, Switzerland) electrode. The titratable acidity was measured with the use of 0.1 $mol\ L^{-1}$ NaOH and to pH 8.2.

3.4. Caffeine Content

Caffeine concentration was determined according to the methodology proposed by the ISO 20481:2008 Standard [40]. The sample of 1 g of milled coffee was mixed with 5 g of magnesium oxide and 200 mL of water, heated to 90 °C and kept at that temperature for 20 min. Then the solution was cooled down to room temperature and filled up with water to 250 mL. After filtration through a 0.22 μm PTFE filter, the sample was ready for HPLC analysis.

The HPLC analysis was performed with the use of a UltiMate 3000 RSLC (Thermo Fisher Scientific, Waltham, MA, USA) equipped with an Accucore XL C18 column (4.6 × 150 mm, 4 μm particle) and a DAD detector. The mobile phase (methanol in water 24% *v*/*v*) flow rate of 1.0 mL/min and isocratic elution were used. UV detection was performed at 272 nm.

3.5. Total Phenolic Content

Determination of the total phenolic content (TPC) was performed with the Folin-Ciocalteu method [3]. A sample of 0.1 mL of coffee solution (3 mg/mL) was diluted with deionized water to 7.5 mL. Subsequently, 0.3 mL of 0.9 M Folin-Ciocalteu reagent and 1 mL of 20% Na_2CO_3 solution were added, and the volume was filled up to 10 mL with deionized water. The solutions were kept at room temperature for one hour, and then the absorbance at 765 nm was measured with a Nanocolor UV/VIS II spectrophotometer (MACHEREY-NAGEL, Düren, Germany). Standard solutions of gallic acid were used to create the calibration curve. The results were therefore expressed in grams of gallic acid equivalents per 1 g of coffee.

3.6. Antiradical Activity against DPPH

The DPPH assay was performed following the methodology presented by Vignoli, Bassoli, and Benassi [3]. In brief, a 10 μL of sample solution (3 mg/mL) was mixed with 1 mL of 0.1 M acetate buffer (pH 5.5), 1 mL of ethanol, and 0.5 mL of 250 μM ethanolic DPPH solution. The control solution was prepared without using the coffee solution. The blank solution was prepared as above, except or the DPPH solution. The absorbance was measured with a Nanocolor UV/VIS II spectrophotometer at 517 nm and after 10 min of solution preparation. The inhibition ratio was calculated with the use of the following equation [41]:

$$\text{Inhibition ratio } (\%) = ((\text{Ac} - \text{As})/\text{Ac}) \times 100$$

where

Ac—absorbance of a control
As—absorbance of a sample

The IC50 was determined using regression model following the procedure of Shimamura et al. [41].

3.7. Statistical Analysis

The data from the analysis, that was performed in triplicate, underwent statistical evaluation. The comparison of quantitative variables in the four groups was performed using the Kruskal–Wallis test. After detecting statistically significant differences, post hoc analysis was performed using the Dunn's test to identify statistically significant groups. The analysis adopted a significance level of 0.05 and was performed in R software, version 4.1.0 [42].

4. Conclusions

The study results show that the choice of the method of coffee cherry processing to some extent affects the functional properties of the coffee beverage. Natural and anaerobic methods allow the highest caffeine concentration to be retained, indicating that coffee beverages obtained from beans subjected to those processing methods are characterized by better properties of central nervous system stimulation. Nevertheless, the selection of processing method does not have a significant influence on the total phenolic content and antiradical activity of coffee. However, it may still affect the phenolic profile of the coffee, and further research in that respect therefore needs to done.

The identified differences concerning water content and pH, along with the lack of significant discrepancies in titratable acidity, may correspond to the influence on the taste and aroma of coffee. Therefore, further research concerning the influence of the indicated processing methods on the sensory profile of coffee is needed.

The research results add new data to knowledge about the influence of coffee fruit processing on the final functional characteristics of coffee. However, there are some limitations to this study. They include: specific place of coffee plant cultivation, which may affect the chemical composition of coffee fruits and fermentation microbiota; and processing conditions, which may vary slightly between different producers and time of the year. Therefore, further research is needed to confirm the outcomes of this study.

Author Contributions: Conceptualization, M.H. and P.O.; methodology, M.H.; formal analysis, M.H.; investigation, P.O. and M.H.; resources, P.O.; writing—preparation of original draft, M.H. and P.O.; writing—review and editing, M.H.; supervision, M.H.; project administration, M.H.; funding acquisition, M.H. All authors have read and agreed to the published version of the manuscript.

Funding: The publication was financed from a subsidy granted to the Krakow University of Economics—Project number: ADN/WOFP/2023/000065.

Institutional Review Board Statement: Not applicable.

Informed Consent Statement: Not applicable.

Data Availability Statement: The data presented in this study are available on request from the corresponding author.

Acknowledgments: The authors would like to thank the farmers form the village of Beiposo, Indonesia for their hard work and help with preparation of research material. The authors would also like to thank Ayrton Wibowo and Kinga Wojtczak (owners of the company Podkawa and the Indonesian coffee plantation) for their assistance in the importing and preparing of the coffee beans for the study.

Conflicts of Interest: The authors declare no conflict of interest.

References

1. Kim, C.H.; Park, S.J.; Yu, J.S.; Lee, D.Y. Interactive effect of post-harvest processing method, roasting degree, and brewing method on coffee metabolite profiles. *Food Chem.* **2022**, *397*, 133749. [CrossRef] [PubMed]
2. Odžakovic, B.; Džinic, N.; Kukric, Z.; Grujic, S. Effect of roasting degree on the antioxidant activity of different Arabica coffee quality classes. *Acta Sci. Pol. Technol. Aliment.* **2016**, *15*, 409–417. [CrossRef] [PubMed]
3. Vignoli, J.A.; Bassoli, D.G.; Benassi, M.T. Antioxidant Activity, Polyphenols, Caffeine and Melanoidins in Soluble Coffee: The Influence of Processing Conditions and Raw Material. *Food Chem.* **2011**, *124*, 863–868. [CrossRef]
4. Gómez-Ruiz, J.Á.; Leake, D.S.; Ames, J.M. In Vitro Antioxidant Activity of Coffee Compounds and Their Metabolites. *J. Agric. Food Chem.* **2007**, *55*, 6962–6969. [CrossRef]
5. Cheong, M.W.; Tong, K.H.; Ong, J.J.M.; Liu, S.Q.; Curran, P.; Yu, B. Volatile composition and antioxidant capacity of Arabica coffee. *Food Res. Int.* **2013**, *51*, 388–396. [CrossRef]
6. Bobková, A.; Hudáček, M.; Jakabová, S.; Belej, L.; Capcarová, M.; Curlej, J.; Bobko, M.; Árvay, J.; Jakab, I.; Čapla, J.; et al. The effect of roasting on the total polyphenols and antioxidant activity of coffee. *J. Environ. Sci. Health Part B* **2020**, *25*, 2574–2588. [CrossRef]
7. Yashin, A.; Yashin, Y.; Wang, J.Y.; Nemzer, B. Antioxidant and Antiradical Activity of Coffee. *Antioxidants* **2013**, *2*, 230–245. [CrossRef]
8. Wei, F.; Furihata, K.; Koda, M.; Hu, F.; Miyakawa, T.; Tanokura, M. Roasting process of coffee beans as studied by nuclear magnetic resonance: Time course of changes in composition. *J. Agric. Food Chem.* **2012**, *60*, 1005–1012. [CrossRef]
9. Esquivel, P.; Jiménez, V.M. Functional properties of coffee and coffee by-products. *Food Res. Int.* **2012**, *46*, 488–495. [CrossRef]
10. Del Castillo, M.D.; Ames, J.M.; Gordon, M.H. Effect of roasting on the antioxidant activity of coffee brews. *J. Agric. Food Chem.* **2002**, *50*, 3698–3703. [CrossRef]
11. Ludwig, I.A.; Sanchez, L.; Caemmerer, B.; Kroh, L.W.; de Peña, M.P.; Cid, C. Extraction of coffee antioxidants: Impact of brewing time and method. *Food Res. Int.* **2012**, *48*, 57–64. [CrossRef]
12. Sunarharum, W.B.; Yuwono, S.S.; Pangestu, N.B.S.W.; Nadhiroh, H. Physical and sensory quality of Java Arabica green coffee beans. In Proceedings of the 1st International Conference on Green Agro-Industry and Bioeconomy (ICGAB 2017), Malang, Indonesia, 24–25 October 2017.

13. Figueroa Campos, G.A.; Sagu, S.T.; Saravia Celis, P.; Rawel, H.M. Comparison of Batch and Continuous Wet-Processing of Coffee: Changes in the Main Compounds in Beans, By-Products and Wastewater. *Foods* **2020**, *9*, 1135. [CrossRef]
14. Pereira, L.L.; Guarçoni, R.C.; Pinheiro, P.F.; Osório, V.M.; Pinheiro, C.A.; Moreira, T.R.; ten Caten, C.S. New Propositions about Coffee Wet Processing: Chemical and Sensory Perspectives. *Food Chem.* **2020**, *310*, 125943. [CrossRef]
15. Clarke, R.J. Green Coffee Processing. In *Coffee*; Clifford, M.N., Willson, K.C., Eds.; Springer: Boston, MA, USA, 2012.
16. Oliveira, P.D.; Biaggioni, M.A.M.; Borém, F.M.; Isquierdo, E.P.; Vaz Damasceno, M.D.O. Quality of natural and pulped coffee as a function of temperature changes during mechanical drying. *Coffee Sci.* **2018**, *13*, 415–425. [CrossRef]
17. Alves, R.C.; Rodrigues, F.; Nunes, M.A.A.; Vinha, A.F.; Oliveira, M.B.P.P. State of the art in coffee processing by-products. In *Handbook of Coffee Processing By-Products: Sustainable Applications*; Galanakis, C., Ed.; Academic Press: Cambridge, MA, USA; Elsevier: Amsterdam, The Netherlands, 2017; pp. 1–26.
18. Tesfa, M. Review on Post-Harvest Processing Operations Affecting Coffee (*Coffea arabica* L.) Quality in Ethiopia. *J. Environ. Earth Sci.* **2019**, *9*, 30–39.
19. de Melo Pereira, G.V.; de Carvalho Neto, D.P.; Magalhães Júnior, A.I.; Vásquez, Z.S.; Medeiros, A.B.P.; Vandenberghe, L.P.S.; Soccol, C.R. Exploring the impacts of postharvest processing on the aroma formation of coffee beans—A review. *Food Chem.* **2019**, *272*, 441–452. [CrossRef] [PubMed]
20. Silva, C.F.; Batista, L.R.; Abreu, L.M.; Dias, E.S.; Schwan, R.F. Succession of bacterial and fungal communities during natural coffee (*Coffea arabica*) fermentation. *Food Microbiol.* **2008**, *25*, 951–957. [CrossRef] [PubMed]
21. Duarte, S.M.S.; Abreu, C.M.P.; Menezes, H.C.; Santos, M.H.; Gouvêa, C.M.C.P. Effect of processing and roasting on the antioxidant activity of coffee brews. *Food Sci. Technol.* **2005**, *25*, 387–393. [CrossRef]
22. Rotta, N.M.; Curry, S.; Han, J.; Reconco, R.; Spang, E.; Ristenpart, W.; Donis-Gonzalez, I.R. A comprehensive analysis of operations and mass flows in postharvest processing of washed coffee. *Resour. Conserv. Recycl.* **2021**, *170*, 105554. [CrossRef]
23. Batista da Mota, M.C.; Dias, N.D.N.; Schwan, R.F. Impact of microbial selfinduced anaerobiosis fermentation (SIAF) on coffee quality. *Food Biosci.* **2022**, *47*, 101640. [CrossRef]
24. Mazzafera, P.; Robinson, S.P. Characterization of polyphenol oxidase in coffee. *Phytochemistry* **2000**, *55*, 285–296. [CrossRef] [PubMed]
25. Orchard, M. Experimental Fermentation in Coffee Processing. 2019. Available online: https://plotroasting.com/blogs/news/experimental-fermentation-in-coffee-processing (accessed on 24 August 2023).
26. Partida-Sedas, J.G.; Muñoz Ferreiro, M.N.; Vázquez-Odériz, M.L.; Romero-Rodríguez, M.Á.; Pérez-Portilla, E. Influence of the postharvest processing of the "Garnica" coffee variety on the sensory characteristics and overall acceptance of the beverage. *J. Sens. Stud.* **2019**, *34*, e12502. [CrossRef]
27. da Silva Vale, A.; Balla, G.; Rodrigues, L.R.S.; de Carvalho Neto, D.P.; Soccol, C.R.; de Melo Pereira, G.V. Understanding the Effects of Self-Induced Anaerobic Fermentation on Coffee Beans Quality: Microbiological, Metabolic, and Sensory Studies. *Foods* **2023**, *12*, 37. [CrossRef]
28. Shen, X.; Zi, C.; Yang, Y.; Wang, Q.; Zhang, Z.; Shao, J.; Zhao, P.; Liu, K.; Li, X.; Fan, J. Effects of Different Primary Processing Methods on the Flavor of *Coffea arabica* Beans by Metabolomics. *Fermentation* **2023**, *9*, 717. [CrossRef]
29. Sunarharum, W.B.; Yuwono, S.; Nadhiroh, H. Effect of different post-harvest processing on the sensory profile of Java Arabica coffee. *Adv. Food Sci. Sustain. Agric. Agroind. Eng.* **2018**, *1*, 9–13. [CrossRef]
30. Baggenstoss, J.; Rainer, P.; Escher, F. Water content of roasted coffee: Impact on grinding behaviour, extraction, and aroma retention. *Eur. Food Res. Technol.* **2008**, *227*, 1357–1365. [CrossRef]
31. De Paula, J.; Farah, A. Caffeine Consumption through Coffee: Content in the Beverage, Metabolism, Health Benefits and Risks. *Beverages* **2019**, *5*, 37. [CrossRef]
32. Eshetu, E.F.; Tolassa, K.; Mohammed, A.; Berecha, G.; Garedew, W. Effect of processing and drying methods on biochemical composition of coffee (*Coffea arabica* L.) varieties in Jimma Zone, Southwestern Ethiopia. *Cogent Food Agric.* **2022**, *8*, 2121203. [CrossRef]
33. Guyot, B.; Manez, J.C.; Perriot, J.J.; Giron, J.; Villain, L. Influence de l'altitude et de l'ombrage sur la qualité des cafés arabica. *Plant Rech. Dév.* **1996**, *3*, 272–280.
34. Mintesnot, A.; Dechassa, N. Effect of altitude, shade, and processing methods on the quality and biochemical composition of green coffee beans in Ethiopia. *East Afr. J. Sci.* **2018**, *12*, 87–100.
35. Jeszka-Skowron, M.; Sentkowska, A.; Pyrzyńska, K.; De Peña, M.P. Chlorogenic acids, caffeine content and antioxidant properties of green coffee extracts: Influence of green coffee bean preparation. *Eur. Food Res. Technol.* **2016**, *242*, 1403–1409. [CrossRef]
36. Mesfin, H.; Bae, H.M.; Kang, W.H. Comparison of the Antioxidant Activities and Volatile Compounds of Coffee Beans Obtained Using Digestive Bio-Processing (Elephant Dung Coffee) and Commonly Known Processing Methods. *Antioxidants* **2020**, *9*, 408.
37. Yusibani, E.; Woodfield, P.L.; Rahwanto, A.; Surbakti, M.S.; Rajibussalim, R.; Rahmi, R. Physical and Chemical Properties of Indonesian Coffee Beans for Different Postharvest Processing Methods. *J. Eng. Technol. Sci.* **2023**, *55*, 1–11. [CrossRef]
38. *PN-A-76100:2009*; Roasted Coffee—Requirements and Test Methods. PKN: Warszawa, Poland, 2009.
39. AOAC. *Official Methods of Analysis*, 18th ed.; International Association of Official Analytical Chemists: Washington, DC, USA, 2010.
40. *ISO 20481:2008*; Coffee and Coffee Products. Determination of the Caffeine Content Using High Performance Liquid Chromatography (HPLC). Reference Method; ISO: Geneva, Switzerland, 2008.

41. Shimamura, T.; Sumikura, Y.; Yamazaki, T.; Tada, A.; Kashiwagi, T.; Ishikawa, H.; Matsui, T.; Sugimoto, N.; Akiyama, H.; Ukeda, H. Applicability of the DPPH Assay for Evaluating the Antioxidant Capacity of Food Additives—Inter-Laboratory Evaluation Study. *Anal. Sci.* **2014**, *30*, 717–721. [CrossRef] [PubMed]
42. R Core Team R. *A Language and Environment for Statistical Computing*; R Foundation for Statistical Computing: Vienna, Austria, 2013; Available online: https://www.R-project.org/ (accessed on 10 July 2023).

molecules

MDPI

Article

Peptides as Potentially Anticarcinogenic Agent from Functional Canned Meat Product with Willow Extract

Karolina M. Wójciak [1], Paulina Kęska [1,*], Monika Prendecka-Wróbel [2] and Karolina Ferysiuk [1]

[1] Department of Animal Food Technology, Faculty of Food Science and Biotechnology, University of Life Sciences in Lublin, Skromna 8, 20-704 Lublin, Poland
[2] Chair and Department of Human Physiology, Medical University of Lublin, Radziwiłłowska 11, 20-080 Lublin, Poland
* Correspondence: paulina.keska@up.lublin.pl; Tel.: +48-81-4623340; Fax: +48-81-4623345

Abstract: The aim of the study was to demonstrate canned pork as a functional meat product due to the presence of potentially anti-cancer factors, e.g., (a) bioactive peptides with potential activity against cancer cells; (b) lowering the content of sodium nitrite and with willow herb extract. In silico (for assessing the anticancer potential of peptides) and in vitro (antiproliferation activity on L-929 and CT-26 cell lines) analysis were performed, and the obtained results confirmed the bioactive potential against cancer of the prepared meat product. After 24 h of incubation with peptides obtained from meat product containing lyophilized herb extract at a concentration of 150 mg/kg, the viability of both tested cell lines was slightly decreased to about 80% and after 72 h to about 40%. On the other hand, after 72 h of incubation with the peptides obtained from the variant containing 1000 mg/kg of freeze-dried willow herb extract, the viability of intestinal cancer cells was decreased to about 40%, while, by comparison, the viability of normal cells was decreased to only about 70%.

Keywords: nitrite replacement; health; *E. angustifolium*; cell lines

Citation: Wójciak, K.M.; Kęska, P.; Prendecka-Wróbel, M.; Ferysiuk, K. Peptides as Potentially Anticarcinogenic Agent from Functional Canned Meat Product with Willow Extract. *Molecules* **2022**, *27*, 6936. https://doi.org/10.3390/molecules27206936

Academic Editors: Andrea Salvo and Adele Papetti

Received: 4 September 2022
Accepted: 12 October 2022
Published: 16 October 2022

1. Introduction

Meat is rich in essential nutrients, including protein, fat, fatty acids, vitamins, and minerals, and is, hence, considered as an important nutritional component of the human diet. However, it may also contain harmful chemical compounds, such as heterocyclic amines, polycyclic aromatic hydrocarbons, nitrates, and N-nitroso compounds, which have been shown to increase the risk of cancer-related morbidity. This poses a significant challenge to producers and consumers, as malignant neoplasms, which are one of the most common noncommunicable diseases, have been increasing worldwide. According to the report of the International Agency for Research on Cancer (IARC) published in 2015 [1], more than 800 studies from the last 20 years have pointed out the relationship between meat consumption and the incidence of cancer. As its regular consumption has been linked with cancer development, and in line with the IARC opinion, red meat has been identified as potentially carcinogenic and classified under Group 2A. Processed meat has been found to be a much more potent carcinogen and is, therefore, classified as Group 1, which includes factors with compelling evidence of human carcinogenicity. As a result, the World Cancer Research Fund (WCRF) recommended that the consumption of red meat should be limited to <500 g per week and processed meat should be avoided as much as possible [2]. The relationship between the consumption of red meat and processed meat and cancer development has been widely described in the literature. Red meat consumption has been shown to increase the risk of bladder cancer, breast cancer, colorectal cancer, endometrial cancer, esophageal cancer, stomach cancer, lung cancer, nasopharyngeal cancer, non-Hodgkin's lymphoma, and overall cancer mortality. Similarly, the consumption of processed meat has been associated with a higher risk of bladder cancer, breast cancer, colorectal cancer, esophageal cancer, gastric cancer, nasopharyngeal cancer, non-Hodgkin's lymphoma, mouth and oropharyngeal cancer, prostate cancer, and overall cancer mortality [3].

Colorectal cancer, in particular, is a major concern, accounting for 10% of all cancer diagnoses (ranking third among the most commonly diagnosed cancers) and 9.4% of all cancer deaths in 2020 (ranking second for cancer mortality). Singh and Fraser [4] indicated that individuals who consumed pork or beef once a week had a 90% higher risk of colon cancer, in comparison to those who were on a meat-free diet. English et al. [5] proved that participants who consumed red meat almost every day (>6.5 times a week) had a 40% higher risk of colorectal cancer and 130% higher risk of rectal cancer, compared to those who consumed less meat (<3 times per week). Furthermore, Norat et al. [6] highlighted that consumption of 100 g of red meat everyday increased the risk of colorectal cancer by 21%. Currently, a wide range of treatment options are available for colorectal cancer. However, these often lead to trauma and side effects, cause damage to healthy cells, and are expensive. Thus, there is a search for new treatments for colon cancer. Recent studies show that functional foods loaded with natural compounds can prevent the risk of developing colon cancer. Therefore, in this study, a meat product with functional properties was developed, in which (a) the content of sodium nitrite was reduced to 50 mg/kg to reduce the risk of formation of carcinogenic compounds and (b) phytochemicals in the form of extracts rich in polyphenols obtained from willow herb, which is known for its strong bioactive properties, including as an antioxidant, anti-inflammatory, anti-cancer. The strategy to halve sodium nitrate is an action to reduce its potentially carcinogenic nature. Eliminating sodium nitrate or reducing the recommended amount to half (from 100 mg/kg to 50 mg/kg) is associated with losses in the quality and safety of canned pork [7], but it turned out to be possible by fortifying low-nitrite canned meat with a lyophilized extract from willow herbs [8]. In addition, our previous studies [9] have shown that the addition of E. angustifolium extracts has an impact on the peptide profile and their various bioactivities. In particular, the addition of herb at the level of 150 mg/kg enhanced the antioxidant effect of peptides from canned pork. Indeed, meat and meat products are a good source of biologically active peptides with various health-promoting effects, including anticarcinogenic properties, and thus can contribute to improving the effectiveness of anticancer therapy. The activity of such peptides may be related to their ability to inhibit angiogenesis and initiate necrosis or apoptosis, distort peptides needed for tumor cell proliferation, delay the activity of enzymes involved in tumor development, or increase immunity against tumor cells [10]. Protein hydrolysates or peptides derived from milk, eggs, fish, crabs, shrimps, sea cucumber, oysters, clams, chlorella (algae), spirulina, rice, soybeans, corn, beans, chickpeas, and rapeseed have been proven to have anticarcinogenic potential [11–23]. Furthermore, previous studies have shown that sausages and whole-muscle products (loins, sirloins, neck) contain biologically active peptides with primarily antioxidant and/or antihypertensive (in vitro) and anticarcinogenic (in silico) properties. Thus far, no study has been performed that investigates the anticarcinogenic potential of canned sterilized pork.

It should be noted that proteolytic digestion of food proteins may involve the formation of anticarcinogenic peptides, which can contribute to strengthening the body's natural defense barrier. Moreover, such food proteins have protective effects against various types of pathogens, as indicated by scientific reports on the anticancer and antimicrobial effects of food peptides [21,24,25]. On the other hand, enzymatic hydrolysis can lead to a loss of the bioactive effect of the peptides due to too extensive degradation of the peptide chains. Therefore, in this study, the anticancer potential of peptides was determined in two stages: (a) the peptides obtained from canned pork were identified by spectrophotometric methods, and their anticarcinogenic effect was analyzed in silico, and (b) in vitro analysis of the antiproliferative properties of protein and peptide hydrolysates with enzymes of the gastrointestinal tract (pepsin and pancreatin) using L-929 and CT-26 cell lines.

The aim of this study was to demonstrate the functional nature of canned meat having reduced the content of sodium nitrite and to evaluate the potentially anticarcinogenic properties of canned meat peptides containing an aqueous, freeze-dried extract of willow herb at an amount of 150 mg/kg (E1) or 1000 mg/kg (E2). The potential of peptides as anticancer agent in the products were evaluated after 6 months of refrigerated (4 °C) storage based on in silico and in vitro analysis.

2. Materials and Methods

2.1. Preparation of *Epilobium angustifolium* L. extract and Lyophilization

Crushed, dried leaves of willow herb were used (herbal enterprise "Polskie Zioła") to prepare water extracts. First, the plant material was immersed in redistilled water (90 °C ± 5 °C) in a ratio of 1:10 (g of sample: mL of solvent). The infusions were placed in an ultrasonic bath for extraction using Sonic 6D equipment (Polsonic Palczynki Sp. J., Poland). The ultrasound frequency was set as 40 kHz and sound intensity as 320 W/cm^2, temperature as 30 °C, and extraction time as 10 min. After 30 min of extraction of the sample with water, the extracts were filtered through a funnel with filter paper. The filter cake was re-extracted under the same conditions, and the collected filtrates were frozen. Then, the frozen extracts were lyophilized using a Labconco FreeZone freeze dryer (Kansas City, MO, USA) for 72 h at −80 °C and 0.04 mbar. The lyophilized material was stored in tightly closed containers.

2.2. Preparation of Canned Pork Meat

Canned pork used in the study was prepared from pork dewlap and pork shoulder in a ratio of 2:8. The muscle was excised in a slaughterhouse (Zakład Mięsny Wasąg sp. J., organic certificate no. PL-EKO-093027/18) after a 48 h slaughter. The raw meat was precrushed with a knife and finely minced using a meat grinder (Ø 5 mm). The minced meat was divided into three individual variants, to which sodium chloride (2%), water (5%), or sodium nitrite (50 or 100 mg/kg) was added. The test variants were as follows: variant added with sodium nitrite at an amount of 50 mg/kg of stuffing and variant added with lyophilized willow herb extract at an amount of 150 mg/kg (E1) or 1000 mg/kg (E2). The control sample (C) contained twice the amount of sodium nitrate (100 mg/kg) and no plant extract. The remaining ingredients were added in equal amounts. The meat stuffing, thus, prepared was mixed using a universal machine-type KU2-3E device (Mesko-AGD, Poland; 5 min/variant) and transferred to metal cans. The batch weight was approximately 250 g. The metal cans were closed and placed in a vertical steam sterilizer (TYP-AS2) and subjected to sterilization (121 °C). The sterilized cans were cooled with cold water and stored in a refrigerator (4 °C) for 180 days.

2.3. Extraction and Spectrometric Identification of Peptides

Peptides were isolated from canned meat samples using the method proposed by Escudero et al. [26]. The resulting supernatant was dried in a vacuum evaporator (Rotavapor R-215, Büchi Labortechnik AG, Flawil, Switzerland). The dried extract was dissolved in a 0.01 N HCl solution, and the mixture was filtered through a 0.45 µM nylon membrane filter (Millipore, Bedford, MA, USA) and stored at −60 °C until further use. Peptide analysis was performed by liquid chromatography coupled with tandem electrospray mass spectrometry, as described in our previous study.

2.4. Sample Extraction and Gastrointestinal Digestion

After 180 days of storage, the meat samples were treated with proteases (pepsin and pancreatin), which are equivalent to human digestive enzymes. Briefly, proteins were obtained from samples by homogenizing them (10 mg) with a phosphate-buffered solution (15.6 mM Na_2HPO_4, 3.5 mM KH_2PO_4, pH 7.5) in a homogenizer (IKA T25, Staufen, Germany) at 8000 rpm for 5 min in a bath. The obtained homogenate was centrifuged at 5000× *g* for 20 min at 4 °C. The protein content of the extracts was determined by the Bradford method, using bovine serum albumin as reference. The resulting supernatant was hydrolyzed, as described by Escudero et al. [26]. Before hydrolysis, the pH of the extracts was adjusted to 2.0–2.5 with 1 M HCl. Then, pepsin was added (enzyme:protein solution 1:100) to the extracts for 2 h at 37 °C under constant stirring. The solution was neutralized to pH 7.0 (with 1 M NaOH), and then pancreatin was added (enzyme:protein solution 1:50) under the above-mentioned conditions. After hydrolysis, the samples were heated at 100 °C for 10 min. Peptides were isolated from the hydrolysates by ultrafiltration (7 kDa molecular weight cut off; Spectra/Por®; 1:4 dilution, phosphate-buffered saline (pH 7.4)). The collected samples were filtered through a 0.45 μM nylon membrane filter (Millipore, Bedford, MA, USA) and stored at −60 °C. The frozen peptides were lyophilized using a LabconcoFreeZone freeze dryer for 72 h at −80 °C and 0.04 mbar. The lyophilized material was stored in tightly closed containers. This material was assessed for antiproliferative activity using cell lines.

2.5. *Anticancer Activity Prediction—In Silico Study*

The peptides isolated from meat products after 180 days of refrigerated storage were subjected to in silico analysis. Peptide sequences were identified by spectrometric methods and then screened for fragments with anticancer activity using the iACP tool, which is a sequence-based tool available online for identifying anticancer peptides (http://lin-group.cn/server/iACP (accessed on 1 December 2021)). The amino acid composition of the peptides was determined using ProtParam (https://web.expasy.org/protparam/ (accessed on 1 December 2021)). Selected physicochemical properties, including hydrophobicity and net charge, as well as toxic properties of peptides were determined using ToxinPred (http://crdd.osdd.net/raghava/toxinpred/ (accessed on 1 December 2021)), and allergenicity was tested using AllerTOP v. 2.0 (https://www.ddg-pharmfac.net/AllerTOP/ (accessed on 1 December 2021)). The spatial structures of the peptides were predicted using the PEP-Fold3 tool (https://bioserv.rpbs.univ-paris-diderot.fr/services/PEP-FOLD3 (accessed on 1 December 2021)).

2.6. *Antiproliferative Activity Prediction—In Vitro Study*

2.6.1. Cell Lines and Culture Conditions

Mouse fibroblast cell line NCTC clone 929 (L cell, L-929, derivative of Strain L; (ATCC® CCL-1™) and colon carcinoma cell line CT-26 (N-nitroso-N-methylurethane-induced, undifferentiated colon carcinoma cells, cloned to generate the designated CT-26. WT cell line; ATCC® CRL-2638™) were obtained from ATCC and cultured with the manufacturer's instructions. Complete Eagle's Minimal Essential medium (Pan-Biotech) and RPMI-1640 medium (Sigma-Aldrich) supplemented with 10% fetal bovine serum (Good HI, Sigma-Aldrich) and antibiotics (100 IU/mL penicillin, 10 mg/mL streptomycin, 25 μg/mL amphotericin B; Pan-Biotech) were used for culturing L-929 and CT-26 cell lines, respectively. Culturing was carried out in a Galaxy 170R incubator under controlled growth conditions, with constant humidity and 5% CO_2. After multiplication and stabilization of the cells (approximately 7–14 days), when the culture reached at least 75% confluence, the cells were cultured with prepared variants in different concentrations.

2.6.2. MTT Assay

This assay is based on the ability of mitochondrial succinate dehydrogenase to reduce the yellow tetrazolium salt (MTT) to purple formazan crystals. As this activity is carried out by living cells, the measured absorbance is directly proportional to the quantity of live cells [27,28]. After multiplication and stabilization of cells (approximately 7–14 days), the culture reached at least 75% of confluence; after reaching confluence, the tested preparations were added to the cell culture in the final concentration (2.5, 25, 50, 100, and 250 μg/mL) and incubated for 24, 48, and 72 h. After this time, the cells were subjected to the MTT assay (MTT Cell Proliferation/Viability Cell Assay: BIOKOM analysis) in triplicate. The control well contained only cells (L-929 or CT-26) without the test preparation. The absorbance was measured using a microplate reader, and absorbance was measured at 570 nm (MULTISCAN FC, Thermo Scientific).

2.6.3. Statistical Analysis

Statistical analysis was carried out using Statistica 13.1 (StatSoft, Cracov, Poland). The differences between mean values were found to be significant if their *p*-value was <0.05 based on Tukey's range t-test. The results are expressed as mean ± standard deviation graphically. In vitro analyses were performed in triplicate.

3. Results

3.1. Spectrometric Identification and In Silico Prediction of Potentially Anticancer Properties

The spectrometric analysis revealed the presence of 1626 peptide sequences in the canned meat extracts. The identified peptide sequences had a chain length of 7–37 amino acids with a lot of Lys, Ala, and Ile, and devoid of Cys and Trp. All the identified peptides were screened for fragments characterized by potentially anticancer activity. In E1 samples (containing extract at a concentration of 150 mg/kg), 21.46% of potentially anticancer fragments were determined (170 out of 792 sequences in total), while in E2 (containing extract at a concentration of 1000 mg/kg) 27.28% of potentially anticancer fragments were identified in the total pool of peptides (203 out of 744 sequences). In the control sample, 25.74% of potentially anticancer peptides were identified (269 out of 1045 sequences). Such a high percentage of potentially anticancer peptides found in canned pork after 180 days of storage indicates that the product is promising and may serve as a functional food for improving human health. Table 1 presents the sequences of potentially anticancer peptides identified (53 fragments) in all analyzed samples (common for E1, E2, and C). A qualitative comparison of unique and common peptides identified in the three study variants, as shown in Figure 1, indicated a greater similarity between the samples containing the willow lyophilizate (E1 and E2) than the control samples.

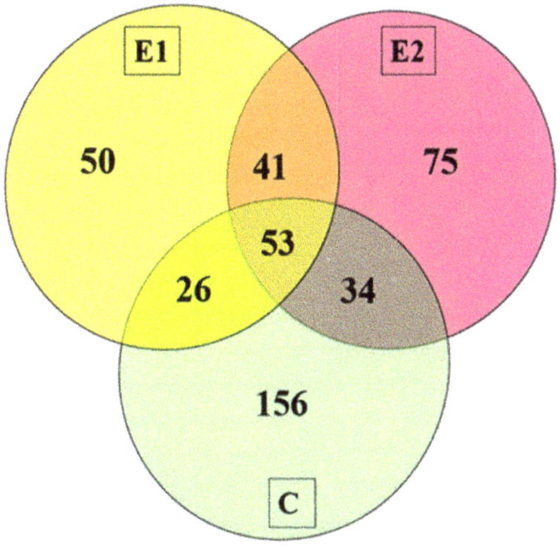

Figure 1. Venn diagram (E1 and E2—samples added with sodium nitrite at an amount of 50 mg/kg of stuffing and lyophilized willow herb extract at an amount of 150 and 1000 mg/kg, respectively; C—control sample added with sodium nitrite at an amount of 100 mg/kg of stuffing but not willow herbs.

Figure 2 shows the values of the net charge distribution and the hydrophobicity of analyzed molecules. The net charge of peptides derived from the tested samples ranged from −5 to 4 (or 3 for E2), while most of the peptides had a net charge of around 0.0. For example, the net charge of LEDEVYAQKMKYKAISEELDNALNDITSL peptide from E1 was on average 0.017 with a minimum value of −5 and that of EALGDKAVFAGRKFRNPKAK was 4. The highest number of potentially anticancer peptides had a net charge of 0.4–2.2, which indicates that they were cationic peptides.

Table 1. List of potentially anticancer peptide sequences found in all the analyzed variants.

No.	Peptide Sequences	Protein Source	Anticancer Probability *	No.	Peptide Sequences	Protein Source	Anticancer Probability *
1	PFGNTHN	Creatine kinase M-type	0.9596	28	VLSAADKANVKAAWGKVGGQAG	Hemoglobin subunit alpha	0.9946
2	LVKAGFAGD	Actin, alpha skeletal muscle	0.9085	29	PPFEVRGANQWIKFKSIS	Pyruvate dehydrogenase E1 component subunit alpha	0.9560
3	SKEYFSKHN	Peroxiredoxin-2	0.9934	30	SEIQNIKSELKYVPRAEQ	Small muscular protein	0.5636
4	VSTVLTSKYR	Hemoglobin subunit alpha	0.5404	31	VQAAFQKVVAGVANALAHKYH	Hemoglobin subunit beta	0.6455
5	VLSAADKANVKA	Hemoglobin subunit alpha	0.9699	32	GELAKHAVSEGTKAVTKYTSSK	Histone H2B type 3-B	0.8512
6	PFGNTHNKYK	Creatine kinase M-type	0.9959	33	VTGNLDYKNLVHIITHGEEKD	Myosin regulatory light chain 2	0.8801
7	VITHGDAKDQE	Myosin regulatory light chain	0.9596	34	PIIQDRHGGYKPTDKHKTDLN	Creatine kinase M-type	0.9860
8	DSKEYFSKHN	Peroxiredoxin-2	0.9944	35	EVYAQKMKYKAISEELDNALNDITSL	Tropomyosin beta chain	0.9880
9	PEDVITGAFKVL	Myosin regulatory light chain 2	0.7486	36	ELYAQKLKYKAISEELDHALNDMTSI	Tropomyosin alpha-3 chain	0.9991
10	KVEELKKKYGI	Acyl-CoA-binding protein	0.9708	37	LEDEVYAQKMKYKAISEELDNALNDITSL	Tropomyosin beta chain	0.9110
11	LVHIITHGEEKD	Myosin regulatory light chain 2	0.8112	38	LEDELYAQKLKYKAISEELDHALNDMTSI	Tropomyosin alpha-3 chain	0.9577
12	PEDVITGAFKVLD	Myosin regulatory light chain 2	0.8566	39	DLEDEVYAQKMKYKAISEELDNALNDITSL	Tropomyosin beta chain	0.9228
13	SREVHTKIISEE	Myosin-1	0.9965	40	DLEDELYAQKLKYKAISEELDHALNDMTSI	Tropomyosin alpha-3 chain	0.9625
14	PFGNTHNKYKLN	Creatine kinase M-type	0.9965	41	ILAADESTGSIAKRLQSIGTEN	Fructose-bisphosphate aldolase	0.9424
15	NLVHIITHGEEKD	Myosin regulatory light chain 2	0.7555	42	PVVQAAYQKVVAGVANALAHKYH	Hemoglobin subunit beta	0.5353
16	PEETILNAFKVFD	Myosin regulatory light chain 2	0.9940	43	PEDVITGAFKVLDPEG	Myosin regulatory light chain 2	0.7505
17	PFGNTHNKYKLNF	Creatine kinase M-type	0.8819	44	KYKAISEELDNALNDITSL	Tropomyosin beta chain	0.7309
18	SADTLWGIQKDLKDL	L-lactate dehydrogenase B chain	0.9896	45	DKEEFVKAKIISREG	Myosin-7	0.9997
19	AISEELDHALNDMTSI	Tropomyosin alpha-3 chain	0.8959	46	DKEEFVKAKILSRE	Myosin-7	0.9808
20	YKNLVHIITHGEEKD	Myosin regulatory light chain 2	0.6486	47	VAGDEESYVVFKDLFD	Creatine kinase M-type	0.5149
21	PEDVITGAFKVLDPEGK	Myosin regulatory light chain 2	0.8019	48	PEDVITGAFKVLDPEGKGT	Myosin regulatory light chain 2	0.8671
22	VLSAADKANVKAAWGKVGG	Hemoglobin subunit alpha	0.9934	49	LAKHAVSEGTKAVTKYTSSK	Histone H2B type 1-N	0.9295
23	FQKVVAGVANALAHKYH	Hemoglobin subunit beta	0.7132	50	DEVYAQKMKYKAISEELDNALNDITSL	Tropomyosin beta chain	0.9898
24	PEDVITGAFKVLDPEGKG	Myosin regulatory light chain 2	0.8395	51	AKLKEIVTNFLAGFEA	ATP synthase subunit alpha	0.5994

Table 1. *Cont.*

No.	Peptide Sequences	Protein Source	Anticancer Probability *	No.	Peptide Sequences	Protein Source	Anticancer Probability *
25	FGPTGIGFGGLTHQVEKKE	Cysteine- and glycine-rich protein 3	0.5000	52	AMKAYINKVEELKKKYGI	Acyl-CoA-binding protein	0.7171
26	KAKDIEHAKKVSQQVSKV	Nebulin	0.5793	53	VLSAADKANVKAAWGKVGGQAGAH	Hemoglobin subunit alpha	0.9858
27	LDYKNLVHIITHGEEKD	Myosin regulatory light chain 2	0.8578				

* Analysis carried out using the iACP tool, in which the probability of biological activity was determined on a scale of 0 (0% activity) to 1 (100% activity).

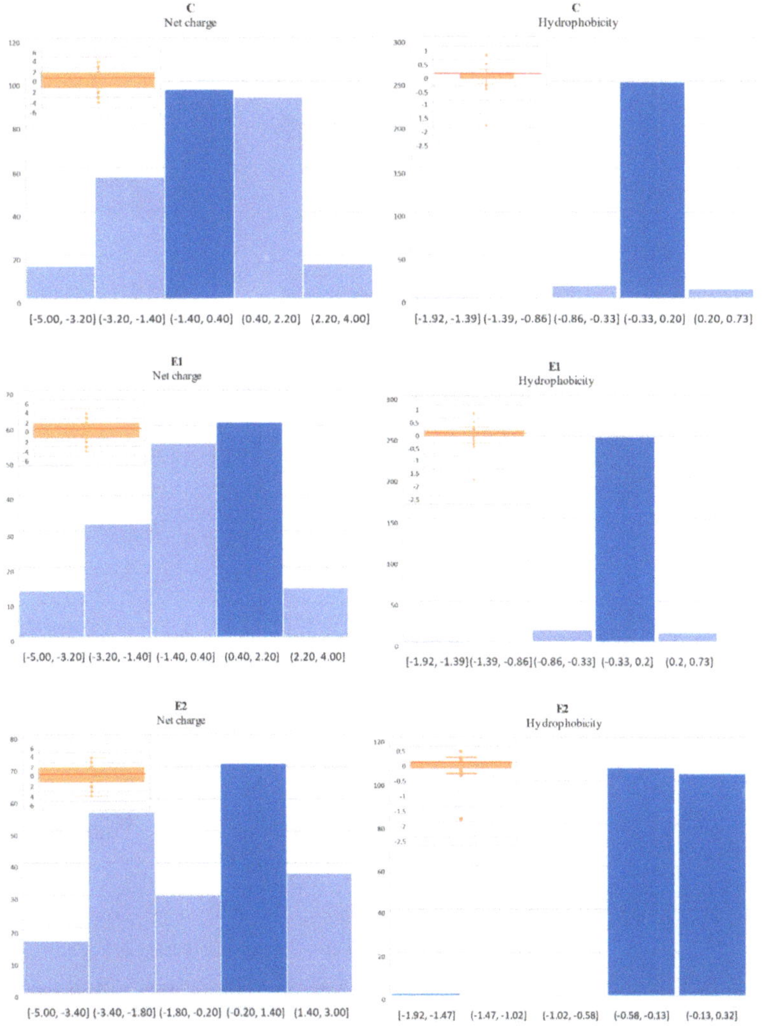

Figure 2. Summary of hydrophobic properties and net charge of all sequences of peptides with potentially anticancer activity derived from canned pork after 6 months of refrigerated storage (the y-axis shows the number of peptides with characteristic net charge or hydrophobicity values, whose specific ranges are represented by the x-axis in the above figures).

Taking into account hydrophobicity, most of the sequences detected in this study had a hydrophobicity value below 0, and most often less than −0.33 (Figure 2). This tendency was observed for all the analyzed variants. The hydrophobicity index values determined in this study did not allow the unequivocal classification of peptides, as well soluble or as poorly soluble, because the obtained values were both additive (but below +1) and negative (even below almost −2), and, in the majority of the cases, the value was close to 0 (zero) in each of the analyzed variants (Figure 2).

To examine the effect of particles' structure on their biological activity on cancer cells, putative antitumor peptides were visualized, and 10 of them with the highest index of "anticancer probability" are shown in Figure 3. All selected molecular structures of the peptides were predicted using the PEP-Fold3 tool. The conformation of the peptides indicated different structures, irrespective of the sequence size. The bioactive sequences were embedded only in the random coil (as was the case with PFGNTHNKYK), but most of the potentially anticancer peptides assumed the α-helical conformation (Figure 3).

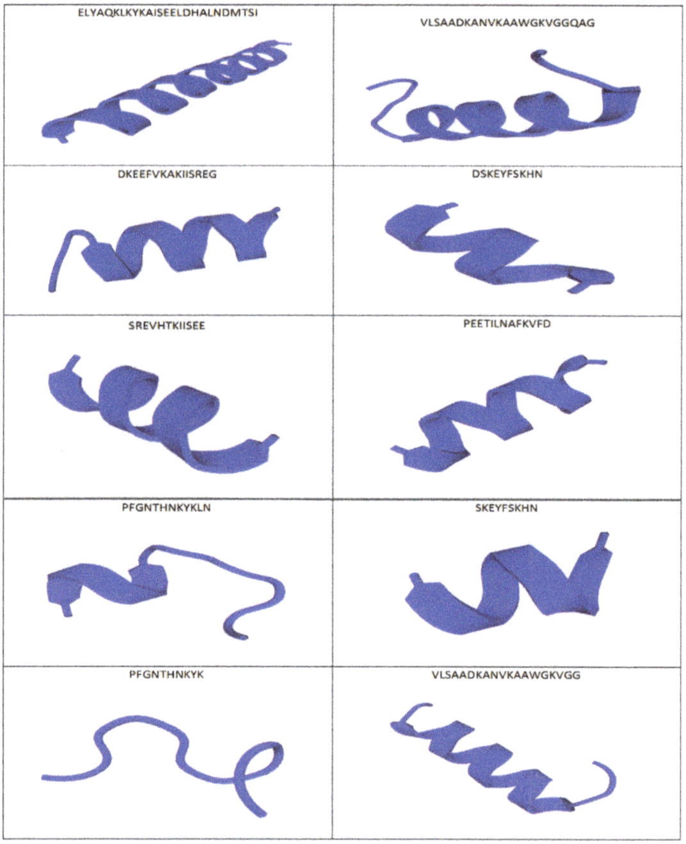

Figure 3. De novo 3D models of selected sequences of peptides with potentially anticancer activity derived from canned pork after 6 months of refrigerated storage irregular sequence—, helix sequence—.

In this study, an in silico analysis of the toxicity and allergenicity of the selected sequences was also performed. None of them were toxic. In contrast, half (51.4%) of the sequences included in the analysis were classified as "possibly allergen" and the other half (48.6%) as "possibly non-allergen".

3.2. *In Vitro Prediction of Antiproliferation Properties*

For the in vitro analysis, the identified sequences were hydrolyzed with pepsin and pancreatin, to simulate the conditions of the human digestive system. The obtained sequences were assessed to ensure whether they were potentially antiproliferation candidates by determining their specificity of action (action against normal cells) and effectiveness against cancer cells. Therefore, after confirming by in silico analysis that canned meat can be a potential source of anticancer peptides, the activity of the experimental variants was examined in vitro by cell viability study using cell lines. The cultures of healthy fibroblast L-929 cells and colorectal carcinoma CT-26 cells were grown until confluence (above 75%) and then treated with the above-mentioned variants containing the tested extract at different concentrations (2.5, 25, 50, 100, and 250 μg/mL). The viability of cells was tested after incubating them with the preparations for 24, 48, and 72 h. As could be expected for the control sample, no particularly negative effect of the pure canned pork preparation on both cell types was noted, regardless of the concentration used. Only in neoplastic cells, a significant decrease in viability was observed after 72 h. In the case of the variant E1, after 24 h of incubation, a slight decrease in cell viability to about 80% was observed in both cell lines, for all concentrations of the preparation used, except for 250 μg/mL, which reduced the viability of CT-26 cells below 80%. The viability of normal L-929 cells remained relatively constant, i.e., about 80%, throughout the experiment after treatment with the E1 variant. On the other hand, cancer cells showed reduced viability regardless of the concentration of the E1 variant used. A decrease in cell viability to about 60% was observed after 48 h of incubation and to about 40% after 72 h, which is very promising (Figure 4). The data obtained for the E2 variant revealed its interesting effect on L-929 cells. After 24 h, an improvement in cell viability, up to around 110%, was noted for the E2 variant at the concentration of 250 μg/mL; at the same time, the appropriate concentration of E2 reduced the viability of CT-26 cells to about 80%. After 48 h of incubation, in both tested cell lines, the results were found to be analogous to the E1 variant. On the other hand, 72 h incubation with E2 reduced the viability of intestinal cancer cells to about 40%, and normal cells to only about 70%.

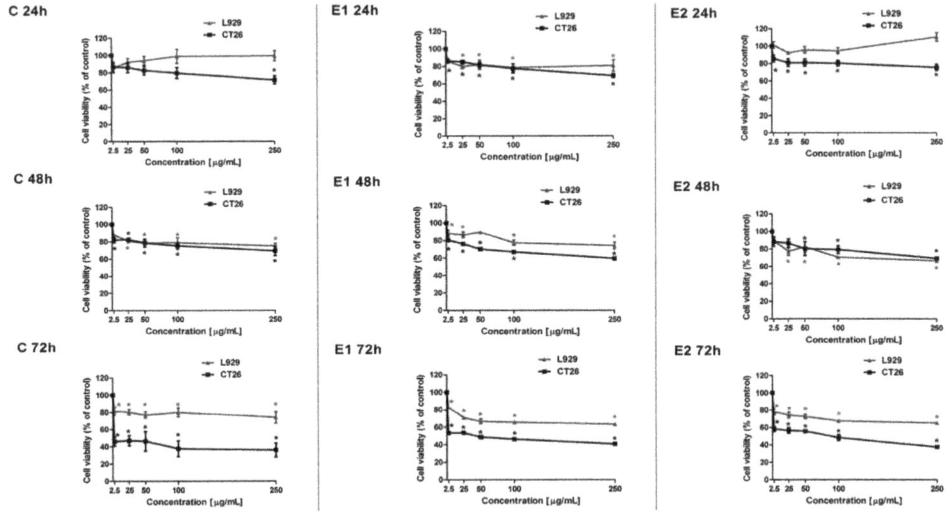

Figure 4. Results of antiproliferation activity in the in vitro analysis (E1 and E2—samples added with sodium nitrite at an amount of 50 mg/kg of stuffing and lyophilized willow herb extract at an amount of 150 and 1000 mg/kg, respectively; C—control sample added with sodium nitrite at an amount of 100 mg/kg of stuffing, but not willow herbs) (sign * means that the differences were statistically significant at $p < 0.05$).

4. Discussion

4.1. Spectrometric Identification and In Silico Prediction of Potentially Anticancer Properties

The biological activity of food-derived peptides, including activity against cancer cells, is determined by their structural characteristics, such as amino acid composition, sequence, length, and total charge/hydrophobicity. Most food-derived anticancer peptides contain short amino acid sequences with residues ranging from 3 to 25 [29] or even longer [30], although peptides of lower molecular weight seem to have higher cytotoxicity [31]. In this study, the identified peptide sequences had a chain length of 7–37 amino acids, with the largest portion of fragments having 13–16 amino acids. Moreover, the presence of amino acids, such as Pro, Leu, Gly, and Ala, as well as one or more Lys, Arg, Ser, Glu, Thr, and Tyr residues, in the anticancer peptides can enhance their interactions with the tumor molecule, thus, exerting a cytotoxic effect on cancer cells [29,32]. Therefore, this study analyzed the amino acid composition of potentially anticancer peptides using the ProtParam bioinformatics tool. The analysis revealed the following relationships: Lys (11%) > Ala (10%) > Ile (9%) > Glu, Leu (8%) > Gly (7%) > Val, Asp, Ser (6%) > Thr (5%) > Pro, Asn (4%) > Phe, Met, Tyr (3%) > His, Arg, Gln (2%). The peptide sequences determined in the study contained neither Cys nor Trp. They consisted of hydrophobic residues, such as Ala, Ile, and Leu, but some polar residues, such as Lys (highest frequency) or Glu, were also found. The presence of Glu and Pro has been linked with anticancer properties of peptides [12,29], which strengthens our assumption about the anticancer activity of the peptides derived from canned meat. Additionally, the hydrophobic amino acids determine the overall hydrophobicity of peptides. Song et al. [31] reported that the Tyr-Ala-Leu-Pro-Ala-His peptide present in the hydrolysate of half-fin anchovy (Setipinnataty) showed antiproliferative activity on PC-3 cells (human prostate cancer), which was attributed to the presence of hydrophobic amino acids (about 50%), as well as the high net charge.

The structure, as well as net charge and hydrophobicity, which depends inter alia on the amino acid composition of the sequence, have an impact on the specificity of targeting and interaction of anticancer peptides with cancer cells. Thus, it is important to analyze these characteristics when assessing the anticancer potential of biological molecules. The net charge is a significant factor in the design of antimicrobial peptides, as it allows their interaction with tumor cells and determines their mechanisms of action against these cells [33]. As shown in Figure 2, the highest number of potentially anticancer peptides were cationic peptides (had a net charge of 0.4–2.2). Cationic peptides are positively charged and are capable of interacting with the phospholipids expressed on the surface of cancer cells, such as phosphatidylserine and phosphatidylglycerol, which carry a net negative charge and provide sites for electrostatic attraction between the peptides and cancer cells. Following electrostatic interaction, the pores of cell membranes are opened, allowing intracellular components to penetrate, which cause cell necrosis [34,35]. According to the literature, peptides that internalize and bind to the mitochondrial membrane destabilize it by activating an apoptotic pathway mediated by caspase release of cytochrome C and, consequently, apoptosis [36]. The net charge values determined in our study are of a wide range, which indicates the presence of cationic peptides (with a positive net charge), although there were also sequences with a negative net charge. In this study, the highest net charge value of +4 was found for peptides from E1 and C samples. This indicates that they may be candidates with anticancer potential, as Ntwasa et al. [37] showed that cationic peptides with charges between +2 and +9 usually interact better with the anionic heads of phospholipids.

Taking into account the hydrophobicity index, the obtained values were quite diverse. This result can be due to the presence of specific amino acids that make up the individual peptide sequences, which included, as described above, both hydrophobic and hydrophilic amino acids.

Moreover, the amino acid composition directly influences the structure of peptide fragments. The literature data suggest that the structural configuration of peptides influences their functions, especially those related to the destruction of pathogenic and/or neoplastic cells [35]. The typical configurations of anticancer peptides were extended and spatial structures (α-helix or β-sheet) [38]. These structures, which are amphipathic in nature, mainly consist of a cationic and hydrophobic surface, which facilitates the interaction of the peptide with the target cell [38]. They were also dominant among the peptides derived from canned pork after 180 days of storage (Figure 3), contributing to their potential anticancer effect.

Despite non-toxicity, an unresolved issue is the allergenicity of the proposed peptides from meat product with dandelion extract. In silico analyses is based on the AllerTOPv. 2.0 tool. This bioinformatics tool is a robust and complementary tool developed based on the k-nearest neighbor method for the classification of allergens and non-allergens. It has an accuracy of 88.7% and is the most effective for in silico allergen prediction [39]. However, it should be noted that in silico methods can only determine whether a new protein is an existing allergen or whether it may cross-react with an existing allergen and not whether the new protein will "become" an allergen [40]. Furthermore, as pointed out by Hayes et al. [41], the predictive value of sequence similarity search should be carefully considered when determining allergenicity potential, as in silico results do not perfectly correlate with the occurrence of food allergy. Moreover, the relatively high degree of amino acid sequence-level identity often observed between cross-reactive IgE proteins cannot guarantee that the protein is a cross-reactive allergen [42]. Thus, a high level of potential allergenicity or no allergenicity determined in this study may not be an indication of allergy, but only draw attention to a potentially health problem. Further biological, biochemical, or in vivo tests should be performed to confirm the allergenicity prediction of peptides derived from canned pork containing freeze-dried willow extract. In particular, the hydrolysates that were proven to be sensitizing by in silico analysis may turn out to be unstable not only in the intestine but also in the vascular system or the liver. Therefore, bioactive peptides must be resistant to enzymes in the human digestive system and remain stable in the further regions of their distribution in the human body up to the site of action. In fact, digestion in the gastrointestinal tract promotes the release of biopeptides from large sequences of food proteins, resulting in the release of biologically active fragments, which, when digested, and absorbed in the small intestine, retain their biological properties. On the other hand, intense hydrolysis can cause the destruction of peptide sequences, reducing their biological activity.

4.2. *In Vitro Prediction of Antiproliferation Properties*

The approach of assessing the viability of peptides against cell lines, as used in this study, has already been successfully applied by other researchers. Various tumor cell lines, such as MCF-7, MDA-MB- 231, and BT549 (human breast cancer); Ca9-22 and CAL 27 (human oral cancer); Hep G2 (human liver cancer); HT-29, RKO, KM12L4, DLD-1, and HCT15 (human colon cancer); Caco-2, TC7, and HCT-116 (human colon cancer); U87 (human glioma); PC3, LNCaP, and DU-145 (human prostate cancer); THP-1 (human monocytic leukemia); Jurkat T cells (human T cell leukemia); AGS (human gastric cancer); A549 and H-1299 (human lung cancer); HeLa (human cervical cancer); 109 (human esophageal cancer); hFOB1.19 (human fetal osteoblastic carcinoma); HL-60 (human acute myeloid leukemia); Kelly, SK-N-DZ, and IMR-32 (human neuroblastoma); L1210 (murine lymphocytic leukemia); P388D1 (mouse monocyte cell line); MC3T3E1 (mouse osteoblastoma), UMR106 (rat osteosarcoma); and vero (monkey kidney cancer cells), have been widely used as model cell culture systems to investigate the antitumor effects of protein hydrolysates prepared from different sources of dietary protein [32]. For example, Ito et al. [43] reported that porcine skin gelatin showed in vitro antitumor effects on murine hepatoma cells (MH134), inducing programmed cell death (apoptosis). Similarly, Castro et al. [44] confirmed the effect of bovine collagen hydrolysates (BCH) on the proliferation of B16F10 melanoma cells. In turn, Daliri et al. [45] analyzed the effect of bioactive peptides isolated from soybean, oyster, and sepia ink hydrolysates, rapeseed protein fermentates, and tuna cooking juice on various cancer cell lines. Other authors [46,47] observed that bioactive peptides derived from bovine sarcoplasmic proteins had a positive effect on breast and gastric cancer cells. Apart from the potential anticancer properties of peptides, the secondary products of willow herb might also be beneficial to human health. Oenothein B is the main substance found in Epilobium species, which is considered to have anticancer properties [48]. Therefore, their application in food may have a positive effect on health [45]. Indeed, in presented study, in vitro analyses on L-929 (mouse fibroblast cell line) and CT-26 (undifferentiated colon carcinoma cell line) confirmed the antitumor activity of peptides obtained from the meat product. The results of own research on cell lines indicate the anticancer potential of peptide from canned meat with willow extract. Particularly noteworthy is the fact that after 72 h of incubation with the peptides obtained from the variant containing 1000 mg/kg of freeze-dried willow herb extract, the viability of intestinal cancer cells was decreased to about 40%, while by comparison the viability of normal cells was decreased to only about 70%. The promising results observed for canned pork added with freeze-dried willow extract suggest the usefulness of the herb in the production of meat products with nutritional, as well as health benefits.

5. Conclusions

This study analyzed the potentially anticancer activity of meat products prepared with freeze-dried willow herb extract. Among the identified peptide sequences, some were cationic and hydrophobic with α-helical or β-sheet conformation, which indicates their anticancer potential. Furthermore, the sequences of peptides present in the proteolytic enzyme hydrolysates showed, on the one hand, the resistance of these peptides to simulated conditions of the human digestive system, and, on the other hand, their preserved biological activity was confirmed by their action on the tested cell lines. Thus, the results of the study shed new light on the value of canned meat as a functional food, due to its complementing nutritional and physiological properties. However, there is a need for further research on anticancer peptides, especially in terms of their selectivity or their half-life in the body and the exclusion of potential allergenicity.

Author Contributions: Conceptualization, K.M.W. and K.F.; methodology, P.K and M.P.-W.; formal analysis, K.M.W. and P.K.; investigation, P.K., M.P.-W. and K.F.; writing—original draft preparation, P.K., M.P.-W. and K.F.; writing—review and editing, K.M.W.; visualization, P.K. All authors have read and agreed to the published version of the manuscript.

Funding: Research was financed under the program of the Minister of Education and Science under the name "Regional Initiative of Excellence" in 2019–2023 project number 029/RID/2018/19 funding amount 11 927 330.00 PLN".

Institutional Review Board Statement: Not applicable.

Informed Consent Statement: Not applicable.

Conflicts of Interest: The authors declare no conflict of interest.

References

1. International Agency for Research on Cancer. Available online: https://www.iarc.who.int/ (accessed on 1 December 2021).
2. World Cancer Research Fund. Available online: https://www.wcrf.org/ (accessed on 1 December 2021).
3. Huang, Y.; Cao, D.; Chen, Z.; Chen, B.; Li, J.; Guo, J.; Dong, Q.; Liu, L.; Wei, Q. Red and processed meat consumption and cancer outcomes: Umbrella review. *Food Chem.* **2021**, *356*, 129697. [CrossRef] [PubMed]
4. Singh, P.; Fraser, G.E. Dietary Risk Factors for Colon Cancer in a Low-risk Population. *Am. J. Epidemiol.* **1998**, *148*, 761–774. [CrossRef] [PubMed]
5. English, D.R.; MacInnis, R.J.; Hodge, A.M.; Hopper, J.L.; Haydon, A.M.; Giles, G.G. Red meat, chicken, and fish consumption andrisk of colorectal cancer. *Cancer Epidemiol. Biomark. Prev.* **2004**, *13*, 1509–1514. [CrossRef]
6. Norat, T.; Bingham, S.; Ferrari, P.; Slimani, N.; Jenab, M.; Mazuir, M.; Overvad, K.; Olsen, A.; Tjønneland, A.; Clavel, F.; et al. Meat, Fish, and Colorectal Cancer Risk: The European Prospective Investigation into Cancer and Nutrition. *J. Natl. Cancer Inst.* **2005**, *97*, 906–916. [CrossRef] [PubMed]
7. Ferysiuk, K.; Wójciak, K.M. The Possibility of Reduction of Synthetic Preservative E 250 in Canned Pork. *Foods* **2020**, *9*, 1869. [CrossRef]

8. Ferysiuk, K.; Wójciak, K.M.; Trząskowska, M. Fortification of low-nitrite canned pork with willow herb (*Epilobium angustifolium* L.). *Int. J. Food Sci. Technol.* **2022**, *57*, 4194–4210. [CrossRef]
9. Ferysiuk, K.; Wójciak, K.M.; Kęska, P. Effect of willow herb (*Epilobium angustifolium* L.) extract addition to canned meat with reduced amount of nitrite on the antioxidant and other activities of peptides. *Food Funct.* **2022**, *13*, 3526–3539. [CrossRef]
10. Wu, M.-L.; Li, H.; Yu, L.-J.; Chen, X.-Y.; Kong, Q.-Y.; Song, X.; Shu, X.-H.; Liu, J. Short-Term Resveratrol Exposure Causes In Vitro and In Vivo Growth Inhibition and Apoptosis of Bladder Cancer Cells. *PLoS ONE* **2014**, *9*, e89806. [CrossRef]
11. Beaulieu, L.; Thibodeau, J.; Bonnet, C.; Bryl, P.; Carbonneau, M.-E. Evidence of Anti-Proliferative Activities in Blue Mussel (*Mytilus edulis*) By-Products. *Mar. Drugs* **2013**, *11*, 975–990. [CrossRef]
12. Kannan, A.; Hettiarachchy, N.S.; Lay, J.O.; Liyanage, R. Human cancer cell proliferation inhibition by a pentapeptide isolated and characterized from rice bran. *Peptides* **2010**, *31*, 1629–1634. [CrossRef]
13. E Kim, S.; Kim, H.H.; Kim, J.Y.; Kang, Y.I.; Woo, H.J.; Lee, H.J. Anticancer activity of hydrophobic peptides from soy proteins. *BioFactors* **2000**, *12*, 151–155. [CrossRef] [PubMed]
14. Picot, L.; Bordenave, S.; Didelot, S.; Fruitier-Arnaudin, I.; Sannier, F.; Thorkelsson, G.; Bergé, J.; Guérard, F.; Chabeaud, A.; Piot, J. Antiproliferative activity of fish protein hydrolysates on human breast cancer cell lines. *Process Biochem.* **2006**, *41*, 1217–1222. [CrossRef]
15. Sheih, I.-C.; Fang, T.J.; Wu, T.-K.; Lin, P.-H. Anticancer and Antioxidant Activities of the Peptide Fraction from Algae Protein Waste. *J. Agric. Food Chem.* **2009**, *58*, 1202–1207. [CrossRef] [PubMed]
16. Xue, Z.; Yu, W.; Wu, M.; Wang, J. In Vivo Antitumor and Antioxidative Effects of a Rapeseed Meal Protein Hydrolysate on an S180 Tumor-Bearing Murine Model. *Biosci. Biotechnol. Biochem.* **2009**, *73*, 2412–2415. [CrossRef] [PubMed]
17. Xue, M.; Ge, Y.; Zhang, J.; Wang, Q.; Hou, L.; Liu, Y.; Sun, L.; Li, Q. Anticancer Properties and Mechanisms of Fucoidan on Mouse Breast Cancer In Vitro and In Vivo. *PLoS ONE* **2012**, *7*, e43483. [CrossRef] [PubMed]
18. Yamaguchi, M.; Takeuchi, M.; Ebihara, K. Inhibitory effect of peptide prepared from corn gluten meal on 7,12-dimethylbenz[a]anthracene-induced mammary tumor progression in rats. *Nutr. Res.* **1997**, *17*, 1121–1130. [CrossRef]
19. Sedighi, M.; Jalili, H.; Ranaei-Siadat, S.-O.; Amrane, A. Potential Health Effects of Enzymatic Protein Hydrolysates from Chlorella vulgaris. *Appl. Food Biotnol.* **2016**, *3*, 160–169. [CrossRef]
20. Sadeghi, S.; Jalili, H.; RanaeiSiadat, S.O.; Sedighi, M. Anticancer and antibacterial properties in peptide fractions from hydrolyzed spirulina protein. *J. Agric. Sci. Technol.* **2018**, *20*, 673–683.
21. Sharma, P.; Kaur, H.; Kehinde, B.A.; Chhikara, N.; Sharma, D.; Panghal, A. Food-Derived Anticancer Peptides: A Review. *Int. J. Pept. Res. Ther.* **2021**, *27*, 55–70. [CrossRef]
22. Sah, B.N.P.; Vasiljevic, T.; McKechnie, S.; Donkor, O.N. Identification of Anticancer Peptides from Bovine Milk Proteins and Their Potential Roles in Management of Cancer: A Critical Review. *Compr. Rev. Food Sci. Food Saf.* **2015**, *14*, 123–138. [CrossRef]
23. Nwachukwu, I.D.; Aluko, R.E. Anticancer and antiproliferative properties of food-derived protein hydrolysates and peptides. *J. Food Bioact.* **2019**, *7*. [CrossRef]
24. Deslouches, B.; Di, Y.P. Antimicrobial peptides with selective antitumor mechanisms: Prospect for anticancer applications. *Oncotarget* **2017**, *8*, 46635–46651. [CrossRef] [PubMed]
25. Montalvo, G.E.B.; Vandenberghe, L.P.D.S.; Soccol, V.T.; Carvalho, J.C.D.; Soccol, C.R. The antihypertensive, antimicrobial and anticancer peptides from Arthrospira with therapeutic potentially: A mini review. *Curr. Mol. Med.* **2020**, *20*, 593–606. [CrossRef]
26. Escudero, E.; Mora, L.; Toldrá, F. Stability of ACE inhibitory ham peptides against heat treatment and in vitro digestion. *Food Chem.* **2014**, *161*, 305–311. [CrossRef] [PubMed]
27. Mosmann, T. Rapid colorimetric assay for cellular growth and survival: Application to proliferation and cytotoxicity assays. *J. Immunol. Methods* **1983**, *65*, 55–63. [CrossRef]
28. van de Loosdrecht, A.; Beelen, R.; Ossenkoppele, G.; Broekhoven, M.; Langenhuijsen, M. A tetrazolium-based colorimetric MTT assay to quantitate human monocyte mediated cytotoxicity against leukemic cells from cell lines and patients with acute myeloid leukemia. *J. Immunol. Methods* **1994**, *174*, 311–320. [CrossRef]
29. Chi, C.; Hu, F.; Wang, B.; Li, T.; Ding, G. Antioxidant and anticancer peptides from the protein hydrolysate of blood clam (Tegillarcagranosa) muscle. *J. Func. Foods* **2015**, *15*, 301–313. [CrossRef]
30. Huang, K.-Y.; Tseng, Y.-J.; Kao, H.-J.; Chen, C.-H.; Yang, H.-H.; Weng, S.-L. Identification of subtypes of anticancer peptides based on sequential features and physicochemical properties. *Sci. Rep.* **2021**, *11*, 13594. [CrossRef]
31. Song, R.; Wei, R.B.; Luo, H.Y.; Yang, Z.S. Isolation and identification of an antiproliferative peptide derived from heated products of peptic hydrolysates of half-fin anchovy (Setipinnataty). *J. Func. Foods* **2014**, *10*, 104–111. [CrossRef]
32. Chalamaiah, M.; Yu, W.; Wu, J. Immunomodulatory and anticancer protein hydrolysates (peptides) from food proteins: A review. *Food Chem.* **2017**, *245*, 205–222. [CrossRef]
33. Liu, X.; Li, Y.; Li, Z.; Lan, X.; Leung, P.H.M.; Li, J.; Yang, M.; Ko, F.; Qin, L. Mechanism of anticancer effects of antimicrobial peptides. *J. Fiber Bioeng. Inform.* **2015**, *8*, 25–36. [CrossRef]
34. Felício, M.R.; Silva, O.N.; Gonçalves, S.; Santos, N.C.; Franco, O.L. Peptides with Dual Antimicrobial and Anticancer Activities. *Front. Chem.* **2017**, *5*, 5. [CrossRef]
35. Kumariya, R.; Sood, S.K.; Rajput, Y.S.; Saini, N.; Garsa, A.K. Increased membrane surface positive charge and altered membrane fluidity leads to cationic antimicrobial peptide resistance in Enterococcus faecalis. *Biochim. Biophys. Acta Biomembr.* **2015**, *1848*, 1367–1375. [CrossRef] [PubMed]

36. Nyström, L.; Malmsten, M. Membrane interactions and cell selectivity of amphiphilic anticancer peptides. *Curr. Opin. Colloid Interface Sci.* **2018**, *38*, 1–17. [CrossRef]
37. Ntwasa, M.; Goto, A.; Kurata, S. Coleopteran Antimicrobial Peptides: Prospects for Clinical Applications. *Int. J. Microbiol.* **2012**, *2012*, 101989. [CrossRef]
38. Hilchie, A.L.; Hoskin, D.W.; Coombs, M.R.P. Anticancer Activities of Natural and Synthetic Peptides. *Antimicrob. Pept.* **2019**, *1117*, 131–147. [CrossRef]
39. Dimitrov, I.; Bangov, I.; Flower, D.R.; Doytchinova, I. AllerTOP v.2—A server for in silico prediction of allergens. *J. Mol. Model.* **2014**, *20*, 2278. [CrossRef]
40. Goodman, R.E. Biosafety: Evaluation and regulation of genetically modified (GM) crops in the United States. *J. Huazhong Agric. Univ.* **2014**, *33*, 85–113.
41. Hayes, M.; Rougé, P.; Barre, A.; Herouet-Guicheney, C.; Roggen, E.L. In silico tools for exploring potential human allergy to proteins. *Drug Discov. Today: Dis. Model.* **2015**, *17*, 3–11. [CrossRef]
42. Aalberse, R.C.; Crameri, R. IgE-binding epitopes: A reappraisal. *Allergy* **2011**, *66*, 1261–1274. [CrossRef]
43. Ito, N.; Kojima, T.; Nagata, H.; Ozeki, N.; Yoshida, Y.; Nonami, T. Apoptosis Induced by Culturing MH134 Cells in the Presence of Porcine Skin Gelatin In Vitro. *Cancer Biother. Radiopharm.* **2002**, *17*, 379–384. [CrossRef]
44. Castro, G.A.; Maria, D.A.; Bouhallab, S.; Sgarbieri, V.C. In vitro impact of a p5whey protein isolate (WPI) and collagen hydrolysates (CHs) on B16F10 melanoma cells proliferation. *J. Dermatol Sci.* **2009**, *56*, 51–57. [CrossRef] [PubMed]
45. Daliri, E.B.M.; Oh, D.H.; Lee, B.H. Bioactive peptides. *Foods* **2017**, *6*, 32. [CrossRef] [PubMed]
46. Albenzio, M.; Santillo, A.; Caroprese, M.; Marino, R.; Della Malva, A. Bioactive Peptides in Animal Food Products. *Foods* **2017**, *6*, 35. [CrossRef] [PubMed]
47. Jang, A.; Jo, C.; Kang, K.-S.; Lee, M. Antimicrobial and human cancer cell cytotoxic effect of synthetic angiotensin-converting enzyme (ACE) inhibitory peptides. *Food Chem.* **2008**, *107*, 327–336. [CrossRef]
48. Vitalone, A.; Allkanjari, O. *Epilobium* spp: Pharmacology and Phytochemistry. *Phytother. Res.* **2018**, *32*, 1229–1240. [CrossRef]

molecules

MDPI

Article

Malvidin and Its Mono- and Di-Glucosides Forms: A Study of Combining Both In Vitro and Molecular Docking Studies Focused on Cholinesterase, Butyrylcholinesterase, COX-1 and COX-2 Activities

Paulina Strugała-Danak [1,*], Maciej Spiegel [2,*] and Janina Gabrielska [1]

1 Department of Physics and Biophysics, Wrocław University of Environmental and Life Sciences, C. K. Norwida 25, 50-375 Wrocław, Poland; janina.gabrielska@upwr.edu.pl

2 Department of Organic Chemistry and Pharmaceutical Technology, Wrocław Medical University, Borowska 211A, 50-556 Wrocław, Poland

* Correspondence: paulina.strugala@upwr.edu.pl (P.S.-D.); maciej.spiegel@umw.edu.pl (M.S.); Tel.: +48-71-320-5295 (P.S.-D.)

Abstract: Malvidin, one of the six most prominent anthocyanins found in various fruits and vegetables, may possess a wide range of health-promoting properties. The biological activity of malvidin and its glycosides is not entirely clear and has been relatively less frequently studied compared to other anthocyanins. Therefore, this study aimed to determine the relationship between the structural derivatives of malvidin and their anti-cholinergic and anti-inflammatory activity. The study selected malvidin (Mv) and its two sugar derivatives: malvidin 3-*O*-glucoside (Mv 3-glc) and malvidin 3,5-*O*-diglucoside (Mv 3,5-diglc). The anti-inflammatory activity was assessed by inhibiting the enzymes, specifically COX-1 and COX-2. Additionally, the inhibitory effects on cholinesterase activity, particularly acetylcholinesterase (AChE) and butyrylcholinesterase (BChE), were evaluated. Molecular modeling was also employed to examine and visualize the interactions between enzymes and anthocyanins. The results revealed that the highest inhibitory capacity at concentration 100 μM was demonstrated by Mv 3-glc in relation to AChE (26.3 ± 3.1%) and BChE (22.1 ± 3.0%), highlighting the crucial role of the glycoside substituent at the C3 position of the C ring in determining the inhibitory efficiency of these enzymes. In addition, the glycosylation of malvidin significantly reduced the anti-inflammatory activity of these derivatives compared to the aglycone form. The IC_{50} parameter demonstrates the following relationship for the COX-1 enzyme: Mv (12.45 ± 0.70 μM) < Mv 3-glc (74.78 ± 0.06 μM) < Mv 3,5-diglc (90.36 ± 1.92 μM). Similarly, for the COX-2 enzyme, we have: Mv (2.76 ± 0.16 μM) < Mv 3-glc (39.92 ± 3.02 μM) < Mv 3.5-diglc (66.45 ± 1.93 μM). All tested forms of malvidin exhibited higher activity towards COX-2 compared to COX-1, indicating their selectivity as inhibitors of COX-2. Theoretical calculations were capable of qualitatively replicating most of the noted patterns in the experimental data, explaining the impact of deprotonation and glycosylation on inhibitory activity. It can be suggested that anthocyanins, such as malvidins, could be valuable in the development of treatments for inflammatory conditions and Alzheimer's disease and deserve further study.

Keywords: cyclooxygenase; cholinesterase; malvidin; molecular docking

Citation: Strugała-Danak, P.; Spiegel, M.; Gabrielska, J. Malvidin and Its Mono- and Di-Glucosides Forms: A Study of Combining Both In Vitro and Molecular Docking Studies Focused on Cholinesterase, Butyrylcholinesterase, COX-1 and COX-2 Activities. *Molecules* **2023**, *28*, 7872. https://doi.org/10.3390/molecules28237872

Academic Editors: Sascha Rohn and Michał Halagarda

Received: 19 September 2023
Revised: 22 November 2023
Accepted: 27 November 2023
Published: 30 November 2023

1. Introduction

Alzheimer's disease (AD) is viewed as a progressive, neurodegenerative pathology. Whereas the origins of AD are not fully explained, it is broadly accepted that it is closely linked to the impairment of cholinergic transmission, in which one of the neurotransmitter, namely acetylcholine, plays a vital role. Furthermore, cholinesterases, including acetylcholinesterase (AChE) and butyrylcholinesterases (BChE), are believed to be key enzymes which are able to hydrolyze acetylcholine [1,2]. As the result of this hydrolysis, there is a significant decrease in acetylcholine which is associated with the loss of basic forebrain cells, which, in turn, are key in order to determine the cognitive impairment of the brain [3]. The process of both enzymes' inhibition, namely AChE and BChE, is the most effective and modern therapeutic approach in cognitive impairment [4,5].

Currently, various therapeutic strategies are being studied in order to develop an effective anti-Alzheimer's drug. Yu et al. [6] recently identified microRNA-485-3p (miR-485-3p) as a biomarker and therapeutic target for AD. Their research primarily focused on miR-485-3p's effects in cell-based in vitro systems, whereas in vivo physiological relevance was not viewed. Research conducted by Koh et al. [7] utilized the miR-485-3p marker in a transgenic mouse model of AD. They demonstrated the overexpression of miR-485-3p in the brain tissues of AD patients, and its antisense oligonucleotide (ASO), effectively reduced Aβ plaque buildup, tau pathology progression, neuroinflammation, and cognitive decline.

Furthermore, there is a vigorous exploration seeking non-toxic natural remedies for Alzheimer's disease, where curcumin, known for its diverse biological properties [8–10], is under close investigation. The antioxidant potential of five mono-carbonyl curcumin analogs was assessed through in vitro tests and an in vivo mouse model focused on the hippocampus [11]. Among these analogs, two with methoxy and chloro-substitutions demonstrated strong DPPH and ABTS free radical scavenging capabilities in the in vitro assays. The authors also revealed a significant decrease in lipid peroxidation and enhanced activities of catalase, superoxide dismutase, and glutathione in the mouse hippocampus through marker analysis, highlighting their antioxidant and memory-enhancing properties. Khan et al. [12] conducted research on the alkaloids nuciferine and norcoclaurine extracted from N. nucifera seeds, highlighting their anti-Alzheimer's and antioxidant properties in a diabetic rat model. The alkaloids notably restored AChE activity in both the blood and brains of the rats, along with the recovery of all antioxidant enzymes. The study showed that nuciferine and norcoclaurine substantially enhance memory and both could be effective phytomedicines against diabetes and Alzheimer's disease (AD). Referring to the compounds from the group of flavonoids, isorhamnetin and quercetin derivatives, which were obtained in the process of isolating them from *Calendula officinalis* L., they are, among others, excellent examples of powerful AChE inhibitors [4,13]. The results of anticholinesterase activities of 24 polyphenolic compounds were interestingly presented in a brief report by Szwajgier [14]. The research was focused on a group of anthocyanins, flavones, flavanols, as well as dihydrochalcone phlorizin and prenylated chalcone xanthohumol. While analyzing the differences in properties in the molecular structure of those compounds, the author suggested that the inhibitory activity decreased in the presence of a 3-hydroxyl group; similarly, it was also stated that aglycons were more effective cholinesterase inhibitors than their corresponding glycosylated forms. Moreover, there are the other studies which proved the inhibitory activities of aqueous fruit mulberry extracts (*Morus* spp.) against AChE and BChE, which are not only rich in anthocyanins, but also in cyanidin, kuromanin, and keracyanin [15].

Oxidative stress and inflammation are both key causative factors for the onset of several diseases, including cancer, metabolic and cardiovascular disorders, and neurodegenerative diseases. Aging is responsible for heightening oxidative stress in the nervous system, leading to impaired nerve regeneration and function [16,17]. Furthermore, inflammation is detected in the human central nervous system and increased levels of inflammatory cytokines were observed in the cerebrospinal fluid of aging individuals [18].

The broad spectrum of potentially anti-inflammatory activity displayed by biologically active compounds can be tested by way of measuring the inhibition of the cyclooxygenase (COX-1, COX-2) activity [19,20]. These two enzymes take part in the synthesis of several mediators of inflammation, including prostaglandin PGE2, leukotriene B4, and thromboxane A2. As a result of the inhibition showed by anthocyanins, and which is generated by those enzymes, the synthesis of the mediators of inflammation, which are responsible for inhibiting leukocyte accumulation and adjusting the vascular system, decreases. Therefore, the overall effects of inflammation are minimalized. The above effects were proved by Fagundes et al. [21], who aimed to compare the anti-inflammatory effect of anthocyanin-rich extract from blueberries (with malvidin-3-galactoside and petunidin-3-arabinoside) with 5-aminosalicylic acid in the TNBS-induced colitis model. It was showed that the anthocyanin extract at a concentration 30 times lower than 5-aminosalicylic acid had higher effectiveness in counteracting intestinal inflammation. Furthermore, it was Seeram et al. [22] who proved, using numerous anthocyanin extracts from berries and cherries, that anthocyanins from raspberries and sweet cherries demonstrated 45% and 47% COX-1 and COX-2 inhibitory activities respectively, referring to assays at 125 μg/mL. Additionally, studies showed that aglycone cyanidin has higher inhibitory activity in comparison to its glycosides, and that COX inhibitory activities increased with a decreasing number of sugar residues, which were attached to the cyanidin moiety.

Malvidin, named 3,5-dimethoxy-3,4,5,7-tetrahydroxyflavylium acid anion, is a type of anthocyanidin cation with a chemical structure similar to delphinidin but with methyl groups attached to positions 3 and 5. In its natural state, it is predominantly encountered in its glycosylated form, with a sugar moiety attached at position 3 on the C-ring. Several studies, both in vitro and in vivo, indicate that these molecules have the potential to mitigate the onset and progression of various disease pathologies, particularly those associated with oxidative stress-related pathogenesis [16]. Previously, we conducted an experiment that proved that Mv and its two glucosides (Mv 3-glc > Mv 3,5-diglc) have a tendency to exhibit in vitro (AAPH$^{\bullet}$ and DPPH$^{\bullet}$ assays) antioxidant activities against liposome membrane in the order illustrated here: Mv > Mv 3-glc > Mv 3,5-diglc. Moreover, we observed that this activity is in line with interaction with this membrane [23]. Despite the fact that the anti-cholinesterase and anti-inflammatory properties of anthocyanins were carefully investigated, there is limited access to information on the effects of such glycosylation on the mechanisms developed by anthocyanins in order to manage/prevent disease conditions. In the experiment that we are describing in this paper, we endeavor to demonstrate the ability of these anthocyanins to inhibit the cholinesterase enzymes and the pro-inflammatory cyclooxygenases COX-1 and COX-2.

As a result, this study is intended to compare the anticholinergic activity and cyclooxygenase (COX-1 and COX-2) inhibitory (as potential anti-inflammatory) properties of Mv and its mono- and di-glucosides. It is important to underline that in this study, for the first time, calculations were carried out with the usage of a computational chemistry approach, which is the molecular docking method, in order to demonstrate the relationship between the anticholinesterase and anti-inflammatory effectiveness of Mv-s and their molecular structure with regard to their interconnections at the atomic level. We hope that the outcome obtained here as a result of these complementary studies will be beneficial in order to indicate what are the potential methods of application for Mv and its mono- and di-glucosides in order to save human lives and treat the early stages of neurodegenerative and anti-inflammatory diseases.

2. Results and Discussion

2.1. Acetylcholinesterase and Buthyrylcholinesterase Inhibition

Anti-cholinergic activity was assessed by inhibiting the enzymes acetylcholinesterase and butyrylcholinesterase specifically. Inhibiting both enzymes is the most effective therapeutic approach for restoring the normal functions of the cholinergic system. Therefore, effective pharmaceuticals are being sought, in particular natural inhibitors of these enzymes. The results of our studies, which were focused on the use of Mv and its two derivatives, as well as the drug neostigmine, are summarized in Table 1, which contains the percentages of AChE and BChE inhibition for 100 μM anthocyanin concentration. There was no dependence on the concentration of anthocyanins within the range 100–200 μM. The highest inhibitory capacity in relation to both enzymes was demonstrated by Mv 3-glc −26.31 ± 3.06% and 22.07 ± 3.00%, in relation to AChE and BChE, respectively. Both Mv and Mv 3,5-diglc were less potent enzyme inhibitors compared to Mv 3-glc. On the one hand, referring to the differences in molecular structure between Mv 3-gluc and Mv, one may suggest the importance of the glycoside substituent in the C3 position of the C ring of the molecule on the inhibitory efficiency of the enzymes of this mono-glycoside. On the other hand, a reduction in this effectiveness occurs after replacing the hydroxyl group at the C5 position with another, second glucoside molecule. In order to reveal fully the relationship between the structure of anthocyanins and the inhibitory efficacy of AChE and BChE enzymes, further research is required. The neostigmine drug used in the treatment of old age disease, AD, showed 50% inhibition efficiency in relation to AChE and BChE at concentration rates of 0.17 ± 0.01 μM and 20.28 ± 1.70 μM, respectively. By relating the inhibition values expressed in % in relation to the results caused by Mv 3-glc, it can be roughly stated that the concentration of this anthocyanin should be increased several times, e.g., 5–10 times in relation to BChE, in order to have an effect similar to neostigmine. Bearing in mind the fact that anthocyanins are natural, non-toxic substances found in red and purple fruits and vegetables, the process of increasing the concentration of these substances several times is possible through supplementation with nutraceuticals. However, there is a serious problem with their low bioavailability and molecular instability under physiological conditions. The core of this problem lies in finding effective solutions proposed by further research, which should be focused on, e.g., the design of encapsulated anthocyanins or extracts with a high content of the compounds required [24–26]

Table 1. Cholinesterase inhibitory activity (against AChE and BChE) of the malvidin (Mv), malvidin 3-*O*-glucoside (Mv 3-glc) and malvidin 3,5-*O*-diglucoside (Mv 3,5-diglc). Concentration of Mv, Mv 3-glc and Mv 3,5-diglc was 100 μM. Neostigmine concentration for AChE inhibition was 0.17 μM, and, for BChE inhibition, it was 20.28 μM. Different letters ($^{a-c}$) in the same row indicate significant differences ($p < 0.05$).

Compound	Inhibition of AChE (%)	Inhibition of BChE (%)
Mv	19.5 ± 2.3 ^{b}	10.2 ± 2.2 ^{b}
Mv 3-glc	26.3 ± 3.1 ^{a}	22.1 ± 3.0 ^{a}
Mv 3,5-diglc	11.9 ± 4.5 ^{c}	13.4 ± 2.4 ^{b}
Neostigmine	50.0 ± 1.2	50.0 ± 4.2

Literature data shows potential AChE inhibitory activity in over 300 natural compounds, including alkaloids (53%), monoterpenes (10%), coumarins (7%), triterpenes (6.5%), flavonoids (5%), simple phenols (5%) and others [13,27]. In the case of compounds obtained from the group of flavonoids, isorhamnetin and quercetin derivatives isolated from *Calendula officinalis* L. are other examples of potential AChE inhibitors [13]. A study conducted by Yusuf et al. [28], which was focused on the inhibition of AChE and BChE by 12 colored carrot varieties, demonstrated the effectiveness of micro purple carrot (MPC) and normal purple carrot (NPC) in inhibiting these enzymes. Whereas IC_{50} values in relation to AChE for purple carrot MPC and NPC were 10.14 mg/mL (i.e., 10140 μg/mL) and 18.96 mg/mL, respectively, in relation to the BChE enzyme they were lower and comparable to each other, at 7.83 mg/mL and 7.85 mg/mL, respectively. The authors of this study also determined the content of anthocyanins, phenolic compounds, and carotenoids in the most active MCP and NCP variants. It was observed that the content of these active compounds was lower in the MPC (namely 53.05 mg/100 g dm, 128.34 mg/100 g dm and 18.91 mg/100 g dm, respectively) than in the NPC, in which these compounds were at a higher level (namely 69.62 mg/100 g dm, 253.3 mg/100 g dm, and 19.07 mg/100 g dm). While comparing these values, it is clear that they did not allow the authors of this study to indicate/explain why micro purple carrot has a higher inhibitory activity in relation to the AChE and BChE enzymes than normal purple carrot, which is richer in these biologically active substances. It is also noticed, by referring to the IC_{50} values determined by Yusuf et al. [28] that the percentage of inhibition obtained in this study, determined for the concentration of Mv-s—100 μM, was about 78 to 190 times higher.

The antioxidant and free radical scavenging and aging-related enzymes, such as AChE and BChE, showed certain properties demonstrated in the study, which proved they are potent compounds, as well as being able to inhibit/delay free radical aging processes observed in the organism. Malvidin and its glycosides have been reported to positively impact neurodegenerative disease [16,29]. Lin et al. [30] demonstrated that preincubation with malvidin increased CAT activity and GSH concentration after hypoxia treatment, and cells also showed higher SOD activities. Zhao et al. [31] in a study on a murine microglial cell line revealed that malvidin prevented mitochondrial dysfunction, reduced ROS accumulation and lipid peroxidation, and increased antioxidant enzyme activity in the cerebrum. Giliani et al. [32] confirmed the neuroprotective effects of orally administered malvidin, regulating antioxidant levels and neuroinflammation in rats exposed to $AlCl_3$. In summary, research clearly supports that malvidin possesses antioxidant activity by inhibiting acetylcholinesterase and managing oxidative stress in neuronal cells.

2.2. *Anti-Inflammatory Activity—COX Inhibition*

The potential anti-inflammatory activity of Mv, Mv 3-glc and Mv 3,5-diglc was proved on the basis of inhibition of the COX activity. Anthocyanins/flavonoids, also known as nature's tender drugs, possess various pharmacological activities, including antioxidant and anti-inflammatory activities. These compounds slow down enzyme activities of arachidonic acid metabolizing enzymes, such as phospholipase PLA2, cyclooxygenase, lipoxygenase and others [33,34]; the inhibition of these enzymes, which is caused by anthocyanins, including cyclooxygenases, is definitely one of the important cellular mechanism of anti-inflammation [35]. The results of our studies, summarized in Table 2, contain the concentration values of IC_{50} for Mv and its glycoside derivatives, which inhibit the activity of COX-1 and COX-2 by 50%. The data presented in Table 2 indicate that Mv had the greatest ability to inhibit both COX-1 and COX-2. The glycosylation of malvidin significantly reduced the anti-inflammatory activity of these derivatives compared to Mv. This reduction was within the range of about 6–7 times in relation to COX-1, and within the range of about 14.5–24 times in the case of COX-2.

In Table 2 the anthocyanin selectivity index, i.e., the COX-2/COX-1 ratio, is also included. In our research, we obtained a COX-2/COX-1 selectivity ratio of 0.22, 0.53, and 0.83 for Mv, Mv 3-glc, and Mv 3,5-diglc, respectively. This indicates that the anthocyanins exhibit approximately 4.5 to 1.4 times higher activity toward COX-2 than COX-1. Medical anti-inflammatory drugs, including NSAIDs, can cause side effects, e.g., gastrointestinal bleeding, cancer promotion, etc. [36]. In our opinion, it is highly likely that anthocyanins/flavonoids have multiple cellular mechanisms acting on multiple sites of cellular machinery; among them, one is responsible for anti-inflammation. which seems to be caused by an effect on eicosanoid generating enzymes (COX1 and COX2 and others. like 5-, 12-LOX, PLA2). It is possible that they can also, similarly to steroidal anti-inflammatory drugs, stimulate protein kinases PKC, PTK and MAPK and down-regulation of the expression of iNOS, TNF-α, IL-1β [37]. Demonstration of such possible modes of action of anthocyanins should be proved in further studies.

Referring to the literature data, which is concentrated on studies of malvidin and its sugar derivatives, and extracts rich in these anthocyanins, it may be concluded that they are connected with numerous anti-inflammatory mechanisms, a fact which was proved in various in vitro models. In the article prepared by Huang et al. [38], the effect of the malvidin-3-glucoside and malvidin-3-galactoside obtained from blueberry *Vaccinium ashei* on inflammatory response in endothelial cells was tested. Particularly, the authors discussed indicated the mechanism of anti-inflammatory activity of malvidin glucosides as a way of restraining tumor necrosis factor-alpha (TNF-α), and by nuclear factor-kappa B (NF-κB). Generally, in these processes, malvidin-3-glucoside exerted a more powerful effect than malvidin-3-galactoside. Fagundes et al. [21] studied the potential of anthocyanidin malvidin, with an intention to protect against and help in peptic ulcer treatment. Expression levels of oxidative and inflammatory genes in the mouse gut in the presence of a 5 $mg \cdot kg^{-1}$

dose were determined in order to investigate the mechanism of malvidin activity. Malvidin, thus. was proved to prevent gastric and duodenal ulcers. As a result, significant anti-inflammatory and anti-oxidative effects on the gastrointestinal tract connected with gene expression modulation and, similarly, an increase in endogenous defense mechanisms were observed. Three anthocyanin di-glucosides from *Syzygium cumini* pulp were isolated in the study conducted by Abdin et al. [39], including delphinidin 3,5-diglucoside, petunidin 3,5-diglucoside, and malvidin 3,5-diglucoside (MDG), with the malvidin derivative isolated at the highest yield. Moreover, MDG showed significant progress in inhibiting nitric oxide release and pro-inflammatory mouse cytokines, such as IL-1 beta, TNF-alpha and IL-6 in lipopolysaccharide-induced RAW264.7 macrophages. This experiment allowed us to better understand the structure–activity correlations of anthocyanin di-glucosides. The aim of yet another study was to examine the impact of malvidin on inflammatory responses and oxidative stress in peripheral blood mononuclear cells (PBMCs), which was generated by lipopolysaccharide (LPS). The most up-to-date studies proved that LPS significantly increased the (gene/cytokine expression) cytokine expression of IL-6, TNF-alpha, IL-1 beta, and COX-2 mRNA, and protein release from PBMCs, 22 h after treatment, whereas the expression of these cytokines (IL-6, TNF-alpha, IL-1 beta) and COX-2 mRNA were induced by LPS, secretion of protein in PBMC, and was much lower as a consequence of pretreatment of malvidin [40]. Finally, it is important to underline that there are also studies focusing on extracts rich in anthocyanins, inter alia malvidins and its sugar derivatives, as well as sugar cyanidin derivatives, which in vitro show their potential significance as drugs and substances which are in support of the treatment of inflammatory processes [41–43].

Table 2. Values of IC_{50} for malvidin (Mv), malvidin 3-*O*-glucoside (Mv 3-glc) and malvidin 3,5-*O*-diglucoside (Mv 3,5-diglc) of cyclooxygenase-1 (COX-1) and cyclooxygenase-2 (COX-2) and, for indomethacin cited as positive control, selectivity index COX-1/COX-2 is included. Different letters ($^{a–d}$) in the same row indicate significant differences ($p < 0.05$).

Compound	IC_{50} (μM)		
	COX-1	**COX-2**	**Selectivity Index COX-2/COX-1**
Mv	12.45 ± 0.70^{d}	2.76 ± 0.16^{d}	0.22
Mv 3-glc	74.78 ± 0.06^{b}	39.92 ± 3.02^{b}	0.53
Mv 3,5-diglc	90.36 ± 1.92^{a}	66.45 ± 1.93^{a}	0.83
Indomethacin *	18.32 ± 0.40^{c}	15.22 ± 1.36^{c}	0.83

* Strugała et al. [44].

2.3. *Computational Outcomes*

With the exception of the cationic structure, which is virtually absent at physiological pH, we conducted docking studies on all other deprotonated species that exist in significant populations. Subsequently, these structures underwent molecular dynamics simulations. We were able to qualitatively reproduce the experimental data to a large extent. Throughout the 10 ns of MD simulations, the RMSD of the complexes under examination remained stable, indicating good structural stability. The RMSF demonstrated similar behaviour, with only marginal deviations associated with the terminal regions, while the remaining ones exhibited highly superimposable behaviour. Additionally, the small extent of the fluctuations in Rg provided assurance of the stability of the formed systems. Supplementary Materials provide detailed information on these topics.

2.3.1. Inhibitory Activity towards AChE and BChE

In the theoretical investigation of AChE inhibition, we were able to replicate the experimental findings on malvidin. Selectivity towards AChE is evident in both the experimental and theoretical approaches, indicated by a greater % of inhibition or lower inhibition constant, respectively. The interaction with the given enzyme appears to be modulated by hydrogen bonds originating from the Tyr338 and Tyr69 amino acids present in the binding pocket. This is different from BChE, where Tyr126, Asp68, Gly113 and Tyr330 participate in binding the ligand. See Figure 1 below.

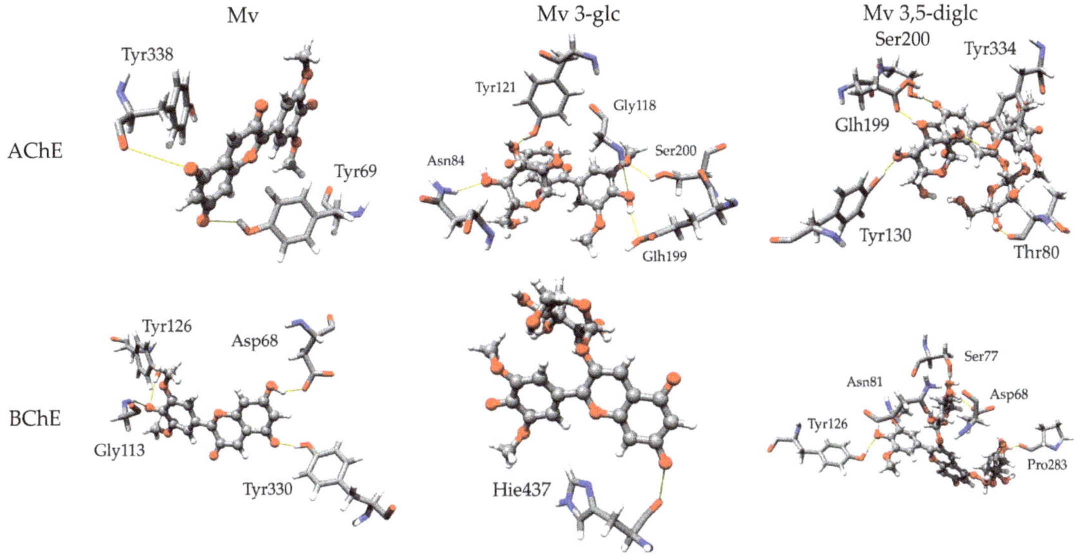

Figure 1. Post-MD 3D representations of the finest docked positions of Mv, Mv 3-glc and Mv 3,5-diglc species interacting with amino acids within the AChE and BChE binding pockets.

For monoglucoside, experimental results indicate more effective inhibition towards AChE, whereas theoretical studies demonstrate lower K_i values for binding towards BChE (see Table 3 below). Nevertheless, selectivity varies only slightly in either case. It is important to note the possibility of the ligand interacting outside the binding pocket, such as via allosteric modulation. These factors could impact the behaviour in this case. This may be expected since, in comparison to malvidin, its glucoside interacts effectively with a greater number of amino acids, thus amplifying the stabilization effect through hydrogen bonding interactions.

Table 3. Theoretically estimated inhibition constant (in M) of each species with a non-negligible population towards the AChE and BChE enzymes.

Compound	${}^fM$	AChE	BChE
Mv	39.56%	2.58×10^{-7}	1.52×10^{-6}
Mv^-	52.15%	1.89×10^{-7}	1.19×10^{-6}
Mv^{2-}	8.08%	1.16×10^{-7}	6.41×10^{-7}
Mv^{3-}	0.13%	8.04×10^{-8}	6.42×10^{-7}
	$K_{i\text{-}overall}$	$\boldsymbol{2.10 \times 10^{-7}}$	$\boldsymbol{1.28 \times 10^{-6}}$
Mv 3-glc	3.49%	2.77×10^{-6}	5.39×10^{-7}
Mv 3-glc$^-$	78.18%	2.42×10^{-6}	5.99×10^{-7}
Mv 3-glc^{2-}	18.33%	2.88×10^{-6}	4.12×10^{-7}
	$K_{i\text{-}overall}$	$\boldsymbol{2.52 \times 10^{-6}}$	$\boldsymbol{5.63 \times 10^{-7}}$
Mv 3,5-diglc	39.19%	4.51×10^{-3}	3.70×10^{-7}
Mv 3,5-diglc$^-$	60.70%	1.63×10^{-4}	8.99×10^{-8}
	$K_{i\text{-}overall}$	$\boldsymbol{1.87 \times 10^{-3}}$	$\boldsymbol{1.99 \times 10^{-7}}$

The experimental data confirm that the diglucoside is preferentially inhibitory towards BChE. However, the computational study suggests significant divergence in inhibition outcomes between AChE and BChE, compared to the experimental data. Similar to monoglucoside, Glh199 and Ser200 participate in this interaction in AChE, while Asp68 and Tyr126 are common between malvidin and its diglucoside in BChE.

Deprotonation appears to result in improved inhibition of AChE by Mv, and even to a greater extent, by diglucoside. However, the effect is less pronounced in the case of monoglucoside. The inhibition of BChE displays a similar deprotonation trend to that of AChE. The addition of a sugar moiety through glycosylation leads to a reduction in inhibitory potential. This effect is more pronounced when two residues are added as opposed to a single glucose. The variations can reach up to three orders of magnitude. However, in the case of BChE, the impact of glycosylation is less noticeable as the diglucoside binds better than the aglycone.

2.3.2. Inhibitory Activity towards COX-1 and COX-2

The assays showed a general preference for COX-1, but computational data only found this behavior for diglucoside. The differences in inhibition constants vary by up to three orders of magnitude for monoglucoside, whereas for aglycone and diglucoside this is only a difference of a single decimal place (see Table 4 below). However, the previously presented IC_{50} is high, which may lead to fluctuation in assessment and make it more prone to errors.

The process of deprotonation seems to improve the inhibitory activity towards COX-1 to some extent, while the reverse effect is observed for COX-2. Therefore, the hydrophobic nature of interactions appears to be of the utmost importance.

Moreover, it is apparent that glycosylation causes a significant increase in inhibition constants, indicating that a higher concentration is required to observe the inhibition process. Certainly, it is evident that glucosides can either shield the formation of stronger, more desirable bonding that retains the ligand in a pocket (see Figure 2 below), or block its further propagation deeper into the binding site, thus indirectly reducing the observed inhibition.

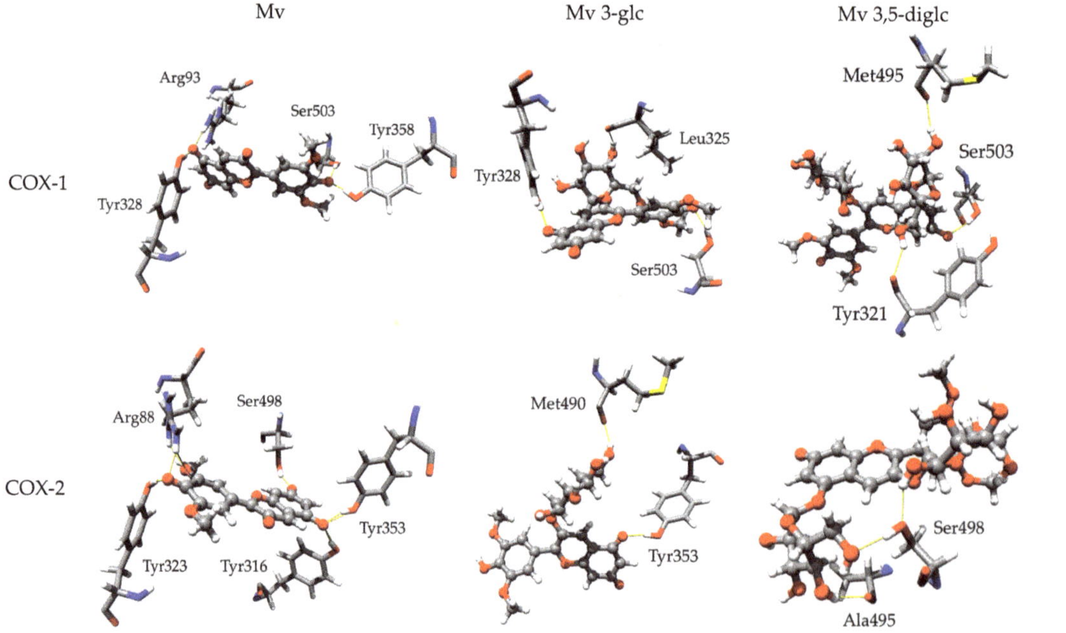

Figure 2. Post-MD 3D representations of the finest docked positions of Mv, Mv 3-glc and Mv 3,5-diglc species interacting with amino acids within the COX-1 and COX-2 binding pockets.

Table 4. Theoretically estimated inhibition constant (in M) of each species with a non-negligible population towards the COX-1 and COX-2 enzymes.

Compound	^{f}M	COX-1	COX-2
Mv	39.56%	9.08×10^{-4}	5.42×10^{-5}
Mv^{-}	52.15%	8.18×10^{-4}	6.68×10^{-5}
Mv^{2-}	8.08%	6.41×10^{-4}	4.85×10^{-5}
Mv^{3-}	0.13%	5.46×10^{-4}	3.92×10^{-5}
	$K_{i\text{-}overall}$	$\mathbf{8.38 \times 10^{-4}}$	$\mathbf{6.02 \times 10^{-5}}$
Mv 3-glc	3.49%	3.43×10^{2}	2.41×10^{-3}
Mv 3-glc^{-}	78.18%	2.89×10^{2}	1.15×10^{-1}
Mv 3-glc^{2-}	18.33%	2.64×10^{2}	2.90×10^{-1}
	$K_{i\text{-}overall}$	$\mathbf{2.87 \times 10^{2}}$	$\mathbf{1.43 \times 10^{-1}}$
Mv 3,5-diglc	39.19%	1.34×10^{11}	1.98×10^{9}
Mv 3,5-$diglc^{-}$	60.70%	1.12×10^{11}	3.20×10^{12}
	$K_{i\text{-}overall}$	$\mathbf{1.21 \times 10^{11}}$	$\mathbf{1.95 \times 10^{12}}$

3. Materials and Methods

3.1. Materials

5,5′-Dithiobis(2-nitrobenzoic acid), (DTNB), acetylthiocholine iodide, butyryl-thiocholine iodide, acetylcholinesterase from Electrophorus electricus, type VI-S (aChE), Butyrylcholinesterase from equine serum (BChE), neostigmine bromide, and sodium dodecyl sulfate (SDS) were purchased from Merck kGaA, Darmstadt, Germany. COX-1 and COX-2 colorimetric inhibitor screening assay kits (Catalog No. 701050) were purchased from Cayman Chemical, Ann Arbor, MI, USA. Malvidin (purity $\geq$ 97%), malvidin 3-*O*-glucoside (purity $\geq$ 95%) and malvidin 3,5-*O*-diglucoside (purity $\geq$ 95%), (Figure 3) were purchased from Extrasynthese (Genay, France), and all other chemicals were of analytical grade.

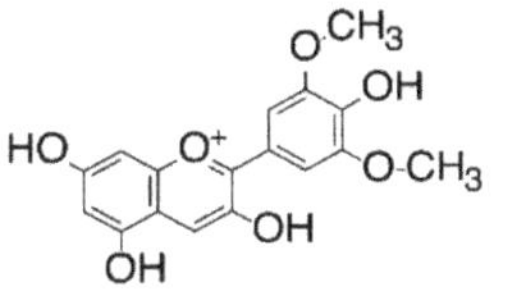

Malvidin

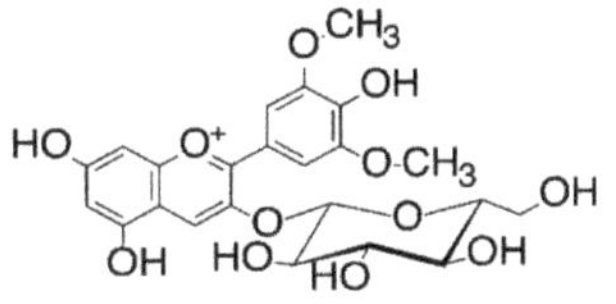

Malvidin-3-*O*-glucoside

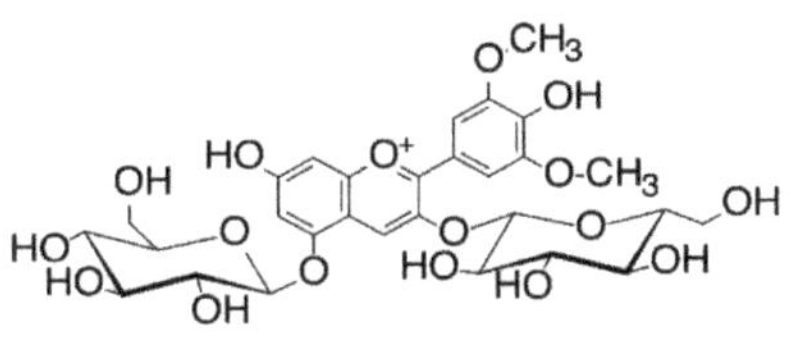

Malvidin-3,5-*O*-diglucoside

Figure 3. Chemical structures of the malvidin group.

3.2. *Acetylcholinesterase and Butyrylcholinesterase Inhibition Assay*

Referring to both the Ellman [45] and Jin et al. [46] methods, we identified acetylcholinesterase/butyrylcholinesterase activity, implementing minor modifications. In brief, the assay was carried out on a 96-well plate, with each well containing 140 µL 0.1 M phosphate buffer (pH 8.0), 10 µL of stock solution of Mv, or Mv 3-glc or Mv 3,5-diglc, 10 µL AChE/BChE (1 units/mL). The plate was incubated for 10 min at 25 °C. Subsequently, 10 µL of a 10 mM DTNB was added to the reaction mixture. In the following step, the reaction was initiated by adding 10 µL of a 14 mM acetylthiocholine iodide/butyryl-thiocholine iodide. The plate was shaken for one minute and finally 20 μL of 5% SDS was added in order to stop the reaction. Control wells, which contained the same composition except for the studied compounds (10 μL, 70% ethanol), were included. Absorbance at λ = 417 nm was recorded using a plate reader (EPOCH, Bio Tech, Dover, MA, USA) after 10 min incubation. All of the reactions were performed in five repetitions. The percent of AChE/BChE inhibition was calculated using the following formula:

$$\%\text{Inhibition} = \frac{A_c - A_s}{A_c} \cdot 100\% \quad (1)$$

where: A_s refers to the absorbance of the sample containing Mvs and Ac refers to the absorbance of the control sample. The studies of AChE and BChE inhibition were also performed depending on the concentration of the drug, neostigmine, as a positive control.

3.3. Cyclooxygenase (COX) Inhibition Assay

We determined the inhibition of both COX-1 and COX-2 activity using a colorimetric inhibitor screening assay kit (Catalog No. 701050), following the manufacturer's instructions and as was previously explained by Jang and Pezzuto [47] and Strugała et al. [48]. The reaction mixtures were prepared in 0.1 M Tris–HCl buffer, pH 7.4.0 containing 10 μL hemin, 10 μL COX-1 (ovine) or COX-2 (human recombinant) and 10 μL Mv, Mv 3-glc Mv 3,5-diglc at final concentrations: 0.9–46 μM. Firstly, the plate was carefully shaken for a few seconds and incubated for 5 min at 25 °C. Secondly, 20 μL of colorimetric substrate solution was added to each plate's well. It was the addition of 20 μL of arachidonic acid (final concentration in reaction mixture 100 μM) which initiated the reaction. Thirdly, the plate was shaken again for a few seconds and incubated for two min at 25 °C. The cyclooxygenase activity was assayed colorimetrically in a way of monitoring the appearance of oxidized TMPD at λ = 590 nm using a plate reader (EPOCH, Bio Tech). The percentage of COX inhibition was calculated using the Equation (1):

The IC_{50} parameter was calculated referring to plots showing the relation between percentage of COX inhibition and concentration of the compounds.

3.4. Computational Studies

Molecular docking was performed using AutoDock Vina 1.2.0 [49,50] on all structures that were present at physiological pH and had been established in the previous publication [23]. To obtain representative polyphenol:protein complexes, the existing bound ligand was used in the docking process.

Enzyme crystal structures were obtained from the RSCB PDB database and consisted of human COX–1 (PDB ID: 6Y3C, res. 3.36 Å) [51] and human COX-2 bound with meclofenamic acid (PDB ID: 5IKQ, res. 2.41 Å) [52], the structure of human acetylcholinesterase (AChE) bound with donepezil (PDB ID: 6O4W, resolution [53] 2.35 Å, and butyrylcholinesterase (AChE) bound with profenamine (PDB ID: 5K5E, resolution 2.80 Å [54]. They were chosen for analysis due to the presence of inhibitors, fair resolution, unmutated *H. sapiens* source of origin, and the absence of any missing loops, except for terminal segments.

To ensure complete residue coverage within 5 Å of the inhibitor's mass center, we employed a docking box. We redocked the initial ligands, and the process succeeded. Unfortunately, we could not find a suitable enzyme inhibitor complex for COX-1, so we overlaid it with COX-2, which shares a binding pocket [51]. Next, we executed the docking process in the relevant box for the meclofenamic acid binding site. As the energy differences and positions of the overall results were not significant between docking outcomes, we conducted molecular dynamics simulations with the best one. Prior to that, we used the corresponding $\Delta G_{\text{binding}}$ values [55] to determine the inhibition constant K_i [M] by following the equation:

$$K_i = \exp\left(\frac{-\Delta G_{binding}}{RT}\right) \quad (2)$$

Before commencing the molecular dynamics, all ligands were parametrized. Gas-phase geometry optimization was carried out with HF/6-31G*. The electrostatic potential was fitted according to the Merz–Singh–Kollman scheme [56] to determine atomic charges, using the RESP procedure. The GAFF2 force field [57] was utilized to extrapolate the parameters. To generate the necessary files, the Antechamber and parmchk modules of Amber16 were utilized [58].

Subsequently, a refinement process was performed on the enzymes to eliminate any superfluous molecules, including crystallographic water and solvents. The protonation states of titratable residues at pH 7.4 were then estimated, using the PDB2PQR server [59,60] and the PropKa algorithm [61]. Each complex was solvated with a TIP3P water box and, in order to maintain system electroneutrality, either sodium or chloride counterions were added. After conducting initial system optimization and heating to 310 K within the NVT and NPT ensembles, 10 ns production simulations were performed for each complex at the equivalent temperature and pressure. The simulations utilized a time constant of 2.0 ps and employed the SHAKE algorithm [62] and Particle Mesh Ewald [63], which made way for an integration step of 2 fs. Further, a cut-off radius of 1.2 nm was utilized. The representative structures were identified from the most populous cluster of the molecular dynamics results.

All simulations were performed using the Gromacs software 2022.2 package [64], whilst the UCSF Chimera software 1.17.3 [65] was employed to analyze the interactions.

3.5. Statistical Analysis

Data are shown as mean values ± standard deviation (SD). p values < 0.05 were considered statistically significant. Statistical analysis was performed using the program Statistica 12.0 (StatSoft, Kraków, Poland).

4. Conclusions

Our in vitro study demonstrates that malvidin, as well as its glucosides, exert on COX-1, COX-2 and AChE, BChE an enzyme inhibitory effect. In addition, it was shown that the presence of sugar residues affected the anti-inflammatory and anti-cholinergic activity properties. This demonstrates that the glycosylation of malvidin substantially diminished its anti-inflammatory activity. The computational protocol led us establish the role of protonation and deprotonation in the inhibitory activity, as well giving insights into the amino acids comprising the given behaviour.

To sum up, despite significant progress being made in the search for natural products with antioxidant properties, understanding of the structure–activity relationships (SARs) among anthocyanins and their effects on enzyme inhibitors remains incomplete, which is why it is necessary to fully elucidate their relationship.

Supplementary Materials: The following supporting information can be downloaded at: https://www.mdpi.com/article/10.3390/molecules28237872/s1.

Author Contributions: Conceptualization, P.S.-D., M.S. and J.G.; methodology, P.S.-D. and M.S.; software, P.S.-D. and M.S.; validation, P.S.-D. and M.S.; investigation, P.S.-D. and M.S.; data curation, P.S.-D. and M.S.; writing—original draft preparation, P.S.-D., M.S. and J.G.; writing—review and editing, P.S.-D.; visualization, P.S.-D. and M.S.; supervision, J.G. All authors have read and agreed to the published version of the manuscript.

Funding: The APC/BPC is co-financed by Wroclaw University of Environmental and Life Sciences and by the National Science Centre, Poland, Grant No. 2017/25/N/NZ9/02915.

Institutional Review Board Statement: Not applicable.

Informed Consent Statement: Not applicable.

Data Availability Statement: Data is contained within the article.

Acknowledgments: Computations were carried out using the computers of Centre of Informatics Tricity Academic Supercomputer and Network.

Conflicts of Interest: The authors declare no conflict of interest.

References

1. Mukherjee, P.K.; Kumar, V.; Mal, M.; Houghton, P.J. Acetylcholinesterase Inhibitors from Plants. *Phytomedicine* **2007**, *14*, 289–300. [CrossRef] [PubMed]
2. Spiegel, M.; Marino, T.; Prejanò, M.; Russo, N. Antioxidant and Copper-Chelating Power of New Molecules Suggested as Multiple Target Agents against Alzheimer's Disease. A Theoretical Comparative Study. *Phys. Chem. Chem. Phys.* **2022**, *24*, 16353–16359. [CrossRef] [PubMed]
3. Colovic, M.B.; Krstic, D.Z.; Lazarevic-Pasti, T.D.; Bondzic, A.M.; Vasic, V.M. Acetylcholinesterase Inhibitors: Pharmacology and Toxicology. *Curr. Neuropharmacol.* **2013**, *11*, 315–335. [CrossRef] [PubMed]
4. Ademosun, A.O.; Oboh, G.; Bello, F.; Ayeni, P.O. Antioxidative Properties and Effect of Quercetin and Its Glycosylated form (Rutin) on Acetylcholinesterase and Butyrylcholinesterase Activities. *J. Evid. Based Complement. Altern. Med.* **2016**, *21*, NP11–NP17. [CrossRef] [PubMed]
5. Pacheco, S.M.; Soares, M.S.P.; Gutierres, J.M.; Gerzson, M.F.B.; Carvalho, F.B.; Azambuja, J.H.; Schetinger, M.R.C.; Stefanello, F.M.; Spanevello, R.M. Anthocyanins as a Potential Pharmacological Agent to Manage Memory Deficit, Oxidative Stress and Alterations in Ion Pump Activity Induced by Experimental Sporadic Dementia of Alzheimer's Type. *J. Nutr. Biochem.* **2018**, *56*, 193–204. [CrossRef] [PubMed]
6. Yu, L.; Li, H.; Liu, W.; Zhang, L.; Tian, Q.; Li, H.; Li, M. MiR-485-3p Serves as a Biomarker and Therapeutic Target of Alzheimer's Disease via Regulating Neuronal Cell Viability and Neuroinflammation by Targeting AKT3. *Mol. Genet. Genom. Med.* **2021**, *9*, e1548. [CrossRef] [PubMed]
7. Koh, H.S.; Lee, S.; Lee, H.J.; Min, J.W.; Iwatsubo, T.; Teunissen, C.E.; Cho, H.J.; Ryu, J.H. Targeting MicroRNA-485-3p Blocks Alzheimer's Disease Progression. *Int. J. Mol. Sci.* **2021**, *22*, 13136. [CrossRef]
8. Rai, M.; Pandit, R.; Gaikwad, S.; Yadav, A.; Gade, A. Potential Applications of Curcumin and Curcumin Nanoparticles: From Traditional Therapeutics to Modern Nanomedicine. *Nanotechnol. Rev.* **2015**, *4*, 161–172. [CrossRef]
9. Arshad, L.; Areeful Haque, M.; Bukhari, S.N.A.; Jantan, I. An Overview of Structure-Activity Relationship Studies of Curcumin Analogs as Antioxidant and Anti-Inflammatory Agents. *Future Med. Chem.* **2017**, *9*, 605–626. [CrossRef]
10. Urošević, M.; Nikolić, L.; Gajić, I.; Nikolić, V.; Dinić, A.; Miljković, V. Curcumin: Biological Activities and Modern Pharmaceutical Forms. *Antibiotics* **2022**, *11*, 135. [CrossRef]
11. Hussain, H.; Ahmad, S.; Shah, S.W.A.; Ullah, A.; Rahman, S.U.; Ahmad, M.; Almehmadi, M.; Abdulaziz, O.; Allahyani, M.; Alsaiari, A.A.; et al. Synthetic Mono-Carbonyl Curcumin Analogues Attenuate Oxidative Stress in Mouse Models. *Biomedicines* **2022**, *10*, 2597. [CrossRef] [PubMed]
12. Khan, S.; Khan, H.U.; Khan, F.A.; Shah, A.; Wadood, A.; Ahmad, S.; Almehmadi, M.; Alsaiari, A.A.; Shah, F.U.; Kamran, N. Anti-Alzheimer and Antioxidant Effects of *Nelumbo Nucifera* L. Alkaloids, Nuciferine and Norcoclaurine in Alloxan-Induced Diabetic Albino Rats. *Pharmaceuticals* **2022**, *15*, 1205. [CrossRef] [PubMed]
13. Olennikov, D.N.; Kashchenko, N.I.; Chirikova, N.K.; Akobirshoeva, A.; Zilfikarov, I.N.; Vennos, C. Isorhamnetin and Quercetin Derivatives as Anti-Acetylcholinesterase Principles of Marigold (*Calendula officinalis*) Flowers and Preparations. *Int. J. Mol. Sci.* **2017**, *18*, 1685. [CrossRef] [PubMed]
14. Szwajgier, D. Anticholinesterase Activities of Selected Polyphenols—A Short Report. *Pol. J. Food Nutr. Sci.* **2014**, *64*, 59–64. [CrossRef]
15. Temviriyanukul, P.; Sritalahareuthai, V.; Jom, K.N.; Jongruaysup, B.; Tabtimsri, S.; Pruesapan, K.; Thangsiri, S.; Inthachat, W.; Siriwan, D.; Charoenkiatkul, S.; et al. Comparison of Phytochemicals, Antioxidant, and In Vitro Anti-Alzheimer Properties of Twenty-Seven *Morus* Spp. Cultivated in Thailand. *Molecules* **2020**, *25*, 2600. [CrossRef] [PubMed]
16. Merecz-Sadowska, A.; Sitarek, P.; Kowalczyk, T.; Zajdel, K.; Jęcek, M.; Nowak, P.; Zajdel, R. Food Anthocyanins: Malvidin and Its Glycosides as Promising Antioxidant and Anti-Inflammatory Agents with Potential Health Benefits. *Nutrients* **2023**, *15*, 3016. [CrossRef]
17. Büttner, R.; Schulz, A.; Reuter, M.; Akula, A.K.; Mindos, T.; Carlstedt, A.; Riecken, L.B.; Baader, S.L.; Bauer, R.; Morrison, H. Inflammaging Impairs Peripheral Nerve Maintenance and Regeneration. *Aging Cell* **2018**, *17*, e12833. [CrossRef]
18. Hu, W.T.; Howell, J.C.; Ozturk, T.; Gangishetti, U.; Kollhoff, A.L.; Hatcher-Martin, J.M.; Anderson, A.M.; Tyor, W.R. CSF Cytokines in Aging, Multiple Sclerosis, and Dementia. *Front. Immunol.* **2019**, *10*, 480. [CrossRef]
19. Salaritabar, A.; Darvishi, B.; HadjiakhoonDi, F.; Manayi, A.; Sureda, A.; Nabavi, S.F.; Fitzpatrick, L.R.; Nabavi, S.M.; Bishayee, A. Therapeutic Potential of Flavonoids in Inflammatory Bowel Disease: A Comprehensive Review. *World J. Gastroenterol.* **2017**, *23*, 5097–5114. [CrossRef]
20. Kim, H.P. The Long Search for Pharmacologically Useful Anti-Inflammatory Flavonoids and Their Action Mechanisms: Past, Present, and Future. *Biomol. Ther.* **2022**, *30*, 117. [CrossRef]
21. Fagundes, F.L.; Pereira, Q.C.; Zarricueta, M.L.; Dos Santos, R.d.C. Malvidin Protects against and Repairs Peptic Ulcers in Mice by Alleviating Oxidative Stress and Inflammation. *Nutrients* **2021**, *13*, 3312. [CrossRef] [PubMed]

22. Seeram, N.P.; Momin, R.A.; Nair, M.G.; Bourquin, L.D. Cyclooxygenase Inhibitory and Antioxidant Cyanidin Glycosides in Cherries and Berries. *Phytomedicine* **2001**, *8*, 362–369. [CrossRef] [PubMed]
23. Strugała-Danak, P.; Spiegel, M.; Hurynowicz, K.; Gabrielska, J. Interference of Malvidin and Its Mono- and Di-Glucosides on the Membrane—Combined in Vitro and Computational Chemistry Study. *J. Funct. Foods* **2022**, *99*, 105340. [CrossRef]
24. Rashwan, A.K.; Karim, N.; Xu, Y.; Xie, J.; Cui, H.; Mozafari, M.R.; Chen, W. Potential Micro-/Nano-Encapsulation Systems for Improving Stability and Bioavailability of Anthocyanins: An Updated Review. *Crit. Rev. Food Sci. Nutr.* **2021**, *63*, 3362–3385. [CrossRef]
25. Strugała, P.; Loi, S.; Bazanów, B.; Kuropka, P.; Kucharska, A.Z.; Włoch, A.; Gabrielska, J. A Comprehensive Study on the Biological Activity of Elderberry Extract and Cyanidin 3-O-Glucoside and Their Interactions with Membranes and Human Serum Albumin. *Molecules* **2018**, *23*, 2566. [CrossRef]
26. Roncato, J.F.F.; Camara, D.; Brussulo Pereira, T.C.; Quines, C.B.; Colomé, L.M.; Denardin, C.; Haas, S.; Ávila, D.S. Lipid Reducing Potential of Liposomes Loaded with Ethanolic Extract of Purple Pitanga (*Eugenia uniflora*) Administered to *Caenorhabditis elegans*. *J. Liposome Res.* **2019**, *29*, 274–282. [CrossRef]
27. Jucá, M.M.; Cysne Filho, F.M.S.; de Almeida, J.C.; Mesquita, D.d.S.; Barriga, J.R.d.M.; Dias, K.C.F.; Barbosa, T.M.; Vasconcelos, L.C.; Leal, L.K.A.M.; Ribeiro, J.E.; et al. Flavonoids: Biological Activities and Therapeutic Potential. *Nat. Prod. Res.* **2020**, *34*, 692–705. [CrossRef]
28. Yusuf, E.; Wojdyło, A.; Oszmiański, J.; Nowicka, P. Nutritional, Phytochemical Characteristics and In Vitro Effect on α-Amylase, α-Glucosidase, Lipase, and Cholinesterase Activities of 12 Coloured Carrot Varieties. *Foods* **2021**, *10*, 808. [CrossRef]
29. Henriques, J.F.; Serra, D.; Dinis, T.C.P.; Almeida, L.M. The Anti-Neuroinflammatory Role of Anthocyanins and Their Metabolites for the Prevention and Treatment of Brain Disorders. *Int. J. Mol. Sci.* **2020**, *21*, 8653. [CrossRef]
30. Lin, Y.C.; Tsai, P.F.; Wu, J.S.B. Protective Effect of Anthocyanidins against Sodium Dithionite-Induced Hypoxia Injury in C6 Glial Cells. *J. Agric. Food Chem.* **2014**, *62*, 5603–5608. [CrossRef]
31. Zhao, P.; Li, X.; Yang, Q.; Lu, Y.; Wang, G.; Yang, H.; Dong, J.; Zhang, H. Malvidin Alleviates Mitochondrial Dysfunction and ROS Accumulation through Activating AMPK-α/UCP2 Axis, Thereby Resisting Inflammation and Apoptosis in SAE Mice. *Front. Pharmacol.* **2023**, *13*, 1038802. [CrossRef] [PubMed]
32. Gilani, S.J.; Bin-Jumah, M.N.; Al-Abbasi, F.A.; Imam, S.S.; Alshehri, S.; Ghoneim, M.M.; Shahid Nadeem, M.; Afzal, M.; Alzarea, S.I.; Sayyed, N.; et al. Antiamnesic Potential of Malvidin on Aluminum Chloride Activated by the Free Radical Scavenging Property. *ACS Omega* **2022**, *7*, 24231–24240. [CrossRef] [PubMed]
33. Hou, D.X.; Yanagita, T.; Uto, T.; Masuzaki, S.; Fujii, M. Anthocyanidins Inhibit Cyclooxygenase-2 Expression in LPS-Evoked Macrophages: Structure-Activity Relationship and Molecular Mechanisms Involved. *Biochem. Pharmacol.* **2005**, *70*, 417–425. [CrossRef]
34. Hämäläinen, M.; Nieminen, R.; Vuorela, P.; Heinonen, M.; Moilanen, E. Anti-Inflammatory Effects of Flavonoids: Genistein, Kaempferol, Quercetin, and Daidzein Inhibit STAT-1 and NF-KappaB Activations, Whereas Flavone, Isorhamnetin, Naringenin, and Pelargonidin Inhibit Only NF-KappaB Activation along with Their Inhibitory Effect on INOS Expression and NO Production in Activated Macrophages. *Mediat. Inflamm.* **2007**, *2007*, 045673. [CrossRef]
35. Baumann, J.; Bruchhausen, F.V.; Wurm, G. Flavonoids and Related Compounds as Inhibition of Arachidonic Acid Peroxidation. *Prostaglandins* **1980**, *20*, 627–639. [CrossRef] [PubMed]
36. Kalita, J.; Dutta, K.; Sen, S.; Dey, B.K.; Gogoi, P. Perceived Benefits and Risk of NSAIDs in Relation to Its Association with Cancer: A Comprehensive Review. *J. Pharm. Res. Int.* **2021**, *33*, 236–244. [CrossRef]
37. Kim, H.P.; Son, K.H.; Chang, H.W.; Kang, S.S. Anti-Inflammatory Plant Flavonoids and Cellular Action Mechanisms. *J. Pharmacol. Sci.* **2004**, *96*, 229–245. [CrossRef]
38. Huang, W.-Y.; Wang, X.-N.; Wang, J.; Sui, Z.-Q. Malvidin and Its Glycosides from Vaccinium Ashei Improve Endothelial Function by Anti-Inflammatory and Angiotensin I-Converting Enzyme Inhibitory Effects. *Nat. Prod. Commun.* **2018**, *13*, 49–52. [CrossRef]
39. Abdin, M.; Hamed, Y.S.; Akhtar, H.M.S.; Chen, D.; Chen, G.; Wan, P.; Zeng, X. Antioxidant and Anti-Inflammatory Activities of Target Anthocyanins Di-Glucosides Isolated from Syzygium Cumini Pulp by High Speed Counter-Current Chromatography. *J. Food Biochem.* **2020**, *44*, 1050–1062. [CrossRef]
40. Bastin, A.R.; Sadeghi, A.; Abolhassani, M.; Doustimotlagh, A.H.; Mohammadi, A. Malvidin Prevents Lipopolysaccharide-Induced Oxidative Stress and Inflammation in Human Peripheral Blood Mononuclear Cells. *IUBMB Life* **2020**, *72*, 1504–1514. [CrossRef]
41. Chao, C.Y.; Liu, W.H.; Wu, J.J.; Yin, M.C. Phytochemical Profile, Antioxidative and Anti-Inflammatory Potentials of Gynura Bicolor DC. *J. Sci. Food Agric.* **2015**, *95*, 1088–1093. [CrossRef] [PubMed]
42. Marchi, P.; Paiotti, A.P.R.; Neto, R.A.; Oshima, C.T.F.; Ribeiro, D.A. Concentrated Grape Juice (G8000^{TM}) Reduces Immunoexpression of INOS, TNF-Alpha, COX-2 and DNA Damage on 2,4,6-Trinitrobenzene Sulfonic Acid-Induced-Colitis. *Environ. Toxicol. Pharmacol.* **2014**, *37*, 819–827. [CrossRef] [PubMed]
43. Bognar, E.; Sarszegi, Z.; Szabo, A.; Debreceni, B.; Kalman, N.; Tucsek, Z.; Sumegi, B.; Gallyas, F. Antioxidant and Anti-Inflammatory Effects in RAW264.7 Macrophages of Malvidin, a Major Red Wine Polyphenol. *PLoS ONE* **2013**, *8*, e65355. [CrossRef] [PubMed]
44. Strugała, P.; Cyboran-Mikołajczyk, S.; Dudra, A.; Mizgier, P.; Kucharska, A.Z.; Olejniczak, T.; Gabrielska, J. Biological Activity of Japanese Quince Extract and Its Interactions with Lipids, Erythrocyte Membrane, and Human Albumin. *J. Membr. Biol.* **2016**, *249*, 393. [CrossRef] [PubMed]

45. Ellman, G.L.; Courtney, K.D.; Andres, V.; Featherstone, R.M. A New and Rapid Colorimetric Determination of Acetylcholinesterase Activity. *Biochem. Pharmacol.* **1961**, *7*, 88–95. [CrossRef]
46. Jin, H. Acetylcholinesterase and Butyrylcholinesterase Inhibitory Properties of Functionalized Tetrahydroacridines and Related Analogs. *Med. Chem.* **2014**, *4*, 688–696. [CrossRef]
47. Jang, M.S.; Pezzuto, J.M. Assessment of Cyclooxygenase Inhibitors Using in Vitro Assay Systems. *Methods Cell Sci.* **1997**, *19*, 25–31. [CrossRef]
48. Strugała, P.; Urbaniak, A.; Kuryś, P.; Włoch, A.; Kral, T.; Ugorski, M.; Hof, M.; Gabrielska, J. Antitumor and Antioxidant Activities of Purple Potato Ethanolic Extract and Its Interaction with Liposomes, Albumin and Plasmid DNA. *Food Funct.* **2021**, *12*, 1271–1290. [CrossRef]
49. Eberhardt, J.; Santos-Martins, D.; Tillack, A.F.; Forli, S. AutoDock Vina 1.2.0: New Docking Methods, Expanded Force Field, and Python Bindings. *J. Chem. Inf. Model.* **2021**, *61*, 3891–3898. [CrossRef]
50. Trott, O.; Olson, A.J. AutoDock Vina: Improving the Speed and Accuracy of Docking with a New Scoring Function, Efficient Optimization, and Multithreading. *J. Comput. Chem.* **2010**, *31*, 455–461. [CrossRef]
51. Blobaum, A.L.; Marnett, L.J. Structural and Functional Basis of Cyclooxygenase Inhibition. *J. Med. Chem.* **2007**, *50*, 1425–1441. [CrossRef] [PubMed]
52. Orlando, B.J.; Malkowski, M.G. Substrate-Selective Inhibition of Cyclooxygeanse-2 by Fenamic Acid Derivatives Is Dependent on Peroxide Tone. *J. Biol. Chem.* **2016**, *291*, 15069–15081. [CrossRef] [PubMed]
53. Gerlits, O.; Ho, K.Y.; Cheng, X.; Blumenthal, D.; Taylor, P.; Kovalevsky, A.; Radić, Z. A New Crystal Form of Human Acetylcholinesterase for Exploratory Room-Temperature Crystallography Studies. *Chem. Biol. Interact.* **2019**, *309*, 108698. [CrossRef] [PubMed]
54. Chen, E.P.; Bondi, R.W.; Michalski, P.J. Model-Based Target Pharmacology Assessment (MTPA): An Approach Using PBPK/PD Modeling and Machine Learning to Design Medicinal Chemistry and DMPK Strategies in Early Drug Discovery. *J. Med. Chem.* **2021**, *64*, 3185–3196. [CrossRef] [PubMed]
55. Spiegel, M.; Krzyżek, P.; Dworniczek, E.; Adamski, R.; Sroka, Z. In Silico Screening and In Vitro Assessment of Natural Products with Anti-Virulence Activity against Helicobacter Pylori. *Molecules* **2021**, *27*, 20. [CrossRef] [PubMed]
56. Bayly, C.I.; Cieplak, P.; Cornell, W.D.; Kollman, P.A. A Well-Behaved Electrostatic Potential Based Method Using Charge Restraints for Deriving Atomic Charges: The RESP Model. *J. Phys. Chem.* **1993**, *97*, 10269–10280. [CrossRef]
57. He, X.; Man, V.H.; Yang, W.; Lee, T.S.; Wang, J. A Fast and High-Quality Charge Model for the next Generation General AMBER Force Field. *J. Chem. Phys.* **2020**, *153*, 114502. [CrossRef]
58. Case Ross, C.; Walker, D.A.; Darden Junmei Wang, T. *Amber 2016 Reference Manual Principal Contributors to the Current Codes*; University of California: San Francisco, CA, USA, 2016.
59. Jensen, L.J.; Kuhn, M.; Stark, M.; Chaffron, S.; Creevey, C.; Muller, J.; Doerks, T.; Julien, P.; Roth, A.; Simonovic, M.; et al. STRING 8—A Global View on Proteins and Their Functional Interactions in 630 Organisms. *Nucleic Acids Res.* **2009**, *37*, D412–D416. [CrossRef]
60. Dolinsky, T.J.; Nielsen, J.E.; McCammon, J.A.; Baker, N.A. PDB2PQR: An Automated Pipeline for the Setup of Poisson–Boltzmann Electrostatics Calculations. *Nucleic Acids Res.* **2004**, *32*, W665–W667. [CrossRef]
61. Søndergaard, C.R.; Olsson, M.H.M.; Rostkowski, M.; Jensen, J.H. Improved Treatment of Ligands and Coupling Effects in Empirical Calculation and Rationalization of pK_a Values. *J. Chem. Theory Comput.* **2011**, *7*, 2284–2295. [CrossRef]
62. Ryckaert, J.P.; Ciccotti, G.; Berendsen, H.J.C. Numerical Integration of the Cartesian Equations of Motion of a System with Constraints: Molecular Dynamics of n-Alkanes. *J. Comput. Phys.* **1977**, *23*, 327–341. [CrossRef]
63. Essmann, U.; Perera, L.; Berkowitz, M.L.; Darden, T.; Lee, H.; Pedersen, L.G. A Smooth Particle Mesh Ewald Method. *J. Chem. Phys.* **1995**, *103*, 8577–8593. [CrossRef]
64. Downloads—GROMACS 2023.3 Documentation. Available online: https://manual.gromacs.org/current/download.html (accessed on 21 November 2023).
65. Pettersen, E.F.; Goddard, T.D.; Huang, C.C.; Couch, G.S.; Greenblatt, D.M.; Meng, E.C.; Ferrin, T.E. UCSF Chimera—A Visualization System for Exploratory Research and Analysis. *J. Comput. Chem.* **2004**, *25*, 1605–1612. [CrossRef] [PubMed]

molecules MDPI

Article

Extraction and Encapsulation of Phytocompounds of Poniol Fruit via Co-Crystallization: Physicochemical Properties and Characterization

N. Afzal Ali [1], Kshirod Kumar Dash [2,*], Vinay Kumar Pandey [3,4], Anjali Tripathi [4], Shaikh Ayaz Mukarram [5], Endre Harsányi [6] and Béla Kovács [5,*]

1 School of Agro and Rural Technology, IIT Guwahati, Guwahati 781039, Assam, India
2 Department of Food Processing Technology, Ghani Khan Choudhury Institute of Engineering and Technology (GKCIET), Malda 732141, West Bengal, India
3 Department of Bioengineering, Integral University, Lucknow 226026, Uttar Pradesh, India
4 Department of Biotechnology, Axis Institute of Higher Education, Kanpur 208001, Uttar Pradesh, India
5 Faculty of Agriculture, Food Science and Environmental Management, Institute of Food Science, University of Debrecen, 4032 Debrecen, Hungary; ayaz.shaikh@agr.unideb.hu
6 Faculty of Agriculture, Food Science and Environmental Management, Institute of Land Utilization, Engineering and Precision Technology, University of Debrecen, 4032 Debrecen, Hungary
* Correspondence: kshirod@tezu.ernet.in (K.K.D.); kovacsb@agr.unideb.hu (B.K.)

Abstract: Poniol (*Flacourtia jangomas*) has beneficial health effects due to its high polyphenolic and good antioxidant activity content. This study aimed to encapsulate the Poniol fruit ethanolic extract to the sucrose matrix using the co-crystallization process and analyze the physicochemical properties of the co-crystalized product. The physicochemical property characterization of the sucrose co-crystallized with the Poniol extract (CC-PE) and the recrystallized sucrose (RC) samples was carried out through analyzing the total phenolic content (TPC), antioxidant activity, loading capacity, entrapment yield, bulk and traped densities, hygroscopicity, solubilization time, flowability, DSC, XRD, FTIR, and SEM. The result revealed that the CC-PE product had a good entrapment yield (76.38%) and could retain the TPC (29.25 mg GAE/100 g) and antioxidant properties (65.10%) even after the co-crystallization process. Compared to the RC sample, the results also showed that the CC-PE had relatively higher flowability and bulk density, lower hygroscopicity, and solubilization time, which are desirable properties for a powder product. The SEM analysis showed that the CC-PE sample has cavities or pores in the sucrose cubic crystals, which proposed that the entrapment was better. The XRD, DSC, and FTIR analyses also showed no changes in the sucrose crystal structure, thermal properties, and functional group bonding structure, respectively. From the results, we can conclude that co-crystallization increased sucrose's functional properties, and the co-crystallized product can be used as a carrier for phytochemical compounds. The CC-PE product with improved properties can also be utilized to develop nutraceuticals, functional foods, and pharmaceuticals.

Keywords: Poniol extract; phytocompounds; sucrose; co-crystallization; encapsulation

Citation: Ali, N.A.; Dash, K.K.; Pandey, V.K.; Tripathi, A.; Mukarram, S.A.; Harsányi, E.; Kovács, B. Extraction and Encapsulation of Phytocompounds of Poniol Fruit via Co-Crystallization: Physicochemical Properties and Characterization. *Molecules* **2023**, *28*, 4764. https://doi.org/10.3390/molecules28124764

Academic Editors: Sascha Rohn and Michał Halagarda

Received: 26 April 2023
Revised: 12 June 2023
Accepted: 12 June 2023
Published: 14 June 2023

1. Introduction

Poniol (local Assamese dialect) (*Flacourtia jangomas* Lour.) belongs to the family Salicaceae and is a fruit-bearing tropical plant. It is an underutilized plant distributed in the Brahmaputra valley and northeast part of India; it probably migrated from upper Myanmar and Bangladesh and has been cultivated as a rare plant [1]. The fruits are palatable, bright in color, and eaten raw during the summer season when they are ripened. The fruit contains magnesium, potassium, calcium, manganese, and aluminum and is high in sodium (139.32 mg/100 g) [2]. The fruit is used to treat bilious conditions and diarrhea and has antidiabetic and cytotoxic properties [3]. According to existing research, the Poniol plant's fruits have astringent, anti-inflammatory, diaphoretic, and antibacterial characteristics, and they are used to treat illnesses such as asthma, diarrhea, jaundice, liver-related diseases, nausea, biliousness, and diabetes [4,5].

Co-crystallization is a new encapsulation process utilizing sucrose as a matrix for incorporating bioactive food components. Co-crystallization is a good alternative as it is flexible and economical for incorporating active compounds into powdered foods. The crystalline structure of sucrose gets modified from perfect shape to irregular agglomerate crystals to form a porous matrix where the target active ingredient is incorporated. The agglomerate appears sponge-like, having void space and a high surface area [6]. Supersaturated sucrose syrup spontaneously crystallizes at elevated temperatures (>120 °C) and low moisture (3–5%, or dissolved solids of 95–97 Brix) [7]. If a second ingredient is added simultaneously, the second ingredient is incorporated into the void spaces within the agglomerates of micro-sized crystals with a size smaller than 30 mm via random crystallization. Co-crystallization enhances the solubility, wettability, homogeneity, dispersibility, hydration, anticaking, stability, and flowability of the encapsulated bioactive compound. It also has advantages such as the ability to transform liquid form (core material) to powdered form without drying, and the agglomerate structure ensures direct tableting characteristics of the substance, which is beneficial to the candy and pharmaceutical industries [8]. The encapsulation of plant components in sucrose through the co-crystallization process has been reported for pomegranate peel extract [9], aqueous ethanol extract of unused Chokeberries [10], carotenoids [11], catechin or curcumin [12], soluble fiber [13], catechin or curcumin pre-emulsified with soybean protein isolate [12], ginger oleoresin [14], and *Securigera securidaca* seed extract [15].

The encapsulation of Jamaica (*Hibiscus Sabdarifa* L) granules through co-crystallization improved the developed product's solubility, dispersibility, and homogeneity. Co-crystallization process increased the entrapment yield and preserved the antioxidant activity of yerba mate (*Ilex paraguariensis*) extract [16]. The co-crystallization of natural antioxidant yerba mate extract with sucrose for the development of compressed tablets showed fast release in an aqueous medium, constituting a useful method for the oral delivery of active compounds with health benefits [17]. Few researchers developed solid dosage forms containing zinc (17 mg/g) obtained through co-crystallization in a sucrose matrix to increase zinc nutrition in high-risk populations. The resulting product had a high encapsulation efficiency (98%) and was stable. Encapsulation of yerba mate extract containing caffeoyl derivatives and flavonoids through co-crystallizing in a supersaturated sucrose solution had a standard cluster-like agglomerate structure with void spaces and sucrose crystal sizes ranging from 2 to 30 mm [18]. Co-crystallization of calcium lactate, magnesium sulfate, and yerba mate extracts in a supersaturated sucrose solution reduced the hygroscopic characteristics of yerba mate extracts without affecting their high solubility [9]. The study showed that the sucrose matrix mainly influenced the flowability, solubility, density, and size distribution of co-crystallized products. The water activity and hygroscopicity depended on the active ingredient added to the co-crystallized product.

Although there are limited scientific reports on the Poniol plant, several studies revealed the fruit extract presented several health benefits. The extracts of the different Poniol plant parts demonstrated cytotoxic, analgesic, antidiabetic, antidiarrheal, antimicrobial, and analgesic properties [1]. Due to the presence of ascorbic acid (24.00 mg 100 g^{-1}) and phenol (1.28 mg 100 g^{-1}), the fruit displayed a considerable quantity of antioxidant activity (8.93% mg^{-1}) [19]. The vitamin C (15.21–223.25 mg 100 g^{-1}), soluble sugar (13.77%), and vitamin B complex of riboflavin (236.84 μg 100 g^{-1}) and thiamine (42.97 μg 100 g^{-1})-reducing sugar (2.15–9.82%) present in the fruit has also been reported [19–21]. The plant fruit's aqueous, methanol, and ethanol extracts showed strong antimicrobial activity against *Klebsiella pneumonia*, *Escherichia coli*, and *Pseudomonas aeruginosa* [22]. Compared to the standard drug, chloramphenicol, the n-butanol extract from the plant fruits had strong antibacterial efficacy against *S. aureus* and *Escherichia coli* [23]. Saikia et al. (2016) observed that the acetone extract of fresh fruits exhibited a total phenolic content of 377.00 mg (mg GAE/100 g) and a total flavonoid content of 6.66 mg (mg QE/100 g). Their study also revealed a metal chelation capability of 18.55% and DPPH radical scavenging activity above 90% [24]. So, developing functional foods via incorporating this underutilized plant fruit extract could be a beneficial approach for valorizing this plant and its health benefits. With the above facts, the objective of this study was to encapsulate the Poniol fruit ethanolic extract using the co-crystallization process and analyze the physicochemical properties of the co-crystalized product.

2. Results and Discussion

2.1. Antioxidant Activity, Entrapment Yield, and Loading Capacity of the Co-Crystallized (Cc-Pe) Sample

The initial phenolic content (L_0) in the crude PE was 38.17 (mg GAE/100 g). The antioxidant activity (DPPH) of co-crystallized powder (Table 1) was 65.10%, and there was a 30.32% reduction in antioxidant activity compared to the crude PE (98.42%). Exposure to heat during the co-crystallization process (60 °C) may lead to the degradation of some polyphenols as well as a lower TPC in the CC-PE powder compared to the PE.

Table 1. Loading capacity, entrapment yield, and antioxidant activity of co-crystallized powder (CC-PE).

Products	Loading Capacity L_C. (mg GAE/100 g)	TPC (mg GAE/100 g of Dried Fruit)	Entrapment Yield (%)	Antioxidant Activity (%)
CC-PE	29.25 ± 0.03	29.25 ± 0.03	76.38 ± 0.07	65.10 ± 0.03
PE	NA	38.17 ± 0.04	NA	98.42 ± 0.37
Control	0.00 ± 0.00	0.00 ± 0.00	0.00 ± 0.00	1.7 ± 0.03

Values are given as mean ± standard deviation. Similar letters within a column indicate no significant difference ($p < 0.05$). NA = Not applicable.

The loading capacity (L_c) is defined as the amount of TPC loaded in one gram of co-crystallized material and was found to be around 29.25 (mg GAE/100 g) for the CC-PE sample (Table 1). The observed lower Lc and the DPPH values show a strong relationship between the phenolic content and antioxidant activity. The entrapment yield of the sucrose CC-PE was 76.38%. Sardar and Singhal (2013) reported an entrapment yield of 35.23% (for 1,8-cineole) and 67.18% (for α-terpinyl acetate) for encapsulating cardamom oleoresin in sucrose [4]. The entrapment yield (%) in our study was found to be reasonably higher due to the fact that the higher L_C resulted from the greater retainment of the TPC and the higher crude extract (PE) quantity (in mL) incorporated during the co-crystallization process. Behnamnik et al. (2019) reported an antioxidant capacity of 56.74% for the product of *Securigera securidaca* (L.) seed extract co-crystallized with sucrose [25], which is lower than our observed antioxidant value. Therefore, the comparatively better entrapment yield and the retainment of antioxidant activity after the co-crystallization process suggest that the CC-PE can be used as a functional food since it carries polyphenolic compounds that promote health. Polyphenolics are very important in promoting health and, at the same time, treating diseases [26–29].

2.2. Bulk Density and Tapped Density

The bulk volume was measured after manually tapping the cylinder two times on a flat tabletop surface. The powder poured into a cylinder will have a particular bulk density. The tapped density denotes the powder's bulk density after a specific compaction process, generally involving the vibration of the container. The bulk density (Table 1) of the CC-PE (0.723 g/cm^3) was found to be slightly higher than dried RC powder, whose value was 0.716 (g/cm^3). The filling up of the interstices/crevices by the PE (where the total soluble solid content is 2%) might increase the solid concentration of the CC-PE agglomerate powder, decreasing the bulk volume and increasing the bulk density. No significant differences ($p > 0.05$) were observed for the bulk densities between the RC and CC-PE powder samples.

Bulk density indicates the material's quantity required to fill a specific package volume, hence providing valuable information regarding handling and logistic applications [30]. Thus, the observed values of the bulk densities of the CC-PE powder confirmed a better packaging performance than the RC powder due to the lesser space requirement for packaging. The parameters that affect it are primarily the material components of the powder product, the humidity and temperature conditions of the environment, and the particle size [31]. The tapped density of the CC-PE (Table 2) powder (0.748 g/cm^3) was slightly lower than the RC powder (0.774 g/cm^3). The more crystalline nature of the PS, which was confirmed via the XRD analysis, could provide more compactness of the RC molecules. Additionally, the regular sugar crystal sizes could provide better alignment/rearrangements of sucrose crystals when tapping and occupying limited spaces between the sucrose powder particles. The more irregular shapes of the CC-PE crystals, which were confirmed via the formation of sucrose crystal agglomerates, made the rearrangement/alignment of the sucrose crystals more difficult and occupied more space between the sucrose powder particles. The SEM micrographs also confirmed the CC-PE powder's irregular crystal size/morphology. The irregular morphology of the CC-PE powder mass might possibly hold more spaces, which would, in turn, occupy relatively more volume than an equivalent RC powder mass. Since the tapped volume is inversely proportional to the tapped density, occupying more volume might reduce the tapped density of the CC-PE powder compared to the RC powder. No significant differences ($p > 0.05$) were observed for the tapped densities between the RC and CC-PE powder samples. A decrease in the tapped density, whose values ranged from 0.77–1.00 (g/cm^3), has been reported for the propolis extract co-crystallized with sucrose when the propolis ethanolic extract amount increased from 10 to 40 mL [32] and soluble fiber co-crystallized with sucrose (0.57 g/cm^3) [13]. The bulk and the tapped density values were significant for the determination of the flowability property of the powder samples.

Table 2. Physical properties of the recrystallized sucrose (RC) and the co-crystallized Poniol extract powder.

Sample	Bulk Density	Tapped Density	Flowability (HR)	Hygroscopicity (%)	Solubilization Time (sec)
RC	0.716 ± 0.034	0.774 ± 0.034	1.034 ± 0.042	12.345 ± 0.036	64.8 ± 0.1
CC-PE	0.723 ± 0.023	0.748 ± 0.022	1.082 ± 0.003	11.672 ± 0.023	62.5 ± 0.1

RC = recrystallized sucrose; CC-PE = sucrose co-crystallized with Poniol extract. Values are given as mean ± standard deviation. Similar letters within a column indicate no significant difference ($p < 0.05$).

2.3. Flowability

The ability of a powder to flow is called powder flowability. Flowability is the result of the combination of material physical properties that affect the material flow. The powder material handling and the caking properties during storage mainly depend on the powder flowability characteristics. The flowability was determined using Hausner's ratio (HR) for the dried co-crystallized powder. Hausner's ratio measures inter-particulate friction and is defined as the ratio of tapped density to the bulk density of the co-crystallized powder. HR < 1.11 is considered excellent flow, while HR > 1.60 is regarded as extremely poor flow; in terms of intermediate values, HR between 1.12–1.18 is considered good flow, HR between 1.19–1.25 is considered passable flow, HR between 1.35–1.45 is considered poor flow, and HR between 1.46–1.54 is considered very poor flow [33,34]. The flowability of the co-crystallized powder (CC-PE) of HR 1.082 was better than the recrystallized sucrose (RC) crystal, which has an HR value of 1.034 (Table 2). The RC sucrose powder hygroscopicity was also slightly higher as compared to the co-crystallized powder. A slight increase in the hygroscopicity of the RC powder would decrease the powder's flowability by increasing the powder's stickiness, resulting in the formation of large sugar agglomerates. This indicates that the co-crystallization process improved the flowability property of the co-crystallized (sucrose + Poniol extract) powder. The improved flowability property (Hausner ratio values ~1) of co-crystallized sucrose with yerba mate extract has been reported [16].

2.4. Hygroscopicity

The tendency of powder to attract and hold water molecules from the surrounding environment is term as hygroscopicity. Hygroscopicity was determined as the weight gained by the products after seven days of their formation reached equilibrium when stored under the environmental conditions of RH = 75% and at 25 °C. This property is an essential factor for defining the product's packaging prerequisites. The hygroscopicity of the dried CC-PE powder (Table 2) was 11.67%. It showed a significant ($p < 0.05$) decrease in the hygroscopicity compared to the recrystallized sucrose (RC), which has a value of 12.34%. The moisture content of any product determines the degree of spoilage or shelf of any product since any biochemical reaction is mainly dependent on the moisture content. The decreased hygroscopicity of the co-crystallized powder indicated that the product would absorb less moisture during the storage period than the RC sucrose powder, thereby prolonging the product's shelf life. Higher hygroscopicity negatively influences the powder's flowability by increasing the powder's stickiness, forming large sugar agglomerates. The sugar agglomerates' chain reaction forms thick masses of sugar powder, which is a very undesirable property of any powder material. Similar findings of the lower hygroscopicity of the sucrose matrix have also been reported when co-crystallized with carotenoids extracted from carrots and found in yerba mate extract [11].

2.5. Solubilization Time

Solubilization time is the time of dissolving the solute (co-crystallized powder) in a solvent (water) to form a homogenous solution (Table 2). The solubilization time of RC powder was found to be around 64.8 s, which signifies that the Brix of the solution did not change (9 Brix), indicating the crystal powder was fully dissolved in the solvent (water). The solubilization time of the CC-PE powder in water was found to be 62.5 s, which was lower than the RC powder. This shows that the co-crystallization process slightly decreased the solubilization time of the product. No significant differences ($p > 0.05$) were observed between the RC and the CC-PE powder samples. Greater amorphousness with reduced crystallinity, confirmed via the XRD analysis, might trigger the reduced solubilization time of the CC-PE. The higher crystalline RC would increase the compactness of the sucrose molecules. Obviously, the more amorphous material could possess more hydration power. This probably could enable quick solvent (distilled water) migration through the agglomerate pores of CC-PE, permitting the active compounds (e.g., PE) residing in the intervening spaces between crystals to release rapidly, thereby increasing solubilization power. A decrease in the solubilization time of sucrose co-crystallized with yerba mate extract (1.33 min) when compared to RC sucrose (1.37 min) has also been reported [31]. A similar report on the decrease in the solubilization time of sucrose matrix co-crystallized with yerba mate extract has also been revealed [16]. It can be concluded that there was a slight tendency for better solubility for the CC-PE, which is beneficial when serving the product.

2.6. X-ray Diffraction Analysis

The XRD diffractograms of the RC sugar and the CC-PE powder are shown in Figure 1. Both the samples showed significant peaks at the 2θ values of 13.2, 18.9, 19.7, 24.8, and 25.3. The graph of the co-crystallized powder did not show any shift in the 2θ values from the RC sucrose values. However, the intensities of the peaks of the RC sucrose were higher than those of the co-crystallized powder sample. The co-crystallized powder did not show any new distinct peaks with higher intensities in the diffraction graph, which confirmed that the extract was not crystallized after the co-crystallization process and was incorporated effectively in the voids of the sucrose. Additionally, the non-crystalline nature of the PE might not influence the crystal structure of the sugar during the co-crystallization process. The degree of crystallinity of the CC-PE was 37.8%, while the RC powder had 51.9% crystallinity. This higher crystallinity of RC sugar was also supported by the DSC analysis that showed that the RC sucrose had a higher endothermic melting peak with a higher intensity as compared to the co-crystallized powder. The XRD diffraction pattern verified that the co-crystallization process reduced the degree of crystallinity of the sucrose, which could be due to the amorphous nature of the extract. Any material had been classified as high crystalline when above 50%, medium crystalline if 20–50%, and low if 20% [35]. Accordingly, the XRD diffractograms proved that the RC sucrose was a crystalline compound while the co-crystalline powder showed medium crystalline properties. The XRD analysis concluded that the co-crystallization treatment did not change the overall crystal structure of the sucrose, although the crystallinity had been reduced. Reports on the non-occurrence of the additional peaks with higher intensities were reported similarly for co-crystallized sucrose with *Basella rubra* extract, yerba mate extract, and cardamom oleoresin [6,16,36]. The decrease in the crystallinity of the co-crystallized product as compared to sucrose has also been reported for vitamin-B12-fortified co-crystallized sugar cubes [37].

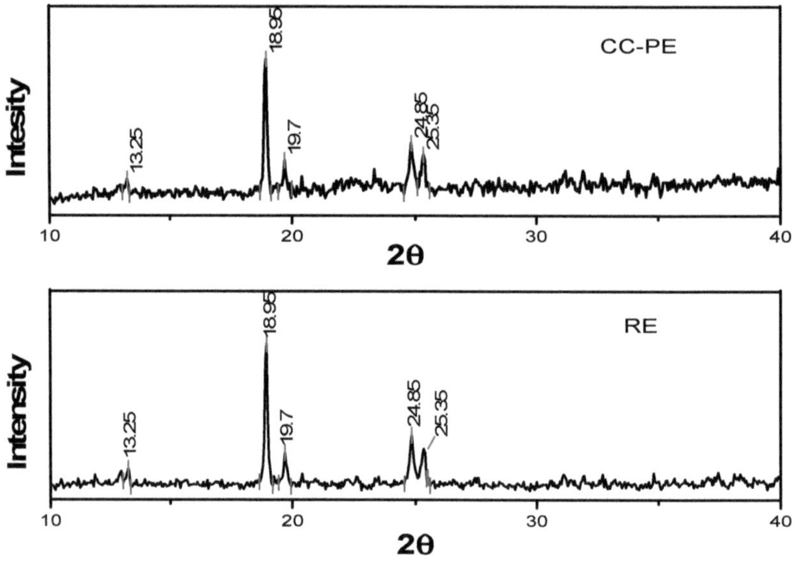

Figure 1. X-ray diffractograms of the RC (recrystallized sucrose) and CC-PE (sucrose co-crystallized with Poniol extract).

2.7. Microstructure of Microencapsulated PONIOL (*F. Jangomas*) Powder after Co-Crystallization

The microcapsules' surface morphology and internal structure were investigated by analyzing the scanning electron microscope (SEM) images (Figure 2). The RC sucrose (Figure 2A,B) showed a relatively regular crystal shape, which could be attributed to the proper alignment of atoms in the RC sucrose molecule. In contrast, the CC-PE powder (Figure 2C,D) exhibited an irregular agglomerated crystal, giving a porous structure that could possibly produce bigger cohesive solids than the original fine RC powder granules.

The porous configuration of the sucrose potentially produced through the co-crystallization process could accept a second ingredient (extract), enabling the incorporation of the second active ingredient [38]. The addition of the Poniol extract during the spontaneous crystallization of the supersaturated sucrose/sugar syrup that had occurred when heated (above 120 °C; 95–97 Brix) resulted in the incorporation of the second ingredient into the void spaces/irregular cavities of the micro-sized crystal agglomerates [35].

The presence of agglomerated clusters in the co-crystallized powder indicated the better entrapment/incorporation of Poniol extract into the recrystallized sucrose. Reports on better entrapment due to the presence of the agglomerated sucrose clusters have also been reported for sucrose crystals that were co-crystallized with yerba mate extract, Basella rubra extract, and carotenoids extracted from carrot [6,11,16]. The SEM analysis provided important pieces of information about the surface morphological characteristics, which are very important for both the products' hygroscopicity and flowability properties.

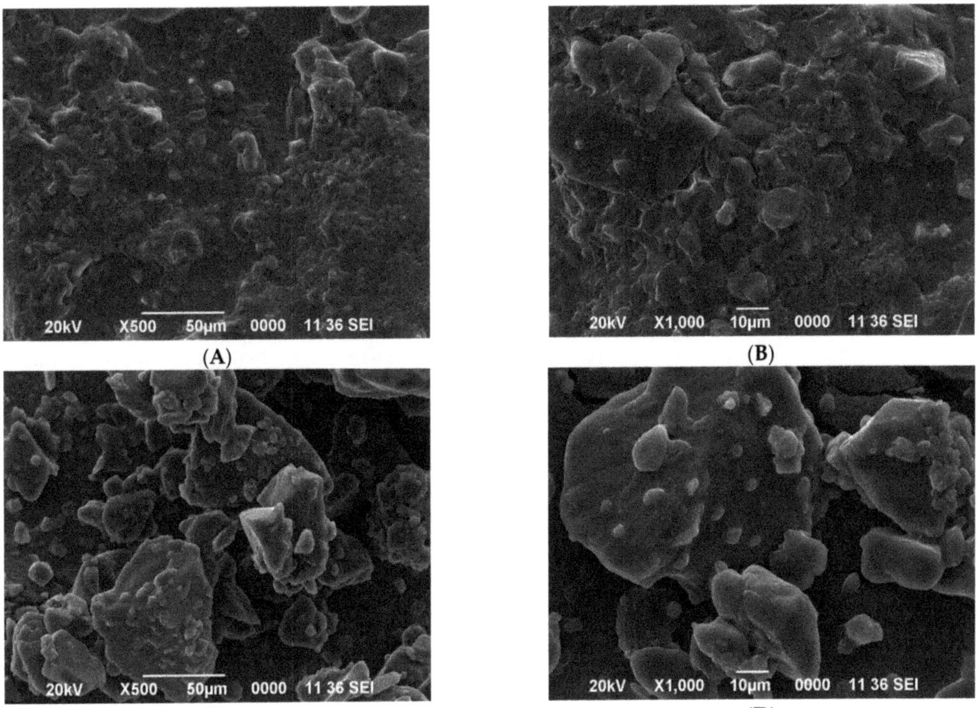

(A) (B) (C) (D)

Figure 2. SEM microphotographs of the RC ((**A**) = 500× and (**B**) = 1000× magnification) and the CC-PE ((**C**) = 500× and (**D**) = 1000× magnification).

2.8. Fourier Transform Infrared (FT-IR) Spectroscopy

The FT-IR analysis analyzed the chemical structure and the functional groups of the RC sucrose and the co-crystallized powder with Poniol extract in the range of 400–4000 cm^{-1} wavenumbers (Figure 3). The significant peaks found at 3568 cm^{-1} and 3395 cm^{-1} were assigned to the -OH groups' vibration (stretching). Bands attributed to the stretching of C-H groups were detected around 3011 cm^{-1} and 2970 cm^{-1}. The characteristic peak for -CH_2 was found at around 2933 cm^{-1}, and for C-O groups, it was found at around 1120 cm^{-1} and 987 cm^{-1}, respectively. The C-O stretching vibration modes were found at around 1084 cm^{-1} and 902 cm^{-1}. The band between 800–1500 cm^{-1} has been called the 'fingerprint' for sucrose (sugars), emergence from the CO-stretching vibrations, and the CH_2 groups' symmetrical deformation, thereby offering high structural information [39–44].

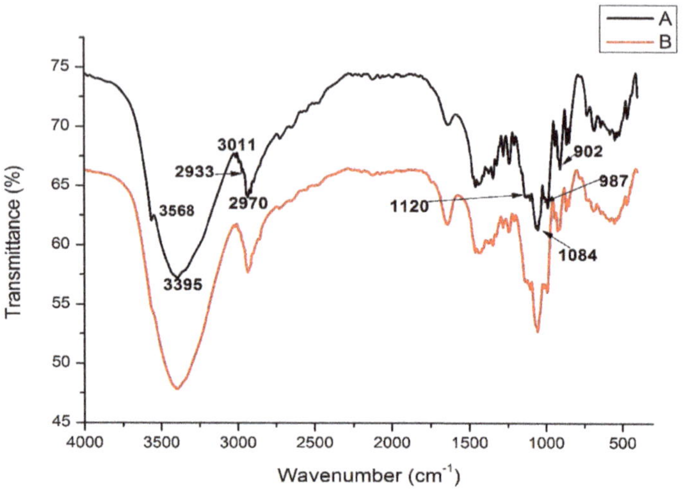

Figure 3. FT-IR spectra of (A) RC and (B) CC-PE.

A similar chemical bonding structure with respect to the wavenumber, as mentioned above, for sucrose has been reported [18,30,32]. The co-crystallized powder had no significant additional peaks compared to the RC sucrose. This means there were no additional conformational changes in the chemical bonding structure during the co-crystallization process. However, there was an increase in the peak intensities at the band of $\sim$1650 cm^{-1}, a wavenumber assigned to the -OH bending vibration of water molecules. It was possible to overlap this transmittance peak ($\sim$1640 cm^{-1}) with antisymmetric (COO−) stretching bands [42]. Therefore, it can be suggested that the overlapping of -OH bending vibration and the stretching vibrations of C=C and C=O of polyphenolic compounds (from the extract) might increase the intensity of the $\sim$1650 cm^{-1} band [32,45]. From the FTIR analysis, it can be suggested that there was a good incorporation/entrapment of the extract to the sucrose matrix since no significant new peaks were observed compared to the RC powder sample.

2.9. Thermal (DSC) Analysis of Co-Crystallized Poniol Extract Powder

Differential scanning calorimetry (DSC) measures the temperatures and heat flows associated with transitions in materials as a function of time and temperature in a controlled atmosphere. These measurements offer information about physical and chemical changes involving endothermic or exothermic processes or changes in heat capacity. Glass transition temperature (Tg) is a characteristic parameter of a material above which it behaves like a liquid (rubbery state). There was a shift in the Tg value of the co-crystallized powder (144.5 °C) from the RC sucrose (145.3 °C). The DSC thermograms for the RC sucrose and the co-crystalized samples are depicted in Figure 4.

The RC sucrose thermogram showed two significant endothermic peaks at approximately 199.9 °C and 221.9 °C, while the co-crystallized powder showed a distinct peak at 175.5 °C. The first peak (199.9 °C) of the RC sucrose was due to the melting point of the sucrose, while the second peak could be assigned to sucrose degradation [31]. The first peak of the co-crystallized powder might be due to the fusion of the sucrose and the extract [46]. Figure 4 shows that the endothermic peaks were shifting. The peaks of the co-crystallized powder sample (both 175 °C and 222.7 °C) were broad with a lower intensity, signifying the effect of the extract entrapped in the powder. The lower crystallinity explained via the XRD property analysis of the co-crystallized power supported this effect. The shifting and total or partial disappearance of the co-crystallized powder's thermal events (melting point) could be taken as proof of its good incorporation into the matrix, indicating good incorporation of the extract into the sucrose matrix. The second endothermic transition at 222.7 °C could be attributed to the degradation of all the components in the CC-PE powder [16,47–49]. The retainment of the crystalline state, as evidenced by the melting endotherms of the DSC analysis, has also been reported in the sucrose co-crystallized with yerba mate extract [50].

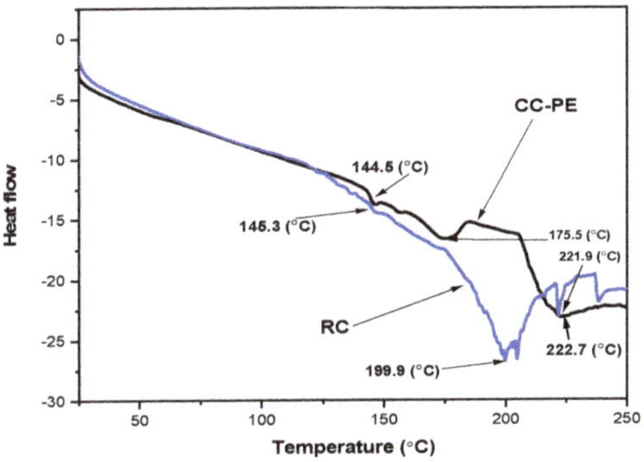

Figure 4. DSC thermograms of the RC and CC-PE samples.

3. Material and Methods

3.1. Raw Material and Chemicals and Reagents

Fresh Poniol (*Flacourtia jangomas*) fruit was obtained from the local market of Udalguri district, Assam, India, and stored at 4 °C for further use. Distilled water was used for washing of the fruit. All the chemicals and reagents were of analytical grade. The ethanol (99.9%) was supplied by Changshu Hongsheng Fine Chemicals Co., Ltd. (Changshu, China). Folin & Ciocalteu's phenol reagent (2.0 N) and gallic acid (98%) were supplied by Sisco Research Laboratories Pvt. Ltd. (Taloja, Maharashtra, India). Sucrose (AR grade), sodium carbonate anhydrous (99.50–100.50%), and 2,2-Diphenyl-1-picrylhydrazyl or DPPH (85.0%) were supplied by Himedia Laboratories Pvt. Ltd. (LBS Marg, Mumbai, India). Ascorbic acid (99.0%) was purchased from Merck Life Science Pvt. Ltd. (Vikroli East, Mumbai, India). Deionized water was used to prepare all the chemical solutions.

3.2. Fruit Extract Preparation

The whole fruit was dried in a domestic microwave oven (IFB, 20BC4, India) at 720 W power, followed by grinding. The dried powdered sample (100 g) was put in a conical flask and dissolved in ethanol (900 mL, 60%) for extraction closed with cotton. The extraction was performed through keeping the solution mixture (ethanol + sample) in a shaker (120 rpm) incubator (Remi RIS 24 plus, India) for 48 h at 37 °C [51,52]. After 48 h, the solution was removed, and filtration was carried out through a cheesecloth to collect the filtrate. The filtrate was again centrifuged (8000 rpm for 10 min), and the supernatant Poniol extract (PE) was collected. The extract was stored at refrigerated conditions (−20 °C) for further applications.

3.3. Preparation of the Co-Crysta–Llization Products

The sucrose (50 g) solution (70 Brix in distilled water) was heated in a glass beaker at 127 ± 4 °C on a hot plate under intermittent manual mixing until the solution reached 95.8 Brix. The obtained solution was cooled, and when it reached 60 °C, the previously prepared PE or pure ethanol (30 mL) was added, followed by covering it with aluminum foil [9,53]. The obtained supersaturated solution was cooled through dipping the beaker in ice-cold water. This cooling was done in order to prevent long-term exposure of the Poniol extract to high temperatures, since this is known to provoke a browning reaction and flavor degradation. Secondly, the product was to increase the supersaturation of the sugars, which facilitates quicker crystallization. The concentrated sucrose syrup was kept for 72 h at room temperature for the formation of crystals. After that, the sucrose co-crystallized with the Poniol extract (CC-PE) was collected and dried in a hot air oven (Universal, JSGW-TC344) at 40 °C for 15 h and then was ground and sieved through a 500 mm mesh. Blends of raw sucrose, distilled water, and ethanol were crystallized as described above for control. This recrystallized sucrose (RC) was referred to as the control sample.

The resultant powder's material characterization was carried out through analyzing phenolic content, scavenging activity, color difference, solubility, hygroscopicity, loading capacity and entrapment yield and flowability, DSC, XRD, and SEM for the CC-PE and the control samples.

3.3.1. Determination of Total Phenolic Content

The total phenolic content (TPC) analysis was conducted according to the Folin–Ciocalteu assay [54] to evaluate and study the influence of the co-crystallization process in the TPC content among the PE, CC-PE, and control samples. The CC-PE powder (5 mg) was dissolved in 5 mL of ethanol (60% *v*/*v*) and vortexed, followed by keeping it at room temperature (25 °C) for 30 min. The concentration of the standard solution of gallic acid ranged from 0.01–0.5 mg/mL. 200 μL of each pure ethanol PE and the solutions of CC-PE and the control powder were pipetted out in different test tubes, followed by adding distilled water (800 μL) and 5 mL Folin–Ciocalteu (10%). The solution mixture was vortexed and kept for 1 min, and then 4 mL sodium carbonate (7.5%) was added. The solution was incubated for 30 min at room temperature in dark conditions. The absorbance was measured at 765 nm wavelength in a UV spectrophotometer (Cary 100, Agilent Technologies, Santa Clara, CA, USA), and the phenolic content was expressed in mg GAE/100 g of dried fruit.

3.3.2. Determination of Antioxidant Activity

Free radical scavenging activity towards the DPPH reagent method was used to determine the antioxidant activity of the samples [55,56]. Co-crystallized powder (0.5 g) was dissolved in 5 mL of ethanol (60% *v*/*v*), followed by vortexing, and was kept at room temperature (25 °C) for 30 min. The RC powder was dissolved via the same procedure as the CC-PE for the control sample preparation. 100μL of each extract/solution and the pure ethanol (60%) were mixed with 3.9 mL of ethanolic solution of DPPH (25 mg/L). Then, the mixture was incubated for 30 min in the dark, and absorbance was measured at 517 nm using a UV-VIS spectrophotometer (Cary 100, Agilent Technologies, Santa Clara, CA, USA). The percentage of inhibition (I %) of the DPPH free radical expressed the antioxidant activity and was calculated using the formula:

$$DPPH\ (\%\ I) = \left(1 - \frac{A_s}{A_o}\right) \times 100 \tag{1}$$

where

A_s = Absorbance of control (DPPH solution without extract)

A_o = Absorbance of sample.

3.3.3. Loading Capacity and Entrapment Yield [11]

Loading capacity (Lc) can be calculated as the total phenolic content of the Poniol extract loaded in 1 g of co-crystallized material expressed as mg GA/g powder.

The entrapment yield (EY) was calculated as follows:

$$EY\ (\%) = \left(\frac{L_c}{L_0}\right) \times 100 \tag{2}$$

where L_0 was the initial TPC of the Poniol extract expressed as mg GA/g dried Poniol fruit.

3.3.4. Bulk Density, Tapped Density, and Flowability

The bulk density (g/cm^3) was calculated through measuring the volume occupied by the known masses of the RC and the RC-PE powder samples that were filled under gravity in a graduated cylinder without compacting. For estimation of tapped density (g/cm^3), the volume measurements were made following 10, 25, and 50 manual taps. Each sample was replicated in triplicate, and the obtained values were averaged.

The flowability of the co-crystallized powders was determined using Hausner's ratio (HR). The value of HR was calculated according to the ratio of the tapped density to the bulk density [3].

The Hausner's ratio was calculated as:

$$HR = \frac{\rho_T}{\rho_B} \tag{3}$$

where ρ_T was the tapped density and ρ_B was the bulk density of the product.

3.3.5. Hygroscopicity

The tendency of powder to attract and hold water molecules from the surrounding environment is termed hygroscopicity. The hygroscopicity of both samples was determined through adopting methods described by [57,58] with slight modifications. Samples (1 g) were kept inside the desiccator, which contains sodium chloride solution (RH 75.3%). The result was calculated as the mass of water absorbed per 100 g of the sample after seven days of storage.

$$Hygroscopicity = \frac{Weight\ of\ moisture\ absorbed(g)}{Initial\ weight\ of\ the\ sample(g)} \tag{4}$$

3.3.6. Solubilization Time

The dried sample (both RC and CC-PE) powder (1 g) was blended with distilled water (10 mL) at 25 °C with continuous stirring [32]. Aliquots of the sugar solutions were taken in a refractometer at regular intervals and the Brix changes were observed. The solubilization time was determined through observing the time (in seconds) when the Brix changes became static.

3.3.7. X-ray Diffraction (XRD)

An X-ray diffractometer (STOE, Darmstadt, Germany) equipped with a Miniflex goniometer in reflection X-ray source: copper anode (Cu Kα = 1.5418 Å), current = 15 mA (fixed), voltage = 30 kV (fixed), provided via a copper Kβ filter was used to analyze the crystallinity of the samples. The samples were scanned between the 2θ values from 10° to 40° [6].

3.3.8. Scanning Electron Microscopy (SEM)

Scanning electron microscopy (JEOLJSM-6390LV, Japan) was used to analyze the microstructure of the RC and CC-PE powder samples. The co-crystallized powder was attached to the SEM stubs using two-sided adhesives tape and then coated with a layer of gold (40–50 nm) and analyzed using a 20 kV acceleration voltage.

3.3.9. FTIR (Fourier Transform Infrared Spectroscopy) Analysis

The chemical bonding and functional groups of the RC and CC-PE powders were analyzed using Fourier transform infrared spectroscopy (Impact −410, Nicolet, Alexandria, USA) and the spectral analysis was performed with the software OMNIC E.S.P.5.0. in the wavenumber range of (400–4000) cm^{-1} at a resolution of 4 cm^{-1} [25,32].

3.3.10. Thermal Analysis (Differential Scanning Calorimetry)

Differential scanning calorimetry (DSC-60, SHIMADZU) with TA-60 WS software was used for the DSC analysis of the RC and CC-PE powder samples. Samples (4–6) mg were placed in aluminum pans and then hermetically sealed. An empty pan was used as the reference. The temperature range for the samples to be heated was (25 to 250) °C and the heating rate was 10 °C/min [30].

3.4. *Statistical Analysis*

The analyses were conducted in triplicate, and the data are presented as mean ± standard deviation. Each parameter underwent an analysis of variance (ANOVA) with Tukey's test to determine the significance of the effects and interactions between them. A $p < 0.05$ was considered to be statistically significant. Origin software was used to draw the graphs.

4. Conclusions

The ethanolic extract of Poniol (*Flacourtia jangomas*) fruit was encapsulated using a co-crystallization process with sucrose, and the physicochemical properties of the co-crystalized product were analyzed in this study. The obtained CC-PE and RC products' material characterization was carried out through analyzing the TPC, antioxidant activity, loading capacity, entrapment yield, bulk and tapped densities, hygroscopicity, solubilization time, flowability, DSC, XRD, FTIR, and SEM. The bulk density and the flowability of the CC-PE sample were 0.723 and 1.082 HR, respectively, which were higher than the RC sample, whose values were 0.716 and 1.034 HR, respectively. The bulk hygroscopicity and solubilization time of the CC-PE sample were 11.672% and 62.5 s, respectively, lower than the RC sample, whose values were 12.345% and 64.8 s, respectively. The result revealed that the CC-PE product had a good entrapment yield (76.38%) and could retain the TPC (29.25 mg GAE/100 g) and antioxidant properties (65.10%) even after the co-crystallization process. The higher flowability bulk density and the lower hygroscopicity and solubilization time of the CC-PE sample compared to the RC sample are desirable for a powder product. The XRD, FTIR, and DSC analyses also showed that there were no changes in the crystal structure, functional group bonding structure, and thermal properties of the sucrose, respectively. The SEM analysis showed that the CC-PE sample has cavities or pores in the sucrose cubic crystals, which suggests that the entrapment was better. From the results, we can conclude that co-crystallization increased the sucrose's functional properties, and the co-crystallized product can be used as a carrier for phytochemical compounds. The obtained CC-PE, which contains bioactive compounds with antioxidant properties, may possibly be used as a natural sweetener with improved properties while considering the verification of consumer acceptance. The CC-PE product with improved properties can also be utilized to develop nutraceuticals, functional foods, and pharmaceuticals.

Author Contributions: Conceptualization, N.A.A.; methodology, K.K.D.; data curation, A.T. and E.H.; writing—original draft preparation, N.A.A., K.K.D. and V.K.P.; writing—review and editing, B.K. and S.A.M.; visualization, V.K.P. and S.A.M.; funding acquisition, B.K. All authors have read and agreed to the published version of the manuscript.

Funding: Project No. TKP2021-NKTA-32 has been implemented with support from the National Research, Development, and Innovation Fund of Hungary, financed under the TKP2021-NKTA funding scheme.

Data Availability Statement: The data presented in this study are available on request from the corresponding author.

Conflicts of Interest: The authors declare no conflict of interest.

Sample Availability: Not Available.

References

1. Sasi, S.; Anjum, N.; Tripathi, Y.C. Ethnomedicinal, Phytochemical and Pharmacological Aspects of *Flacourtia Jangomas*: A Review. *Int. J. Pharm. Pharm. Sci.* **2018**, *10*, 9–15. [CrossRef]
2. Ara, R.; Jahan, S.; Abdullah, A.; Fakhruddin, A.N.M.; Saha, B. Physico-chemical properties and mineral content of selected tropical fruits in Bangladesh. *Bangladesh J. Sci. Ind. Res.* **2015**, *49*, 131–136. [CrossRef]
3. Barua, U.; Das, R.; Gogoi, B.; Baruah, B.G.S. Underutilized Fruits of Assam for Livelihood and Nutritional Security. *Agric. Rev.* **2019**, *40*, 175–184. [CrossRef]
4. Jeyachandran, R.; Mahesh, A. Enumeration of Antidiabetic Herbal Flora of Tamil Nadu. *Res. J. Med. Plant* **2007**, *1*, 144–148. [CrossRef]
5. Lim, T.K. Flacourtia jangomas. In *Edible Medicinal and Non-Medicinal Plants*; Springer: Dordrecht, The Netherlands, 2013; pp. 771–775. [CrossRef]
6. Karangutkar, A.V.; Ananthanarayan, L. Co-crystallization of Basella rubra extract with sucrose: Characterization of co-crystals and evaluating the storage stability of betacyanin pigments. *J. Food Eng.* **2020**, *271*, 109776. [CrossRef]
7. Chezanoglou, E.; Goula, A.M. Co-crystallization in sucrose: A promising method for encapsulation of food bioactive components. *Trends Food Sci. Technol.* **2021**, *114*, 262–274. [CrossRef]
8. Chikodili, I.M.; Chioma, I.I.; Ukamaka, I.A.; Nnenna, O.T.; Ogechukwu, O.D.; Mmesoma, E.E.; Chikodi, E.C.; IfedibaluChukwu, E.I. Phytocompound inhibitors of caspase 3 as beta-cell apoptosis treatment development option: An In-silico approach. *Sci. Phytochem.* **2023**, *2*, 17–37. [CrossRef]
9. Chezanoglou, E.; Kenanidou, N.; Spyropoulos, C.; Xenitopoulou, D.; Zlati, E.; Goula, A.M. Encapsulation of pomegranate peel extract in sucrose matrix by co-crystallization. *Sustain. Chem. Pharm.* **2023**, *31*, 100949. [CrossRef]
10. Tzatsi, P.; Goula, A.M. Encapsulation of Extract from Unused Chokeberries by Spray Drying, Co-crystallization, and Ionic Gelation. *Waste Biomass-Valoriz.* **2021**, *12*, 4567–4585. [CrossRef]
11. Kaur, P.; Elsayed, A.; Subramanian, J.; Singh, A. Encapsulation of carotenoids with sucrose by co-crystallization: Physicochemical properties, characterization and thermal stability of pigments. *LWT* **2021**, *140*, 110810. [CrossRef]
12. Wang, H.; Gao, S.; Zhang, D.; Wang, Y.; Zhang, Y.; Jiang, S.; Li, B.; Wu, D.; Lv, G.; Zou, X.; et al. Encapsulation of catechin or curcumin in co-crystallized sucrose: Fabrication, characterization and application in beef meatballs. *LWT* **2022**, *168*, 113911. [CrossRef]
13. Queiroz, M.B.; Sousa, F.R.; da Silva, L.B.; Alves, R.M.V.; Alvim, I.D. Co-crystallized sucrose-soluble fiber matrix: Physicochemical and structural characterization. *LWT* **2022**, *154*, 112685. [CrossRef]
14. Rai, K.; Chhanwal, N.; Shah, N.N.; Singhal, R.S. Encapsulation of ginger oleoresin in co-crystallized sucrose: Development, characterization and storage stability. *Food Funct.* **2021**, *12*, 7964–7974. [CrossRef]
15. Behnamnik, A.; Vazifedoost, M.; Didar, Z.; Hajirostamloo, B. Evaluation of physicochemical, structural, and antioxidant properties of microencapsulated seed extract from *Securigera securidaca* by co-crystallization method during storage time. *Biocatal. Agric. Biotechnol.* **2021**, *35*, 102090. [CrossRef]
16. López-Córdoba, A.; Deladino, L.; Agudelo-Mesa, L.; Martino, M. Yerba mate antioxidant powders obtained by co-crystallization: Stability during storage. *J. Food Eng.* **2014**, *124*, 158–165. [CrossRef]
17. López-Córdoba, A.; Matera, S.; Deladino, L.; Hoya, A.; Navarro, A.; Martino, M. Compressed tablets based on mineral-functionalized starch and co-crystallized sucrose with natural antioxidants. *J. Food Eng.* **2015**, *146*, 234–242. [CrossRef]
18. López-Córdoba, A.; Gallo, L.; Bucalá, V.; Martino, M.; Navarro, A. Co-crystallization of zinc sulfate with sucrose: A promissory strategy to render zinc solid dosage forms more palatable. *J. Food Eng.* **2016**, *170*, 100–107. [CrossRef]
19. Barua, U.; Das, R.P.; Das, P.; Das, K.; Gogoi, B. Morpho-physiological, proximate composition and antioxidant activity of *Flacourtia jangomas* (Lour.) Raeus. *Plant Physiol. Rep.* **2022**, *27*, 56–64. [CrossRef]
20. Biswas, S.C.; Kumar, P.; Kumar, R.; Das, S.; Misra, T.K.; Dey, D. Nutritional Composition and Antioxidant Properties of the Wild Edible Fruits of Tripura, Northeast India. *Sustainability* **2022**, *14*, 12194. [CrossRef]
21. Hossain, M.; Rahim, A.; Haque, R. Biochemical properties of some important underutilized minor fruits. *J. Agric. Food Res.* **2021**, *5*, 100148. [CrossRef]
22. Srivastava, D.; Prabhuji, S.; Tripathi, A.; Srivastava, R.; Mishra, P. In vitroantibacterial activities of *Flacourtia jungomas* (Lour.) Raeus. fruit extracts. *Med. Plants—Int. J. Phytomed. Relat. Ind.* **2012**, *4*, 98–100. [CrossRef]
23. Das, S.; Dewan, N.; Das, K.J.; Kalita, D. Preliminary Phytochemical, Antioxidant and Antimicrobial Studies of *Flacourtia jangomas* Fruits. *Int. J. Curr. Pharm. Res.* **2017**, *9*, 86–91. [CrossRef]
24. Saikia, S.; Nikhil Kumar Mahnot, N.K.; Charu Lata Mahanta, C.L. Phytochemical content and antioxidant activities of thirteen fruits of Assam, India. *Food Biosci.* **2016**, *13*, 15–20. [CrossRef]
25. BehnamNik, A.; Vazifedoost, M.; Didar, Z.; Hajirostamloo, B. The antioxidant and physicochemical properties of microencapsulated bioactive compounds in *Securigera securidaca* (L.) seed extract by co-crystallization. *Food Qual. Saf.* **2019**, *3*, 243–250. [CrossRef]
26. García-Pérez, P.; Gallego, P.P. Plant Phenolics as Dietary Antioxidants: Insights on Their Biosynthesis, Sources, Health-Promoting Effects, Sustainable Production, and Effects on Lipid Oxidation. In *Lipid Oxidation in Food and Biological Systems*; Bravo-Diaz, C., Ed.; Springer International Publishing: Cham, Switzerland, 2022; pp. 405–426. [CrossRef]

27. Jesus, F.; Gonçalves, A.C.; Alves, G.; Silva, L.R. Health Benefits of *Prunus avium* Plant Parts: An Unexplored Source Rich in Phenolic Compounds. *Food Rev. Int.* **2022**, *38*, 118–146. [CrossRef]
28. Rajashekar, C.B. Dual Role of Plant Phenolic Compounds as Antioxidants and Prooxidants. *Am. J. Plant Sci.* **2023**, *14*, 15–28. [CrossRef]
29. Sun, W.; Shahrajabian, M.H. Therapeutic Potential of Phenolic Compounds in Medicinal Plants—Natural Health Products for Human Health. *Molecules* **2023**, *28*, 1845. [CrossRef] [PubMed]
30. Sarabandi, K.; Mahoonak, A.S.; Akbari, M. Physicochemical properties and antioxidant stability of microencapsulated marjoram extract prepared by co-crystallization method. *J. Food Process. Eng.* **2019**, *42*, e12949. [CrossRef]
31. Deladino, L.; Anbinder, P.S.; Navarro, A.S.; Martino, M.N. Co-crystallization of yerba mate extract (Ilex paraguariensis) and mineral salts within a sucrose matrix. *J. Food Eng.* **2007**, *80*, 573–580. [CrossRef]
32. Irigoiti, Y.; Yamul, D.K.; Navarro, A.S. Co-crystallized sucrose with propolis extract as a food ingredient: Powder characterization and antioxidant stability. *LWT* **2021**, *143*, 111164. [CrossRef]
33. Jeong, Y.; Yoo, B. Physical, Morphological, and Rheological Properties of Agglomerated Milk Protein Isolate Powders: Effect of Binder Type and Concentration. *Polymers* **2023**, *15*, 411. [CrossRef] [PubMed]
34. Jinapong, N.; Suphantharika, M.; Jamnong, P. Production of instant soymilk powders by ultrafiltration, spray drying and fluidized bed agglomeration. *J. Food Eng.* **2008**, *84*, 194–205. [CrossRef]
35. Bhandari, B.; Hartel, R. Co-crystallization of Sucrose at High Concentration in the Presence of Glucose and Fructose. *J. Food Sci.* **2002**, *67*, 1797–1802. [CrossRef]
36. Sardar, B.R.; Singhal, R.S. Characterization of co-crystallized sucrose entrapped with cardamom oleoresin. *J. Food Eng.* **2013**, *117*, 521–529. [CrossRef]
37. Bajaj, S.R.; Singhal, R.S. Enhancement of stability of vitamin B12 by co-crystallization: A convenient and palatable form of fortification. *J. Food Eng.* **2021**, *291*, 110231. [CrossRef]
38. Chen, A.C.; Veiga, M.F.; Rizzuto, A.B. Co-crystallization: An encapsulation process. *Food Technol.* **1989**, *42*, 87–90.
39. Limm, W.; Karunathilaka, S.R.; Mossoba, M.M. Fourier Transform Infrared Spectroscopy and Chemometrics for the Rapid Screening of Economically Motivated Adulteration of Honey Spiked With Corn or Rice Syrup. *J. Food Prot.* **2023**, *86*, 100054. [CrossRef]
40. Riswahyuli, Y.; Rohman, A.; Setyabudi, F.M.C.S.; Raharjo, S. Indonesian wild honey authenticity analysis using attenuated total reflectance-fourier transform infrared (ATR-FTIR) spectroscopy combined with multivariate statistical techniques. *Heliyon* **2020**, *6*, e03662. [CrossRef]
41. Salelign, K.; Duraisamy, R. Sugar and ethanol production potential of sweet potato (*Ipomoea batatas*) as an alternative energy feedstock: Processing and physicochemical characterizations. *Heliyon* **2021**, *7*, e08402. [CrossRef]
42. Sritham, E.; Gunasekaran, S. FTIR spectroscopic evaluation of sucrose-maltodextrin-sodium citrate bioglass. *Food Hydrocoll.* **2017**, *70*, 371–382. [CrossRef]
43. Teklemariam, T.A.; Moisey, J.; Gotera, J. Attenuated Total Reflectance-Fourier transform infrared spectroscopy coupled with chemometrics for the rapid detection of coconut water adulteration. *Food Chem.* **2021**, *355*, 129616. [CrossRef] [PubMed]
44. Wang, J.; Kliks, M.M.; Jun, S.; Jackson, M.; Li, Q.X. Rapid Analysis of Glucose, Fructose, Sucrose, and Maltose in Honeys from Different Geographic Regions using Fourier Transform Infrared Spectroscopy and Multivariate Analysis. *J. Food Sci.* **2010**, *75*, C208–C214. [CrossRef] [PubMed]
45. Fangio, M.F.; Orallo, D.E.; Gende, L.B.; Churio, M.S. Chemical characterization and antimicrobial activity against *Paenibacillus larvae* of propolis from Buenos Aires province, Argentina. *J. Apic. Res.* **2019**, *58*, 626–638. [CrossRef]
46. Raemy, A.; Schweizer, T.F. Thermal behaviour of carbohydrates studied by heat flow calorimetry. *J. Therm. Anal.* **1983**, *28*, 95–108. [CrossRef]
47. Cevallos, P.A.P.; Buera, M.P.; Elizalde, B.E. Encapsulation of cinnamon and thyme essential oils components (cinnamaldehyde and thymol) in β-cyclodextrin: Effect of interactions with water on complex stability. *J. Food Eng.* **2010**, *99*, 70–75. [CrossRef]
48. Pralhad, T.; Rajendrakumar, K. Study of freeze-dried quercetin–cyclodextrin binary systems by DSC, FT-IR, X-ray diffraction and SEM analysis. *J. Pharm. Biomed. Anal.* **2004**, *34*, 333–339. [CrossRef]
49. Vasisht, K.; Chadha, K.; Karan, M.; Bhalla, Y.; Jena, A.K.; Chadha, R. Enhancing biopharmaceutical parameters of bioflavonoid quercetin by co-crystallization. *Crystengcomm* **2016**, *18*, 1403–1415. [CrossRef]
50. Deladino, L.; Navarro, A.S.; Martino, M.N. Microstructure of minerals and yerba mate extract co-crystallized with sucrose. *J. Food Eng.* **2010**, *96*, 410–415. [CrossRef]
51. Bozkir, B.; Acet, T.; Özcan, K. Investigation of the effects of different extraction methods on some biological activities of Dactylorhiza romana subsp. georgica (Klinge) Soó ex Renz & Taubenheim. *S. Afr. J. Bot.* **2022**, *149*, 347–354. [CrossRef]
52. Szalata, M.; Dreger, M.; Zielińska, A.; Banach, J.; Szalata, M.; Wielgus, K. Simple Extraction of Cannabinoids from Female Inflorescences of Hemp (*Cannabis sativa* L.). *Molecules* **2022**, *27*, 5868. [CrossRef]
53. Yang, L.; Shen, S.-Y.; Wang, Z.-N.; Yang, T.; Guo, J.-W.; Hu, R.-Y.; Li, Y.-F.; Burner, D.M.; Ying, X.-M. New Value-Added Sugar and Brown Sugar Products from Sugarcane: A Commercial Approach. *Sugar Technol* **2020**, *22*, 853–857. [CrossRef]
54. Singleton, V.L.; Orthofer, R.; Lamuela-Raventós, R.M. Analysis of total phenols and other oxidation substrates and antioxidants by means of folin-ciocalteu reagent. In *Methods in Enzymology*; Elsevier: Amsterdam, The Netherlands, 1999; pp. 152–178.

55. Brand-Williams, W.; Cuvelier, M.E.; Berset, C. Use of a free radical method to evaluate antioxidant activity. *LWT Food Sci. Technol.* **1995**, *28*, 25–30. [CrossRef]
56. Wu, Y.; Xu, L.; Liu, X.; Hasan, K.F.; Li, H.; Zhou, S.; Zhang, Q.; Zhou, Y. Effect of thermosonication treatment on blueberry juice quality: Total phenolics, flavonoids, anthocyanin, and antioxidant activity. *LWT* **2021**, *150*, 112021. [CrossRef]
57. Cai, Y.; Corke, H. Production and Properties of Spray-dried Amaranthus Betacyanin Pigments. *J. Food Sci.* **2000**, *65*, 1248–1252. [CrossRef]
58. Daza, L.D.; Fujita, A.; Fávaro-Trindade, C.S.; Rodrigues-Ract, J.N.; Granato, D.; Genovese, M.I. Effect of spray drying conditions on the physical properties of Cagaita (*Eugenia dysenterica* DC.) fruit extracts. *Food Bioprod. Process.* **2016**, *97*, 20–29. [CrossRef]

molecules MDPI

Article

Optimization of the Ultrasound Operating Conditions for Extraction and Quantification of Fructooligosaccharides from Garlic (*Allium sativum* L.) via High-Performance Liquid Chromatography with Refractive Index Detector

Muhammad Abdul Rahim [1], Adeela Yasmin [1], Muhammad Imran [1,*], Mahr Un Nisa [2], Waseem Khalid [1], Tuba Esatbeyoglu [3,*] and Sameh A. Korma [4]

1 Department of Food Science, Faculty of Life Sciences, Government College University, Faisalabad 38000, Pakistan
2 Department of Nutritional Sciences, Faculty of Medical Sciences, Government College University, Faisalabad 38000, Pakistan
3 Department of Food Development and Food Quality, Institute of Food Science and Human Nutrition, Gottfried Wilhelm Leibniz University Hannover, Am Kleinen Felde 30, 30167 Hannover, Germany
4 Department of Food Science, Faculty of Agriculture, Zagazig University, Zagazig 44519, Egypt
* Correspondence: imran@gcuf.edu.pk (M.I.); esatbeyoglu@lw.uni-hannover.de (T.E.); Tel.: +92-335-2020050 (M.I.); +49-511-7625589 (T.E.)

Abstract: Dietary interventions have captured the attention of nutritionists due to their health-promoting aspects, in addition to medications. In this connection, supplementation of nutraceuticals is considered as a rational approach to alleviating various metabolic disorders. Among novel strategies, prebiotic-supplemented foods are an encouraging trend in addressing the issue. In the present investigation, prebiotic fructooligosaccharides (FOS) were extracted from garlic (*Allium sativum* L.) powder using ultrasound-assisted extraction (UAE). The response surface methodology (RSM) was used to optimize the independent sonication variables, i.e., extraction temperature (ET, 80, 90, and 100 °C), amplitude level (AL, 70, 80, and 90%) and sonication time (ST, 10, 15 and 20 min). The maximum FOS yield (6.23 ± 0.52%) was obtained at sonication conditions of ET (80 °C), AL (80%) and ST (10 min), while the minimum yield of FOS was obtained at high operating temperatures and time. The optimized FOS yield (7.19%) was obtained at ET (80 °C), AL (73%) and ST (15 min) after model validation. The influence of sonication parameters, i.e., ET, AL and ST, on FOS yield was evaluated by varying their coded levels from −1 to +1, respectively, for each independent variable. High-performance liquid chromatography with refractive index detector (HPLC-RID) detection and quantification indicated that sucrose was present in high amounts (2.06 ± 0.10 g/100 g) followed by fructose and glucose. Total FOS fractions which included nystose present in maximum concentration (526 ± 14.7 mg/100 g), followed by 1-kestose (428 ± 19.5 mg/100 g) and fructosylnystoses (195 ± 6.89 mg/100 g). Conclusively, garlic is a good source of potential prebiotics FOS and they can be extracted using optimized sonication parameters using ultrasound-assisted techniques with maximum yield percentage.

Keywords: garlic (*Allium sativum* L.); FOS; prebiotics; UAE; RSM; HPLC-RID; spice; detection quantification

Citation: Rahim, M.A.; Yasmin, A.; Imran, M.; Nisa, M.U.; Khalid, W.; Esatbeyoglu, T.; Korma, S.A. Optimization of the Ultrasound Operating Conditions for Extraction and Quantification of Fructooligosaccharides from Garlic (*Allium sativum* L.) via High-Performance Liquid Chromatography with Refractive Index Detector. *Molecules* **2022**, *27*, 6388. https://doi.org/10.3390/molecules27196388

Academic Editors: Michał Halagarda and Sascha Rohn

Received: 20 August 2022
Accepted: 22 September 2022
Published: 27 September 2022

1. Introduction

Consumers are cautious about their food in recent times as poor nutritional habits such as more intake of saturated fatty acids and sugar contents and low intake of long-chain polyunsaturated fatty acids, vitamins, minerals and dietary fibers result in heart diseases, metabolic syndrome, chronic anxiety disorders, inflammation and various other maladies both in developed and developing countries. People need those functional foods that not only fulfil the nutritional requirements but also provide bioactive compounds that help in maintaining good health and longevity [1]. Therefore, the utilization of functional ingredients is important to provide the health benefits that ultimately reduce these risk factors due to poor nutritional intake. In this regard, many food industries are more interested in the production of fortified food products using different functional ingredients than actual food. The important functional ingredients in the human diet are prebiotics, probiotics, polyphenols, fatty acids and vitamins [2]. Among these functional ingredients, prebiotics plays an effective role in intestinal health by selectively stimulating the growth and activity of bacteria in the bowel [3]. Prebiotics act as feed for probiotics bacteria and other beneficial microbiota in the small intestine. It produces more health benefits by modulating intestinal microbiota as compared to other techniques such as drug therapy, aging, disease and antibiotics. It helps to promote certain microbial species, which are not present in the gut, for gaining better health benefits [4].

The prebiotics concerned are present in vegetables, roots and tuber crops. Among roots, garlic is an excellent source of natural prebiotics in the form of fructooligosaccharides (FOS). FOS are a diverse group of carbohydrates including fructose residues as prime monomers [5]. FOS, also synonymously called as oligofructose or oligofructan, are oligosaccharide fructans. Short-chain fructooligosaccharides (scFOS or FOS) are a combination of 1F-(1-β-fructofuranosyl) n-1 sucrose oligomers, where n varies from two to four [6]. In nature, they are sucrose molecules (glucose–fructose disaccharide) to which one or more extra fructose units are connected by β, 2-1glycosidic linkages The individual components of FOS contain GF_2 (α-D-glucopyronoside-(1,2)-β-D-fructofuranosyl-(1,2)-β-D-fructofuranosyl or kestose), GF_3 (α-D-glucopyronoside-(1,2)-β-Dfructofuranosyl-(1,2)-β-D-fructofuranosyl-(1,2)-β-D-fructofuranosyl-(1,2)-β-Dfructofuranosyl (nystose) and 1F-fructofuranosyl-nystose (GF_4) [7]. Bioactive compounds 1-kestose (GF_2), nystose (GF_3), quercetin, kaempferol and fructosylnystose (GF_4) have been reported in garlic samples [8,9]. FOS are also responsible for many functions in the human body such as the consumption of non-digestible oligosaccharides that increase gastrointestinal metabolism, improve the activity of bifidobacteria in the large intestine and act as an essential nutrient that must be present in the diet to reduce the risk of heart diseases and maintain gut health. Additionally, FOS also act as an antimicrobial agent, antioxidant, hypoglycemic index and hepatoprotective compounds, lowering alanine aminotransferase and mineral absorption to maintain bone homeostasis and maintain lipid profile levels in the human body [10]. Femia et al. [11] reported that FOS could reduce colonic epithelial cell proliferation in the colon and reduce the number of pills. They also help to reduce plasma cholesterol and hypertriglyceridemia and maintain colon health and gut microflora [12]. This dietary fiber can also help to reduce the effect of hypertriglyceridemia and decrease glucose intolerance in the colon [13]. Furthermore, FOS can also be used as a sweetener in the form of sucrose as it contains low calories and is rich in fiber [10,13].

Prebiotics are commonly obtained through three main techniques, i.e., microbiological synthesis, enzymatic degradation of polysaccharides and isolation from natural resources. These are naturally present in vegetables, roots and tuber crops. Among roots, garlic has been recognized as an excellent source of natural prebiotics in the form of FOS, comprising 70–80% of its total dry matter. Commercial extraction and quantification methods have been employed for the determination of prebiotic oligosaccharides from different vegetables and fruits such as enzymatic extraction, electrophoresis and ion exchange chromatography; most of these methods are technically complicated, require many laborious steps, with many impurities, and are time-consuming and expensive [10,14–16]. In addition, previous research has elaborated that the extraction yield of prebiotics FOS can be increased by 25% using ultrasound-assisted extraction (UAE) technology when compared to traditional enzymatic and solvent extraction methods. High-performance liquid chromatography (HPLC) is the method of choice among the chromatographic techniques for the quantification and detection of FOS and their other structural components [15,17,18]. Optimization of the methods for natural product extraction and quantification may reduce the cost and time consumption with a higher yield. In this connection, the present research aimed to optimize the ultrasound operating conditions and check their impact on FOS extraction from garlic (*Allium sativum* L.) powder using response surface methodology (RSM) and to determine their main sugar contents of FOS by high-performance liquid chromatography with refractive index detection (HPLC-RID) for the first time in Pakistan.

2. Results and Discussion

2.1. Model Fitting and Extraction Yield of Fructooligosaccharides (FOS)

In order to optimize extraction conditions, the combined impact of independent variables (extraction temperatures (ET) 80–100 °C, amplitude level (AL) 70–90% and sonication time (ST) 10–20 min on the extraction of FOS, experiments were performed for different combinations of the independent variables using statistically designed experiments, and the results have been described in Table 1. The total number of the experiment was 16-run to determine their optimum levels. In this study, the highest yield of FOS in garlic powder was obtained at 6.23 ± 0.52% at ET (80 °C), AL (80%), and ST (10 min). The minimum response value in experimental samples was estimated at 4.55 ± 0.40% at ET (100 °C), AL (80%) and ST (20 min). Further, the FOS yield was significantly improved by reducing the ET and ST. The predicted extraction of FOS from garlic powder with the combinations of independent variables such as ET (°C), AL (%) and ST (min) as defined by the regression model were found in the range of 5.75 ± 0.44% to 7.19 ± 0.57% (Table 1).

Table 1. Extraction yield of FOS in garlic as analyzed using Box–Behnken design (BBD) after ultrasound treatment.

Sonication Run	Independent Variables						Response Variable FOS Yield (%)	
	Coded (ET)	Coded (AL)	Coded (ST)	ET (°C)	AL (%)	ST (min)	Experimental Value	Predicted Value
1	+1	−1	0	100	70	15	4.76 ± 0.42 [h]	5.82 ± 0.48 [e]
2(C_1)	0	0	0	90	80	15	5.25 ± 0.47 [fg]	6.25 ± 0.53 [c]
3	−1	0	+1	80	80	20	5.95 ± 0.49 [de]	7.08 ± 0.55 [ab]
4	−1	−1	0	80	70	15	6.06 ± 0.50 [d]	7.19 ± 0.57 [a]
5	+1	+1	0	100	90	15	5.13 ± 0.45 [g]	5.97 ± 0.50 [de]
6	0	+1	−1	90	90	10	4.62 ± 0.41 [hi]	5.88 ± 0.49 [e]
7	−1	0	−1	80	80	10	6.23 ± 0.52 [c]	7.02 ± 0.53 [b]
8	0	−1	+1	90	70	20	5.11 ± 0.44 [g]	5.84 ± 0.48 [e]
9(C_2)	0	0	0	90	80	15	5.27 ± 0.48 [fg]	6.25 ± 0.53 [c]
10	−1	+1	0	80	90	15	6.17 ± 0.51 [cd]	7.11 ± 0.56 [ab]
11(C_3)	0	0	0	90	80	15	5.23 ± 0.46 [fg]	6.25 ± 0.53 [c]
12	+1	0	+1	100	80	20	4.55 ± 0.40 [i]	5.75 ± 0.44 [ef]
13	+1	0	−1	100	80	10	4.98 ± 0.43 [gh]	5.84 ± 0.48 [e]
14	0	−1	−1	90	70	10	5.24 ± 0.47 [fg]	6.30 ± 0.54 [bc]
15	0	+1	+1	90	90	20	5.41 ± 0.48 [f]	6.33 ± 0.54 [bc]
16(C_4)	0	0	0	90	80	15	5.26 ± 0.45 [fg]	6.25 ± 0.53 [c]

C_1, C_2, C_3 and C_4 FOS sonication conditions are set at center points of the Box–Behnken design (BBD); [a–i] Means with different superscripts represent the change in FOS yield; ET, extraction temperature; AL, amplitude level; ST, sonication time; FOS, fructooligosaccharides.

There is very limited published data that provides information to support the extraction of FOS from garlic powder using optimized operating conditions of ultrasound-assisted extraction (UAE). The predicted values of FOS were compared with experimental values in order to evaluate the validity of the model. Table 2 indicates the analysis of variance (ANOVA) obtained from the fitting of the experimental data and the interaction effect of ultrasonic conditions on FOS yield. The model and lack of fit *f*-value showed a significant effect on the dependent variables. The R_2 computed for FOS was found to be 0.92. The analytical results for adjusted and predicted R_2 values were reported as 0.80 and 0.25, respectively.

Table 2. Analysis of variance (ANOVA) for the quadratic model of FOS yield.

Source of Variation	SS	DF	MS	*f*-Value	*p*-Value
Model	3.81	9	0.42	7.84	0.01
ET	3.14	1	3.14	58.18	0.003
AL	0.06	1	0.004	0.052	0.82
ST	0.04	1	0.002	0.003	0.95
ET × AL	0.01	1	0.014	0.27	0.62
ET × ST	0.05	1	0.003	0.10	0.75
AL × ST	0.21	1	0.21	3.84	0.09
ET^2	0.37	1	0.37	6.84	0.03
AL^2	0.01	1	0.0005	0.072	0.79
ST^2	0.06	1	0.066	1.23	0.30
Residual	0.31	6	0.054	–	–
Lack of Fit	0.34	3	0.11	368.77	0.82
Pure Error	0.07	3	0.001	–	–
Total	4.13	16	–	–	–

ET, extraction temperature; AL, amplitude level; ST, sonication time; SS, the sum of squares; DF, degree of freedom; MS, mean square. Level of significance: $p \leq 0.05$.

UAE is based on the propagation of mechanical waves that are capable of destroying the cell walls of the sample. It analyzes the variables involved in the production of high-value compounds and the extraction of prebiotics [19]. The FOS content in garlic powder was found to be 3.34% as described in a research study conducted by Prayogi Sunu et al. [20]. According to the literature of Campbell et al. [14], FOS content in garlic powder was 1.70%. The FOS yield depends on the concentration, temperature, solvent and treatment time. Moreover, the temperature and time comprehensively affected the yield of bioactive compounds [21]. Another research work by Heydari and Darabi Bazvand [22] revealed that the maximum extraction efficiency of vitamin C or ascorbic acid was estimated in various matrices at the lower ultrasonic time (10 s) and higher ultrasonic amplitude (100%). In another similar study, the extraction efficiency of mineral components from edible oils was increased up to 10 min and then decreased, while increasing the optimum ultrasonic bath temperature to 60 °C contributes to an increase in the yield [23]. Furthermore, the maximum yield was obtained at optimizing ultrasonic conditions such as, ultrasound time = 30 min; volume of organic solvent = 2.5 mL; salt concentration = 25% *w*/*v*; and pH = 4 [24]. According to the previous report of Rezaeepour et al. [25], a higher extraction efficiency occurs at the initial ultrasonic time range from 1 to 30 min and then decreases. A comparative study was carried out by Louie et al. [26] and found a significantly higher yield of FOS as compared to other traditional extraction methods. The improvement in yield was noticed as time and temperature decreased [10]. The highest value of yield (112 µg/mL) was determined at 25 °C for 90 min with optimum frequency [21]. The highest withdrawal rate was observed in the first few minutes, which is considered to be the most profitable period [27]. Higher FOS contents can be used in functional products to improve the activity of microbiota and may reduce the attack of pathogens on intestinal cells [28].

2.2. Single Factor Analysis for FOS Yield

The influence of ET, AL and ST on FOS yield was evaluated by varying their coded levels from −1 to +1, respectively, for each independent variable (Figure 1). The mean value of independent variables was set to rotate the model uniformly during the analysis of each individual variable of the process and response. The regression equations for the independent variables are given in Equations (1)–(3), respectively.

$$\text{Regression equation for ET} = (36.492) + (-0.609)\ \text{ET} + (0.003)\ \text{ET}^2 \quad (1)$$

$$\text{Regression equation for AL} = (4.1025) + (0.051875)\ \text{AL} + (-0.003)\ \text{AL}^2 \quad (2)$$

$$\text{Regression equation for ST} = (5.108) + (0.153)\ \text{ST} + (-0.005)\ \text{ST}^2 \quad (3)$$

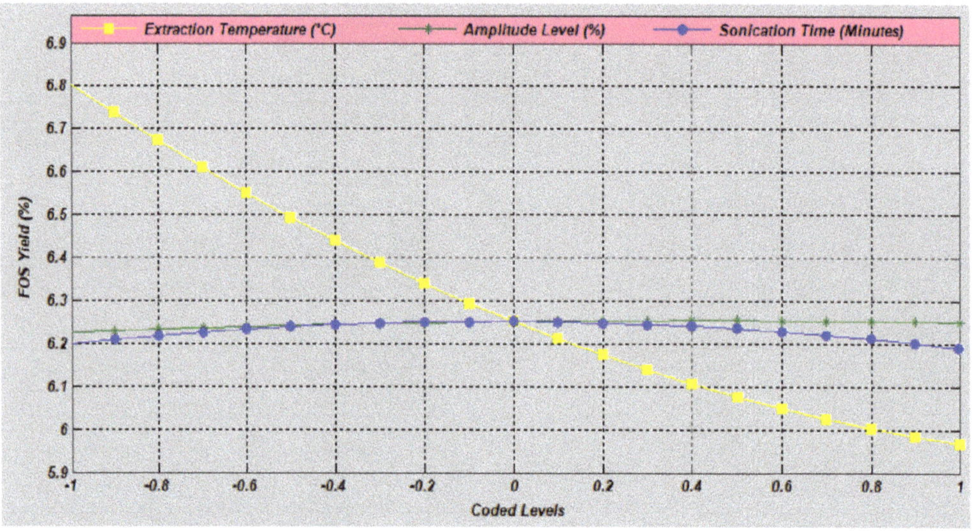

Figure 1. Single factor analysis of independent sonication operating conditions for FOS yield. For interpretation of the references to color in this figure legend, the reader is referred to the Web version of this article.).

The analysis of the single factor showed that the ET had a strong effect on the percentage of FOS yield. The FOS yield was inversely proportional to the level of ET. The FOS yield was increased by lowering the level of ET. The AL and ST level imparts minimum effect on the FOS yield (Figure 1). The regression coefficient was used to calculate the quadratic impact of independent variables. In this quadratic regression model, the regression coefficient between the independent variables and the response variables was high, indicating the best evaluation of the experimental data.

2.3. Mutual Interaction Effect on FOS Yield

The effect of the mutual interaction on the independent variables for the yield of FOS in garlic powder was estimated by rotating two independent factors and fixing the third factor at the coded zero level. The surface and contour plots representing the mutual interaction of sonication independent variables have been shown in Figures 2 and 3. The mutual interaction effect between ET and AL showed that the FOS yield was reduced by increasing the ET and AL (Figure 2a). Furthermore, the correlation between ST and ET indicated that the increase in ST and ET leads to a decline in FOS yield (Figure 2b). Moreover, the FOS yield was improved by lowering the ST and AL levels (Figure 2c). The validation of the model depends on the optimized experimental values and response yield. The Box–Behnken design (BBD) was used to optimize the operational conditions and FOS yield. Based on the above findings, the interaction between ET and AL showed the FOS yield of 7.19% at ET (80 °C), AL (73.34%) and ST (15 min) (Figure 3a). Moreover, the relation between optimized and predicted values of ET and ST indicated the FOS yield as 7.18%, at ET (80 °C), AL (80%) and ST (15.67 min) (Figure 3b). Furthermore, sonication-independent conditions for AL and ST in UAE were determined as ET (90 °C), AL (90%) and ST (19.34 min) for a 6.33% FOS yield (Figure 3a). The optimized sonication conditions for FOS yield validation were again performed with three different replications to confirm the final predicted value and response yield for future recommendations at the commercial scale for discerning food processing industries. Finally, the optimized FOS yield (7.19%) was obtained at ET (80 °C), AL (73%) and ST (15 min) after model validation.

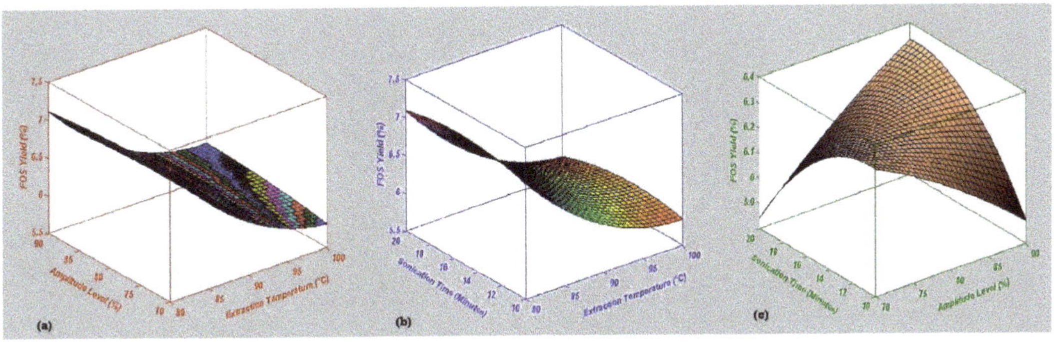

Figure 2. Surface graphical representation of mutual interaction effect of sonication conditions on response yield. (**a**) Extraction temperature (°C) and amplitude level (%) versus FOS yield (%), (**b**) extraction temperature (°C) and sonication time (min) versus FOS yield (%) and (**c**) amplitude level (%) and sonication time (min) versus FOS yield (%). For interpretation of the references to color in this figure legend, the reader is referred to the Web version of this article.

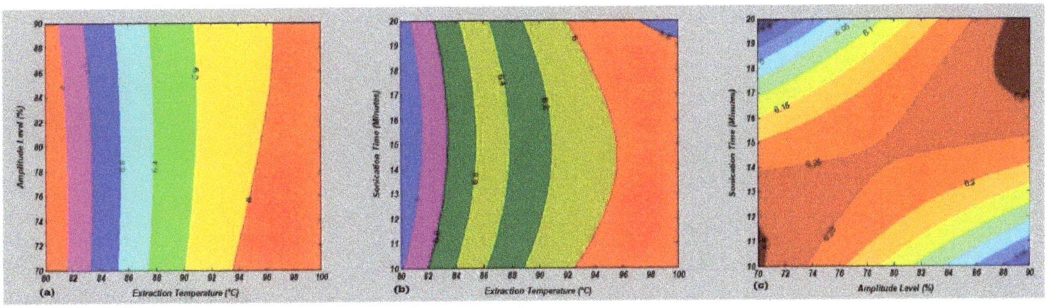

Figure 3. Mutual interaction analysis of sonication operating conditions in terms of contour plots for FOS yield. (**a**) Extraction temperature (°C) versus amplitude level (%), (**b**) extraction temperature (°C) versus sonication time (min) and (**c**) amplitude level (%) versus sonication time (min). For interpretation of the references to color in this figure legend, the reader is referred to the Web version of this article.

The results of the present study are in line with the previous findings of Ahmad et al. [29]. Notably, they used extraction temperature, extraction time and liquid–solid ratio in the representative quadratic model and statistical validation of the polynomial equation was performed. The highest polysaccharide yield (11.56%) was noted at optimum conditions [29]. The mutual interaction between response and predicted values was validated by the RSM model using regression coefficient correlation [30,31]. In a similar fashion, the effect of time, temperature, volume to mass ratio and ultrasound treatment on yield was validated for individual regression coefficients [32]. Moreover, Khumpirapang et al. [33] described the strong correlation between predicted and experimental values obtained at optimal extraction conditions, and such a finding ultimately strengthens the outcomes reported in the present study.

2.4. *Quantification of FOS by High-Performance Liquid Chromatography with Refractive Index Detector (HPLC-RID)*

The quantification of FOS via HPLC-RID in garlic powder is presented in Table 3. It is obvious from the data that sucrose is present in the highest amount (2.06 ± 0.10 g/100 g), followed by fructose and glucose. On the other hand, out of total FOS fractions, nystose (GF_3) with three glucose units was present in maximum concentration (526 ± 14.68 mg/100 g), preceded by 1-kestose (GF_2) at 428 ± 19.45 mg/100 g and fructosylnystose (GF_4) 195 ± 6.89 mg/100 g. The results from intra- and inter-day analysis showed good precision. The results concerning extracted FOS from garlic in the current investigation are in agreement with the observations of Król and Grzelak [34]. They categorized commercially available FOS and observed the values for individual monosaccharides containing sucrose, glucose and fructose as 3.00, 0.40 and 0.30 g/100 g, whilst nystose, kestose, and fructosylnystose were observed as 49.00, 36.00 and 12.00 g/100 g. Nevertheless, the FOS composition varies with the source, degree of polymerization as well as the method of extraction. The obtained results for FOS composition are also corroborated by the findings of Chen et al. [35], who assessed FOS powder for its modulating role in elderly men. According to their observations, HPLC analysis exposed that FOS contained various fractions i.e., sucrose, glucose, fructose, 1-kestose, nystose and fructosylnystose. Out of these individual fractions, the maximum level was noticed for 1-kestose and the minimum for fructosylnystose. The current data regarding FOS characterization are in agreement with the findings of Judprasong et al. [36]; they verified the FOS and inulin composition of numerous fruits and vegetables including garlic and established the presence of the above-mentioned fractions.

Table 3. Quantification of individual fractions (mono- and oligosaccharides) of FOS contents of garlic powder by HPLC-RID.

Sugar	**Concentration (g/100 g)**
Fructose	1.30 ± 0.09
Glucose	0.30 ± 0.02
Sucrose	2.06 ± 0.10
Fructooligosaccharides	**Concentration (mg/100 g)**
1-Kestose (GF_2)	428 ± 19.45
Nystose (GF_3)	526 ± 14.68
Fructosylnystose (GF_4)	195 ± 6.89
Total (FOS)	1149 ± 22.35

Values expressed as means ± standard deviation.

FOS hold quite a lot of characteristics that make them a desirable food additive to augment the shelf life and taste profile of food products [7]. Remarkably, FOS are one of the new functional fibers employed in the food industry that effectually increase the fiber contents of usually non-fibrous foods. Still, it is appropriate to probe the compositional parameters and degree of polymerization of FOS contents of different vegetables as it impacts the rheology of functional food products to which they can be added. Moreover, the chain length also affects the level of fermentation in the intestine and a shorter chain length is preferred [37].

Garlic has a peculiar smell due to which it is not liked by all consumers. Moreover, its pungent taste may not be appropriate for addition to some types of food products. Hence, extraction of FOS and provision in their purified form may improve their overall consumption. The extracted FOS in this study were off-white in color and almost odorless. So, they can be used as an ingredient in the preparation of many ready-to-eat food products with prebiotic properties.

3. Materials and Methods

3.1. Chemicals and Preparation of Sample

All the chemicals and HPLC grade reagents and standards were purchased from Merck (Merck KGaA, Darmstadt, Germany) and Sigma-Aldrich (Sigma-Aldrich, Tokyo, Japan). Spring garlic was purchased from the registered superstore, Punjab, Pakistan. In this study, garlic bulbs were manually separated into cloves. The undesirable components were removed from the garlic cloves. After that, approximately 100 g of samples were randomly selected, cut into small pieces of 20 to 30 mm and dried in a universal hot air oven (Memmert® UN55, Memmert, Schwabach, Germany) for 12 to 24 h at 60 °C. At the end, garlic powder was prepared by grinding the dry bulb through a grinder [38].

3.2. *Ultrasound-Assisted Extraction (UAE) of Fructooligosaccharides (FOS)*

FOS were extracted from the garlic powder samples using the method elaborated by Jovanovic-Malinovska et al. [18] with some modifications. In this method, 100 g of garlic powder was dissolved in 20 mL of ethanol (96%) and prepared into a solution. The sonication apparatus (model VCX 750, Sonic and Materials, Inc., Newtown, CT, USA) was used for the extraction process. The different ratios of prepared solutions were placed in the ultrasound sonication apparatus at extraction temperatures (ET) 80–100 °C, amplitude level (AL) 70–90% and sonication time (ST) 10–20 min. After that, the solution samples were kept for 10 to 20 min to cool at room temperature (25 ± 2 °C). Then, the solution was centrifuged at a low temperature of 10 °C by adjusting the speed at 3000 rpm. After the centrifugation, 10 mL of supernatants were mixed and placed in the vacuum rotary evaporator at 50 °C until the solvent was removed. Finally, the slurry was re-suspended by adding 1.5 mL of deionized water. Whatman No. 1 filter paper was used for filtration.

3.3. *Determination of the FOS Content by HPLC-RID*

The concentration of FOS in the garlic powder sample was estimated using HPLC-RID (PerkinElmer Series 200, PerkinElmer, Shelton, CT, USA) fitted with a refractive index detector (RID-10A) and C18 column (250 mm × 4.6 mm, 5.0 µm particle size). In this study, 1 mL sample was dissolved in HPLC grade water and the solution volume was made up to 50 mL. After that, the prepared solution was kept in a shaking water bath for 5 min at 30 °C and filtered with Whatman No. 1 (pore size 25 µm). Then, 10 µL of the aliquot sample was injected. The mobile phase was HPLC grade water. The flow rate was adjusted to 1 mL/minutes and the amounts of 1-kestose (GF_2), nystose (GF_3), fructosylnystose (GF_4) and total (FOS) were performed by their respective standards of fructose, glucose, sucrose and FOS (Sigma-Aldrich, St. Louis, MO, USA). The precision or accuracy of the presented method was assessed through the intra- and inter-day repeatability of the method of respective standards. The intra-day repeatability study was performed through injection of standard solution six times in one day followed by calculation of the relative standard deviation. Furthermore, inter-day repeatability was assessed by analyzing the same standard solution once a day over a three-day period [39].

3.4. *Experimental Design and Statistical Analysis*

The obtained data for garlic powder were subjected to statistical analysis, and the level of significance was determined at 5% ($p \leq 0.05$) using the software Design Expert® (version 11.1.2.0, E Hennepin Ave, Minneapolis, MN, USA) and MathWorks Matlab® (version 7.5.0.338; R2007a, Natick, MA, USA) software according to the method described in Montgomery [40]. In addition to this, response surface methodology (RSM) was performed to estimate the optimized values of independent variables on which maximum dependent response was obtained using the Box–Behnken design (BBD). Analysis of variance (ANOVA) was employed to check the ampleness of the model. The modeling was started with a quadratic model including linear effects, interaction effects and quadratic effects. Significant terms in the model for dependent variables were found by ANOVA and significance was assessed by the F-statistic intended from the data. The results were evaluated with various descriptive statistical analyses such as the sum of squares (SS), degree of freedom (DF); and mean square (MS). After fitting the value to the model, the generated values were used for contour and surface plots. Table 4 presents the coded and actual values of experimental treatments. The three independent variables at three levels and results obtained for each response were used to evaluate the BBD of RSM explained by non-linear Equations (4)–(6), respectively.

$$Y = \beta_0 + \beta_1\,(ET) + \beta_2\,(AL) + \beta_3\,(ST) + \beta_1 \times \beta_2\,(ET \times AL) + \beta_1 \times \beta_3\,(ET \times ST) + \beta_2 \times \beta_3\,(AL \times ST) + \beta_1\,(T)^2 + \beta_2\,(AL)^2 + \beta_3\,(ST)^2 \quad (4)$$

Table 4. Coded and actual levels of independent sonication variables for optimization of fructooligosaccharides yield determined by BBD of response surface methodology (RSM).

Independent Variables	Units	Coded Levels		
		−1 (Low)	0 (Medium)	+1 (High)
Extraction temperature (ET)	°C	80	90	100
Amplitude level (AL)	%	70	80	90
Sonication time (ST)	min	10	15	20

The regression equation in terms of coded and actual factors is given below:

$$Y_1 = +\,4.25 - 0.63\,\beta_1 + 0.019\,\beta_2 - 0.005\,\beta_3 + 0.060\,\beta_1 \times \beta_2 - 0.037\,\beta_1 \times \beta_3 + 0.23\,\beta_2 \times \beta_3 + 0.30\,\beta_{12} - 0.031\,\beta_{22} - 0.13\,\beta_3 \quad (5)$$

$$Y_2 = +\,39.96 - 0.64\,ET - 0.070\,AL - 0.14\,ST + 0.006\,ET \times AL - 0.007\,ET \times ST + 0.004\,AL \times ST + 0.003\,ET^2 - 0.003\,AL^2 - 0.005\,ST^2 \quad (6)$$

where Y, Y_1, Y_2 = Dependent variables; β_0 = Intercept; β_1 = Extraction Temperature (ET); β_2 = Amplitude Level (AL); β_3 = Sonication Time (ST); β_1 to β_3 = Regression coefficients; ET, AL, ST = Independent variables.

4. Conclusions

This study has provided detailed information regarding the extraction of fructooligosaccharides (FOS) from the local garlic variety consumed in Pakistan. It may be concluded that garlic may be one of the major sources of FOS and the maximum yield of fructooligosaccharides (FOS) from garlic can be obtained by using the optimized conditions of ultrasound green technology. Moreover, the HPLC-RID quantification revealed the presence of 1-kestose (GF_2), nystose (GF_3) and fructosylnystose (GF_4) in higher concentrations as individual sugar fractions. The results may provide a good basis for optimized extraction parameters as well as the composition of FOS in garlic. Furthermore, the present study also recommended that the extracted FOS may be explored as a functional food ingredient to formulate prebiotics-supplemented food products. Additionally, long-term storage quality and biological evaluation of such FOS-fortified food products should be considered in future studies.

Author Contributions: Conceptualization, M.A.R., M.I., M.U.N. and T.E.; methodology, M.A.R., M.I. and M.U.N.; software, M.A.R., M.I. and M.U.N.; validation, M.A.R., M.I., S.A.K. and M.U.N.; formal analysis, M.A.R., M.I. and M.U.N.; investigation, M.A.R., A.Y., M.I., M.U.N., S.A.K. and T.E.; resources, M.A.R., M.I., M.U.N., S.A.K. and T.E.; data curation, M.A.R., M.I., M.U.N., S.A.K. and T.E.; writing—original draft preparation, M.A.R., M.I. and M.U.N.; writing—review and editing, M.U.N., A.Y., M.A.R., W.K., S.A.K. and T.E.; visualization, M.A.R., M.I., M.U.N. and S.A.K.; supervision, M.A.R., M.I., M.U.N., S.A.K. and T.E.; project administration, M.A.R., M.I., M.U.N., S.A.K. and T.E.; funding acquisition, M.A.R., M.I., M.U.N., S.A.K. and T.E. All authors have read and agreed to the published version of the manuscript.

Funding: The publication of this article was funded by the Open Access Fund of Leibniz Universität Hannover.

Institutional Review Board Statement: Not applicable.

Informed Consent Statement: Not applicable.

Data Availability Statement: Data are available from the authors upon request.

Acknowledgments: The authors thank the Library Department, Government College University Faisalabad (GCUF) and IT Department, Higher Education Commission (HEC) for access to journals, books and valuable databases. The authors are also thankful to the Innovative Processing Technologies Lab, Department of Food Science, Government College University, Faisalabad for providing access to sonication processing.

Conflicts of Interest: The authors have declared no conflict of interest for this article.

References

1. Alamgir, K.; Sami, U.K.; Salahuddin, K. Nutritional complications and its effects on human health. *J. Food Sci. Nutr.* **2018**, *1*, 17–20.
2. Peng, M.; Tabashsum, Z.; Anderson, M.; Truong, A.; Houser, A.K.; Padilla, J.; Akmel, A.; Bhatti, J.; Rahaman, S.O.; Biswas, D. Effectiveness of probiotics, prebiotics, and prebiotic-like components in common functional foods. *Compr. Rev. Food Sci. Food Saf.* **2020**, *19*, 1908–1933. [CrossRef] [PubMed]
3. Davani-Davari, D.; Negahdaripour, M.; Karimzadeh, I.; Seifan, M.; Mohkam, M.; Masoumi, S.J.; Berenjian, A.; Ghasemi, Y. Prebiotics: Definition, types, sources, mechanisms, and clinical applications. *Foods* **2019**, *8*, 92. [CrossRef] [PubMed]
4. Markowiak, P.; Śliżewska, K. Effects of probiotics, prebiotics, and synbiotics on human health. *Nutrients* **2017**, *9*, 1021. [CrossRef] [PubMed]
5. Bonnema, A.L.; Kolberg, L.W.; Thomas, W.; Slavin, J.L. Gastrointestinal tolerance of chicory inulin products. *J. Am. Diet Assoc.* **2010**, *110*, 865–868. [CrossRef] [PubMed]
6. Birkett, A.; Francis, C. Short-chain fructooligosaccharides: A low molecular weight fructan. In *Handbook of Prebiotics and Probiotics Ingredients: Health Benefits and Food*; Cho, S., Ed.; CRC Press: Boca Raton, FL, USA, 2009; pp. 14–38. [CrossRef]
7. Kolida, S.; Gibson, G. Prebiotic capacity if inulin-type fructans. *J. Nutr.* **2007**, *137*, 2503–2506. [CrossRef] [PubMed]
8. Mabrok, H.; Soliman, M.; Mohammad, M.; Hussein, L. HPLC profiles of onion fructooligosaccharides and inulin and their prebiotic effects on modulating key markers of colon function, calcium metabolism and bone mass in rat model. *J. Biochem. Physiol.* **2018**, *1*, 10.
9. Rahim, M.A.; Saeed, F.; Khalid, W.; Hussain, M.; Anjum, F.M. Functional and nutraceutical properties of fructo-oligosaccharides derivatives: A review. *Int. J. Food Prop.* **2021**, *24*, 1588–1602. [CrossRef]
10. Jovanovic-Malinovska, R.; Kuzmanova, S.; Winkelhausen, E. Oligosaccharide profile in fruits and vegetables as sources of prebiotics and functional foods. *Int. J. Food Prop.* **2014**, *17*, 949–965. [CrossRef]
11. Femia, A.P.; Luceri, C.; Dolara, P.; Giannini, A.; Biggeri, A.; Salvadori, M.; Clune, Y.; Collins, K.J.; Paglierani, M.; Caderni, G. Antitumorigenic activity of the prebiotic inulin enriched with oligofructose in combination with the probiotics *Lactobacillus rhamnosus* and *Bifidobacterium lactis* on azoxymethane-induced colon carcinogenesis in rats. *Carcinogenesis* **2002**, *23*, 1953–1960. [CrossRef] [PubMed]
12. Kherade, M.; Solanke, S.; Tawar, M.; Wankhede, S. Fructooligosaccharides: A comprehensive review. *J. Ayurvedic Herb. Med.* **2021**, *7*, 193–200. [CrossRef]
13. Hills, R.D.; Pontefract, B.A.; Mishcon, H.R.; Black, C.A.; Sutton, S.C.; Theberge, C.R. Gut microbiome: Profound implications for diet and disease. *Nutrients* **2000**, *11*, 1613. [CrossRef] [PubMed]
14. Campbell, J.M.; Bauer, L.L.; Fahey, G.C.; Hogarth, A.J.C.L.; Wolf, B.W.; Hunter, D.E. Selected fructooligosaccharide (1-kestose, nystose, and 1F-β-fructofuranosylnystose) composition of foods and feeds. *J. Agric. Food Chem.* **1997**, *45*, 3076–3082. [CrossRef]
15. Wichienchot, S.; Thammarutwasik, P.; Jongjareonrak, A.; Chansuwan, W.; Hmadhlu, P.; Hongpattarakere, T.; Itharat, A.; Ooraikul, B. Extraction and analysis of prebiotics from selected plants from southern Thailand. Songklanakarin. *J. Sci. Technol.* **2011**, *33*, 517–523.
16. Shalini, R.; Krishna, J.; Sankaranarayanan, M.; Antony, U. Enhancement of fructan extraction from garlic and fructooligosaccharide purification using an activated charcoal column. *LWT* **2021**, *148*, 111703.
17. Benkeblia, N. Fructooligosaccharides and fructans analysis in plants and food crops. *J. Chromatogr. A* **2013**, *1313*, 54–61. [CrossRef]
18. Jovanovic-Malinovska, R.; Kuzmanova, S.; Winkelhausen, E. Application of ultrasound for enhanced extraction of prebiotic oligosaccharides from selected fruits and vegetables. *Ultrason. Sonochem.* **2015**, *22*, 446–453. [CrossRef]
19. Scudino, H.; Guimarães, J.; Lino, D.; Duarte, M.; Esmerino, E.; Freitas, M. Ultrasound for probiotic and prebiotic foods. In *Probiotics and Prebiotics in Foods*; Elsevier: Amsterdam, The Netherlands, 2021; pp. 293–307. [CrossRef]
20. Prayogi Sunu, D.S.; Mahfudz, L.D.; Yunianto, V.D. Prebiotic activity of garlic (*Allium sativum*) extract on *Lactobacillus acidophilus*. *Vet. World* **2019**, *12*, 2046–2051. [CrossRef]
21. Mathialagan, R.; Mansor, N.; Shamsuddin, M.R.; Uemura, Y.; Majeed, Z. Optimisation of ultrasonic-assisted extraction (UAE) of allicin from garlic (*Allium sativum* L.). *Chem. Eng. Trans.* **2017**, *56*, 1747–1752.
22. Heydari, R.; Darabi Bazvand, M.R. Ultrasound-assisted matrix solid-phase dispersion coupled with reversed-phase dispersive liquid–liquid microextraction for determination of vitamin C in various matrices. *Food Anal. Methods* **2019**, *12*, 1949–1956. [CrossRef]
23. Mohebbi, M.; Heydari, R.; Ramezani, M. Determination of Cu, Cd, Ni, Pb and Zn in edible oils using reversed-phase ultrasonic assisted liquid–liquid microextraction and flame atomic absorption spectrometry. *J. Anal. Chem.* **2018**, *73*, 30–35. [CrossRef]

24. Ismaili, A.; Heydari, R.; Rezaeepour, R. Monitoring the oleuropein content of olive leaves and fruits using ultrasound-and salt-assisted liquid–liquid extraction optimized by response surface methodology and high-performance liquid chromatography. *J. Sep. Sci.* **2016**, *39*, 405–411. [CrossRef]
25. Rezaeepour, R.; Heydari, R.; Ismaili, A. Ultrasound and salt-assisted liquid–liquid extraction as an efficient method for natural product extraction. *Anal. Methods* **2015**, *7*, 3253–3259. [CrossRef]
26. Louie, K.B.; Kosina, S.M.; Hu, Y.; Otani, H.; de Raad, M.; Kuftin, A.N.; Nigel, J.M.; Benjamin, P.B.; Northen, T.R. Mass spectrometry for natural product discovery. In *Comprehensive Natural Products III*, 3rd ed.; Elsevier: Amsterdam, The Netherlands, 2020; Volume 6, pp. 263–306. [CrossRef]
27. Esclapez, M.D.; García-Pérez, J.V.; Mulet, A.; Cárcel, J.A. Ultrasound-assisted extraction of natural products. *Food Eng. Rev.* **2000**, *3*, 108–120. [CrossRef]
28. Green, M.; Arora, K.; Prakash, S. Microbial medicine: Prebiotic and probiotic functional foods to target obesity and metabolic syndrome. *Int. J. Mol. Sci.* **2000**, *21*, 2890. [CrossRef]
29. Ahmad, A.; Rehman, M.U.; Wali, A.F.; El-Serehy, H.A.; Al-Misned, F.A.; Maodaa, S.N.; Aljawdah, H.M.; Mir, T.M.; Ahmad, P. Box–Behnken response surface design of polysaccharide extraction from rhododendron arboreum and the evaluation of its antioxidant potential. *Molecules* **2020**, *25*, 3835. [CrossRef]
30. Tomšik, A.; Pavlić, B.; Vladić, J.; Ramić, M.; Brindza, J.; Vidović, S. Optimization of ultrasound-assisted extraction of bioactive compounds from wild garlic (*Allium ursinum* L.). *Ultrason. Sonochem.* **2016**, *29*, 502–511. [CrossRef]
31. Rahim, M.A.; Imran, M.; Khan, M.K.; Ahmad, M.H.; Ahmad, R.S. Impact of spray drying operating conditions on encapsulation efficiency, oxidative quality, and sensorial evaluation of chia and fish oil blends. *J. Food Process. Preserv.* **2022**, *46*, 16248. [CrossRef]
32. Medlej, M.K.; Cherri, B.; Nasser, G.; Zaviska, F.; Hijazi, A.; Li, S.; Pochat-Bohatier, C. Optimization of polysaccharides extraction from a wild species of Ornithogalum combining ultrasound and maceration and their anti-oxidant properties. *Int. J. Biol. Macromol.* **2020**, *161*, 958–968. [CrossRef]
33. Khumpirapang, N.; Srituptim, S.; Kriangkrai, W. Optimization for powder yield of spray-dried garlic extract powder using box-behnken experimental design. *Key Eng. Mater.* **2020**, *859*, 301–306. [CrossRef]
34. Król, B.; Grzelak, K. Qualitative and quantitative composition of fructooligosaccharides in bread. *Eur. Food Res. Technol.* **2006**, *223*, 755–758. [CrossRef]
35. Chen, H.L.; Lu, Y.H.; Ko, L.Y. Effects of fructooligosaccharide on bowel function and indicators of nutritional status in constipated elderly men. *Nutr. Res.* **2000**, *20*, 1725–1733. [CrossRef]
36. Judprasong, K.; Tanjor, S.; Puwastien, P.; Sungpuag, P. Investigation of Thai plants for potential sources of inulin-type fructans. *J. Food Compost. Anal.* **2011**, *24*, 642–649. [CrossRef]
37. Stewart, M.L.; Timm, D.A.; Slavin, J.L. Fructooligosaccharides exhibit more rapid fermentation than long-chain inulin in an in vitro fermentation system. *Nutr. Res.* **2008**, *28*, 329–334. [CrossRef]
38. Ratti, C.; Araya-Farias, M.; Mendez-Lagunas, L.; Makhlouf, J. Drying of garlic (*Allium sativum*) and its effect on allicin retention. *Dry Technol.* **2007**, *25*, 349–356. [CrossRef]
39. Petkova, N.; Vrancheva, R.; Denev, P.; Ivanov, I.; Pavlov, A. HPLC-RID method for determination of inulin and fructooligosacharides. *Acta Nat. Sci.* **2014**, *1*, 107.
40. Montgomery, D.C. *Design and Analysis of Experiments*; John Wiley & Sons: Hoboken, NJ, USA, 2017.

molecules

Article

The Impact of Furfural on the Quality of Meads

Paweł Sroka *, Tomasz Tarko * and Aleksandra Duda

Department of Fermentation Technology and Microbiology, University of Agriculture in Krakow, ul. Balicka 122, 30-149 Kraków, Poland; aleksandra.duda@urk.edu.pl
* Correspondence: pawel.sroka@urk.edu.pl (P.S.); tomasz.tarko@urk.edu.pl (T.T.); Tel.: +48-12662-47-90 (P.S.); +48-12662-47-92 (T.T.)

Abstract: Furfural is a naturally occurring compound in bee honey, classified as a fermentation inhibitor. The aim of this study was to ascertain the concentration of furfural in mead worts, prepared at room temperature (unsaturated) and heated to boiling for 10 to 70 min (saturated), with an extract of 25 to 45°Brix. Moreover, the impact of the furfural on the fermentation course of mead wort was assessed. For this purpose, fermentation tests were conducted using mead wort (30°Brix) to which furfural was added at concentrations ranging from 1 to 100 mg/L. HS-SPME-GC-TOF-MS analysis revealed that the furfural concentration in mead worts varied between 2.3 and 5.3 mg/L. In saturated worts, the concentration increased by 2.8 to 4.5 times. Acidification of mead wort prior to boiling led to further increase in furfural concentration. The greatest changes occurred in the least concentrated worts, having the lowest buffer capacity. The addition of furfural to the mead wort did not inhibit fermentation, and an increase in attenuation was observed in the samples containing 2 mg/L of furfural compared to the control. Throughout the fermentation most of the furfural was reduced to furfuryl alcohol.

Keywords: mead; honey wine; fermentation; furfural; furfuryl alcohol; *Saccharomyces cerevisiae*

1. Introduction

Mead is a fermented beverage with an ethanol content ranging from 8% to 18% *v*/*v* produced through the fermentation of mead wort obtained from diluted bee honey [1]. Depending on the ratio of water to honey in the mead wort, the drinks are called: półtorak (1:0.5), dwójniak (1:1), trójniak (1:2) and czwórniak (1:3). These names have been registered as Traditional Specialities Guaranteed (TSGs) and are protected throughout the European Union to preserve traditional production methods and recipes [2].

Mead worts are characterized by a high sugar content, which is one of the factors that inhibit the fermentation process. The high osmotic pressure resulting from the high concentration of the extract contributes to the relatively long adaptation time of the yeast [3]. This increases the synthesis of acetic acid [4,5]. Bee honey is characterized by low acidity and low content of substances necessary for yeast metabolism. Mead wort is usually additionally acidified with citric acid. It is also supplemented with salts containing ammonium ions, phosphates(V) and complex preparations based on yeast extracts [6–10]. A deficiency of compounds containing elements such as calcium, magnesium, phosphorus, nitrogen and vitamins, mainly from the B group, can lead to stuck and slow fermentation [1,7,11]. Acidification of mead worts is aimed at lowering the pH of the solution, reducing the risk of development of harmful microbiota that may lead to spoilage of the product, and balancing the sweet taste resulting from the high concentration of carbohydrates [6,12].

Problems with low attenuation of mead wort are also caused by the presence of compounds with antimicrobial activity in bee honey [13]. This group includes polyphenolic compounds, essential oils and amphiphiles, i.e., compounds with both hydrophilic and hydrophobic properties. The latter group includes medium molecular weight aliphatic acids, e.g., hexanoic, octanoic and decanoic acids, which in high concentrations and at

Citation: Sroka, P.; Tarko, T.; Duda, A. The Impact of Furfural on the Quality of Meads. *Molecules* **2024**, *29*, 29. https://doi.org/10.3390/molecules29010029

Academic Editors: Sascha Rohn and Michał Halagarda

Received: 17 November 2023
Revised: 13 December 2023
Accepted: 18 December 2023
Published: 20 December 2023

low pH can contribute to fermentation inhibition [6,14]. These compounds have surface-active properties and are absorbed by suspensions and colloids present in the fermentation medium [15,16]. In the case of mead wort, yeast hulls, pollen, water-insoluble polymers and hydrocolloids can be added to accelerate fermentation. A similar effect can be achieved by immobilizing yeast cells on organic or inorganic supports [15–20].

According to the method of wort preparation, meads can be divided into saturated and unsaturated. Saturated mead is made from wort that is heated to boiling before fermentation [21]. Unsaturated meads are richer in aromatic substances of raw material origin, and partially retain the enzymatic activity of honey, which results in the formation of hydrogen peroxide and gluconolactone in the worts [22]. As a result of lactone hydrolysis, gluconic acid is produced, which lowers the pH of the wort environment. Increasing acidity increases the effective concentration of undissociated medium molecular weight fatty acids. These compounds are classified as fermentation inhibitors, which can also negatively affect the fermentation rate and significantly reduce the degree of sugar attenuation [23].

When saturated meads are obtained, microorganisms and enzymes present in mead wort are deactivated by heating, while proteins coagulate and some polyphenolic compounds and waxes are precipitated [3,6]. Saturated worts are characterized by faster fermentation and clarification and higher microbiological stability [24]. Heating mead worts above 70 °C deactivates glucose oxidase [22]. During heating, valuable aromatic compounds characteristic for the type of bee honey are evaporated, but also new substances are formed, which significantly affect the aroma [25]. Maillard reactions occur in heated worts, contributing to increased concentration of furan derivatives, including 5-hydroxymethylfurfural (HMF) and furfural [24]. These compounds occur naturally in bee honey, especially when stored at elevated temperatures [26]. HMF is formed by the dehydration of hexoses in an acidic environment, while furfural is formed by the dehydration of pentoses and the decomposition of ascorbic acid [27]. These reactions also take place during the preparation and fermentation of mead wort. The rate of formation of furan derivatives depends mainly on the substrate concentration, temperature and pH [26]. The concentrations of the mentioned substances depend on the type and storage conditions of the honey and are inversely proportional to the freshness of the product. At the same time, the method of mead wort production can significantly affect the concentration of furan derivatives in the fermented liquids [24].

Furfural is classified as a fermentation inhibitor, i.e., a factor that inhibits the fermentation process and reduces the attenuation of sugars in the wort. Furan compounds inhibit the growth of *S. cerevisiae* yeast and reduce the concentration of ethanol in the final products [27–29]. Furfural inhibits glycolysis and the fermentation rate, especially alcohol dehydrogenase, aldehyde dehydrogenase and pyruvate dehydrogenase [30–33]. In a study by Palmqvist et al. [34], a negative effect of a mixture of furfural and acetic acid on biomass growth and ethyl alcohol production was found. Allen et al. [35] showed that under aerobic conditions, the presence of furfural inhibits the growth of yeast, despite the presence of sufficient amounts of appropriate nutrients. Under anaerobic conditions, yeast cells reduce furfural to furfuryl alcohol [27]. The degree of furfural reduction is proportional to the inoculum density [34]. The furfural reducing agent is NADH, and the reaction leads to a reduction in the furfural concentration in the solution within a few to several dozen hours [36].

After inoculation of the mead worts with *S. cerevisiae* yeast, alcoholic fermentation lasts from several days to up to three months [1]. Once the fermentation process is complete, young meads are aged for several months to several years. To shorten the fermentation time of mead wort, one strategy is to select appropriate yeast strains that are adapted to the specific high-sugar environment [37]. Research was conducted on the adaptation of yeast cells to high sugar concentration solutions containing substances classified as fermentation inhibitors. The data showed that adaptation to a furfural-enriched environment led to a significant reduction in fermentation time. The adapted strains also showed an increased ability to degrade HMF and a significant increase in the conversion of furfural to furfuryl alcohol [38].

During the fermentation of mead worts, a synergistic effect is often observed as a result of high osmotic pressure, high ethanol concentration, and numerous substances classified as fermentation inhibitors on yeast cells. The objectives of this work were: (i) to evaluate the quantity of furfural produced during the preparation and saturation of mead worts and (ii) to determine the impact of various concentrations of furfural on the fermentation course and the parameters of young honey. Additionally, the amount of furfuryl alcohol generated in the young meads was monitored.

2. Results

2.1. The Influence of Heating Mead Wort on the Furfural Formation

As shown in Figures 1–3 the unheated control worts contained furfural at concentrations ranging from 2.3 (mead wort with an extract of 25°Brix, Figure 1) to 5.2 mg/L (45°Brix, Figure 3). The differences in furfural concentrations resulted from the different amount of bee honey mixed with water, i.e., the dilution of the wort, which ranged approximately from two (1:1 in dwójniak mead, 45°Brix) to four (1:3 in czwórniak, 25°Brix). In heated mead worts, an increase in furfural concentration proportional to the heating time was observed (Figures 1–3). In more diluted samples the percentage increase in furfural was higher. In the wort of unacidified czwórniak (25°Brix), the concentration of furfural increased by 242% after 70 min of heating compared to the unheated wort (Figure 1). In the acidified worts heated for 70 min, the concentration, compared to the corresponding unacidified worts, was even higher by between 14% (30°Brix, Figure 2) and 30% (25°Brix, Figure 1).

Since dwójniak mead, with the highest extract of 45°Brix, is most often produced from wort with an extract of 30°Brix by adding honey in portions during fermentation, further experiments were conducted only on trójniak mead worts with an extract of 30°Brix.

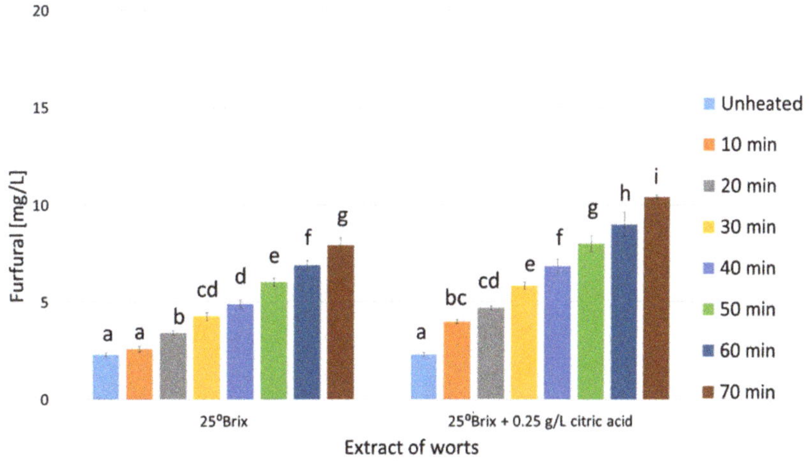

Figure 1. Changes in the concentration of furfural in mead wort with an extract of 25°Brix as a function of acidification and time of boiling. Values are expressed as the mean ± standard deviation. Means with different letters (a–i) are statistically different ($p < 0.05$).

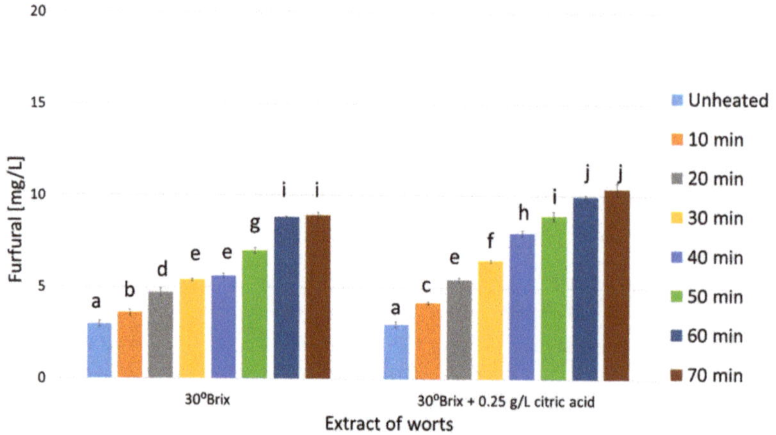

Figure 2. Changes in the concentration of furfural in mead wort with an extract of 30°Brix as a function of acidification and time of boiling. Values are expressed as the mean ± standard deviation. Means with different letters (a–j) are statistically different ($p < 0.05$).

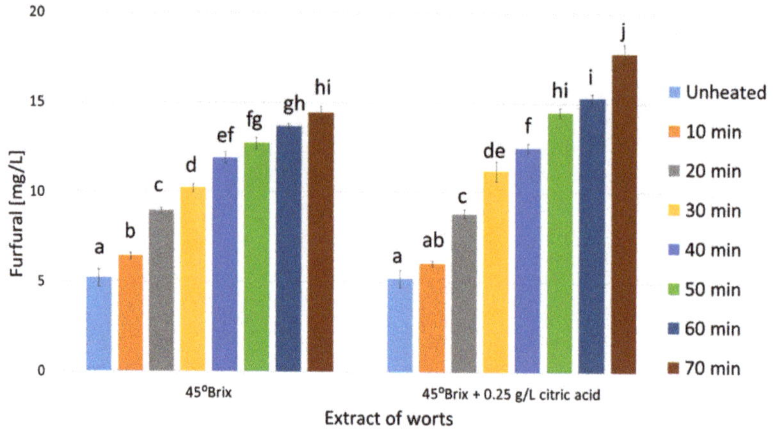

Figure 3. Changes in the concentration of furfural in mead wort with an extract of 45°Brix as a function of acidification and time of boiling. Values are expressed as the mean ± standard deviation. Means with different letters (a–j) are statistically different ($p < 0.05$).

2.2. Fermentation of Mead Worts Supplemented with Furfural

During the fermentation of mead worts (30°Brix) with different concentrations of furfural (ranging from 1 to 100 mg/L), the weight of the samples was monitored. The loss of weight was due to the release of carbon dioxide produced during fermentation (Figure 4). During the first few days of the process, no changes were observed as a result of yeast adaptation to a high-sugar environment. Samples supplemented with furfural (from 1 to 5 mg/L) were generally characterized by a greater CO_2 release as early as the fifth day of the process, proving a faster adaptation of yeast cells in the supplemented wort. After about one week, all samples were fermenting vigorously, with intense foaming and carbon dioxide release. In the third week, the violent fermentation ended and the post-fermentation period, characterized by low carbon dioxide release, began. The greatest CO_2 release was found in samples supplemented with furfural at a concentration of 2 mg/L. The experiment indicates that even relatively high concentrations of furfural (100 mg/L) do not inhibit the fermentation of mead wort.

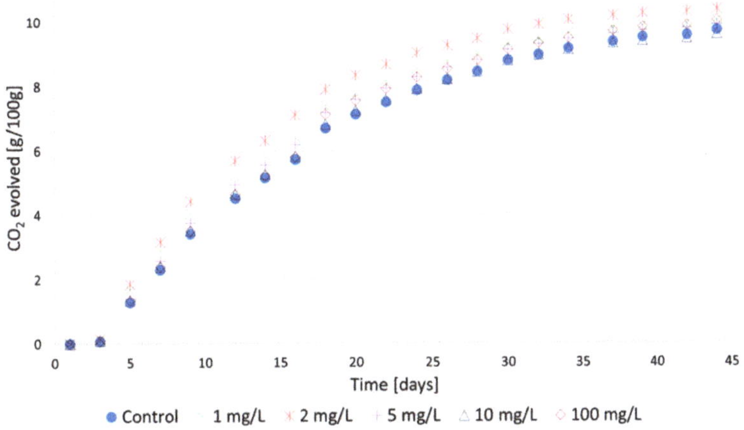

Figure 4. Carbon dioxide (g/100 g) released during the fermentation of mead worts without (control) and with furfural addition (1 to 100 mg/L).

2.3. *Young Meads' Parameters*

After 7 weeks of fermentation, young meads contained only 0.68 to 1.08 mg/L furfural (Figure 5), which means that the vast majority of the initial furfural content was decomposed during fermentation. The main product of furfural transformations was furfuryl alcohol (Figure 6). Its concentration in the fermented worts ranged from 2.3 to 61.3 mg/L, depending on the dose of furfural added. In samples supplemented with furfural at concentrations higher than 50 mg/L, more than 70% of this compound was reduced to furfuryl alcohol (Figure 6).

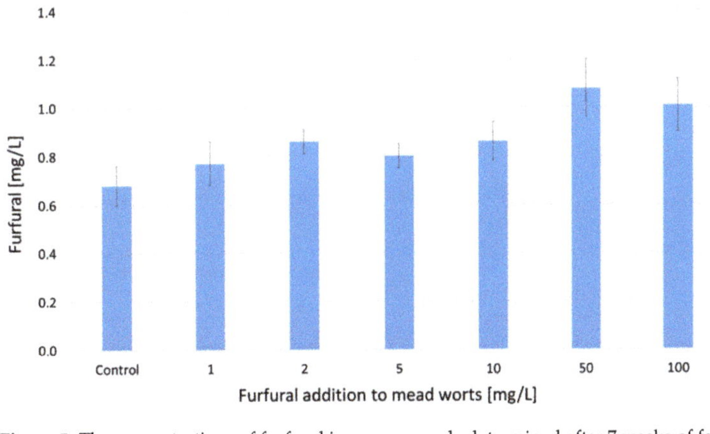

Figure 5. The concentrations of furfural in young meads determined after 7 weeks of fermentation of 30°Brix mead worts, with furfural added at concentrations ranging from 1 to 100 mg/L. Values are expressed as the mean ± standard deviation. The differences were not statistically significant.

The ethanol concentration in the fermented samples ranged from 13.0 to 14.3% *v*/*v* (Table 1), and only samples with 2 mg/L furfural addition had higher ethanol concentrations after fermentation compared to the control sample. In this case, a slightly (more than 1% *v*/*v*) higher ethanol concentration was obtained. The alcohol degree obtained in the fermented samples was correlated with the amount of carbon dioxide released during fermentation (Figure 4).

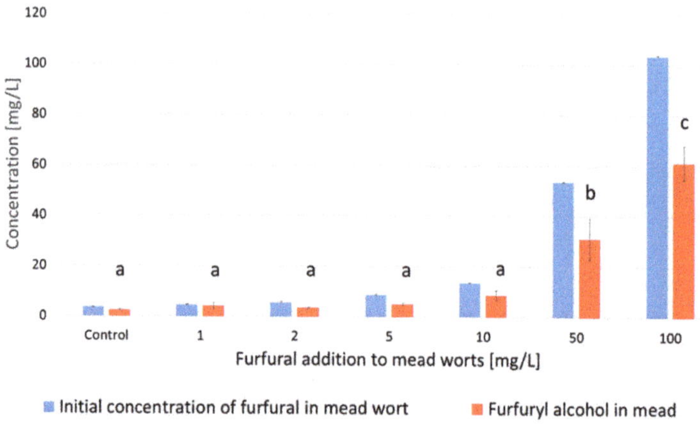

Figure 6. Initial concentrations of furfural in mead worts before fermentation and furfuryl alcohol concentrations in young meads, for samples with starting extracts of 30° Brix with furfural addition before fermentation at concentrations from 1 to 100 mg/L. Values are expressed as the mean ± standard deviation. Means with different letters (a–c) are statistically different ($p < 0.05$).

The total extract content ranged from 146 to 163 g/L in young mead (Table 1), and no significant differences were found among the samples analyzed. Based on the results obtained, it can be concluded that furfural has no significant effect on the total acidity and volatile acidity of young meads. The total acidity ranged from 3.4 to 3.7 g/L, which is relatively low and the resulting meads require additional correction after maturation with the addition of an appropriate amount of citric acid. The pH during fermentation is particularly important for the proper course of the process. A low pH prevents the development of undesirable microbiota. Wine yeast is adapted to low pH, but values below 3.0 can slow down or inhibit fermentation. In the fermented mead worts, the pH was between 3.3 and 3.5. Fermentation of high-sugar mixtures contributes to osmotic stress acting on yeast cells, which under these conditions increase the synthesis of acetic acid, which is the main compound influencing volatile acidity. The volatile acidity values of the fermented samples did not differ significantly and ranged from 1.2 to 1.4 g/L, which was relatively high.

Table 1. Parameters of fermented mead worts (30° Brix) with furfural added at concentrations ranging from 1 to 100 mg/L.

		Amount of Furfural Added to the Wort (mg/L)						
		0	1	2	5	10	50	100
Ethanol	(% v/v)	13.0 ± 0.4 a [1]	13.4 ± 0.2 ab	14.3 ± 0.3 b	13.7 ± 0.4 ab	13.6 ± 0.2 ab	13.6 ± 0.7 ab	13.6 ± 0.8 ab
Extract	(g/L)	158 ± 3 a	152 ± 8 a	146 ± 5 a	152 ± 2 a	157 ± 1 a	147 ± 13 a	163 ± 9 a
Titratable acidity	(g/L)	3.4 ± 0.1 a	3.4 ± 0.1 a	3.6 ± 0.1 a	3.6 ± 0.1 a	3.7 ± 0.1 a	3.6 ± 0.1 a	3.6 ± 0.1 a
pH		3.32 ± 0.04 a	3.37 ± 0.03 a	3.34 ± 0.03 a	3.46 ± 0.11 a	3.52 ± 0.04 a	3.38 ± 0.01 a	3.38 ± 0.01 a
Volatile acidity	(g/L)	1.3 ± 0.1 a	1.2 ± 0.1 a	1.4 ± 0.1 a	1.4 ± 0.1 a	1.4 ± 0.1 a	1.2 ± 0.2 a	1.3 ± 0.1 a

[1] Values are expresses as the mean ± standard deviation. Means with different letters (a,b) are statistically different ($p < 0.05$).

3. Discussion

As Figures 1–3 display, heating mead worts significantly changes the contents of furfural. As a result of the Maillard reaction, furan derivatives are formed [24]. The conducted experiments showed that boiling the worts for 70 min increases the concentration of furfural. The growth was between 2.8 times in the 45° Brix wort (Figure 3) and about 4.5 times in the acidified 25° Brix wort sample (Figure 1). Unacidified worts contained between 7.9 and 14.4 mg/L of furfural (for 25° Brix and 45° Brix boiled worts, respectively), while the addition of citric acid before heating contributed to the increase in furfural concentration to values between 10.3 (25°Brix) and 17.7 mg/L (45°Brix), depending on the wort extract (Figures 1–3). The relatively large impact of wort acidification on the concentration of furfural formed during heating is probably due to the low buffer capacity of the solutions and the lower pH of the worts, which catalyzes the reaction of formation of furan compounds [26].

It has been reported that furfural inhibits alcohol production by *Saccharomyces cerevisiae* in solutions containing 10 mg/L of this compound [27]. Previous studies have shown that furfural at a concentration of 4 g/L inhibits the yeast cell growth by 80% and ethanol production by 97% [30]. In our experiments, unheated wort contained from 2.3 (25°Brix) to 5.2 mg/L (45°Brix) of furfural, which should not significantly affect the fermentation process. However, boiled worts contained furfural at concentrations up to 18 mg/L, which, according to research [27], should significantly reduce the concentration of ethanol produced. In the case of fermentation of concentrated worts, the yeast is additionally subjected to osmotic stress, which may result in an increased inhibitory effect on the fermentation process.

Knowing the concentration of furfural produced during the heating of mead worts (Figures 1–3), trójniak worts with an extract of 30°Brix supplemented with furfural were prepared in order to determine its effect on the fermentation process. However, our experiments showed that even high (over 100 mg/L) concentrations of furfural did not inhibit alcohol fermentation (Figure 4), and low concentrations (2 mg/L) of added furfural contributed to increased alcohol production (Table 1), probably as a result of better adaptation of cells to the environment [38].

Furfural inhibits the fermentation process and also significantly inhibits the growth of yeast [33]. It should be noted, that relatively large amounts of yeast are added during mead production, usually twice as much as in the case of fermentation of wine musts. Therefore, inhibition of yeast multiplication by furfural during the adaptation of mead worts may not be as important as in the case of wine production. In addition, the reduction of furfural to furfuryl alcohol occurs within a few tens of hours after fermentation initiation [34,39–41]. A relatively large number of yeast cells and an extended adaptation time of honey worts, up to several days [16], contribute to an increase in the rate of reduction of furfural to furfuryl alcohol. The obtained experimental results could be influenced by two factors: the relatively rapid reduction of furfural to furfuryl alcohol and the absence of an inhibitory effect of this compound on the concentration of ethyl alcohol in the turbulent phase of fermentation, which commenced in the second week following the addition of yeast to the wort (Figure 4).

S. cerevisiae cells rapidly decompose the furfural in the solution, yielding furfuryl alcohol as the main product (Figure 6). The reaction employs the reduced form of nicotinamide adenine dinucleotide (NADH) as the reducing agent, with the catalytic participation of alcohol dehydrogenase (ADH) [41].

$$\text{Furfural} + \text{NADH} + \text{H}^+ = \text{furfuryl alcohol} + \text{NAD}^+$$

A small portion of furfural may have been converted into 1-(2-furyl)-1-hydroxy-2-propanone and 1-(2-furyl)-propane-1,2-diol [42] or oxidized into furoic acid [33,39,41]. Furfural can also react with amino acids present in mead wort and produce colored compounds. According to Murata et al. [43], furfural reacts with lysine to generate the yellow dye furpipatide. All these processes lead to the final concentration of furfural in the fermented worts not exceeding 1.1 mg/L (Figure 5).

It has been demonstrated that adding a low concentration of furfural (2 mg/L) can speed up fermentation and increase the ethanol concentration by 1% v/v in the fermented samples (Table 1). This suggests that low furfural concentrations have a beneficial effect on yeast adaptation processes or cell metabolism aimed at reducing furfural to furfuryl alcohol [38].

Fermenting solutions with high initial wort extracts results in strong osmotic stress. Yeasts respond to high sugar concentrations by producing more acetic acid, as a result of interference with the reduction of acetaldehyde [5]. During the production of mead, significant amounts of acetic and succinic acids are produced. An increase in acidity lowers the pH and, in extreme cases, may contribute to the inhibition of fermentation [44]. Volatile acidity was high in all of the analyzed samples (Table 1). Nonetheless, the addition of furfural did not result in any significant changes in volatile acidity. This can be attributed to the rapid decomposition of furfural, most likely occurring during the yeast adaptation phase.

4. Materials and Methods

4.1. Biological Material

The mead worts were prepared from buckwheat honey (Sądecki Bartnik, Stróże, Poland) by appropriate dilution with potable water (96.0 mg/L Ca^{2+}, 11.0 mg/L Mg^{2+}, 37.4 mg/L Cl^-, 49.0 mg/L SO_4^{2-}, pH 7.7).

Commercial wine yeast *Saccharomyces cerevisiae* Vintage White from Enartis (Navarrete, Spain) was used in the fermentation experiments. To prepare the yeast for fermentation, a yeast slurry was created by adding 10 g of dry yeast to 100 mL of sterile tap water. The resulting suspension, known as starter yeast, was left to settle for 30 min at 30 °C.

4.2. Experimental Design

4.2.1. Effect of Heating on Furfural Concentration

Buckwheat honey was diluted with water at room temperature to obtain worts with an extract of 25°Brix (czwórniak), 30°Brix (trójniak) and 45°Brix (dwójniak). The wort was divided into two parts: unacidified (no treatment) and acidified (with added 0.25 g/L of citric acid monohydrate pure p.a. (Avantor Performance Materials, Gliwice, Poland)). A volume of 200 mL of the wort was measured in 500 mL round-bottom flasks and heated to boiling under reflux for various time ranges (from 10 to 70 min) to obtain the saturated worts. The unheated mead wort was used as a negative control (unsaturated wort). The wort samples were then cooled in a stream of cold water and the furfural content was determined using the HS-SPME-GC-MS method. Three replicates of the experiment were performed.

4.2.2. Effect of Furfural Concentration on Mead Fermentation

Buckwheat honey was mixed with potable water in a ratio 1:2 (*v*/*v*), heated and gently boiled for 10 min, then topped up with diammonium hydrogen phosphate(V) (pure p.a., Avantor Performance Materials, Poland; 0.4 g/L) and acidified with citric acid (pure p.a., Avantor Performance Materials, Poland; 0.25 g/L). The wort extract was checked and corrected to 30°Brix after mixing. The hot wort was poured into flasks (0.2 L of wort into 0.5 L flasks) closed with a stopper with sterile fermentation trap tubes filled with glycerin. After cooling the wort to about 20 °C, furfural (Sigma-Aldrich, St. Louis, MO, USA) was added in amounts from 1 to 100 mg/L and a precisely defined amount of starter yeast was added (0.5 g/L calculated on dry substance). Worts without the addition of furfural were treated as control samples.

4.2.3. Control of the Fermentation Process

Fermentation of mead worts with and without furfural was conducted at 20 ± 1 °C and the samples were weighed on a balance PS 4500/X (Radwag company, Radom, Poland), three times per week. The weight loss resulted from the release of carbon dioxide produced during fermentation [45]. The fermentation process was continued until two consecutive weights of the samples, did not differ by more than 1 g. The weight loss was converted into the amount of CO_2 released in g/100 g. All fermentation experiments were conducted three times.

4.3. Analytical Methods

4.3.1. Physicochemical Parameters

Ethanol content, pH and the total and volatile acidity were determined using official methods recommended by O.I.V. [46]. The pH value was measured using a CP-505 pH-meter (Elmetron, Zabrze, Poland). The density of the samples and sample distillates was determined using an oscillating density meter DMA 4500 M (Anton Paar GmbH, Graz, Austria).

4.3.2. Determination of Furfural and Furfural Alcohol Content Using HS-SPME-GC-MS

The analysis was performed using gas chromatography (GC) coupled to time-of-flight mass spectrometry (TOF-MS) after headspace solid-phase microextraction (HS-SPME) [47].

Sample Preparation

Headspace solid-phase microextraction (HS-SPME) was performed using a Multi-Purpose Autosampler (MPS Dual Head, Gerstel GmbH and Co.KG, Mulheim, Germany). Samples of mead wort and young mead after 7 weeks of fermentation were stored at −20 °C prior to analysis. After heating to 20 °C, 1 mL of the sample was transferred to a 15 mL glass vial, 1 mL of saturated sodium chloride (pure p.a., Avantor Performance Materials, Gliwice, Poland) solution was added and the vial was capped and transferred to the autosampler tray. HS-SPME microextraction (30 min, 40 °C) was performed on PDMS fiber (100 μm, polydimethylsiloxane, SUPELCO, Sigma-Aldrich, St. Louis, MO, USA). Desorption was carried out in the inlet of the gas chromatograph at a temperature of 260 °C.

Calibration curves were prepared in the same way as the analyzed samples. Furfural and furfuryl alcohol (Sigma-Aldrich, St. Louis, MO, USA) were used to prepare calibration solutions.

TOF-MS Chromatographic Separation Conditions

The experiments were carried out using an Agilent Technologies gas chromatograph model 7890 B, coupled with time-of-flight mass spectrometer (TOF-MS Pegasus HT, LECO Corporation, St. Joseph, MI, USA). Chromatographic separation was performed on a Restek Stabilwax (cross-linked poly(ethylene glycol) column, dimensions: 30 m; 0.25 mm; 0.25 μm), at the programmed GC column temperature: initial at 35 °C (5 min), ramp 5 °C/min to 110 °C and 40 °C/min up to 230 °C, and final heating for 5 min. SPME fiber desorption was performed in a septumless inlet at a temperature of 260 °C, for 60 s in splitless mode. Helium (6.0, Linde Gaz Polska sp. z o.o., Kraków, Poland) at a flow rate of 1 mL/min was used as the carrier gas. TOF-MS detector parameters: scanning frequency: 20 Hz, acquisition voltage: 1500 V, ionization energy: −70 eV, ion source temperature: 250 °C. The transfer line temperature was 250 °C.

4.4. Statistical Analysis

All analytical results were selected for an analysis of variance (ANOVA). The post-hoc analysis of means was performed with Tukey's test at 5% error probability using Statistica 13.3 software (TIBCO Software Inc., Palo Alto, CA, USA), the graphs were generated using Microsoft Office Professional Plus 2013 software.

5. Conclusions

The performed study proved that heating the mead wort to boiling before fermentation correlated with the significantly increased amount of furfural. This should be taken into account when deciding on the length of wort saturation and the method and rate of wort cooling prior to fermentation. Most of the furfural is reduced to furfuryl alcohol during fermentation.

Fermentation of mead worts with 30°Brix extract supplemented with furfural in amounts ranging from 1 to 100 mg/L showed an acceleration of fermentation and an increase in ethanol concentration only in samples with 2 mg/L added furfural. Higher furfural concentrations did not significantly affect the fermentation rate or the basic parameters of the fermented worts. In particular, there was no inhibition of fermentation and no increase in acetic acid synthesis by yeast and volatile acidity, even in samples containing furfural concentrations ten times higher than those naturally occurring in mead worts heated for over one hour. The lack of influence of furfural on the course of fermentation does not mean that the heating of honey worts does not affect the quality of the obtained meads. As a result of an increase in temperature and a decrease in the pH of mead worts, furfural is formed, which, after reduction during fermentation, contributes to an increase in the concentration of furfuryl alcohol in the finished product.

The obtained results can be applied to practical mead production. It should be noted that the heating time is a critical parameter enhancing the furfural production. The boiling time of the mead wort should be restricted, and the solution should be cooled rapidly to minimize Maillard reactions, as these reactions proceed even while the boiled wort cools. Acidification of the wort with citric acid should be performed after cooling the solution. Additionally, it may be beneficial to prepare wort with a higher extract and dilute it after saturation, as this method produces solutions with lower concentrations of furfural and saves the energy required to heat a larger volume of water.

Author Contributions: Conceptualization, P.S.; methodology, P.S., T.T. and A.D.; software, P.S.; validation, P.S., T.T. and A.D.; formal analysis, P.S. and T.T.; investigation, P.S.; resources, P.S.; data curation, P.S.; writing—original draft preparation, P.S.; writing—review and editing, T.T. and A.D.; visualization, P.S.; supervision, A.D.; project administration, T.T.; funding acquisition, T.T. All authors have read and agreed to the published version of the manuscript.

Funding: This research was financed by The Ministry of Science and Higher Education of Poland, as a part of the Science Subsidy number 072000-D020.

Institutional Review Board Statement: Not applicable.

Informed Consent Statement: Not applicable.

Data Availability Statement: Data are contained within the article.

Conflicts of Interest: The authors declare no conflicts of interest.

References

1. Iglesias, A.; Pascoal, A.; Choupina, A.B.; Carvalho, C.A.; Feás, X.; Estevinho, L.M. Developments in the fermentation process and quality improvement strategies for mead production. *Molecules* **2014**, *19*, 12577–12590. [CrossRef] [PubMed]
2. European Commission. *Regulation (EU) No 729/2008 of 28 July 2008 Entering Certain Designations in the Register of Traditional Specialities Guaranteed (Czwórniak (TSG), Dwójniak (TSG), Półtorak (TSG), Trójniak (TSG))*; European Commission: Brussels, Belgium, 2008.
3. Ramalhosa, E.; Gomes, T.; Pereira, A.P.; Dias, T.; Estevinho, L.M. Mead production: Tradition versus modernity. In *Advances in Food and Nutrition Research*; Jackson, R.S., Ed.; Academic Press: Burlington, VA, USA, 2011; pp. 101–118.
4. Martini, S.; Ricci, M.; Bonechi, C.; Trabalzini, L.; Santucci, A.; Rossi, C. In vivo 13C-NMR and modelling study of metabolic yield response to ethanol stress in a wild-type strain of *Saccharomyces cerevisiae*. *FEBS Lett.* **2004**, *564*, 63–68. [CrossRef] [PubMed]
5. Erasmus, D.J.; Cliff, M.; Van Vuuren, H.J.J. Impact of yeast strain on the production of acetic acid, glycerol, and the sensory attributes of icewine. *Am. J. Enol. Vitic.* **2004**, *55*, 371–378. [CrossRef]
6. Sroka, P.; Tuszyński, T. Changes in organic acid contents during mead wort fermentation. *Food Chem.* **2007**, *104*, 1250–1257. [CrossRef]
7. Gupta, J.K.; Sharma, R. Production technology and quality characteristics of mead and fruit-honey wines: A review. *Nat. Prod. Radiance* **2009**, *8*, 345–355.
8. Mendes-Ferreira, A.; Cosme, F.; Barbosa, C.; Falco, V.; Ines, A.; Mendes-Faia, A. Optimization of honey-must preparation ad alcoholic fermentation by *Saccharomyces cerevisiae* for mead production. *Int. J. Food Microbiol.* **2010**, *144*, 193–198. [CrossRef] [PubMed]
9. Pereira, A.P.; Mendes-Ferreira, A.; Oliveira, J.M.; Estevinho, L.M.; Mendes-Faia, A. Mead production: Effect of nitrogen supplementation on growth, fermentation profile and aroma formation by yeasts in mead fermentation. *J. Inst. Brew.* **2015**, *121*, 122–128. [CrossRef]
10. Pereira, A.P.; Mendes-Ferreira, A.; Oliveira, J.M.; Estevinho, L.M.; Mendes-Faia, A. Improvement of mead fermentation by honey-must supplementation. *J. Inst. Brew.* **2015**, *121*, 405–410. [CrossRef]
11. O'Connor-Cox, E.S.C.; Ingledew, W.M. Alleviation of the effects of nitrogen limitation in high gravity worts through increased inoculation rates. *J. Ind. Microbiol.* **1991**, *7*, 89–96. [CrossRef]

12. McConnell, D.S.; Schramm, K.D. Mead success: Ingredients, processes and techniques. *Zymurgy* **1995**, *4*, 33–39.
13. Alvarez-Suarez, J.M.; Tulipani, S.; Romandini, S.; Bertoli, E.; Battino, M. Contribution of honey in nutrition and human health: A review. *Med. J. Nutrition. Metab.* **2010**, *3*, 15–23. [CrossRef]
14. Cabral, M.G.; Viegas, C.A.; Sá-Correia, I. Mechanisms underlying the acquisition of resistance to octanoic-acid-induced-death following exposure of *Saccharomyces cerevisiae* to mild stress imposed by octanoic acid or ethanol. *Arch. Microbiol.* **2001**, *175*, 301–307. [CrossRef] [PubMed]
15. Sroka, P.; Satora, P. The influence of hydrocolloids on mead wort fermentation. *Food Hydrocoll.* **2017**, *63*, 233–239. [CrossRef]
16. Sroka, P.; Satora, P.; Tarko, T.; Duda-Chodak, A. The influence of yeast immobilization on selected parameters of young meads. *J. Inst. Brew.* **2017**, *123*, 289–295. [CrossRef]
17. Roldán, A.; Muiswinkel, G.C.J.; Lasanta, C.; Palacios, V.; Caro, I. Influence of pollen addition on mead elaboration: Physicochemical and sensory characteristics. *Food Chem.* **2011**, *126*, 574–582. [CrossRef]
18. Qureshi, N.; Tamhane, D.V. Production of mead by immobilized cells of *Hansenula anomala*. *Appl. Microbiol. Biotechnol.* **1987**, *27*, 27–30. [CrossRef]
19. Pereira, A.P.; Mendes-Ferreira, A.; Estevinho, L.M.; Mendes-Faia, A. Mead's production: Fermentative performance of yeasts entrapped in different concentrations of alginate. *J. Inst. Brew.* **2014**, *120*, 575–580.
20. Pereira, A.P.; Mendes-Ferreira, A.; Oliveira, J.M.; Estevinho, L.M.; Mendes-Faia, A. Effect of *Saccharomyces cerevisiae* cells immobilisation on mead production. *LWT* **2014**, *56*, 21–30. [CrossRef]
21. Bednarek, M.; Szwengiel, A. Distinguishing between saturated and unsaturated meads based on their chemical characteristics. *LWT* **2020**, *133*, 109962. [CrossRef]
22. Kretavičius, J.; Kurtinaitienė, B.; Račys, J.; Čeksterytė, V. Inactivation of glucose oxidase during heat-treatment de-crystallization of honey. *Žemdirbystė=Agriculture* **2010**, *97*, 115–122.
23. Lambert, R.J.; Stratford, M. Weak-acid preservatives: Modelling microbial inhibition and response. *J. Appl. Microbiol.* **1999**, *86*, 1157–1164. [CrossRef] [PubMed]
24. Czabaj, S.; Kawa-Rygielska, J.; Kucharska, A.Z.; Kliks, J. Effects of mead wort heat treatment on the mead fermentation process and antioxidant activity. *Molecules* **2017**, *22*, 803. [CrossRef] [PubMed]
25. Starowicz, M.; Granvogl, M. Effect of wort boiling on volatiles formation and sensory properties of mead. *Molecules* **2022**, *27*, 710. [CrossRef] [PubMed]
26. Kowalski, S.; Łukasiewicz, M.; Duda-Chodak, A.; Zięć, G. 5-Hydroxymethyl-2-furfural (HMF)—Heat-induced formation, occurrence in food and biotransformation—A Review. *Pol. J. Food Nutr. Sci.* **2013**, *63*, 207–225. [CrossRef]
27. Abalos, D.; Vejarano, R.; Morata, A.; González, C.; Suárez-Lepe, J.A. The use of furfural as a metabolic inhibitor for reducing the alcohol content of model wines. *Eur. Food Res. Technol.* **2011**, *232*, 663–669. [CrossRef]
28. Boyer, L.J.; Vega, J.L.; Klasson, K.T.; Clausen, E.C.; Gaddy, J.L. The effects of furfural on ethanol production by *Saccharomyces cerevisiae* in batch culture. *Biomass Bioenergy* **1992**, *3*, 41–48. [CrossRef]
29. Klinke, H.B.; Thomsen, A.B.; Ahring, B.K. Inhibition of ethanol-producing yeast and bacteria by degradation products produced during pre-treatment of biomass. *Appl. Microbiol. Biotechnol.* **2004**, *66*, 10–26. [CrossRef] [PubMed]
30. Banerjee, N.; Bhatnagar, R.; Viswanathan, L. Inhibition of glycolysis by furfural in *Saccharomyces cerevisiae*. *Eur. J. Microbiol. Biotechnol.* **1981**, *11*, 226–228. [CrossRef]
31. Pienkos, T.; Zhang, M. Role of pretreatment and conditioning processes on toxicity of lignocellulosic biomass hydrolysates. *Cellulose* **2009**, *16*, 743–762. [CrossRef]
32. Chandel, A.K.; Silva, S.S.; Singh, O.V. Detoxification of lignocellulose hydrolysates: Biochemical and metabolic engineering toward white biotechnology. *Bioenerg. Res.* **2013**, *6*, 388–401. [CrossRef]
33. Modig, T.; Liden, G.; Taherzadeh, M.J. Inhibition effects of furfural on alcohol dehydrogenase, aldehyde dehydrogenase and pyruvate dehydrogenase. *Biochem. J.* **2002**, *363*, 769–776. [CrossRef] [PubMed]
34. Palmqvist, E.; Almeida, J.S.; Hahn-Hagerdal, B. Influence of furfural on anaerobic glycolytic kinetics of *Saccharomyces cerevisiae* in batch culture. *Biotechnol. Bioeng.* **1998**, *62*, 447–454. [CrossRef]
35. Allen, S.A.; Clark, W.; McCaffery, J.M.; Cai, Z.; Lanctot, A.; Slininger, P.J.; Liu, Z.L.; Gorsich, S.W. Furfural induces reactive oxygen species accumulation and cellular damage in *Saccharomyces cerevisiae*. *Biotechnol. Biofuels* **2010**, *3*, 2. [CrossRef] [PubMed]
36. Ask, M.; Bettiga, M.; Mapelli, V.; Olsson, L. The influence of HMF and furfural on redox-balance and energy-state of xylose-utilizing *Saccharomyces cerevisiae*. *Biotechnol. Biofuels* **2013**, *6*, 22. [CrossRef] [PubMed]
37. Srimeena, N.; Gunasekaran, S.; Murugesan, R. Screening of yeast from honey for mead production. *Madras Agric. J.* **2013**, *100*, 858–861.
38. Almeida, J.R.M.; Bertilsson, M.; Gorwa-Grauslund, M.F.; Gorsich, S.; Lidén, G. Metabolic effects of furaldehydes and impacts on biotechnological processes. *Appl. Microbiol. Biotechnol.* **2009**, *82*, 625–638. [CrossRef] [PubMed]
39. Villa, G.P.; Bartoli, R.; Lopez, R.; Guerra, M.; Enrique, M.; Penas, M.; Rodriquez, E.; Redondo, D.; Iglesias, I.; Diaz, M. Microbial transformation of furfural to furfuryl alcohol by *Saccharomyces cerevisiae*. *Acta Biotechnol.* **1992**, *12*, 509–512. [CrossRef]
40. Villegas, M.E.; Villa, P.; Guerra, M.; Rodríguez, E.; Redondo, D.; Martínez, A. Conversion of furfural into furfuryl alcohol by *Saccharomyces cerevisiae*. *Acta. Biotechnol.* **1992**, *12*, 351–354. [CrossRef]
41. Taherzadeh, M.J.; Gustaffson, L.; Niklasson, C.; Lidén, G. Conversion of furfural on aerobic and anaerobic batch fermentation of glucose by *Saccharomyces cerevisiae*. *J. Biosci. Bioeng.* **1999**, *87*, 169–174. [CrossRef]

42. Mochizuki, N.; Kitabatake, K. Analysis of 1-(2-furyl)propane-1,2-diol, a furfural metabolite in beer. *J. Ferment. Bioeng.* **1997**, *83*, 401–403. [CrossRef]
43. Murata, M.; Totsuka, H.; Ono, H. Browning of furfural and amino acids, and a novel yellow compound, furpipate, formed from lysine and furfural. *Biosci. Biotechnol. Biochem.* **2007**, *71*, 1717–1723. [CrossRef]
44. Taherzadeh, M.J.; Niklasson, C.; Lidén, G. Acetic acid—Friend or foe in anaerobic batch conversion of glucose to ethanol by *Saccharomyces cerevisiae*? *Chem. Eng. Sci.* **1997**, *52*, 2653–2659. [CrossRef]
45. Adamenko, K.; Kawa-Rygielska, J.; Kucharska, A.Z.; Piórecki, N. Characteristics of biologically active compounds in Cornelian cherry meads. *Molecules* **2018**, *23*, 2024. [CrossRef]
46. International Organisation of Vine and Wine. *Compendium of International Methods of Wine and Must Analysis*; International Organisation of Vine and Wine: Paris, France, 2023.
47. Plutowska, B.; Chmiel, T.; Dymerski, T.; Wardencki, W. A headspace solid-phase microextraction method development and its application in the determination of volatiles in honeys by gas chromatography. *Food Chem.* **2011**, *126*, 1288–1298. [CrossRef]

molecules

Article

Potato Resistant Starch Type 1 Promotes Obesity Linked with Modified Gut Microbiota in High-Fat Diet-Fed Mice

Weiyue Zhang [1,2,†], Nana Zhang [2,†], Xinxin Guo [2], Bei Fan [2], Shumei Cheng [1,*] and Fengzhong Wang [2,*]

[1] College of Food Science and Technology, Hebei Agricultural University, Baoding 071000, China; zhangwy921@163.com
[2] Key Laboratory of Agro-Products Processing, Ministry of Agriculture, Institute of Food Science and Technology, Chinese Academy of Agricultural Sciences, Beijing 100090, China; zhangnn16@163.com (N.Z.); guoxx26@163.com (X.G.); fanbei@caas.cn (B.F.)
* Correspondence: 13483279169@163.com (S.C.); wangfengzhong@sina.com (F.W.)
† These authors contributed equally to this work.

Abstract: Obesity has become a major disease that endangers human health. Studies have shown that dietary interventions can reduce the prevalence of obesity and diabetes. Resistant starch (RS) exerts anti-obesity effects, alleviates metabolic syndrome, and maintains intestinal health. However, different RS types have different physical and chemical properties. Current research on RS has focused mainly on RS types 2, 3, and 4, with few studies on RS1. Therefore, this study aimed to investigate the effect of RS1 on obesity and gut microbiota structure in mice. In this study, we investigated the effect of potato RS type 1 (PRS1) on obesity and inflammation. Mouse weights, as well as their food intake, blood glucose, and lipid indexes, were assessed, and inflammatory factors were measured in the blood and tissues of the mice. We also analyzed the expression levels of related genes using PCR, with 16S rRNA sequencing used to study intestinal microbiota changes in the mice. Finally, the level of short-chain fatty acids was determined. The results indicated that PRS1 promoted host obesity and weight gain and increased blood glucose and inflammatory cytokine levels by altering the gut microbiota structure.

Keywords: potato resistant starch type 1; obesity; inflammation; gut microbiota; short-chain fatty acids

Citation: Zhang, W.; Zhang, N.; Guo, X.; Fan, B.; Cheng, S.; Wang, F. Potato Resistant Starch Type 1 Promotes Obesity Linked with Modified Gut Microbiota in High-Fat Diet-Fed Mice. *Molecules* **2024**, *29*, 370. https://doi.org/10.3390/molecules29020370

Academic Editors: Lucia Panzella, Sascha Rohn and Michał Halagarda

Received: 3 November 2023
Revised: 22 December 2023
Accepted: 3 January 2024
Published: 11 January 2024

1. Introduction

Obesity, a primary disease, is a danger to human health globally [1]. According to epidemiological data, obesity has become the fifth leading cause of death worldwide. In 2016, approximately 13% of the global adult population was obese, with a prevalence rate of 11% for males and 15% for females [2]. The International Diabetes Federation reported that there were 451 million adults with diabetes worldwide, with a projected increase to 693 million by 2045 [3]. Obesity and diabetes affect a wide range of people and a large age range, thereby becoming important diseases endangering human health. Interestingly, many studies have shown that dietary interventions can reduce the prevalence of obesity and diabetes [4].

Resistant starch (RS) refers to a special type of starch that cannot be digested in the small intestine. Rather, it ferments with volatile fatty acids in the colon. Research has shown that RS affects the content of certain metabolic products, such as short-chain fatty acids (SCFAs), by altering the structure and abundance of the gut microbiota. RS has beneficial effects on weight loss, thereby lowering blood glucose levels and alleviating inflammation [5,6]. This type of starch may also affect the occurrence or progression of cancer by altering circulating hormone levels and other factors [7]. RS has been shown to affect endogenous intestinal hormone release and improve appetite control and blood glucose control [8].

The differences in RS types arise from differences in food sources, molecular structures, and physicochemical properties. Therefore, their physiological functions also differ. Type 2 RS (RS2) is a dietary fiber composed solely of glucose. Research has shown that RS2 treatment can reverse weight gain, liver steatosis, inflammation, and increased intestinal permeability caused by a high-fat diet (HFD)-influenced changes in the gut microbiota and metabolites [9]. Indeed, RS2 intervention increased the α-diversity of the gut microbiota and promoted *Brucella*, *Bifidobacterium*, and other organisms in the gut microbiota. Meanwhile, RS2 reduced the abundance of *Desulfovibrio*, *Helicobacter*, and *Enterococcus*, which are associated with obesity, inflammation, and aging [10,11]. Research has shown that intervention with RS type 3 (RS3) can reduce host weight and food intake as well as lower blood lipid levels and liver fat accumulation [12]. RS3 has a dose-dependent regulatory effect on HFD-induced obesity-related metabolic syndrome by promoting the proliferation of intestinal cells and expression of tight junction proteins such as occludin and (ZO)-1 [13]. In addition, RS3 promotes the production of microbial metabolites such as propionic acid and acetic acid by regulating the relative abundance of certain gut microbiota, including *Bifidobacterium*, *Ruminococcus*, and *Bacteroidetes* and reducing the abundance of harmful bacteria such as *Escherichia coli* and *Shigella* [14]. Research has shown that RS type 4 (RS4) significantly increases the abundance of *Actinobacteria*, *Bacteroidetes*, and *Bifidobacteria* and reduces the abundance of *Firmicutes* [15]. RS4 intervention reduced the level of cholesterol, fasting blood glucose, and proinflammatory factors in the blood and increased the content of fasting fatty acids such as butyrate, propionate, and valerate in feces [16]. Finally, research has shown that RS type 5 (RS5) has a significant effect in alleviating postprandial hyperglycemia in mice [17]. At the same time, RS5 was shown to alleviate weight gain, reduce the fasting blood glucose level, reduce triglyceride and cholesterol levels in the serum, and increase high-density lipoprotein levels in diabetic mice [18]. These findings suggest that RS has beneficial physiological functions, but significant gaps in research on RS type 1 (RS1) remain.

Because potatoes contain high RS levels, this study examined the effect of potato RS1 (PRS1) on mouse body weights, blood glucose levels, and inflammatory responses. Furthermore, the gut microbiota and metabolites were analyzed to improve our understanding of RS1. This study aimed to provide insights into antagonistic starch and offer a more theoretical basis for dietary intervention.

2. Results

2.1. *Effect of PRS1 on Body Weight, Tissue Weight, and Food Intake*

The body weight and fat weight of mice are important indicators for judging obesity. We examined how PRS1 affected diet-induced obesity in C57BL6J mice that were fed a control diet or high-fat diet supplemented with three doses of PRS1 (5, 15, or 25 g/100 g diet) according to previous research [14]. The high-fat diet led to a marked increase in body weight compared to the control diet (Figure 1a). Throughout 11 weeks of treatment, there was no difference in weight gain between PRL receivers and HFD receivers, as shown in Figure 1a. Body weight was slightly, yet not significantly, increased in PRM mice ($p = 0.1$), while the highest dose of PRS1 promoted weight gain significantly ($p < 0.05$) at week 9. At the end of the experiment, PRH receivers still showed a stronger weight gain ($p = 0.0007$) than PRM receivers ($p = 0.05$) in comparison with HFD-fed control mice. These data suggested that PRH-treated mice had a strong tendency for higher body weight gain than that of PRM-treated mice in the experiment. It is reasonable to suppose that the supplementation of 25% PRS1 would result in a strong degree of the indices of metabolic syndrome that would be amenable to the study of the mechanism. As a result, we sought to select the highest dose of PRS1 to study how PRS1 is impacting obesity.

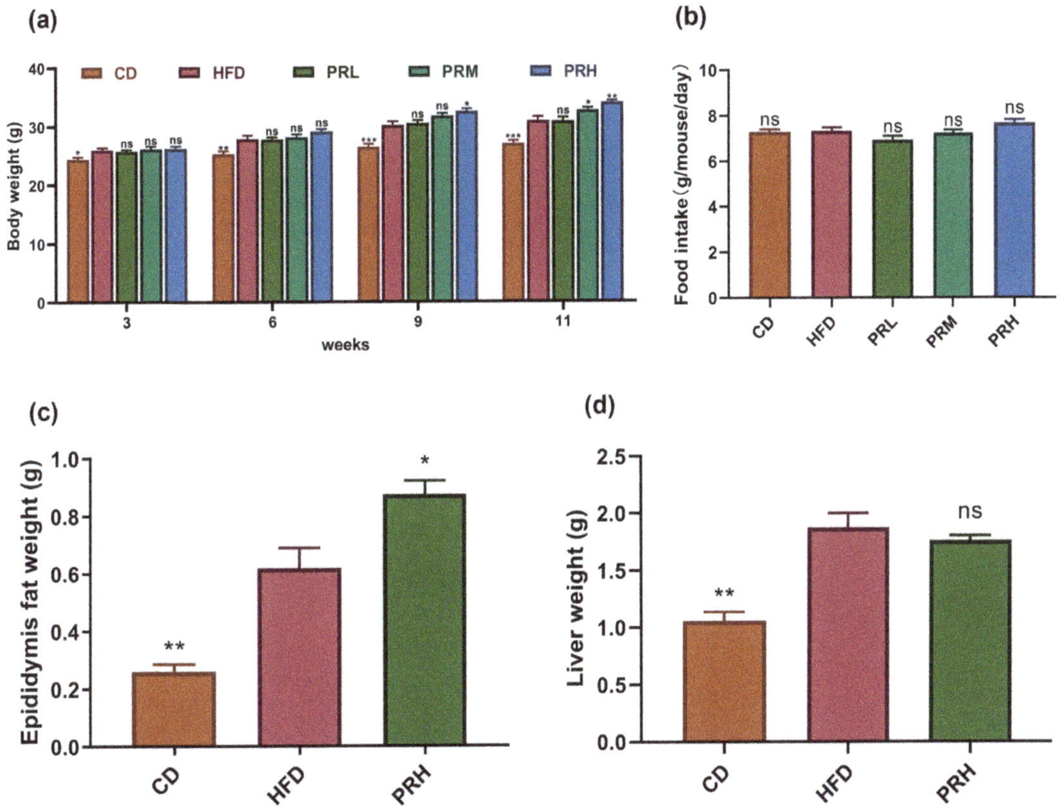

Figure 1. Effect of PRS1 on mouse body weight (**a**), food intake (**b**), epididymal fat weight (**c**), liver weight (**d**). CD: common diet group; HFD: high-fat diet group; PRL: PRS1 low-dose group; PRM: PRS1 medium-dose group; PRH: PRS1 high-dose group. Compared with the HFD group, * $p < 0.05$; ** $p < 0.01$, *** $p < 0.001$; ns $p > 0.05$.

No significant differences were found in the daily food intake for each group (Figure 1b). This result indicated that even with the same level of food intake, the PRS1 intervention groups gained weight more efficiently than the HFD group and that this effect was significant. The most obvious effect occurred in the PRH group. We subsequently chose the PRH group as the treatment group for our research. The results for the epididymal and brown adipose tissue weights showed that the epididymal adipose and liver tissue weights (Figure 1c,d) were significantly increased in the HFD group compared with the CD group ($p < 0.01$). The epididymal fat weights were significantly higher in the PRH group than in the HFD group ($p < 0.05$), and there was an upward trend in the liver weight. These results indicated that PRS1 intervention increased mainly the adipose tissue weight and that its effect on the liver was not obvious.

2.2. Effect of PRS1 on Glucose Metabolism in HFD-Fed Mice

The insulin content in the blood of mice was detected, and the oral glucose tolerance test (OGTT) and insulin tolerance test (ITT) were evaluated to determine the effect of HFD on blood glucose levels and insulin secretion in mice. Our results indicated that the HFD group had significantly increased blood insulin levels. Compared with the HFD group, the PRH group had a more significant increase in blood insulin levels (Figure 2a). OGTT was performed with the assumption that there were no significant differences in fasting blood glucose levels between the groups (Figure 2b). The OGTT results showed that in each group, the peak blood glucose level was reached at the 15th min after glucose stimulation. Compared with the HFD group, the PRH group had a more significant glucose increase (Figure 2c). After the 15th min, the blood glucose level of all three groups decreased. Compared to the CD group, both the HFD and PRH groups had higher area under the curve (AUC) values (Figure 2d). Furthermore, the AUC was significantly higher in the PRH group than in the HFD group ($p < 0.01$). The insulin tolerance test (ITT) results showed that at 60 min after insulin injection, the blood glucose level of each group gradually increased (Figure 2e). In particular, the PRH group had a higher AUC than the HFD group (Figure 2f). These results indicated that PRS1 intervention weakened the ability of the mice to control their blood glucose levels, which led to insulin resistance.

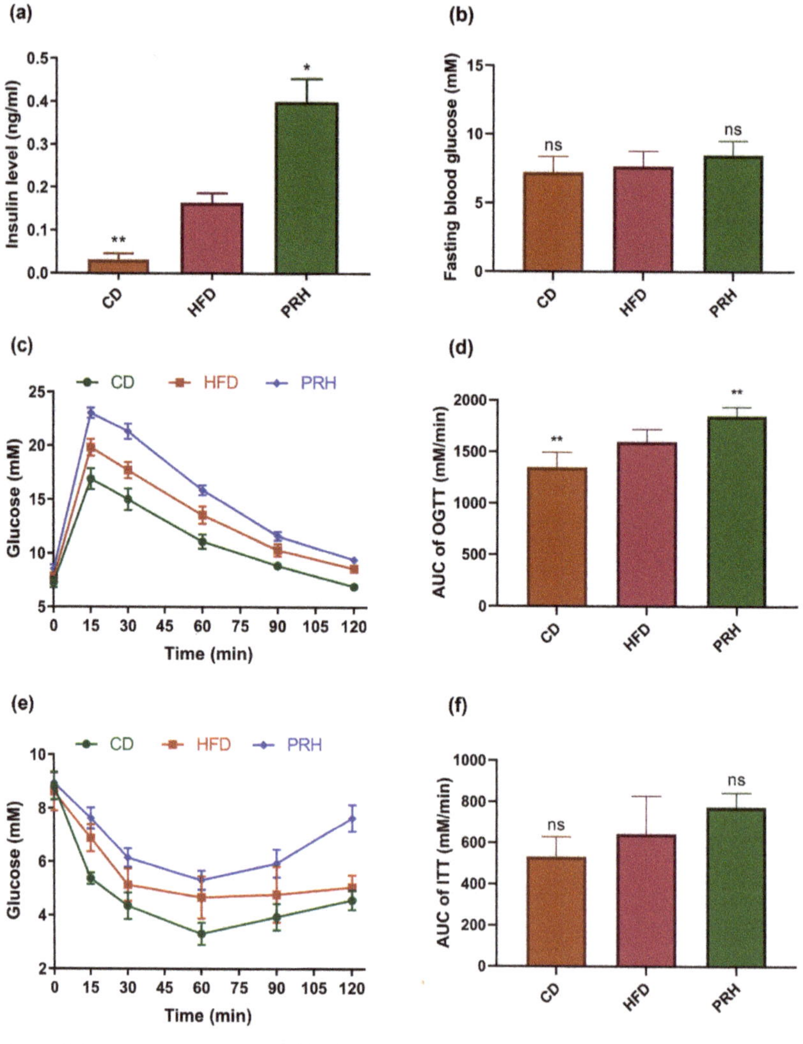

Figure 2. Effect of PRS1 on mouse blood insulin (**a**), fasting blood glucose (**b**), OGTT (**c**), AUC of OGTT (**d**), ITT (**e**), and AUC of ITT (**f**). CD: common diet group; HFD: high-fat diet group; PRH: PRS1 high-dose group. Compared with the HFD group, * $p < 0.05$; ** $p < 0.01$; ns $p > 0.05$.

2.3. Effect of PRS1 on Lipid Profile and Serum Inflammatory Factors

The effect of PRS1 on obesity in mice was evaluated by detecting the level of blood glucose, lipid-related indicators, and inflammatory factors in the blood. The results showed that serum total glyceride (TG) and free fatty acid (FFA) levels were significantly higher in the PRH group than in the HFD group (Figure 3a,b). It is known that adipose tissue growth in mice can cause an increase in serum leptin levels [19,20]. In this study, leptin levels in the blood also showed a significant upward trend (Figure 3c). Compared with the HFD group, the PRH group had a significantly increased serum level of IL-6 (Figure 3e) and significantly increased serum levels of IL-1β and TNF-α (Figure 3d,f). These results indicated that PRS1 not only increased blood TG and FFA levels in mice but also caused an inflammatory response and leptin resistance.

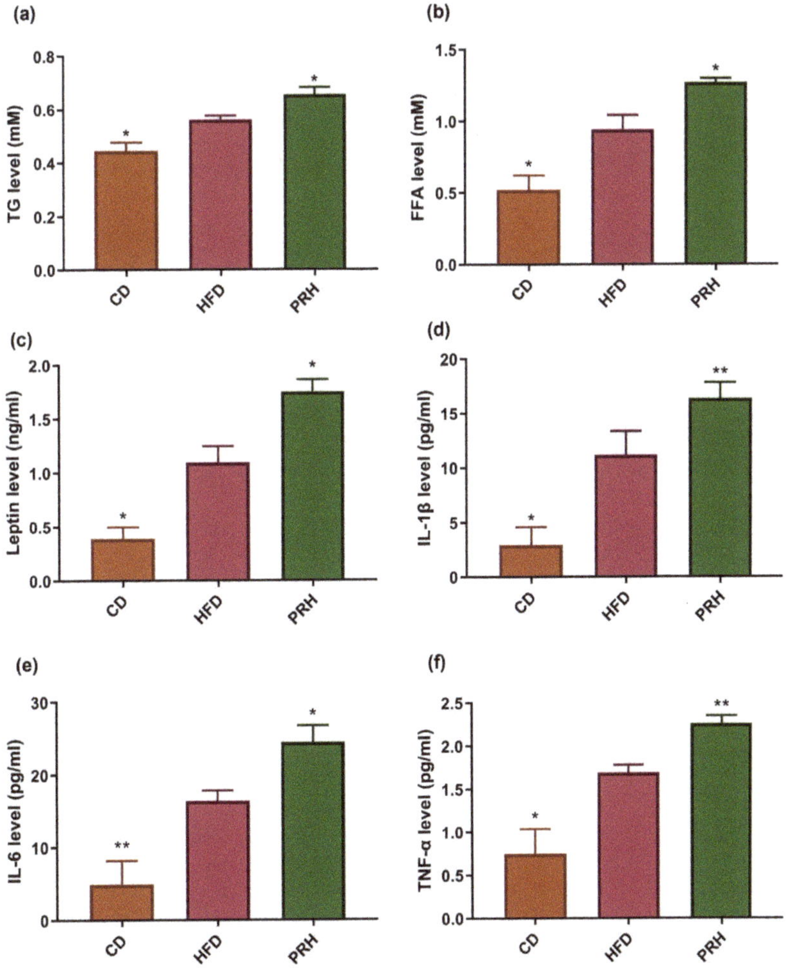

Figure 3. Effect of PRS1 on mouse serum level of TG (**a**), FFA (**b**), leptin (**c**), IL-1β (**d**), IL-6 (**e**), and TNF-α (**f**). CD: common diet group; HFD: high-fat diet group; PRH: PRS1 high-dose group. Compared with the HFD group, * $p < 0.05$; ** $p < 0.01$.

2.4. Effect of PRS1 on the Micromorphology of White Adipose, Brown Adipose, and Liver Tissues

H&E staining was performed to visually assess the effect of PRS1 intervention on mouse adipose tissue, liver tissue, and colon tissue. Compared with the model group, the HFD group exhibited an increased average size of white adipocytes and a reduced number of brown adipocytes (Figure 4a,b). PRS1 intervention alleviated the HFD-induced damage to the mouse colon tissue structure (Figure 4c). These findings indicated that PRS1 intervention accelerated the growth of white adipose tissue, weakened the thermogenic capacity of the original brown adipose tissue, and probably had adverse effects on liver health.

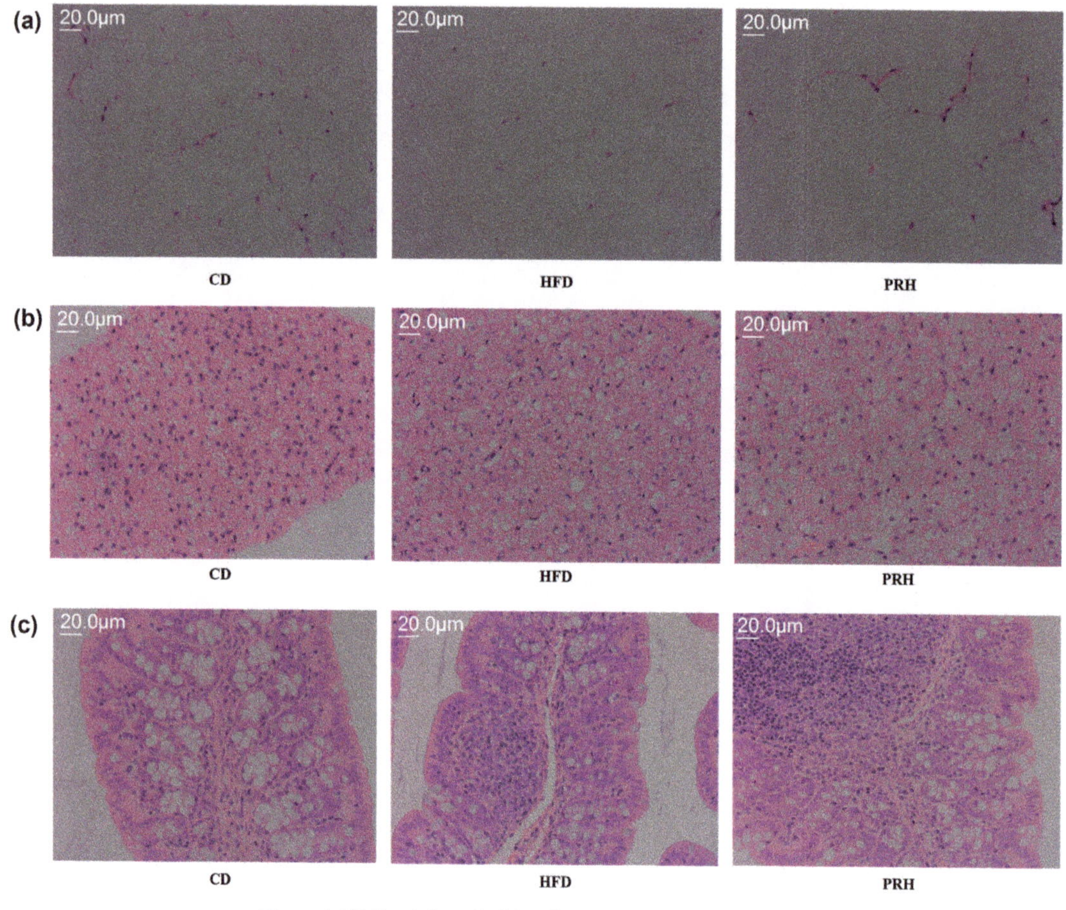

Figure 4. H&E staining of white adipose tissue (**a**), brown adipose tissue (**b**), and colon tissue (**c**). CD: common diet group; HFD: high-fat diet group; PRH: PRS1 high-dose group.

2.5. PRS1 Increases Inflammatory Factor Expression and Intestinal Permeability in Colon Tissue

Inflammatory factors in the colon tissue were investigated to explore the effect of PRS1 on colonic inflammatory response. The results of inflammatory factor levels in the colon and the expression of genes related to intestinal permeability are shown in Figure 5. Compared with the CD group, the HFD group had lower levels of IL-6, MCP-1, GRO-α, and IL-1β. The inflammatory factor levels in the PRH group showed an upward trend, and there were significant differences compared to the HFD group (Figure 5a–d). The gene expression level of occludin, which is related to intestinal permeability, was significantly decreased in the PRH group compared to the HFD group ($p < 0.05$) (Figure 5e–g). Additionally, ZO-1 and Muc2 gene expression levels were significantly decreased ($p < 0.01$). The above results indicated that PRS1 intervention did not alleviate the inflammatory response induced by HFD. In contrast, the intervention damaged the intestinal colon barrier and increased its permeability.

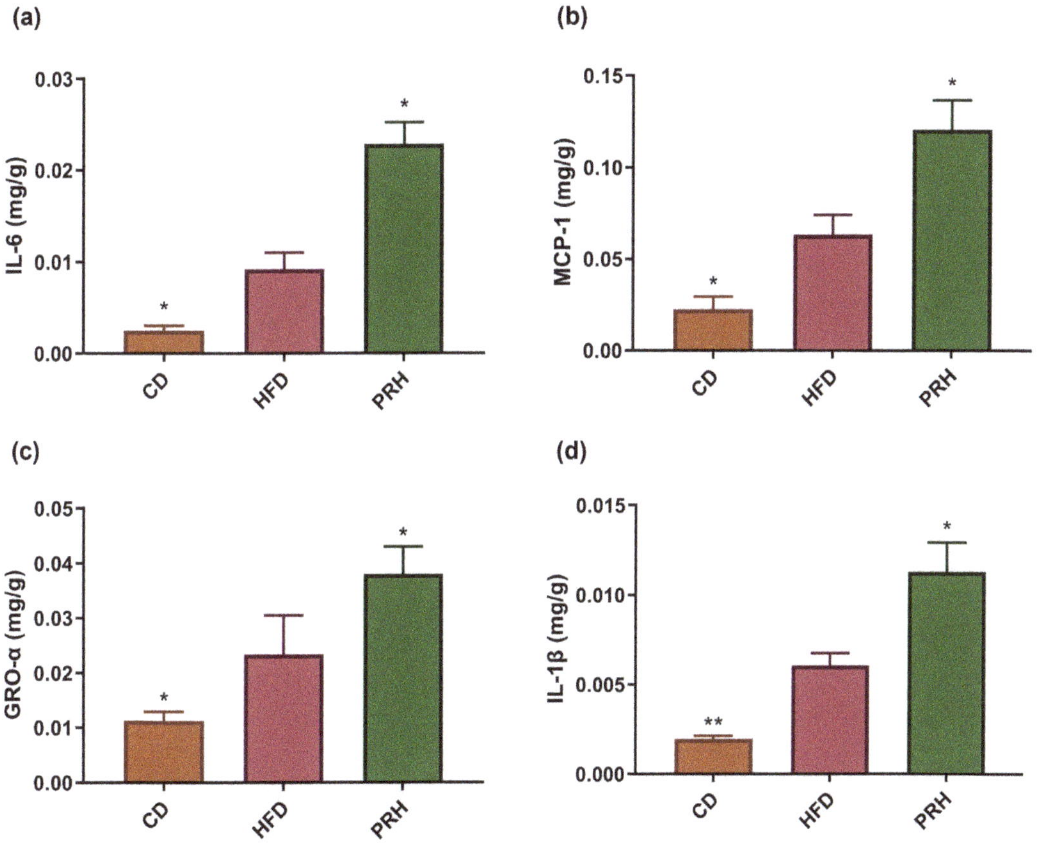

Figure 5. *Cont.*

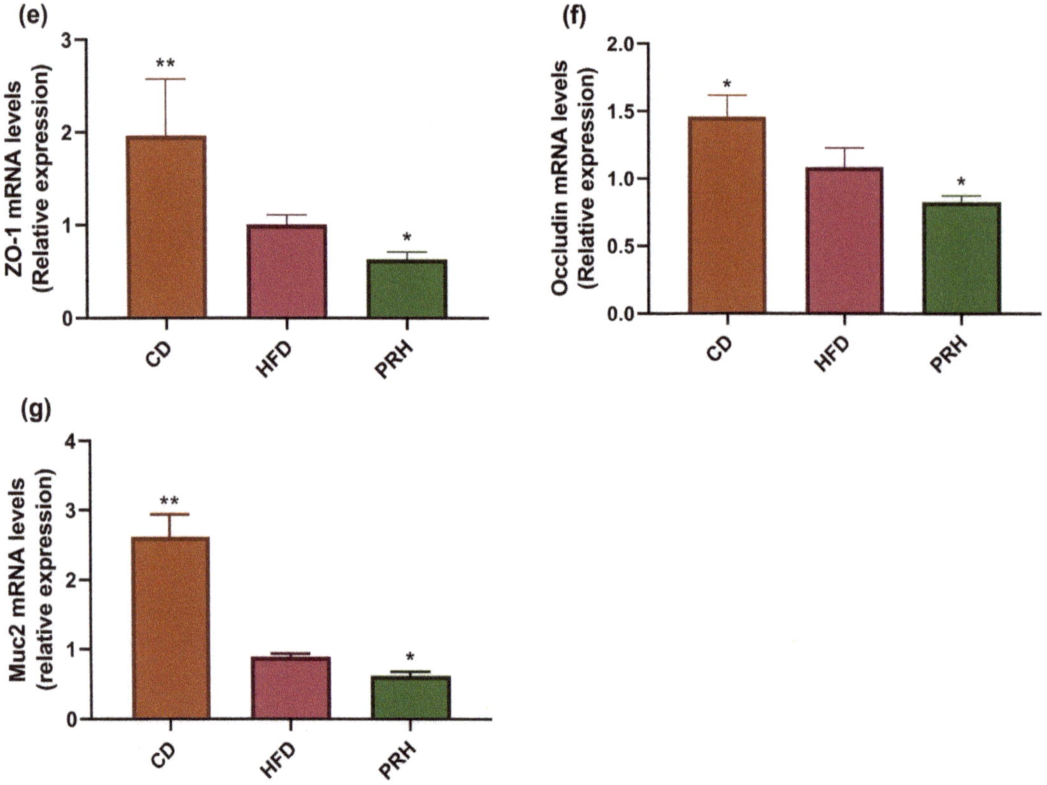

Figure 5. Level of IL-6 (**a**), MCP-1 (**b**), GRO-α (**c**), IL-1β (**d**), ZO-1 relative expression (**e**), occludin relative expression (**f**), and Muc2 relative expression (**g**). CD: common diet group; HFD: high-fat diet group; PRH: PRS1 high-dose group. Compared with the HFD group, * $p < 0.05$; ** $p < 0.01$.

2.6. PRS1 Changes the Abundance and Diversity of the Intestinal Flora in Mice

RS is fermented and broken down by the colonic bacterial community. Therefore, the gut microbiota in the colon was analyzed to explore its effect on obesity in mice.

2.6.1. Alpha Diversity Analysis

Rarefaction curves can directly reflect the rationality of sequencing data and indirectly reflect the richness of species in a sample [21]. Our results (Figure 6a) showed that the species diversity was far lower in the PRH group than in the CD and HFD groups. In addition, the dilution curves of the three groups indicated that Group C tended to be flat at 22,000 OTUs, Group M tended to be flat at 14,000 OTUs, and Group PRH tended to be flat at 12,000 OTUs. These findings indicated that the sequencing data volume was reasonable and that a greater volume of data can only produce a small number of new OTUs.

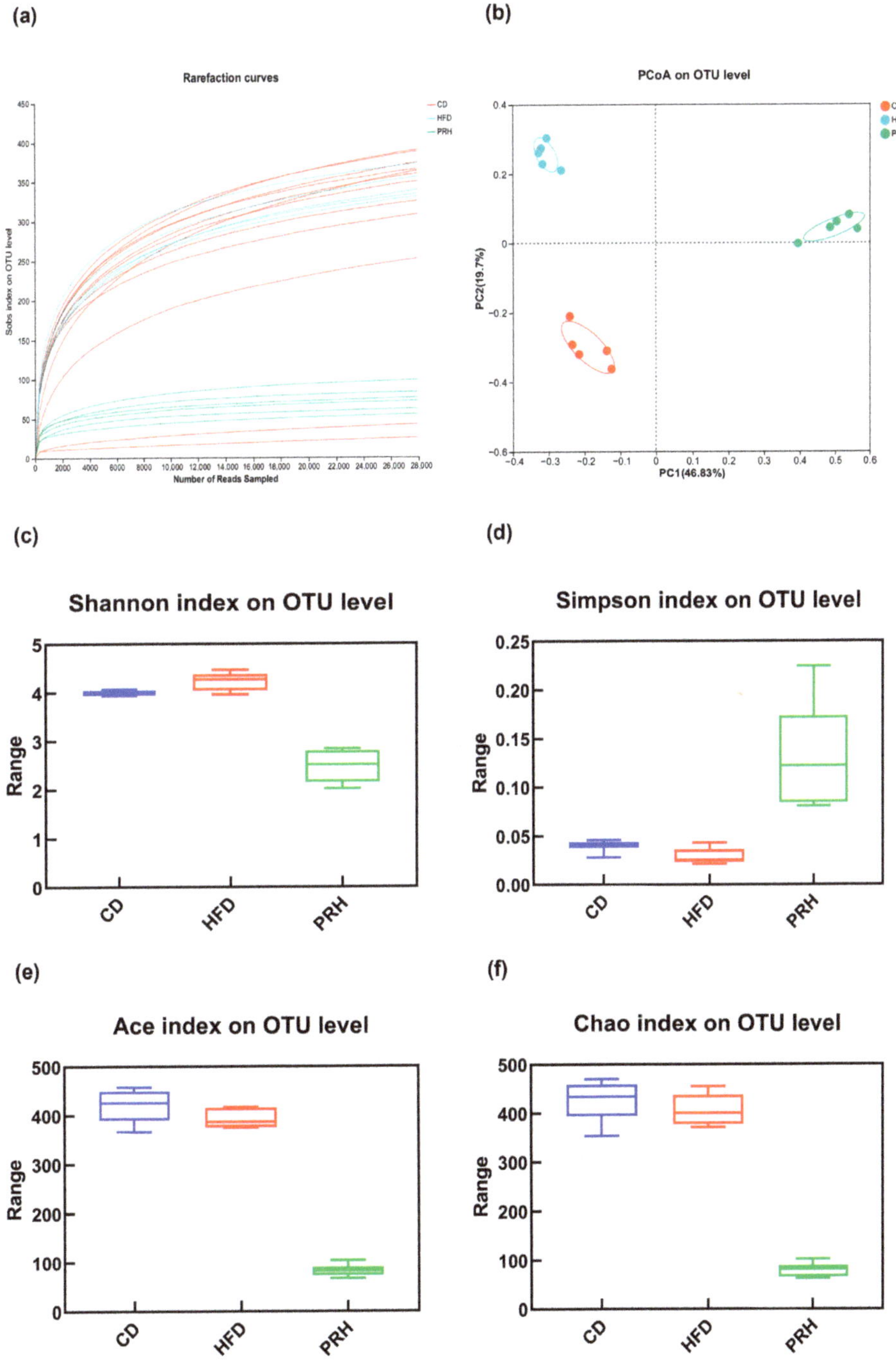

Figure 6. *Cont.*

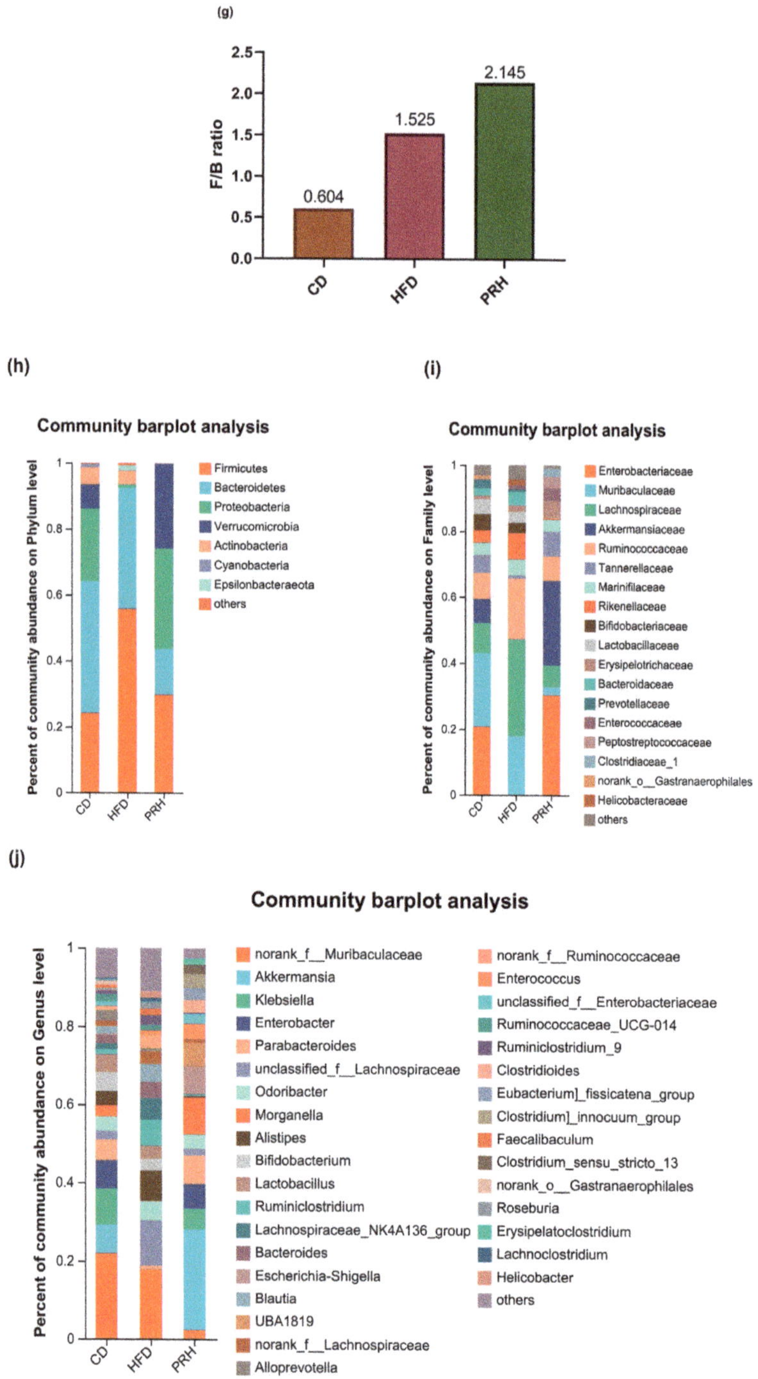

Figure 6. Rarefaction curve (**a**), PCoA (**b**), Shannon index (**c**), Simpson index (**d**), Ace index (**e**), Chao index (**f**), Firmicutes/*Bacteroidetes* (F/B) ratio (**g**), phylum level species composition (**h**), family level species composition (**i**), and genus level species composition (**j**). CD: common diet group; HFD: high-fat diet group; PRH: PRS1 high-dose group.

Alpha diversity is an ecological indicator that shows how many taxonomic groups exist in each sample and whether the abundance of these groups is evenly distributed [22]. The dietary interventions of functional foods that increase alpha diversity and change the abundance of specific bacteria can significantly change the fecal microbiota, independent of antidiabetes drugs [23]. We used Mothur's method to calculate and evaluate the gut microbiota abundance after PRS1 dietary intervention. The alpha diversity and community diversity indexes were analyzed by principal coordinate analysis (PCoA) using the Bray–Curtis distance, and the microbial community composition of each group before and after the intervention was compared at the phylum, family, and genus levels [24].

This experiment analyzed the abundance (Ace and Chao indexes) and diversity of the gut microbiota (Shannon and Simpson indexes). The Shannon index was higher for the PRH group than the HFD group, but the Simpson index was lower for the PRH group than the HFD group (Figure 6c,d). These data indicated that PRS1 intervention improved the richness of the gut microbiota community in the mice but reduced the uniformity of the community. At the same time, the Ace and Chao indexes were significantly lower for the PRH group than the HFD group (Figure 6e,f). We thus concluded that PRS1 intervention reduced the evenness of the intestinal microbial community of mice and ultimately reduced the species diversity of the flora.

2.6.2. Beta Diversity Analysis

A PCoA was conducted for the gut microbiota to examine the mechanism of microbial community composition changes between the different groups. The results showed that compared with no intervention in the HFD group, increased PRS1 administration resulted in significant changes in the intestinal microbiota composition (Figure 6b). The changes observed for PC1 and PC2 were 46.83% and 19.7%, respectively.

2.6.3. Gut Microbiota Composition

Sample level clustering analyses at the phylum, family, and genus classification levels revealed the following (all compared to the HFD group). At the phylum level, PRS1 intervention increased the ratio of *Firmicutes* to *Bacteroidetes* (F/B ratio) and the abundance of *Verrucomimicrobia* (Figure 6g,h). At the family level, the abundance of *Muribacillaceae, Lachnospiraceae, Tannellaceae, Marinifilaceae, Rikenellaceae, Lactobacillaceae,* and *Bacteroidaceae* significantly decreased in the PRH group and that of *Akkermansiaceae, Enterobacteriaceae, Erysipelotrichaceae, Peptostreptococcaceae,* and *Clostridiaceae_1* significantly increased (Figure 6h). At the genus level, the abundance of *norank_f_Muribaculaceae, unclassified_f_Lachnospiraceae, Odoribacter, Bifidobacterium, Alistipes, Ruminiclostridium, Lactobacillus,* and *Bacteroides* decreased in the PRH group, while the abundance of *Akkermansia, Morganella, Escherichia-Shigella, Enterobacter, UBA1819, Klebsiella,* and *Clostridioides* increased (Figure 6i).

2.7. PRS1 Reduces SCFA Secretion

The fermentation and decomposition of PRS1 by gut microbiota can produce various metabolites. Previous studies have shown that SCFAs are closely related to obesity in mice [25,26]. According to previous research, RS enters the intestine and is fermented by microorganisms, thereby producing a series of metabolic products [27,28]. Our SCFA results are shown in Figure 7. The level of acetic acid, propionic acid, butyric acid, and isobutyric acid followed a decreasing trend in the HFD group compared with the CD group. Compared with the HFD group, the PRH group had lower levels of SCFAs. This result indicated that PRS1 intervention reduced SCFA secretion, promoted weight gain, and caused inflammation in mice.

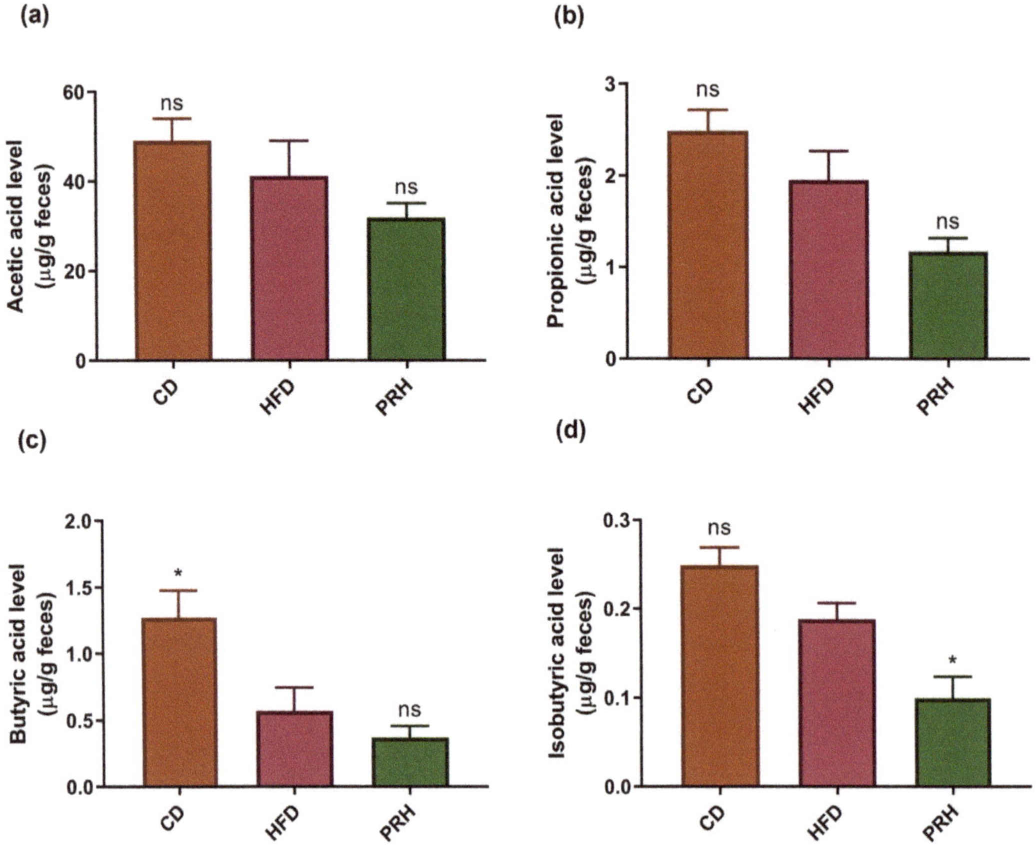

Figure 7. Level of acetic acid (**a**), propionic acid (**b**), butyric acid (**c**), and isobutyric acid (**d**). CD: common diet group; HFD: high-fat diet group; PRH: PRS1 high-dose group. Compared with the HFD group, * $p < 0.05$; ns $p > 0.05$.

3. Discussion

3.1. Effect of PRS1 on HFD-Induced Obesity and Inflammatory Responses

This study showed that PRS1 intervention led to increases in body and fat weights. The average weekly food intake of each mouse was nearly the same, and these results indicated that PRS1 increased the body and fat tissue weights of mice consuming HFD. Additionally, this intervention had a dose-dependent effect. H&E staining results showed that PRS1 accelerated the accumulation of white adipose tissue and reduced the amount of brown adipose tissue. PRS1 weakened the fat consumption and heat production of the mice and led to weight gain and obesity. However, PRS1 may have adverse effects on liver health. The OGTT and ITT results showed that PRS1 reduced the ability of the mice to control their blood glucose levels, as evidenced by increased blood glucose levels and insulin resistance.

3.2. Effect of PRS1 on Blood Lipid Levels, Inflammatory Response, and Intestinal Barrier in HFD Mice

The PRS1 intervention increased the serum TG and FFA levels in the mice. Leptin, which is positively correlated with fat content, was also significantly increased. The serum levels of the inflammatory factors IL-1β, TNF-α, and IL-6 increased as well as the levels of the inflammatory cytokines IL-6, MCP-1, GRO-α, and IL-1β in the colon. PCR amplification analysis showed that the expressions of ZO-1, occludin, and Muc2, which are related to epithelial cell junction and intestinal barrier integrity, all decreased. These results indicated that inflammation occurred in the colon. PRS1 disrupted the integrity of the colon epithelium, causing damage to the intestinal barrier.

3.3. Effect of PRS1 on the Gut Microbiota Composition of HFD-Fed Mice

According to previous research findings, disrupted glucose and lipid metabolism and inflammatory responses in mice are closely related to the intestinal microbiota composition. This study investigated gut microbiota changes at three taxonomic levels: phylum, family, and genus. At the phylum level, previous studies have shown that dietary fiber and probiotics can increase the relative abundance of *Actinobacteria* and *Bacteroidetes* and reduce the relative abundance of Firmicutes, thereby preventing obesity. However, the results of this study showed a decreased abundance of *Firmicutes*, *Bacteroidota*, and *Actinobacteria*, indicating that PRS1 did not play a role in preventing obesity.

The *Firmicutes/Bacteroidetes* (F/B) ratio is considered a biomarker of weight loss [29–31], and the human gut microbiota is composed mainly of these two dominant phyla, which account for over 90% of the total microbiota, as well as some other subdominant phyla [32]. Many studies have shown that the F/B proportion was higher in obese individuals than in lean individuals and that this ratio decreased with weight loss. Further studies have shown a positive correlation between the fecal concentration of Bacteroides and body mass index [33], indicating that high F/B ratio values are often associated with obesity. In this study, the F/B ratio was significantly higher in the PRH group than in the HFD group, and the overall abundance was decreased. In addition, the abundance of Proteobacteria, a biomarker of ecological imbalance and risk, also followed an increasing trend [34]. These data indicated that PRS1 not only had no preventive effect on obesity but also caused a certain degree of damage to the original microbial community, thereby increasing the risk of obesity in HFD-fed mice.

At the genus level, studies have shown a negative correlation between intestinal inflammation and Lactobacillus and a positive correlation with *Ruminococcus* [35,36]. *Ruminococcus* helps alleviate diarrhea and reduces the risk of diabetes and colon cancer [37–39]. Studies have shown that *Akkermansia* can control HFD-induced inflammation and body weight in mice and that adding RS to the diet can increase the relative abundance of *Akkermansia* [9,40]. *Bifidobacterium* is an important bacterial genus closely related to human health [41,42]. It can break down RS and accelerate growth, which is negatively correlated with obesity, glucose intolerance, and SCFA production [43].

Research has shown that RS can increase the relative abundance of Bacteroides, which can reduce body weight by fermenting RS and promoting SCFA production, especially acetic acid, in the intestine [44]. Our study indicated that PRS1 intervention increased *Akkermansia* abundance, which is consistent with the previous results documented in the literature. However, the abundance of *Bifidobacterium* decreased, indicating that the PRS1-consuming mice exhibited the symptoms of glucose intolerance and had an increased risk of obesity. By contrast, consuming PRS1 decreased the abundance of Bacteroides, indicating that PRS1 could not indirectly reduce weight or promote intestinal SCFA production through Bacteroides.

3.4. Effect of PRS1 Intervention on SCFA Content

According to previous research, RS enters the intestine and is fermented by microorganisms, producing a series of metabolic products. The level of certain SCFAs, such as acetic acid, propionic acid, and butyric acid, is negatively correlated with obesity-related parameters. SCFAs promote physiological benefits in the intestinal epithelium of the host by increasing the weight of the large intestine and cecal tissue. In this study, the levels of acetic acid, propionic acid, butyric acid, and isobutyric acid were decreased. Based on these metabolite results, we assert that PRS1 intervention did not promote weight loss in HFD mice and that its effect on alleviating obesity was not significant.

4. Materials and Methods

4.1. Mouse Experimental Environment and Groups

PRS1 with a purity of >90% was purchased from Louis Francois Co. (Paris, France). Animal experiments were approved by the Basic Medical Animal Care and Use Committee of Inner Mongolia Medical University (Inner Mongolia, China). The ethical review batch number is "YKD202101103". For this study, 60 C57BL/6 male mice 5–6 weeks of age were purchased from Vital River Laboratory Animal Technology (Beijing, China). All mice were fed in the animal room, which was maintained at 25 °C ± 1 °C and 55% ± 5% humidity. The animals were exposed to light for 12 h and darkness for 12 h in a day [14]. Prior to the start of the experiment, all mice were allowed to acclimate for 1 week with access to normal food and water. One week later, the mice were divided into six groups, with 10 mice in each group, and each group was placed in a separate cage [45]. The mice were fed with control diet, high-fat diet (HFD), and HFD diet supplemented with 5%, 15%, or 25% PRS1/kg of food mass for 11 weeks. The experimental groups were as follows: CD group (ccontrol diet; 19.2% protein, 4.3% fat, 67.3% carbohydrate, total energy 4057 kcal/kg), HFD group (high-fat diet group; 23.7% protein, 23.6% fat, 41.4% carbohydrate, total energy 4057 kcal/kg), PRL group (PRS1 low-dose group; HFD with 5% PRS1; 23.9% protein, 23.8% fat, 41.8% carbohydrate, total energy 4057 kcal/kg;), PRM group (PRS1 medium-dose group; HFD with 15% PRS1; 22.6% protein, 22.5% fat, 39.6% carbohydrate, total energy 4057 kcal/kg), and PRH (PRS1 high-dose group; HFD with 25% PRS1; 18.8% protein, 18.8% fat, 33.0% carbohydrate, total energy 4057 kcal/kg). The diet composition of all groups used in this study is shown in Table 1. In this study, we first studied the effect of PRS1 on diet-induced weight gain with doses of 5%, 15%, and 25% according to previous research [14]. The supplementation dose of PRS1 that would result in the strongest degree of weight gain was then used to investigate the effect on other indices of metabolic syndrome and the underlying mechanism. The mouse weights and dietary intakes were measured each week.

Table 1. The composition of the diet used in this study.

	CD		HFD		PRL		PRM		PRH	
Ingredient	**gm**	**kcal**	**gm**	**kcal**	**gm**	**kcal**	**gm**	**kcal**	**gm**	**kcal**
Casein	200	800	200	800	200	800	200	800	200	800
L-Cystine	3	12	3	12	3	12	3	12	3	12
Maltodextrin 10	150	600	100	400	100	400	100	400	100	400
Sucrose	0	0	172.8	691.2	172.8	691.2	172.8	691.2	172.8	691.2
PRS1	0	0	0	0	42.5	0	90	0	269	0
Soybean Oil	25	225	25	225	25	225	25	225	25	225
Lard	20	180	177.5	1597.5	177.5	1597.5	177.5	1597.5	177.5	1597.5
Mineral Mix S10026	10	0	10	0	10	0	10	0	10	0
Dicalcium Phosphate	13	0	13	0	13	0	13	0	13	0
Calcium Carbonate	5.5	0	5.5	0	5.5	0	5.5	0	5.5	0
Potassium Citrate, 1 H_2O	16.5	0	16.5	0	16.5	0	16.5	0	16.5	0
Vitamin Mix V10001	10	40	10	40	10	40	10	40	10	40
Choline Bitartrate	2	0	2	0	2	0	2	0	2	0
Total	1055.1	4057	858.15	4057	850.65	4057	898.15	4057	1077.1	4057

PRS1, potato-resistant starch type 1.

4.2. Mouse Adipose, Colon, and Liver Tissue Samples

The mice were euthanized via the cervical dislocation method. Epididymal adipose, brown adipose, colon, and liver tissues were removed from the mice and frozen in liquid nitrogen. To collect and process colon tissues, 2 mm colon segments were cleaned of their intestinal contents with PBS. Next, the samples were homogenized using a grinder, and after centrifuging the sample at 1500× *g* and 4 °C for 15 min, the supernatant was collected. ELISA kits (Mercodia, Uppsala, Sweden) were used to measure the content of the inflammatory factors in the colon tissues.

4.3. Blood Biochemical Indexes

Retro-orbital blood samples were collected from the mice and centrifuged at 1800× *g* for 20 min. The upper serum portions were removed and stored at −80 °C. Total triglyceride (TG), free fatty acid (FFA), insulin, and leptin levels in the blood were detected using ELISA kits (Mercodia, Uppsala, Sweden) according to the manufacturer's instructions.

4.4. OGTT and ITT

Before experiments, the mice were placed in clean cages and fasted with free access to only water for 16 h. Then, the mice were weighed and gavaged with 1 g/kg glucose. Blood was collected from the tail vein at 0, 15, 30, 60, 90, and 120 min after the glucose was administered. The blood glucose concentrations were measured using a glucometer (Roche, Basel, Switzerland) to determine the OGT.

Prior to the next experiment, the mice were fasted for 6 h with free access to only water. Then, the mice were injected intraperitoneally with 0.5 U/kg insulin (Sigma-Aldrich, St. Louis, MO, USA). At 0, 15, 30, 60, 90, and 120 min after insulin administration, tail vein blood samples were collected, and the glucose concentrations were measured as described above to determine the IT.

4.5. Hematoxylin and Eosin Staining (H&E)

Mouse tissues were fixed in 4% paraformaldehyde at 4 °C for 24 h, embedded in paraffin, and cut into 4 mm sections. The paraffin sections were stained with hematoxylin for 5 min, rinsed with water, and placed in 1% acetic acid 5 times for 30 s each time. The samples were then rinsed with water for 5 min and stained with eosin [46], followed by rinsing with water for 5 min. A microscope (Olympus, Tokyo, Japan) was used to observe the staining results.

4.6. 16S rRNA Gene Sequence Analysis

On the fourth week of the experiment, the mice were placed in metabolic cages, and feces were collected for 24 h. A total of 300–400 mg of feces was collected from each group. The samples were placed in 1-mL sterile centrifuge tubes and quickly transferred to a −80 °C ultra-low temperature freezer for storage. Total microbiome DNA was extracted from the feces using the QIAamp DNA Stool Mini Kit (Qiagen, Hilden, Germany), and the extracted samples were stored at −20 °C. The DNA concentration and purity were assessed using the A260 nm/A280 nm ratio. The ABI GeneAmp® 9700 PCR System (Applied Biosystems, Foster City, CA, USA) was used for PCR amplification [47], and the QuantiFluorTM-ST Handheld Fluorometer with UV/Blue Channels (Promega Corporation, Madison, WI, USA) was used to quantify the amplification product. The collected samples were quantified and sequenced using the Major Bio Pharm Technology Co., Ltd. platform (Shanghai, China). Universal primers 27F (5′-AGAGTTTGATCCTGGCTCAG-3′) and 533R (5′-TTACCGCGGCTGCTGGCAC-3′) were used to amplify the hypervariable V3–V4 regions via the polymerase chain reaction (PCR) of the 16S rDNA gene. The gene sequence was filtered, spliced, and classified into operational taxon units (OTUs) to ensure that the sequencing error rate was <1% and the percentage of bases < 0.01 was >97% [48].

4.7. Real Time-qPCR

Colon tissue samples (1 cm in length) were rinsed with PBS buffer (Invitrogen, Carlsbad, CA, USA) to remove the colon contents. After lysis buffer (Solarbio, Beijing, China) was added, the samples were homogenized with a grinder (Servicebio, Wuhan, China). The homogenates were centrifuged at 1500× *g* and 4 °C for 15 min; the supernatants were then collected, and RNA was extracted using a kit (Aidlab, Beijing, China). The DNA concentration and purity were assessed using the A260 nm/A280 nm ratio. Reverse transcription was performed according to the manufacturer's instructions. cDNA was obtained using a transcription Kit (Aidlab, Beijing, China), and the ABI GeneAmp® 9700 PCR System

(Applied Biosystems, Foster City, CA, USA) was used for gene expression detection [49–51]. Table 2 lists the primer sequences used in this study.

Table 2. Reaction primers.

Gene	Forward and Reverse Primers
ZO-1	F: 5′-TTTGAGACGACTCGGGGGAT-3′ R: 5′-TCTCGTTTTCTGGTTGGCAGT-3′
Occludin	F: 5′-CGCGTGCACACACACAATAA-3′ R: 5′-TAGTAACGGAAAGGACCCCC-3′
Muc2	F: 5′-GTTTGGACACGCACAAGGAC-3′ R: 5′-CTCGGGTAGCTTCCACTGTT-3′

4.8. Statistical Analysis

SPSS statistics 23 (IBM, Armonk, NY, USA) and GraphPad Prism 8.0 (GraphPad Software Inc., San Diego, CA, USA) were used to conduct one-way analysis of variance (ANOVA) and Duncan multiple comparison analysis. The threshold value of significance was set as $p < 0.05$, and $p < 0.01$ was considered extremely significant. Excel 2016 (Microsoft, Washington, DC, USA) was used to calculate the average values of the data, which are expressed as the mean ± standard deviation.

5. Conclusions

PRS1 intervention can lead to obesity and weight gain in mice. PRS1 increased blood glucose and lipid levels and led to both insulin and leptin resistance. Furthermore, PRS1 increased serum inflammatory factor levels in mice. It also increased the permeability of the colon, destroyed the intestinal barrier, and increased the level of inflammatory factors in the colon. PRS1 promoted white fat accumulation, reduced the production of brown adipose tissue, and inhibited the transformation and consumption of fat. The abundance of Bacteroides as well as level of SCFA decreased after PRS1 intervention. This suggests that PRS1 does not reduce body weight indirectly or promote SCFA production in the gut through Bacteroides. In this study, physiological indexes, gene expression, intestinal microbiota, and metabolites were analyzed. The results indicated that PRS1 indirectly affected SCFA production by altering the structure of the intestinal flora, thereby promoting host obesity and weight gain.

This study evaluated the effect of RS1 on HFD-induced obesity in mice to provide a more theoretical basis for alleviating obesity and dietary interventions. Differences in the molecular structure and physicochemical properties of different RS types, and their physiological functions are also different. It is needed to address the physiological functional differences between different types of RS in the future.

Author Contributions: Conceptualization, N.Z. and W.Z.; methodology, N.Z.; software, N.Z.; validation, N.Z., W.Z. and X.G.; formal analysis, N.Z.; investigation, W.Z.; resources, F.W.; data curation, N.Z.; writing—original draft preparation, W.Z.; writing—review and editing, W.Z. and N.Z.; visualization, W.Z.; supervision, S.C.; project administration, B.F.; funding acquisition, F.W. All authors have read and agreed to the published version of the manuscript.

Funding: This research was funded by the Agricultural Science and Technology Innovation Program of the Institute of Food Science and Technology, Chinese Academy of Agricultural Sciences (No. CI2021A05031, China) and the National Natural Science Foundation of China (No. 82073837, China).

Institutional Review Board Statement: Animal experiments were approved by the Basic Medical Animal Care and Use Committee of Inner Mongolia Medical University (Inner Mongolia, China). Animal feeding and handling were conducted to minimize the discomfort and pain of the animals in accordance with the national standard guidelines outlined in the Requirements for Experimental Animals in Environment and Housing Facilities (GB 14925-2010).

Informed Consent Statement: Not applicable.

Data Availability Statement: Data are contained within this article.

Conflicts of Interest: The authors declare no conflict of interest.

References

1. Flegal, K.M.; Kruszon-Moran, D.; Carroll, M.D.; Fryar, C.D.; Ogden, C.L. Trends in Obesity among Adults in the United States, 2005 to 2014. *JAMA* **2016**, *315*, 2284–2291. [CrossRef] [PubMed]
2. NCD Risk Factor Collaboration (NCD-RisC) Trends in Adult Body-Mass Index in 200 Countries from 1975 to 2014: A Pooled Analysis of 1698 Population-Based Measurement Studies with 19.2 Million Participants. *Lancet* **2016**, *387*, 1377–1396. [CrossRef]
3. Kolarić, V.; Svirčević, V.; Bijuk, R.; Zupančič, V. Chronic Complications of Diabetes and Quality of Life. *Acta Clin. Croat.* **2022**, *61*, 520–527. [CrossRef] [PubMed]
4. Gallagher, D.; Heshka, S.; Kelley, D.E.; Thornton, J.; Boxt, L.; Pi-Sunyer, F.X.; Patricio, J.; Mancino, J.; Clark, J.M.; MRI Ancillary Study Group of Look AHEAD Research Group. Changes in Adipose Tissue Depots and Metabolic Markers Following a 1-Year Diet and Exercise Intervention in Overweight and Obese Patients with Type 2 Diabetes. *Diabetes Care* **2014**, *37*, 3325–3332. [CrossRef] [PubMed]
5. Deng, J.; Wu, X.; Bin, S.; Li, T.-J.; Huang, R.; Liu, Z.; Liu, Y.; Ruan, Z.; Deng, Z.; Hou, Y.; et al. Dietary Amylose and Amylopectin Ratio and Resistant Starch Content Affects Plasma Glucose, Lactic Acid, Hormone Levels and Protein Synthesis in Splanchnic Tissues. *J. Anim. Physiol. Anim. Nutr.* **2010**, *94*, 220–226. [CrossRef]
6. Venkataraman, A.; Sieber, J.R.; Schmidt, A.W.; Waldron, C.; Theis, K.R.; Schmidt, T.M. Variable Responses of Human Microbiomes to Dietary Supplementation with Resistant Starch. *Microbiome* **2016**, *4*, 33. [CrossRef]
7. Effects of Dietary Beef and Chicken with and without High Amylose Maize Starch on Blood Malondialdehyde, Interleukins, IGF-I, Insulin, Leptin, MMP-2, and TIMP-2 Concentrations in Rats—PubMed. Available online: https://pubmed.ncbi.nlm.nih.gov/20432166/ (accessed on 7 December 2023).
8. Nilsson, A.C.; Johansson-Boll, E.V.; Björck, I.M.E. Increased Gut Hormones and Insulin Sensitivity Index Following a 3-d Intervention with a Barley Kernel-Based Product: A Randomised Cross-over Study in Healthy Middle-Aged Subjects. *Br. J. Nutr.* **2015**, *114*, 899–907. [CrossRef]
9. Barouei, J.; Bendiks, Z.; Martinic, A.; Mishchuk, D.; Heeney, D.; Hsieh, Y.-H.; Kieffer, D.; Zaragoza, J.; Martin, R.; Slupsky, C.; et al. Microbiota, Metabolome, and Immune Alterations in Obese Mice Fed a High-Fat Diet Containing Type 2 Resistant Starch. *Mol. Nutr. Food Res.* **2017**, *61*, 1700184. [CrossRef]
10. Vital, M.; Howe, A.; Bergeron, N.; Krauss, R.M.; Jansson, J.K.; Tiedje, J.M. Metagenomic Insights into the Degradation of Resistant Starch by Human Gut Microbiota. *Appl. Environ. Microbiol.* **2018**, *84*, e01562-18. [CrossRef]
11. Hughes, R.L.; Horn, W.H.; Finnegan, P.; Newman, J.W.; Marco, M.L.; Keim, N.L.; Kable, M.E. Resistant Starch Type 2 from Wheat Reduces Postprandial Glycemic Response with Concurrent Alterations in Gut Microbiota Composition. *Nutrients* **2021**, *13*, 645. [CrossRef]
12. Chen, X.; Wang, Z.; Wang, D.; Kan, J. Effects of Resistant starch III on the Serum Lipids Levels and Gut Microbiota of Kunming Mice under High-Fat Diet. *Food Sci. Hum. Health* **2023**, *12*, 9. [CrossRef]
13. Zhang, X.; Liang, D.; Liu, N.; Qiao, O.; Gao, W.; Li, X. Preparation and Characterization of Resistant Starch Type 3 from Yam and Its Effect on the Gut Microbiota. *Tradit. Med. Res.* **2022**, *7*, 11–19. [CrossRef]
14. Liang, D.; Zhang, L.; Chen, H.; Zhang, H.; Hu, H.; Dai, X. Potato Resistant Starch Inhibits Diet-Induced Obesity by Modifying the Composition of Intestinal Microbiota and Their Metabolites in Obese Mice. *Int. J. Biol. Macromol.* **2021**, *180*, 458–469. [CrossRef]
15. Zhang, Y.; Chen, L.; Hu, M.; Kim, J.J.; Lin, R.; Xu, J.; Fan, L.; Qi, Y.; Wang, L.; Liu, W.; et al. Dietary Type 2 Resistant Starch Improves Systemic Inflammation and Intestinal Permeability by Modulating Microbiota and Metabolites in Aged Mice on High-Fat Diet. *Aging* **2020**, *12*, 9173–9187. [CrossRef] [PubMed]
16. Upadhyaya, B.; McCormack, L.; Fardin-Kia, A.R.; Juenemann, R.; Nichenametla, S.; Clapper, J.; Specker, B.; Dey, M. Impact of Dietary Resistant Starch Type 4 on Human Gut Microbiota and Immunometabolic Functions. *Sci. Rep.* **2016**, *6*, 28797. [CrossRef]
17. Lau, E.; Zhou, W.; Henry, C.J. Effect of Fat Type in Baked Bread on Amylose-Lipid Complex Formation and Glycaemic Response. *Br. J. Nutr.* **2016**, *115*, 2122–2129. [CrossRef]
18. Guo, J.; Ellis, A.; Zhang, Y.; Kong, L.; Tan, L. Starch-Ascorbyl Palmitate Inclusion Complex, a Type 5 Resistant Starch, Reduced in Vitro Digestibility and Improved in Vivo Glycemic Response in Mice. *Carbohydr. Polym.* **2023**, *321*, 121289. [CrossRef]
19. Härle, P.; Straub, R.H. Leptin Is a Link between Adipose Tissue and Inflammation. *Ann. N. Y. Acad. Sci.* **2006**, *1069*, 454–462. [CrossRef]
20. Blum, W.F. Leptin: The Voice of the Adipose Tissue. *Horm. Res.* **1997**, *48* (Suppl. S4), 2–8. [CrossRef]
21. Liu, S.; Li, T.; Yu, S.; Zhou, X.; Liu, Z.; Zhang, X.; Cai, H.; Hu, Z. Analysis of Bacterial Community Structure of Fuzhuan Tea with Different Processing Techniques. *Open Life Sci.* **2023**, *18*, 20220573. [CrossRef]
22. Independence of Alpha and Beta Diversities—PubMed. Available online: https://pubmed.ncbi.nlm.nih.gov/20715617/ (accessed on 31 May 2023).
23. Medina-Vera, I.; Sanchez-Tapia, M.; Noriega-López, L.; Granados-Portillo, O.; Guevara-Cruz, M.; Flores-López, A.; Avila-Nava, A.; Fernández, M.L.; Tovar, A.R.; Torres, N. A Dietary Intervention with Functional Foods Reduces Metabolic Endotoxaemia and

Attenuates Biochemical Abnormalities by Modifying Faecal Microbiota in People with Type 2 Diabetes. *Diabetes Metab.* **2019**, *45*, 122–131. [CrossRef]
24. Schloss, P.D.; Westcott, S.L.; Ryabin, T.; Hall, J.R.; Hartmann, M.; Hollister, E.B.; Lesniewski, R.A.; Oakley, B.B.; Parks, D.H.; Robinson, C.J.; et al. Introducing Mothur: Open-Source, Platform-Independent, Community-Supported Software for Describing and Comparing Microbial Communities. *Appl. Environ. Microbiol.* **2009**, *75*, 7537–7541. [CrossRef]
25. You, H.; Tan, Y.; Yu, D.; Qiu, S.; Bai, Y.; He, J.; Cao, H.; Che, Q.; Guo, J.; Su, Z. The Therapeutic Effect of SCFA-Mediated Regulation of the Intestinal Environment on Obesity. *Front. Nutr.* **2022**, *9*, 886902. [CrossRef] [PubMed]
26. Machate, D.J.; Figueiredo, P.S.; Marcelino, G.; Guimarães, R.D.C.A.; Hiane, P.A.; Bogo, D.; Pinheiro, V.A.Z.; Oliveira, L.C.S.D.; Pott, A. Fatty Acid Diets: Regulation of Gut Microbiota Composition and Obesity and Its Related Metabolic Dysbiosis. *Int. J. Mol. Sci.* **2020**, *21*, 4093. [CrossRef]
27. Wang, A.; Guo, T.; An, R.; Zhuang, M.; Wang, X.; Ke, S.; Zhou, Z. Long-Term Consumption of Resistant Starch Induced Changes in Gut Microbiota, Metabolites, and Energy Homeostasis in a High-Fat Diet. *J. Agric. Food Chem.* **2023**, *71*, 8448–8457. [CrossRef]
28. Liu, H.; Zhang, M.; Ma, Q.; Tian, B.; Nie, C.; Chen, Z.; Li, J. Health Beneficial Effects of Resistant Starch on Diabetes and Obesity via Regulation of Gut Microbiota: A Review. *Food Funct.* **2020**, *11*, 5749–5767. [CrossRef] [PubMed]
29. Yamei, Y.; Yujia, P.; Jilong, T.; Jia, M.; Lu, L.; Xiaoying, L.; Linwu, R.; Xiaoxiong, Z.; Youlong, C. Effects of Anthocyanins from the Fruit of Lycium Ruthenicum Murray on Intestinal Microbiota. *J. Funct. Foods* **2018**, *48*, 533–541. [CrossRef]
30. Magne, F.; Gotteland, M.; Gauthier, L.; Zazueta, A.; Pesoa, S.; Navarrete, P.; Balamurugan, R. The Firmicutes/Bacteroidetes Ratio: A Relevant Marker of Gut Dysbiosis in Obese Patients? *Nutrients* **2020**, *12*, 1474. [CrossRef]
31. Koliada, A.; Syzenko, G.; Moseiko, V.; Budovska, L.; Puchkov, K.; Perederiy, V.; Gavalko, Y.; Dorofeyev, A.; Romanenko, M.; Tkach, S.; et al. Association between Body Mass Index and Firmicutes/Bacteroidetes Ratio in an Adult Ukrainian Population. *BMC Microbiol.* **2017**, *17*, 120. [CrossRef]
32. Qin, J.; Li, R.; Raes, J.; Arumugam, M.; Burgdorf, K.S.; Manichanh, C.; Nielsen, T.; Pons, N.; Levenez, F.; Yamada, T.; et al. A Human Gut Microbial Gene Catalogue Established by Metagenomic Sequencing. *Nature* **2010**, *464*, 59–65. [CrossRef]
33. Ignacio, A.; Fernandes, M.R.; Rodrigues, V.a.A.; Groppo, F.C.; Cardoso, A.L.; Avila-Campos, M.J.; Nakano, V. Correlation between Body Mass Index and Faecal Microbiota from Children. *Clin. Microbiol. Infect.* **2016**, *22*, 258.e1-8. [CrossRef] [PubMed]
34. Spain, A.M.; Krumholz, L.R.; Elshahed, M.S. Abundance, Composition, Diversity and Novelty of Soil Proteobacteria. *ISME J.* **2009**, *3*, 992–1000. [CrossRef]
35. Zhang, W.; Xu, J.-H.; Yu, T.; Chen, Q.-K. Effects of Berberine and Metformin on Intestinal Inflammation and Gut Microbiome Composition in Db/Db Mice. *Biomed. Pharmacother.* **2019**, *118*, 109131. [CrossRef] [PubMed]
36. Bai, Y.; Ma, K.; Li, J.; Ren, Z.; Zhang, J.; Shan, A. Lactobacillus Rhamnosus GG Ameliorates DON-Induced Intestinal Damage Depending on the Enrichment of Beneficial Bacteria in Weaned Piglets. *J. Anim. Sci. Biotechnol.* **2022**, *13*, 90. [CrossRef]
37. Xu, H.; Wang, S.; Jiang, Y.; Wu, J.; Chen, L.; Ding, Y.; Zhou, Y.; Deng, L.; Chen, X. Poria Cocos Polysaccharide Ameliorated Antibiotic-Associated Diarrhea in Mice via Regulating the Homeostasis of the Gut Microbiota and Intestinal Mucosal Barrier. *Int. J. Mol. Sci.* **2023**, *24*, 1423. [CrossRef] [PubMed]
38. Díaz-Perdigones, C.M.; Muñoz-Garach, A.; Álvarez-Bermúdez, M.D.; Moreno-Indias, I.; Tinahones, F.J. Gut Microbiota of Patients with Type 2 Diabetes and Gastrointestinal Intolerance to Metformin Differs in Composition and Functionality from Tolerant Patients. *Biomed. Pharmacother.* **2022**, *145*, 112448. [CrossRef] [PubMed]
39. Wang, C.-S.-E.; Li, W.-B.; Wang, H.-Y.; Ma, Y.-M.; Zhao, X.-H.; Yang, H.; Qian, J.-M.; Li, J.-N. VSL#3 Can Prevent Ulcerative Colitis-Associated Carcinogenesis in Mice. *World J. Gastroenterol.* **2018**, *24*, 4254–4262. [CrossRef]
40. Du, H.; Zhao, A.; Wang, Q.; Yang, X.; Ren, D. Supplementation of Inulin with Various Degree of Polymerization Ameliorates Liver Injury and Gut Microbiota Dysbiosis in High Fat-Fed Obese Mice. *J. Agric. Food Chem.* **2020**, *68*, 779–787. [CrossRef]
41. Zeng, H.; Huang, C.; Lin, S.; Zheng, M.; Chen, C.; Zheng, B.; Zhang, Y. Lotus Seed Resistant Starch Regulates Gut Microbiota and Increases Short-Chain Fatty Acids Production and Mineral Absorption in Mice. *J. Agric. Food Chem.* **2017**, *65*, 9217–9225. [CrossRef]
42. Louis, P.; Hold, G.L.; Flint, H.J. The Gut Microbiota, Bacterial Metabolites and Colorectal Cancer. *Nat. Rev. Microbiol.* **2014**, *12*, 661–672. [CrossRef]
43. Koh, A.; De Vadder, F.; Kovatcheva-Datchary, P.; Bäckhed, F. From Dietary Fiber to Host Physiology: Short-Chain Fatty Acids as Key Bacterial Metabolites. *Cell* **2016**, *165*, 1332–1345. [CrossRef] [PubMed]
44. Lee, E.-S.; Song, E.-J.; Nam, Y.-D.; Nam, T.G.; Kim, H.-J.; Lee, B.-H.; Seo, M.-J.; Seo, D.-H. Effects of Enzymatically Modified Chestnut Starch on the Gut Microbiome, Microbial Metabolome, and Transcriptome of Diet-Induced Obese Mice. *Int. J. Biol. Macromol.* **2020**, *145*, 235–243. [CrossRef]
45. Duan, M.; Sun, X.; Ma, N.; Liu, Y.; Luo, T.; Song, S.; Ai, C. Polysaccharides from Laminaria Japonica Alleviated Metabolic Syndrome in BALB/c Mice by Normalizing the Gut Microbiota. *Int. J. Biol. Macromol.* **2019**, *121*, 996–1004. [CrossRef] [PubMed]
46. Andrés-Manzano, M.J.; Andrés, V.; Dorado, B. Oil Red O and Hematoxylin and Eosin Staining for Quantification of Atherosclerosis Burden in Mouse Aorta and Aortic Root. *Methods Mol. Biol.* **2015**, *1339*, 85–99. [CrossRef]
47. Clarridge, J.E. Impact of 16S rRNA Gene Sequence Analysis for Identification of Bacteria on Clinical Microbiology and Infectious Diseases. *Clin. Microbiol. Rev.* **2004**, *17*, 840–862. [CrossRef] [PubMed]

48. Watts, G.S.; Youens-Clark, K.; Slepian, M.J.; Wolk, D.M.; Oshiro, M.M.; Metzger, G.S.; Dhingra, D.; Cranmer, L.D.; Hurwitz, B.L. 16S rRNA Gene Sequencing on a Benchtop Sequencer: Accuracy for Identification of Clinically Important Bacteria. *J. Appl. Microbiol.* **2017**, *123*, 1584–1596. [CrossRef]
49. Heid, C.A.; Stevens, J.; Livak, K.J.; Williams, P.M. Real Time Quantitative PCR. *Genome Res.* **1996**, *6*, 986–994. [CrossRef]
50. McKinzie, P.B.; Myers, M.B. A Brief Practical Guide to PCR. *Methods Mol. Biol.* **2023**, *2621*, 3–13. [CrossRef]
51. Maren, N.A.; Duduit, J.R.; Huang, D.; Zhao, F.; Ranney, T.G.; Liu, W. Stepwise Optimization of Real-Time RT-PCR Analysis. *Methods Mol. Biol.* **2023**, *2653*, 317–332. [CrossRef]

MDPI
St. Alban-Anlage 66
4052 Basel
Switzerland
www.mdpi.com

Molecules Editorial Office
E-mail: molecules@mdpi.com
www.mdpi.com/journal/molecules

www.ingramcontent.com/pod-product-compliance
Lightning Source LLC
LaVergne TN
LVHW070459100526
838202LV00014B/1756